T0185347

More information about this series at http://www.springer.com/series/7899

Zongben Xu · Xinbo Gao
Qiguang Miao · Yunquan Zhang
Jiajun Bu (Eds.)

Big Data

6th CCF Conference, Big Data 2018
Xi'an, China, October 11–13, 2018
Proceedings

Springer

Editors
Zongben Xu
School of Mathematics and Statistics
Xi'an Jiaotong University
Xi'an, China

Xinbo Gao
Xidian University
Xi'an, China

Qiguang Miao
Xidian University
Xi'an, Shaanxi, China

Yunquan Zhang
Chinese Academy of Sciences
Beijing, China

Jiajun Bu
Zhejiang University
Hangzhou, China

ISSN 1865-0929 ISSN 1865-0937 (electronic)
Communications in Computer and Information Science
ISBN 978-981-13-2921-0 ISBN 978-981-13-2922-7 (eBook)
https://doi.org/10.1007/978-981-13-2922-7

Library of Congress Control Number: 2018957485

This Springer imprint is published by the registered company Springer Nature Singapore Pte Ltd.
The registered company address is: 152 Beach Road, #21-01/04 Gateway East, Singapore 189721, Singapore

Preface

This volume constitutes the proceedings of the 6th CCF Academic Conference on Big Data (CCF Big Data 2018), which was held during October 11–13, 2018, in Xi'an, China, with the theme "Challenges, Innovations, and Achievements."

In the past decade, research on big data has made great progress accompanied by many achievements. Currently, we have entered a new decade of big data, in which we need to continue our efforts to further explore all theoretical and technical aspects of big data, but with more attention to the trend toward ever-increasing real-world demands on big data applications.

CCF Big Data 2018 achieved its aim to further explore the theory, techniques, and applications of big data. The submitted papers covered a broad range of research topics on big data, including natural language processing and text mining, big data analysis and smart computing, data quality control and data governance, big data applications, the application of big data in machine learning, social networks and recommendation systems, parallel computing and storage of big data, data science theory and methods, big data systems and management, and network security.

In this volume, we are not able to include all topics related to big data in an exhaustive way, but we believe that the coverage is wide enough to show the research and development promises that the subject area holds for the future. With the technical programs and forums, CCF Big Data 2018 provided all participants with valuable opportunities for communication and cooperation.

There were 304 submissions (in English and Chinese) from about 700 authors in different organizations, countries, and regions. Each submission was reviewed by at least two reviewers, and on average by 2.9 reviewers. Finally, 37 high-quality papers written in English were selected and are included in this volume of proceedings. The acceptance rate is 12.2%.

We would like to express our sincere thanks to the China Computer Federation (CCF), China Computer Federation Task Force on Big Data, Xidian University, and many other sponsors for their support and sponsorship. We thank all members of our Steering Committee, Program Committee, and Organizing Committee for their advice and contributions. Also many thanks to all reviewers for reviewing the papers in such a short time.

We are especially grateful to, Springer for publishing the proceedings in the CCIS series (*Communications in Computer and Information Science*). Moreover, we greatly appreciate all our keynote speakers, sessions chairs, authors, conference attendees, and student volunteers for their participation and contribution.

September 2018

Zongben Xu
Xinbo Gao
Qiguang Miao
Yunquan Zhang
Jiajun Bu

Organization

CCF Big Data 2018 was organized by the China Computer Federation and co-organized by the China Computer Federation Task Force on Big Data and Xidian University.

Conference Honorary Chair

Guojie Li   Institute of Computing Technology Chinese Academy of Sciences, Academician of Chinese Academy of Engineering, China

Steering Committee Chair

Hong Mei   Beijing Institute of Technology, Academic Divisions of the Chinese Academy of Sciences, China

Steering Committee Secretary-General

Xueqi Cheng   Institute of Computing Technology Chinese Academy of Sciences, China

General Chairs

Zongben Xu   Xi'an Jiaotong University, Academic Divisions of the Chinese Academy of Sciences, China

Xinbo Gao   Xidian University, China

Steering Committee

Guojie Li	Institute of Computing Technology Chinese Academy of Sciences, China
Hong Mei	Beijing Institute of Technology, China
Yunsheng Hua	The Chinese University of Hong Kong, China
Jianzhong Li	Harbin Institute of Technology, China
Enhong Chen	University of Science and Technology of China, China
Jianmin Wang	Tsinghua University, China
Xueqi Cheng	Institute of Computing Technology Chinese Academy of Sciences, China
Bin Hu	Lanzhou University, China
Yihua Huang	Nanjing University, China
Philip Yu	University of Illinois at Chicago, USA

Program Chairs

Qiguang Miao	Xidian University, China
Yunquan Zhang	Institute of Computing Technology Chinese Academy of Sciences, China
Jiajun Bu	Zhejiang University, China

Publicity Chairs

Xiaolong Jin	Institute of Computing Technology Chinese Academy of Sciences, China
Zhixin Ma	Lanzhou University, China
Laizhong Cui	Shenzhen University, China

Publication Chairs

Yang Gao	Nanjing University, China
Li Wang	Taiyuan University of Technology, China
Peiyi Shen	Xidian University, China
Rui Mao	Shenzhen University, China

Publication Vice-chairs

Yue Wu	Xidian University, China
Wen Lu	Xidian University, China
Ruyi Liu	Xidian University, China

Sub-forum Chairs

Yihua Huang	Nanjing University, China
Enhong Chen	University of Science and Technology of China, China
Liang Wang	Institute of Automation, Chinese Academy of Sciences, China

Industry Forum Chairs

Xiaosheng Tan	Qihoo 360 Technology Co., Ltd., China
Lin Mei	The Third Research Institute of Ministry of Public Security
Tao Wang	iQIYI, China

Award Chairs

Xiaoyang Wang	Fudan University, China
Yi Xiao	Sun Yat-sen University, China

Organizing Co-chairs

Quan Wang	Xidian University, China
Yanning Zhang	Northwestern Polytechnical University, China
Xingjun Zhang	Xi'an Jiaotong University, China
Jinye Peng	Northwest University, China
Huansheng Song	Chang'an University, China
Xiaoming Wang	Shaanxi Normal University, China
Xinhong Hei	Xi'an University of Technology, China
Zhongmin Wang	Xi'an University of Posts and Telecommunications, China
Zhongxing Duan	Xi'an University of Architecture and Technology, China
Tao Xue	Xi'an Polytechnic University, China

Outstanding Youth Scholars Forum Chairs

Maoguo Gong	Xidian University, China
Huawei Shen	Institute of Computing Technology Chinese Academy of Sciences, China

Finance Chairs

Yining Quan	Xidian University, China
Jianfeng Song	Xidian University, China

Sponsorship Chair

Yuan Su	NowledgeData, China

Program Committee

Shuo Bai	Shanghai Stock Exchange, China
Jiajun Bu	Zhejiang University, China
Li Zha	Institute of Computing Technology Chinese Academy of Sciences, China
Jun Chen	China Defense Science and Technology Information Center, China
BaoQuan Chen	Shandong University, China
Jidong Chen	Ant Financial Services Group, China
Shangyi Chen	Baidu, China
Wei Chen	Microsoft Research Asia, China
Yixin Chen	Washington University in St. Louis, USA
Laizhong Cui	Shenzhen University, China
Lizhen Cui	Shandong University, China
Peng Cui	Tsinghua University, China
Bo Deng	Beijing System Engineering Institute, China
Zhiming Ding	Beijing University of Technology, China

Junping Du	Beijing University of Posts and Telecommunications, China
Xiaoyong Du	Renmin University of China, China
Yuejin Du	Alibaba, China
Liang Fang	National University of Defense Technology, China
Shicong Feng	Mininglamp
Changshui Gao	Center for International Economic and Technological Cooperation Ministry of Industry and Information Technology, China
Hong Gao	Harbin Institute of Technology, China
Sheng Gao	Beijing University of Posts and Telecommunications, China
Xinbo Gao	Xidian University, China
Yang Gao	Nanjing University, China
Ning Gu	Fudan University, China
Yanbo Han	North China University of Technology, China
Jieyue He	Southeast University, China
Jun He	Renmin University of China, China
Liwen He	Jiangsu Industrial Technology Research Institute, China
Xiaofei He	Zhejiang University, China
Yalou Huang	Nankai University, China
Zhexue Huang	Shenzhen University, China
Tongkai Ji	Guangdong Electronics Industry Institute, China
Bin Jiang	Hunan University, China
Jie Jiang	Tencent, China
Hai Jin	Huazhong University of Science and Technology, China
Bo Jin	The Third Research Institute of Ministry of Public Security, China
Tao Lei	BeagleData, China
Chonggang Li	Beijing Jinxinwangyin Financial Information Service Company, China
Cuiping Li	Renmin University of China, China
Jie Li	Youku Tudou Inc., China
Ru Li	Inner Mongolia University, China
Jianxin Li	Beihang University, China
Shanshan Li	National University of Defense Technology, China
Dongsheng Li	National University of Defense Technology, China
Guangya Li	Wonders Information System Co., Ltd., China
Jianzhong Li	Harbin Institute of Technology, China
Keqiu Li	Dalian University of Technology, China
Qing Li	City University of Hong Kong, SAR China
Ruixuan Li	Huazhong University of Science and Technology, China
Tianrui Li	Xi'an Jiaotong University, China
Xiaoming Li	Peking University, China
Jianhui Li	Computer Network Information Center, Chinese Academy of Sciences, China
Jiye Liang	Shanxi University, China

Hui Liu	China Information Technology Security Evaluation Center, China
Tao Xue	Xi'an Polytechnic University, China
Xueping Su	Xi'an Polytechnic University, China
Quanli Gao	Xi'an Polytechnic University, China
Jinguang Chen	Xi'an Polytechnic University, China
Xueqing Zhao	Xi'an Polytechnic University, China
Su-Nan Ge	Xi'an Polytechnic University, China
Xinjuan Zhu	Xi'an Polytechnic University, China
Bo Yang	Xi'an Polytechnic University, China
Liang Chen	Xi'an Polytechnic University, China
Tao Wu	Xi'an Polytechnic University, China
Weiqian Li	Xi'an Polytechnic University, China
Ting Li	Xi'an Polytechnic University, China
Linming Gong	Xi'an Polytechnic University, China
Yanping Chen	Xi'an University of Posts and Telecommunications, China
Hao Chen	Xi'an University of Posts and Telecommunications, China
Xiaoyan Xie	Xi'an University of Posts and Telecommunications, China
Yang Jia	Xi'an University of Posts and Telecommunications, China
Kui Xing	Xi'an University of Posts and Telecommunications, China
Fuxi Wen	Xi'an University of Posts and Telecommunications, China
Hengshan Zhang	Xi'an University of Posts and Telecommunications, China
Weiwei Kong	Xi'an University of Posts and Telecommunications, China
Jingtao Sun	Xi'an University of Posts and Telecommunications, China
Hanlin Sun	Xi'an University of Posts and Telecommunications, China
Jinwei Zhao	Xi'an University of Technology, China
Xinhong Hei	Xi'an University of Technology, China
Xiaofeng Lu	Xi'an University of Technology, China
Zhiqiang Zhao	Xi'an University of Technology, China
Zhurong Wang	Xi'an University of Technology, China
Qindong Sun	Xi'an University of Technology, China
Yichuan Wang	Xi'an University of Technology, China
Junhuai Li	Xi'an University of Technology, China
Yuan Qiu	Xi'an University of Technology, China
Minghua Zhao	Xi'an University of Technology, China
Zhenghao Shi	Xi'an University of Technology, China
Guo Xie	Xi'an University of Technology, China
Zhongbin Sun	Xi'an Jiaotong University, China
Yuanqi Su	Xi'an Jiaotong University, China
Peng Zhang	Xi'an Jiaotong University, China
Jian An	Xi'an Jiaotong University, China
Xuebin Ren	Xi'an Jiaotong University, China
Longxiang Wang	Xi'an Jiaotong University, China
Minnan Luo	Xi'an Jiaotong University, China
Jie Lin	Xi'an Jiaotong University, China
Junpeng Bao	Xi'an Jiaotong University, China

Xingjun Zhang	Xi'an Jiaotong University, China
Xi Zhao	Xi'an Jiaotong University, China
Hao Li	Xi'an Jiaotong University, China
Fei Hao	Shaanxi Normal University, China
Wangyang Yu	Shaanxi Normal University, China
Liang Wang	Shaanxi Normal University, China
Lichen Zhang	Shaanxi Normal University, China
Xiaoming Wang	Shaanxi Normal University, China
Liang Wang	Northwestern Polytechnical University, China
Mengchi Liu	Wuhan University
Qi Liu	University of Science and Technology of China, China
Song Liu	Alibaba, China
Wei Liu	EMC, China
Xinran Liu	National Computer Network Emergency Response Technical Team/Coordination Center of China, China
YuBao Liu	Sun Yat-sen University, China
Zheng Liu	SAS, China
Xiaodong Lin	Rutgers University, USA
Xuemin Lin	East China Normal University, China
Chao Liu	Beijing golden dike Technology Co., Ltd., China
Yilei Lu	AdMaster, China
Jiaheng Lu	University of Helsinki, Finland
Ping Lu	Zhongxing Telecommunication Equipment Corporation, China
Shengmei Luo	Zhongxing Telecommunication Equipment Corporation, China
Jiawei Luo	Hunan University, China
Huadong Ma	Beijing University of Posts and Telecommunications, China
Jun Ma	Shandong University, China
Shuai Ma	Beihang University, China
Zhixin Ma	Lanzhou University, China
Rui Mao	Shenzhen University, China
Wenji Mao	Institute of Automation, Chinese Academy of Sciences, China
Luoming Meng	Beijing University of Posts and Telecommunications, China
Xiaofeng Meng	Renmin University of China, China
Kaixiang Miao	Cloudera, China
Qiguang Miao	Xidian University, China
Hong Mu	Huawei Technologies Co., Ltd., China
Zhuting Pan	Beijing Venustech Co., Ltd., China
Shaoliang Peng	National University of Defense Technology, China
Zhiyong Peng	Wuhan University, China
Hongwei Qi	Datatang, China
Zheng Qin	Hunan University, China
Jiadong Ren	Yanshan University, China
Tong Ruan	East China University of Science and Technology, China
Shuo Shen	China Internet Network Information Center, China

Shuicai Shi   TRS, China
Yong Shi   CAS Research Center on Fictitious Economy and Data
                          Science, China
Jiwu Shu   Tsinghua University, China
Huaiming Song   Sugon, China
Shaoling Sun   China Mobile Communications Corporation, Suzhou R&D
                          Center, China
Yunchuan Sun   Beijing Normal University, China
Chang Tan   iFLYTEK, China
Jin tang   Anhui University, China
Tuergen Yibulayin   Xinjiang University, China
Guoren Wang   Northeastern University, China
Guoyin Wang   Chongqing University of Posts and Telecommunications,
                          China
Hao Wang   Qihoo 360 Technology Co., Ltd., China
Hongzhi Wang   Harbin Institute of Technology, China
Jianmin Wang   Tsinghua University, China
Jianzong Wang   Ping An Technology, China
Li Wang   Taiyuan University of Technology, China
Liang Wang   Institute of Automation, Chinese Academy of Sciences, China
Tengjiao Wang   Peking University, China
Wenjun Wang   Tianjin University, China
Xiaoyang Wang   Fudan University, China
Xin Wang   Fudan University, China
Yijie Wang   National University of Defense Technology, China
Yuanzhuo Wang   Institute of Computing Technology Chinese Academy
                          of Sciences, China
Zhiping Wang   Huawei Technologies Co., Ltd., China
Jirong Wen   Renmin University of China, China
Gansha Wu   UISEE, China
Wushouer Silamu   Xinjiang University, China
Yongwei Wu   Tsinghua University, China
Nong Xiao   National University of Defense Technology, China
Limin Xiao   Beihang University, China
Weidong Xiao   National University of Defense Technology, China
Xianghui Xie   The State Key Laboratory of Engineering Mathematics
                          and Advanced Computing
Cunxiao Xing   Tsinghua University, China
Hui Xiong   Rutgers, The State University of New Jersey, USA
Gang Xiong   Institute of Automation, Chinese Academy of Sciences, China
Yun Xiong   Fudan University, China
Guirong Xue   Alibaba, China
Zhenghua Xue   Baidu, China
Dongri Yang   CSIP, China
Ming Yang   Nanjing Normal University, China
Jian Yin   Sun Yat-sen University, China

Sponsoring Institutions

NVIDIA Corporation   INSPUR Group Co., Ltd.   Huawei Technologies Co., Ltd.

SenseTime   Qihoo 360 Technology Co., Ltd.

Contents

Natural Language Processing and Text Mining

Big Data Analytics and Smart Computing

Big Data Applications

The Application of Big Data in Machine Learning

Data Quality Control and Data Governance

Big Data System and Management

Natural Language Processing and Text Mining

A Method to Chinese-Vietnamese Bilingual Metallurgy Term Extraction Based on a Pivot Language

Shengxiang Gao, Haodong Zhu, Zhengtao Yu$^{(\boxtimes)}$, Xiaoxu He, and Yunlong Li

School of Information Engineering and Automation,
Kunming University of Science and Technology, No. 727 South Jingming Rd.,
Chenggong District, Kunming 650500, China
ztyu@hotmail.com

Abstract. To settle resource scarcity problem for Chinese-Vietnamese bilingual aligned corpus in metallurgy field, a method to Chinese-Vietnamese bilingual term extraction in metallurgy field based on a pivot language is proposed. Firstly, term-unit and term-hood features are selected and inputted to the trained CRFs model to identify and extract Chinese metallurgy terminology. Secondly, the phrase-based statistical machine translation model is used to generate the Chinese-English phrase table and English-Vietnamese phrase table. With the pivot mapping idea, A Chinese-Vietnamese phrase table will be inferred out through pivot English. Finally, the former extracted Chinese metallurgy terms are used to filter the Chinese-Vietnamese phrase table, a Chinese-Vietnamese bilingual metallurgy term base, therefore, will be built. Experiments show that the proposed method achieved an accuracy rate at 69.45%. The method, under the resource absence of Chinese-Vietnamese bilingual alignment corpus, is validated as an effective solution to the difficult problem for Chinese-Vietnamese bilingual metallurgy term extraction.

Keywords: Bilingual term extraction · Pivot language
Chinese-English-Vietnamese · Metallurgy domain

1 Introduction

Metallurgical industry has become a pillar industry of our country. It performs an important role in national economy. As One Belt, One Road initiative continues, cooperation and exchange of knowledge and technology in metallurgy field between China and Vietnam are more urgent and frequent. Metallurgical field terminology as the core of metallurgical science and technology knowledge, its mutual translation remains one of the biggest language obstacles between the two countries in metallurgical knowledge, technical exchanges. Preparing terminology translation pairs manually not only spends a lot of manpower, material resources and time, but also will get limited quantities. Therefore, Automatic

© Springer Nature Singapore Pte Ltd. 2018
Z. Xu et al. (Eds.): Big Data 2018, CCIS 945, pp. 3–20, 2018.
https://doi.org/10.1007/978-981-13-2922-7_1

extraction of Chinese-Vietnamese bilingual terms in metallurgical field is of great significance for cross language retrieval and construction of bilingual dictionaries in metallurgical field. In addition to, that bilingual terminology in metallurgical field, as bilingual dictionaries, are integrated into Chinese-Vietnamese Machine Translation system for scientific and technical documentations in metallurgy domain, will improve translation performance to a certain extent [1].

At present, the study of automatic bilingual term extraction technology is mainly two steps, extracting monolingual terms and then calculating probabilities of bilingual terminology translation [2–4]. Firstly, two monolingual term candidate lists, source side and target side, are generated through monolingual term extraction. Then, based on co-occurrence probability of candidate terms in Parallel Corpora [5,6] or based on translation probability of bilingual term pairs in bilingual dictionary [7,8], the candidate results whose translation probability satisfies a certain threshold can be served as bilingual term.

Monolingual term extraction method is mainly divided into three categories. The first one is based on linguistic rules. It realizes heuristic term extraction by using rule templates and terminology dictionary authored by professionals [9–11]. This method is of high precision. Writing rules, which depend on language environment and domain themes, are difficult to transplant. The second one is based on statistical features. Based on the assumption that words in a term have higher degree of adhesion, it implements term extraction by using statistical features, such as Chi square test, log likelihood test, mutual information and Cvalue/NC-Value and so on [12,13]. This kind of method has good portability, but the accuracy depends on the size of corpus and word frequency of candidate terms. The third one is based on machine learning. Term extraction task is transformed into a classification problem or an annotation problem with the aid of decision tree (DT), support vector machine (SVM) [14], hidden Markov model (HMM) and conditional random field model (CRF) [15] etc. This method has better portability, but both training and testing model needs a lot of labeled data.

The calculation of bilingual term translation probability is mainly based on statistical features or bilingual dictionary. Liu et al. calculated the bilingual term translation probability through bilingual dictionary, and algorithm accuracy rate reached 76.14% [2]. Sun et al. calculated the bilingual term translation probability based on parallel corpora. Extracting 72 bilingual term pairs from 5792 pairs of Chinese-English parallel sentences, algorithm accuracy rate reached 86.2% [3]. Sun et al. calculated the bilingual term translation probability based on phrase-base statistical machine translation model. Using conditional random field component analysis technology and domain subject information to do filtering, the accuracy reached up to 94% [16].

However, these bilingual term extraction methods, whatever extracting multilingual terms and then aligning them, or extracting monolingual term and then recognizing its translation equivalent in another language, are based on a rich corpus. For Chinese-Vietnamese, data collection and annotation are very difficult, and aligned corpuses are scarce. Especially in metallurgical field, accessible Chinese-Vietnamese bilingual resources are scarce on the Internet. Under the circumstances, it is very difficult to extract Chinese-Vietnamese bilingual terms in metallurgical field. Fortunately, English as an international language, Chinese to English and English to Vietnamese bilingual corpus resources are relatively abundant. For example, Chinese and English abstracts of papers in CNKI, Chinese-English inter-translatable word alignment corpus and English-Vietnamese discourse corpus of entry alignment in Wikipedia. Therefore, this paper proposes a Chinese-Vietnamese bilingual metallurgical term extraction method based on English pivot. First, extract Chinese metallurgical terminology. Then, get Chinese-English phrase list, English-Vietnamese phrase list by using the phrase-based statistical machine translation model. Mapping by English pivot, get Chinese-Vietnamese bilingual phrase list. Final, filter the Chinese-Vietnamese phrase list with the Chinese metallurgical terms, and construct a bilingual term library in the field of metallurgy.

2   Chinese-Vietnamese Bilingual Metallurgy Term Extraction Based on English Pivot

Bilingual terminology extraction requires large-scale bilingual alignment corpus. Model trained by small-scale corpus has some problems including low coverage, sparse data and poor quality of translation. For Chinese-Vietnamese bilingual term extraction, aligned Chinese-Vietnamese data are very scarce, so it is difficult to collect bilingual alignment corpus. However, English as an international language, Chinese to English and English to Vietnamese bilingual corpus resources are relatively abundant. For example, Chinese and English abstracts of scientific and technical documents in CNKI, English and Vietnamese aligned discourse corpus of entries in Wikipedia. From these resources, a lot of Chinese-English and English-Vietnamese bilingual parallel corpus can be acquired. From the corpus, Chinese Metallurgical terminologies are extracted out by CRF model. Then, Chinese-English phrase list, English-Vietnamese phrase list are acquired by using the phrase-based statistical Machine Translation model. Mapping by English pivot, Chinese-Vietnamese bilingual phrase list can be obtained. In the end, the Chinese-Vietnamese phrase list is filtered by the Chinese metallurgical terminologies, and a bilingual term library in the field of metallurgy will be constructed.

2.1   Chinese-English, English-Vietnamese Bilingual Metallurgical Text Acquisition

The literature in metallurgy field has strong specialty. CNKI journal database of China included abundant Sci-tech literatures from all walks of life with Chinese and English aligned abstract. According to the website features of CNKI journal database, we design professional Crawler to search and download metallurgy Sci-tech literature from the CNKI database. Extracting their Chinese and English abstracts, we construct a certain scale of Chinese-English bilingual paragraph alignment corpus. Wikipedia as a global, multi-language, complete, accurate and neutral network encyclopedia, it has nearly 50000 Vietnamese entries. Designing professional crawler according to Wikipedia entries' multi language structure relationship, we crawl Vietnamese and English Wikipedia equivalent entries in the field of metallurgy. After the downloaded Web pages are preprocessed, the page's content is extracted out to construct an aligned English-Vietnamese bilingual discourse corpus in metallurgy. Using the method mentioned in reference [17], sentence alignment on Chinese-English discourse alignment corpus and English-Vietnamese discourse alignment corpus are done to construct Chinese-English sentence alignment corpus and English-Vietnamese sentence alignment corpus.

2.2 Chinese Metallurgy Term Extraction Based on Conditional Random Field

In our work, on the basis of analyzing the characteristics of terminology in the field of metallurgy, term-unit and term-hood are selected as main features of domain terms, the corpus is manually annotated, the term extraction task is transformed into sequence labeling problem, and the conditional random field is used as our machine learning model [18]. From former-mentioned Chinese abstract corpus of CNKI scientific literature, Chinese metallurgical terms are identified and extracted.

(1) Feature Selection

Metallurgical terminology has two characteristics of term-unit and term-hood. Term-unit means that a term must be a correct string as a complete meaning. Term-unit can measure the tightness of internal combination of the string. Here, word, part of speech, mutual information, left & right information entropy are selected to measure the term-unit. Term-hood is also called domain property, indicating that a term represents the extent of its particular domain. We take economics field as the background corpus, select the frequency of the current word in the field corpus, the frequency in the background corpus, and the frequency difference between the two corpus to measure the term-hood. Thus, the selected features are summarized in Table 1:

Here, 'State' is the tagging status of the word. It identifies whether the word is a domain term. For a given input sentence, after Chinese word segmentation, BMEO is used to mark terms. The tagging rule is shown in Table 2.

MI, named mutual information, calculates the internal bonding strength of a string. For an example, a sting c to be treated, '钢包渣结壳' (Ladle slag crust), it has two longest strings, $a =$ '钢包熔渣结' (ladle slag knot) and $b =$ '包熔渣结壳' (package slag crust). $f(c)$ means co-occurrence frequency of C in the corpus. $p(c)$ means co-occurrence probability of C in the corpus. According to maximum likelihood estimation, if the corpus size is large enough, we suppose $p(c) = f(c)$. Mutual information is defined as Eq. (1) for String c.

$$MI = \log_2 \frac{p(c)}{p(a)\,p(b)} = \log_2 \frac{f(c)}{f(a)\,f(b)} \tag{1}$$

Table 1. Feature selection in domain term recognition

No	Feature name	Feature description
1	Word	The word itself
2	POS	Part of speech
3	State	whether the word is a domain term
4	MI	Mutual information
5	LE	Left information entropy
6	RE	right information entropy
7	DomainFreq	Frequency in domain corpus
8	ContrastFreq	Frequency in Background corpus
9	ΔFreq	Frequency difference between the two domain corpus

Table 2. Tagging rule

Tag	Position of tagging
TB	The first word of a term
TM	The middle words of a term
TE	The last word of a term
O	Others

If c is well-bonded, there is little difference between $f(c)$ and $f(a)$ or $f(b)$. Its mutual information calculated by Eq. (1) is relatively large. Otherwise, $f(a)$ and $f(b)$ will be much larger than $f(c)$. Its mutual information calculated will be relatively small. The mutual information will be floating point value. It is divided into 20 categories by their size: 1, 2, 3, 4, 5, 6, 7, 8, 9, 10, 11, 12, 13, 14, 15, 16, 17, 18, 19, 20.

LE represents left information entropy. LR represents right informa-tion entropy. They are used to measure the uncertainty of left and right boundaries of a string. The more uncertain the string bound-ary, the higher the information entropy, the more likely it is to be a complete word. For example, in Chinese metallurgical corpus "烧结作业是烧结生产的中心环节，它包括布料、点火、烧结等主要工序。(The sintering operation is the key process of sintering production, which includes the main processes such as cloth, ignition, sintering and so on.)", the sintering operation is the key process of sintering production. "烧结(Sintering)" appears 3 times, its left adjacent words are "ε", "是" and "、", and each appears 1 times. Its right adjacent words are "作", "生" and "等", and each emerges 1 times. Left adjacent words of "烧" are "ε", "是" and "、", each emerges 1 times. Right adja-cent word of "烧" is "结", which appears 3 times. These indicate that the word "烧结" have not fixed left and right adjacent words, and "烧结" is likely to be a term in the metallurgy field. The word "烧" has a relatively fixed right adjacent word "结", and so the word "烧" is probably not a term in the metallurgical field.

S is a candidate string. S' left information entropy is labeled as LE(s), S' right information entropy is labeled as RE(s). They can be seen as Eqs. (2) (3)

$$LE\,(s) = -\sum_{l \in L} p\,(ls|s) \log_2 p\,(ls|s) \tag{2}$$

$$RE\,(s) = -\sum_{r \in R} p\,(sr|s) \log_2 p\,(sr|s) \tag{3}$$

Here, LS is a string consisted of 's' and its left adjacent word 'l'. In the case of 's' appearing in the corpus, p(ls|s) represents the probability that 'l' is the left adjacent word of 's'. 'sr' is a string consisted of 's' and its right adjacent word 'r'. In the case of 's' appearing in the corpus, p(sr|s) represents the probability that 'r' is the right adjacent word of 's'. The larger LE(s) and RE(s), the more unfixed the left and right adjacent word, and so the more likely 's' is to be a individual term. Set a same threshold for LE (s) and RE (s) to filter candidate words which cannot be a term alone. As shown in the Eq. (4).

$$RE\,(s) \geq E_{\min} \text{ and } E\,(s) \geq E_{\min} \tag{4}$$

The left and right information entropy is a floating point value. They are divided into 20 categories by size. Their values are 1, 2, 3, 4, 5, 6, 7, 8, 9, 10, 11, 12, 13, 14, 15, 16, 17, 18, 19 and 20 individually.

DomainFreq is the domain corpus frequency. It represents the frequency of a word in domain corpus.

$$DomainFreq = f\,(w)/C \tag{5}$$

C is the total number of words in the corpus. $F(W)$ represents the number of times that the word 'w' appears in the corpus. The calculated value is floating point which can not be directly used as the feature value of the CRFs model. According to the value, then, it is divided into five categories 1, 2, 3, 4, 5.

ContrastFreq represents the background corpus frequency, which calculates the word's frequency in the background corpus. Choose economics as background corpus, The computation of the background corpus frequency is the same as that of the domain corpus frequency.

Freq is difference of the two domain frequency, which calculates the word's frequency difference between the metallurgical domain and the background corpus.

$$\Delta Freq = DomainFreq - ContrastFreq \tag{6}$$

(2) Feature template

After selecting features, we need to develop a feature template. According to the feature template, the system needs to extract features from the training corpus. The basic feature template is defined as Table 3:

Table 3. Define feature template

Template type	Templates	Template description
Template 1	$Word(n)\{n=-2,-1,0,1,2\}$	the current word and two words before and after it
Template 2	$POS(n)\{n=-2,-1,0,1,2\}$	Part of speech of the current word and the two words before and after it
Template 3	$State(n)\{n=-2,-1,0\}$	Tag state of the current word and the two words before it
Template 4	$MI(n-1,n)\{n=0\}$	Mutual information of the current word
Template 5	$LE(n)\{n=0\}$	Left information entropy of the current word
Template 6	$RE(n)\{n=0\}$	Right information entropy of the current word
Template 7	$DomainFreq(n)\{n=0\}$	Domain frequency of the current word
Template 8	$ContrastFreq(n)\{n=0\}$	Background domain frequency of the current word
Template 9	$\Delta Freq(n)\{n=0\}$	Domain Frequency difference of the current word
Template 10	$Word(n-1)Word(n)\{n=-1,0,1\}$	Two adjacent words
Template 11	$Word(n)POS(n)\{n=0\}$	the current word and its Part of speech

Table 4 shows an example of term features extracted from Chinese corpus "冶金法制备太阳能级硅工艺中湿法提纯及半工业化研究 (Study on wet purification and semi industrialization of solar grade silicon by metallurgical method)".

Table 4. Examples of term feature extraction in metallurgical field

Word	POS	State	MI	LE	RE	DomainFreq	ContrastFreq	ΔFreq
冶金(metallurgical)	n	TB	18	19	15	5	1	4
法(method)	n	TE	20	2	11	4	2	2
制备(preparation)	v	TB	15	11	13	4	2	2
太阳(solar)	n	TB	19	12	4	4	1	3
能(power)	n	TM	20	2	7	4	1	3
级(grade)	n	TM	20	3	2	4	2	2
硅(silicon)	n	TE	20	5	15	5	1	4
工艺(craft)	n	TB	19	14	12	4	1	3
中(in)	f	O	20	17	14	1	3	1
湿(wet)	a	TB	20	11	8	3	1	2
法(method)	n	TM	20	2	11	4	2	2
提纯(purification)	vn	TE	12	13	11	5	1	4
及(and)	cc	O	20	16	17	3	3	1
半(semi)	m	TB	20	4	3	2	2	1
工业化(industrialized)	vn	TE	16	12	18	3	2	1
研究(study)	vn	O	18	15	17	4	4	1

(3) Chinese metallurgy term extraction based on Conditional Random Field
According to the features selected above, calculating their eigenvalues and manually tagging corpus, we train conditional random field model on the tagged corpus. Then we extract the Chinese metallurgy terminology on the test data.

Conditional random field [19,20], is an Undirected Probabilistic Graphic Classifying Model. It is able to calculate the conditional probability of output value to input value of the designated node. The training objective is to maximize the conditional probability. A observation sequence $X(X1, X2, X3, \ldots, XN)$ is of length N. The probability of its output sequence $Y(Y1, Y2, \ldots, YN)$ is defined as follows.

$$P(Y/X) = \frac{1}{Zx} \exp \left(\sum_i \sum_j \lambda_j f_j(y_{i-1}, y_i, X, i) \right) \tag{7}$$

In Eq. (7), Z is a normalized constant. It makes the probabilistically sum of all state sequences 1. The calculation formula of Zx is as follows.

$$Zx = \sum_y \exp \left(\sum_i \sum_j \lambda_j f_j(y_{i-1}, y_i, X, i) \right) \tag{8}$$

In Eq. (8), $FJ(yi-1, yi, X, i)$ is the feature function for sequence X whose tag is located in i and $i-1$. The feature function is a bi-valued function, namely Boolean value, the value set being $\{0, 1\}$. λ_j obtained in the training, is weight coefficient associated with each feature f_j.

From an input sequence x, the goal of tagging is to find out the most probable annotation result sequence $\overline{y}$. That is

$$\overline{y} = \arg \max p_\lambda(y|x) \tag{9}$$

Because $Z_\lambda(X)$ has not reliance on y, there is $\overline{y} = \arg \max p_\lambda(y|x) = \arg \max \lambda \cdot F(y, x)$. Similar to hidden Markov models, CRFs uses Viterbi decoding method [21] to get the best annotation result sequence.

Based on the training model, the string "冶金(metallurgy)" and "法(method)" in the Corpus "冶金法制备太阳能级硅工艺中湿法提纯及半工业化研究 (Study on wet purification and semi industrialization of solar grade silicon by metallurgical method)" are merged into a term "冶金法(metallurgical method)" because left information entropy of "法(method)" is relatively small. Similarly, the metallurgical domain term sheet extracted from this corpus: 冶金法 (Metallurgical method) /n TB, 制备 (preparation) /v TB, 太阳能级硅 (solar grade silicon) /n TB, 工艺 (process) /n TB, 湿法提纯 (wet purification) /n TB, 半工业化 (semi industrialized) /n TB.

2.3   Chinese-Vietnamese Bilingual Phrase Table Generation Based on English Pivot

After the Chinese metallurgical terminologies was obtained, on the hand, direct Chinese-Vietnamese bilingual aligned corpus is very lack, on the contrary, English as an international communication language, the Chinese-English and English-Vietnamese bilingual aligned corpus is relatively abundant. Therefore, we will make full use of English as pivot Language, and on the basis of pivotal thinking put forward the Chinese-Vietnamese bilingual terminology extraction method in metallurgy field Firstly, in Sect. 2.1, we will obtain a large-scale Chinese-English, English-Vietnamese bilingual sentence pairs based on the aligned corpus and use phrase-based statistical machine translation method to respectively construct Chinese-English phrase table and English-Vietnamese phrase table; And then, by mapping English between the Chinese-English table and the English-Vietnamese phrase table, the Chinese-Vietnamese Bilingual Phrase Table is formed.

When constructing Chinese-English and EnglishVietnamese phrase translation model, we will use three features: the phrase translation probability, the lexical weight in phrase pair and the reorder probability According to Koehn's research [22], the reorder feature is independent of the special phrase modeling. Therefore, in our study based on pivot language, we focus on the analysis of phrase translation probability and lexical weights.

(1) Phrase translation probability

Phrase translation Probability: The way we construct the phrase translation probability is the same as a generating probability process and computing condition exists in three languages: Chinese phrase, English phrase and Vietnamese phrase. As shown in Eq. (10).

$$\phi(\bar{s}|\bar{t}) = \sum_{\bar{p}} \phi(\bar{s}|\bar{p})\phi(\bar{p}|\bar{t}) \tag{10}$$

In the Eq. (10), and respectively represents a Chinese phrase, a English phrase and a Vietnamese phrase The phrase translation probabilities and are respectively estimated from Chinese-English and English-Vietnamese parallel sentences. From the two translation probabilities, the probability of Chinese-Vietnamese phrase translation can be calculated, namely.

In order to obtain a valid probability of phrase rule translation, when estimating the probabilities of phrase translation, we need to establish the correct connection between Chinese-English-Vietnamese phrases. Because of the ambiguity of the phrase, the ambiguity problem will be confronted when the intermediate language connects the source language and the target language for corresponding relationship. The disjunction of the connected phrases is taken into account by the context information of the pivot language to establish the Chinese-English-Vietnamese phrase corresponding relationship. We adopt the phrase alignment method based on language context characteristics. At first, we construct a lexical feature space in Chinese and Vietnamese three languages, then construct eigenvectors for the phrase context respectively, and calculate the similarity between the phrase vector and the source language and target language phrase vector respectively. And to establish the correct connection of the Chinese-English-Vietnamese phrases according to the similarity value and to adjust the translation probability of the phrase rule according to the number of correct connections.

(2) Calculation of Lexical Probability Weighting

Lexicalization weight: Given a pair of phrases and the 'a' represents a word-aligned positional relationship between the source language and the target language. And then the lexicalization weight can be estimated by Eq. (11).

$$p_w(\bar{s}|\bar{t},a) = \prod_{i=1}^{n} \frac{1}{|j|(i,j) \in a|} \sum_{\forall(i,j) \in a} w(s_i|t_j) \tag{11}$$

In order to estimate lexicalization weights, firstly, we need to obtain the alignment information of two phrases s and t, and then we will estimate the probability of lexical translation from the alignment information. And the alignment information of the phrase pair can be deduced from the two phrase pairs a1 and a2.

The a1 and a2 respectively represents the alignment of words in the phrase pairs. The word-alignment relation 'a' in the phrase pair can be deduced from Eq. (12).

$$a = \{(s,t)|\exists p : (s,p) \in a1 \& (p,t) \in a2\} \tag{12}$$

Figure 1 is an example of a word alignment method based on a pivot-based phrase. The left side of the figure is the Chinese-English (solar 1 to 2 3 silicon 4, solar1, 2 null2 grade3 silicon4), the word with the right is English-Vietnamese (solar1, 2 null2 grade3 silicon4, silicon4 cấp3 năng_luong2 mặt_trời1) word-alignment relationship. Based on these two pairs of word alignment relationship, we can get the Chinese-Vietnamese word alignment relationship (Sun 1 2 3 3 4, Silicon4 cấp3 năng_luong2 mặt_trời1). Due to the ambiguity of word meaning, the problem of word meaning ambiguity exists in the process of word translation. In the lexicalization weight calculation, the lexical context information should be considered to determine the vocabulary correspondence. Therefore, the similarity relation between Chinese and English is established by lexical-context cross-linguistic similarity calculation model and lexical-level proper alignment relation according to Chinese-English-Vietnamese correspondence.

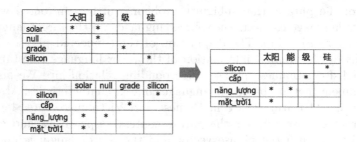

Fig. 1. A phrase alignment process based on pivot language.

We will obtain the alignment of the words by the pivot language phrases. The number of the deduced phrase pairs is represented by k. And the co-occurrence frequency of the word pair(s,t) can be estimated according to Eq. (13).

$$count(s,t) = \sum_{k=1}^{K} \varphi_k(\bar{s}|\bar{t}) \sum_{i=1}^{n_k} \delta(s,s_i)\delta(t,t_{ai}) \tag{13}$$

Where $\phi_k(\bar{s}|\bar{t})$ is the translation probability of the phrase pairs k, n_z is the length of the source language phrase $\bar{s}$;The function$\delta(x,y)$ represents the correct alignment function for the words in the phrase, the words have an alignment relation of 1 and do not have an alignment relationship of 0. After calculating $count(s,t)$, it is possible to calculate the probability of translation of the word$w(s|t)$ by the Eq. (14)

$$w(s|t) = \frac{count(s,t)}{\sum_{s'} count(s',t)} \qquad (14)$$

After obtaining the probabilities of lexical translation, the bilingual lexicalization weights of Chinese and Vietnamese are obtained by Eq. (11).

In order to use the probabilistic phrase translation rules generated by different corpora, the linear translation method is used to integrate the translation model obtained from the pivot language and the Chinese-Vietnamese standard translation model obtained from the small-scale Chinese-Vietnamese corpus. The probabilities of phrase translation and lexicalization weights are given by Eqs. (15) and (16).

$$\phi(\bar{s}|\bar{t}) = \sum_{i=0}^{1} \alpha_i \phi_i(\bar{s}|\bar{t}) \qquad (15)$$

$$p_w(\bar{s}|\bar{t},a) = \sum_{i=0}^{1} \beta_i p_{w,i}(\bar{s}|\bar{t},a) \qquad (16)$$

In the Eqs. (15) and (16). A and B respectively represent the probabilities of lexical translation and lexicalization weights of the Chinese-Vietnamese standard translation model. A and B respectively represent the translation probabilities and lexical weights of the pivot model, of which A and B represent the insertion coefficients. Using the above-mentioned translation model of Chinese-Vietnamese phrases, we can construct the Chinese-English-Vietnamese three-language metallurgical domain glossary from the example of "metallurgical preparation of solar-grade silicon technology in the wet-purification and semi-industrial research", and the information is shown in Table 5.

When we construct the translation Model of Chinese - Vietnamese Phrases, we can translate the Chinese sentences into "nghiên cúu hydrometallurgy sach và Bán công nghiêp trong do thu công Luyên kim su chuan bi Silicon cấp năng_luong mǎt_tròi".

2.4   Chinese-Vietnamese Bilingual Phrase Filter

After the Chinese-Vietnamese bilingual phrasal table was obtained in Sect. 2.3, the bilingual phrasal vocabulary of Chinese-Vietnamese metallurgical domain was constructed by filtering the Chinese-Vietnamese bilingual phrase table using the Chinese metallurgical terminology extracted in Sect. 2.2.

3   Experiment and Analysis

3.1   Experimental Data

In order to verify the effect of the method proposed in this chapter, we extracted 3027 academic papers from the metallurgical field (IPC classification number TF1) from the CNKI journal library, and we extracted these abstracts of Chinese and English to form an alignment document. The sentence alignment tool extracts a total of 24216 pairs of Chinese and English parallel sentences as the data set for the Chinese-English phrase table. At the same time, 791 English and Vietnamese aligned documents were retrieved from Wikipedia and 31640 pairs of English and Vietnamese sentences were extracted from sentence pairs by using the sentence alignment tool as the data set for English-Vietnamese phrases.

For the Chinese metallurgical terminology, we selected 500 training Chinese corpora as conditional random field model from the above 3027 CNKI Chinese abstracts, and the others as test corpora. At the same time, we extracted 1,500 papers from the CNKI periodicals which field is about economics (IPC classification number F11), and the Chinese abstracts as the background domain corpus.

3.2   Evaluation Index

In order to verify the effectiveness of the method, we select the Accuracy rate (P), Recall rate(R), F value to be the evaluation indices:

The Accuracy rate (P)=The number of domain terms correctly identified by the system/The number of domain terms identified by the system*100%

The Recall rate (R)=The number of domain terms correctly identified by the system/The total number of domain terms in the document*100%

F1 value=2*P*R/(P+R)*100%

3.3   Experimental Results and Analysis

Experiment 1: A comparison of automatic Chinese metallurgical term extraction with different features

In order to verify the effectiveness of the feature, the experiment sets up the same Chinese corpus case, and adopts different characteristics. The experiment

Table 5. An example for Chinese-English-Vietnamese metallurgy domain terminology

Metallurgical Terminology in Chinese	Metallurgical Terminology in English	Metallurgical Terminology in Vietnamese
冶金法	Metallurgical Method	Luyn kim
制备	Preparation	s chun b
太阳能级硅	Solar Grade Silicon	Silicon c?p nng_l?ng m?t_tr?i
工艺	Craft	? th? công
湿法提纯	Hydrometallurgy Purification	hydrometallurgy s?ch
半工业化	Semi-industrialization	Semi-công nghi?p

Table 6. Copus statistics table

Corpus	Thesis of CNKI The Metallurgical	Chinese-English parallel pairs (pairs)	Wikipedia English-Vietnamese aligned documents (articles)	English-Vietnamese parallel pairs (pairs)	CNKI Economic Papers (articles)
Total corpus	3027	24216	791	31640	1500
Chinese term extraction training corpus	500				
Chinese term extraction test corpus	2527				
The corpus of Economic background					1500
Phrases machine translation training corpus		24000		31000	
Phrase Machine Translation Test Corpus		216		640	

Table 7. The experimental results for Chinese metallurgy domain terminology

Characteristics	Accuracy rate (%)	Recall rate (%)	F1 (%)
Word, POS	16.34	18.27	17.25
Word, POS, MI	42.76	36.54	39.4
Word, POS, MI, LE, RE	58.29	54.71	56.44
Word, POS, MI, LE, RE, DomainFreq	71.66	69.35	70.48
Word, POS, MI, LE, RE, DomainFreq, ContrastFreq, ΔFreq	84.23	79.48	81.78
Word, POS, MI, LE, RE, DomainFreq, ContrastFreq, DomainFreq, ContrastFreq,	85.16	81.69	83.39

based on the conditional random field proposed in this paper is used to compare the term extraction in Chinese metallurgy field (Table 6).

At first, The experiment selects different characteristics. Based on the training data of 500 Chinese abstracts, a term extraction model of Chinese metallurgy based on conditional random field was trained, and this model will be used. In the remaining 2527 Chinese abstract test corpus, we will extract the Chinese metallurgical field terms. And then, we will compare the Accuracy rate (P), Recall rate (R), and F value in the first 500 terms. The experimental results are shown in Table 7.

As can be seen from the experimental results from Table 7, if only words and parts of speech are selected as features, the accuracy rate, recall rate and F-value of the Chinese metallurgical terminology which is extracted are the worst.

The value is 16.34%, 18.27%, 17.25% respectively. With the addition of mutual information, the information entropy, domain frequency, the frequency of the background field and the composite features, the evaluation index would gradually increase. Finally the Accuracy is 85.16% and the Recall is 81.69%. It shows that mutual information, left and right information entropy, domain frequency and so on have a very positive effect on the metallurgical term extraction.

Experiment 2: A comparison in Automatic Chinese-Vietnamese bilingual metallurgical term extraction on different corpus sizes

In order to validate the influence of the terminology extraction method of Chinese and Vietnamese bilingual metallurgy based on the pivot language, the training corpus and test corpus scale of the data set are changed, and the number of bilingual terminology pairs is extracted.

In this paper, we firstly select all the features defined in training data onto 500 Chinese abstracts to train Chinese metallurgical terminology extraction model based on conditional random field, and then use this model. In the remaining 2527 Chinese abstract test corpus, we will extract 6738 terms of Chinese metallurgical terminology. In 24216 pairs of English-Chinese sentence pairs and 31640 pairs of English and Vietnamese sentence pairs on the closed corpus, we will change the data set of training data and test data and use Niutrans open source tools based on the phrase statistical machine translation model. We will respectively construct the Chinese-English bilingual sentences and English-Vietnamese sentences which bases on the English and Chinese bilingual phrases and the English and Vietnamese bilingual phonetic tables. And we will obtain the Chinese-Vietnamese bilingual phrase alignment tables through the mapping of the phrase table. Finally, the Chinese-Vietnamese bilingual phrasal table is filtered by using the Chinese metallurgical terminology extracted in step 1, and the number of terms in Chinese-Vietnamese bilingual metallurgical field obtained from different languages is compared. Experimental results shown in Table 8 and the trend shown in Fig. 2.

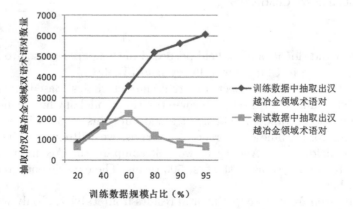

Fig. 2. Chinese-Vietnamese bilingual metallurgy terminology pairs on different corpus sizes.

Table 8. The experimental results for automatic bilingual terminology extraction

Proportion of Training data (%)	Proportion Test data (%)	Number of Chinese-Vietnamese phrase pairs extracted from training data	Number of Chinese-Vietnamese phrase pairs extracted from test data
20	80	831	657
40	60	1739	1651
60	40	3576	2243
80	20	5194	1189
90	10	5628	772
95	5	6065	673

On the selected closed corpus, we can get some information from Table 8: (1) With the increasing of the corpus of Chinese-English and English-Vietnamese sentence-aligned training data, the term pairs of Chinese and Vietnamese bilingual metallurgical fields extracted from the training data would gradually increase, while the Chinese-Vietnamese bilingual metallurgical terms extracted from the test data would gradually decrease. The reason is that the more corpus, the number of terms included in the more; (2) The Chinese-Vietnamese bilingual metallurgical field terms with to 6738 pairs which are extracted from the training data and test data; (3) It can be seen from Fig. 2 that the number of terms extracted from the training data and the test data has not changed greatly since the scale of the training data reaches 80%, because the corpus scale has covered most of the term pairs.

Experiment 3: A Comparison of Chinese-Vietnamese metallurgical term extraction based on different alignment

In order to compare the validity of bilingual term extraction based on phonetic pivots and bilingual terminology extraction based on word alignment pivots, this contrast experiment is set up.

In Experiment 1, all the domain terms are selected and the Chinese metallurgical term extraction method based on conditional random field is adopted. And Chinese metallurgical terminology is extracted from Chinese corpus of test. After that, we choose the word alignment method of GIZA ++ on Chinese-English and English-Vietnamese parallel sentences to align the closed corpus in the experiment 2. Firstly, the Chinese-English and English-Vietnamese sentences aligned text are used to construct Chinese-English word alignment table and English-Vietnamese word alignment table.

The Chinese-Vietnamese word alignment table is obtained through the English pivoting mapping. In the end, we use the previously extracted Chinese metallurgical terminology to filter the Chinese-Vietnamese word alignment table, and finally get the bilingual terminology of Chinese-Vietnamese metallurgy domain which is based on GIZA ++ word alignment pivots method. At the same time, the bilingual terminology extraction method in Chinese-Vietnamese metallurgical field based on the phrasal pivot of English phrase is used to extract the term pairs in Chinese and Vietnamese bilingual metallurgical fields, in which is based on the phrase alignment pivots method. The results of the two methods are shown in Table 9:

Table 9. Bilingual terminology extraction results based on different alignment

Terminology extraction method	Chinese-Vietnamese bilingual metallurgy terminology pairs
Pivot method based on GIZA ++ word alignment	3898
Pivot method based on the phrase alignment	6738

As can be seen from Table 9, in the closed corpus, the method based on the phrase alignment of the pivot can extract more Chinese-Vietnamese metallurgical bilingual terms compared with GIZA ++ based word alignment pivot method. Analyzing the reasons, we will find that word alignment to extract the majority of the term is short and the long term extraction is not good.

Experiment 4: A comparison of Chinese-Vietnamese bilingual metallurgical terminology extraction based on different methods

In this paper, we accomplish a method to extract terminology from Chinese metallurgy based on statistics, for instance, the chi-square test, the Mutual Information Law and the Information entropy method. Firstly, we will extract the terminology of Chinese metallurgical domain. And then, we use the English as the pivot based on the phrase statistical machine translation model, so that we can construct the Vietnamese bilingual phrases. And then, we will use the Chinese metallurgical terminology to filter the Chinese-Vietnamese bilingual phrase table in the first step and we can obtain the bilingual terminology pairs. The experimental results are shown in Table 10:

Table 10. Bilingual terminology extraction results based on different methods

Algorithm	Accuracy rate (%)	Recall rate (%)	F-value (%)
Chi-square test and pivot	49.56	41.23	45.01
Mutual Information Law and pivot	52.14	49.97	51.03
Information entropy and pivot	54.73	51.48	53.06
CRFs and pivot	69.45	61.74	65.37

As can be seen from Table 10, the accuracy rate is only 49.56%, and the recall rate is only 41.23% which fuse chi-square tests and English pivot of Chinese-Vietnamese metallurgy bilingual terminology extraction method. But the extracting method fuses the mutual information or information entropy and the English pivot of Chinese-Vietnamese and metallurgy field bilingual terminology which accuracy and recall rate should be higher. Analyzing the reason, we can find that the unit and the terminology characteristics of the mutual information, left and right information entropy as well as domain frequency and background domain frequency terms have a positive contribution to the Chinese-Vietnamese and metallurgy field bilingual terms.

4 Conclusion

The acquisition of terminology in the bilingual field is a difficult task. In order to solve the problem of Chinese and Vietnamese bilingual corpus scarcity, using English as a pivotal language, using the pivotal idea to extract probability phrases. And we will get the bilingual glossary of Chinese and Vietnamese metallurgy fields through the identification and filtering of terms. Experiments show that the proposed method is effective. With the help of pivot language, it can solve the problem of bilingual terminology extraction in the certain extent.

Acknowledgment. This work was supported by National Natural Science Foundation of China (Grant Nos. 61761026, 61732005, 61672271, 61472168, 61762056), Science and technology leading talents in Yunnan, and Yunnan high and new technology industry project (Grant No. 201606), Natural Science Fundation of Yunnan Province (Grant No. 2018FB104), and Talent Fund for Kunming University of Science and Technology (Grant No. KKSY201703005).

References

1. Li, X.: Terminology and machine translation: experimental results and construction of terminological databank. Res. Explor. Lab. **27**(11), 51–56 (2008). (in Chinese)
2. Liu, L.: Research on automatic bilingual term extraction technology for patents. Shenyang Institute of Aeronautical Engineering, Shenyang (2009). (in Chinese)
3. Sun, L., Jin, Y.: Automatic extraction of bilingual term lexicon from parallel corpora. J. Chin. Inf. Process. **14**(6), 33–39 (2000). (in Chinese)
4. Erdmann, M., Nakayama, K., Hara, T., Nishio, S.: An approach for extracting bilingual terminology from wikipedia. In: Haritsa, J.R., Kotagiri, R., Pudi, V. (eds.) DASFAA 2008. LNCS, vol. 4947, pp. 380–392. Springer, Heidelberg (2008). https://doi.org/10.1007/978-3-540-78568-2_28
5. Aker, A., Paramita, M., Gaizauskas, R.: Extracting bilingual terminologies from comparable corpora. In: Proceedings of the 51st Annual Meeting of the Association for Computational Linguistics (ACL 2013), pp. 402–411. ACL (2013)
6. Liu, S., Zhu, D.: Automatic term alignment based on advanced multi-strategy and Giza++ integration. J. Softw. **3**, 1650–1661 (2015). (in Chinese)
7. Haque, R., Penkale, S., Way, A.: Bilingual termbank creation via log-likelihood comparison and phrase-based statistical machine translation. In: Proceedings of the 4th International Workshop on Computational Terminology (COLING 2014), pp. 42–51(2014)
8. Zhang, J., Cao, C.G., Wang, S.: Web-based term translation extraction and verification method. Comput. Sci. **39**(3), 170–174 (2012)
9. Bourigault, D.: Surface grammatical analysis for the extraction of terminological noun phrases. In: Proceedings of the 14th Conference on Computational Linguistics, Nantes, France, vol. 3, pp. 977–981. Association for Computational Linguistics (1992)
10. Justeson, J.S., Katz, S.M.: Technical terminology: some linguistic properties and an algorithm for identification in text. Nat. Lang. Eng. **1**(1), 9–27 (1995)
11. Ananiadou, S.: A methodology for automatic term recognition. In: Proceedings of the 15th Conference on Computational Linguistics, Kyoto, Japan, vol. 2, pp. 1034–1038. Association for Computational Linguistics (1994)

12. Frantzi, K., Ananiadou, S., Mima, H.: Automatic recognition of multi-word terms: the C-value/NC-value method. Int. J. Digit. Libr. **3**(2), 115–130 (2000)
13. Zhan, Q., Wang, C.: A hybrid strategy for chinese domain-specific terminology extraction. In: Proceedings of International Joint Conference on Neural Networks (IEEE 2016), pp. 217–221. IEEE (2016)
14. Takeuchi, K., Collier, N.: Use of support vector machines in extended named entity recognition. In: Proceedings of the 6th Conference on Natural Language Learning, Stroudsburg, PA, vol. 20, 119–125. Association for Computational Linguistics (2002)
15. Lafferty, J., Mccallum, A., Pereira, F.C.: Conditional random fields: probabilistic models for segmenting and labeling sequence data. In: Proceedings of the 18th International Conference on Machine Learning, pp. 282–289. Morgan Kaufmann Publishers, San Francisco (2001)
16. Sun, M., Li, L., Liu, Z.: Unsupervised bilingual terminology extraction algorithm for Chinese-English parallel patents. J. Tsinghua Univ. (Sci. Technol.), **54**(10), 1339–1343. (in Chinese)
17. Li, P., Sun, M., Xue, P.: Fast-champollion: a fast and robust sentence alignment algorithm. In: Proceedings of International Conference on Computational Linguistics (COLING 2010), pp. 710–718 (2010)
18. Li, D.: Chinese term extraction in specific domain. Dalian University of Technology, Dalian (2011). (in Chinese)
19. Wallach H.: Efficient Training of conditional random fields. Effic. Train. Cond. Random Fields (2002). http://www.cogsci.ed.ac.uk
20. Sutton, C., McCallum, A.: Dynamic conditional random fields: factorized probabilistic models for labeling and segmenting sequence data. J. Mach. Learn. Res. **8**, 693–723 (2007)
21. Guo, J., Xue, Z., Zhengtao, Y., Zhang, Z., Zhang, Y., Yao, X.: Named entity recognition for the tourism domain based on cascaded conditional random fields. J. Chin. Inf. Process. **23**(2), 47–52 (2009). (in Chinese)
22. Koehn, P., Och, F.J., Marcu, D.: Statistical phrase-based translation. In: Proceedings of the Human Language Technology and North American Association for Computational Linguistics Conference, Edmonton, Alberta, pp. 48–54 (2003)

Big Data Analytics and Smart Computing

Preprocessing and Feature Extraction Methods for Microfinance Overdue Data

Jiahao Wang[1] , Liang Zhang[1], Peiyi Shen[1(✉)], Guangming Zhu[1], and Yuhuai Zhang[2]

[1] Xidian University, Xi'an, China
jhwangl@stu.xidian.edu.cn,
{liangzhang, pyshen}@xidian.edu.cn
[2] Xi'an University, Xi'an, China

Abstract. With rapid development of the microfinance industry, the number of customs has surged and the bad debt rate has risen dramatically. Increase of the overdue customers has led to a substantial augment in business volume in the collection industry. However, under the current policy of protecting customer privacy, the lack of credit information, as well as the constraints of collection's cost and scale is two major issues that the collection industry comes across. This paper proposes a repayment probability forecasting system that does not rely on credit information, but can improve the collection efficiency. The proposed system focuses on preprocessing more than one hundred thousand overdue data, using word2vec to locate the keyword, extracting features of the data according to their types. Our system also depends on mature machine learning models to predict the customers' ability of repayment, including LR, GBDT, XGBoost and RF. Meanwhile, we not only use AUC but also design a new evaluation index that can be adapted to the business background to evaluate the system's performance. Experiments results show that, in the case of a surge in business volume and around 1.5% of the overdue costumers' repayment, through our system, collection on only the first half of the customers with high scores can increase the repayment rate by at least 1.2%, which greatly increases the work efficiency and reduces manual labor for collection.

Keywords: Repayment probability forecasting · Preprocessing
Feature extraction · Machine learning models

1 Introduction

In the economic downturn, the scale and ratio of non-performing loans [1] of commercial banks have all increased by a certain amount. The "City Regulatory Indicators Status Table (Quarterly)" published by the China Banking Regulatory Commission shows that in late 2016 the balance of non-performing loans increased from 1274.4 billion Yuan to 1512.2 billion Yuan in late 2015, increasing by 18.66%. The non-performing loan ratio also increased from 1.67% to 1.74%. In 2016, compared to 2015, "substandard", "doubtful" and "loss" loans respectively increased by 2.84%, 25.69% and 55.36%.

© Springer Nature Singapore Pte Ltd. 2018
Z. Xu et al. (Eds.): Big Data 2018, CCIS 945, pp. 23–43, 2018.
https://doi.org/10.1007/978-981-13-2922-7_2

China has implemented a five-category loan classification system [2] since 1998, dividing loans into "performing", "special mention", "substandard", "doubtful" and "loss", especially the latter three cases are non-performing loans. Considering that some unfavorable factors existing in "special mention" loans may affect the repayment ability, we should combine non-performing loans with "special mention" loans to analyze risk.

According to the annual reports issued by 25 listed banks in late 2016, the total number of non-performing loans of listed banks was 1185.29 billion Yuan, with an increase of 167 billion Yuan and a 16.40% growth rate over the end of 2015 (in Fig. 1). The overall non-performing loan ratio of listed banks in late 2016 was 1.70%. Compared to 2015, it rose 0.07 percentage points. In late 2016, the listed banks' total amount of loans for "special mention" was 2393.056 billion Yuan. There is an increase of 261.8 billion Yuan and a 12.28% growth rate over the end of 2015. The total risky exposure including non-performing loans and "special mention" loans was 3578.34 billion Yuan, with an increase of 428.8 billion Yuan and a 13.61% growth rate over 2015 [3].

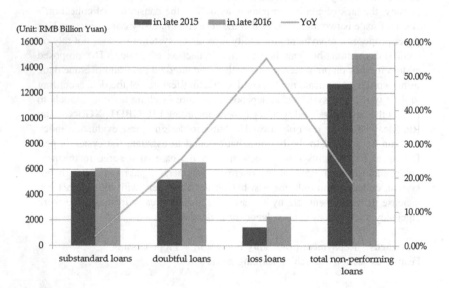

Fig. 1. Non-performing loans of commercial banks in late 2016 comparing to 2015. YoY (Year-on-year percentage) refers to the current period's growth rate over the same period of the previous year.

In the current non-performing assets market, the bad debt ratios of banks and consumer finances range from 2% to 3%, while the bad debt ratios of P2P and small loan companies are even more than 10%. It is expected that the actual annual non-performing assets will exceed 2.5 trillion Yuan. Therefore, for a lending institution, a small increase in the repayment rate of non-performing asset can bring considerable benefits. Once the repayment rate of one hundred million non-performing assets has increased by 1%, it will recover at least 1 million. Thus, non-performing assets are also valuable.

In the collection field, the big data technologies are widely used for many applications including but not limited to: (1) Predicting the difficulty of collection, (2) Guiding the formulation of collection and incentive strategies, (3) Intelligently distributing cases, (4) Assessing and tracking the performance of cases in process of collection, and (5) Effectively monitoring the legalization and regularization of collection.

Another application of big data technologies in the collection field is to make repayment forecasts and then price credit [4]. According to reports, currently the industry's big data model has included dozens of dimensions. For a collection platform with hundreds of thousands of cases, its accuracy of forecasting repayment rate model based on big data has reached a tolerance of plus and minus 5%.

In recent years, illegal access to personal information of citizens, providing personal information of citizens, and other serious circumstances are clearly defined. Further emphasis is placed on protecting the security of citizens' personal information and their legitimate rights and interests.

In fact, the collection industry cannot obtain the customer's credit information. The lack of detailed customers' information makes many prediction models based on credit ratings unusable. Under these circumstances, we need a technology to forecast the probability of re-payment according to limited collection data.

The rest of the paper is organized as follows. Section 2 discusses related work. In Sect. 3, we present our proposed preprocessing and feature extraction methods in order to construct our own datasets in financial collection field. Section 4 presents comprehensive evaluation results by comparing several Machine Learning methods. Finally, we conclude the paper in Sect. 5.

2 Related Work

From the perspective of Data mining, mathematical tools that can perform default analysis or personal credit evaluation mainly include four types: *discriminant analysis* [5], *logistic regression* [6], *neural network* [7] *and classification decision tree* [8]. In recent years, *ensemble learning* [9] technologies like Bagging and Boosting also become stable and effective tools for improving classification accuracy or regression accuracy of machine learning models.

- **Discriminant Analysis.** Discriminant analysis is a classification method that measures the degree of importance of factors in a specific category. For example, it can examine the main factors causing the customer's breach of contract. As long as all possible influencing factors are determined, the model can use these factors to make a discriminant analysis between the main and the secondary factors of the breach of contract. On the premise that the probability of misclassification is the smallest or the misjudgment loss is the minimum, a calculation criterion is established. According to the criterion, whether or not the contract is breached is determined for a given sample. The calculation of the probability of customer default is a multivariate discriminant analysis. Specifically, the existing customer default data are grouped by corresponding customer credit classification. Select the corresponding independent variables for statistical analysis of each grouped samples and obtain combined covariance matrix. Then Mahalanobis distance is calculated by substituting the corresponding variable in the new sample data. The smallest distance indicates that the new sample data is most similar to the sample, and thus falls into this category (default or non-default). And according to the distance, we can find out the default probability of new customers.

- **Logistic Regression.** This kind of model is a traditional tool for calculating the probability. Its basic principle is to make a 0-1 classification of default and non-default samples of existing customers. According to business rules, certain indicators are selected as explanatory variables. After obtaining a sample of the priori data, set P as the customer default probability, $(1-P)$ as the probability that the customer does not default, and take the ratio $P/(1-P)$ as the natural logarithm $Ln(P/(1-P))$, that is, P for LOGIT conversion, thus establishing a linear regression equation for analysis. Practice has shown that this model has a good effect on judging the relationship between two classification variables. The default event belongs to the dichotomy category, so this model has a good applicability in calculating the probability of default.

- **Neural Network.** The neural network model is a credit analysis model developed in recent years. It is very similar to non-linear discriminant analysis. It discards that the assumption that the variables of the crisis prediction function is linear and independent of each other. The neural network model can deeply mine the "hidden" relationship between the predictors and is becoming an important basis for the nonlinear default prediction function. In the human brain, whether the electronic signals traveling between neurons are suppressed or activated depends on what the neuron network has learned in the past. Similarly, the behavior of artificial neurons constructed using hardware or software has a similar manner to biological neurons. The behavior of neural networks comes from the collective behavior of interrelated units. The association between neurons is not fixed which can be modified during the learning process generated by the interaction between the neural network and the outside world.

- **Classification Decision Tree.** In the classification problem, Classification decision tree indicates the process of classifying samples based on features. It can be consider as a set of if-then rules, or as a conditional probability distribution defined in feature space and class space. Compared to Naive Bayes classification [10], the advantage of decision tree is that the construction process does not require any domain knowledge or parameter settings. Therefore, in practical applications, decision trees are more applicable to the detection of knowledge discovery. The classification decision tree model is a tree structure that describes the classification of samples. The decision tree consists of nodes and directed edges. There are two types of nodes: internal nodes and leaf nodes. Internal nodes represent a feature or attribute and leaf nodes represent a class. The process of classification is starting from the root node where a feature of the sample is tested and the sample is assigned to the node's child node according to the test result. At this time, each child node corresponds to a value of the feature. This process recursively moves down until the sample reaches the leaf node, and finally assigns the sample to the leaf node's class. The principle and operation of the decision tree model is relatively simple, and the system development is less difficult. It is mainly used in the lack of mature statistics and measurement analysis capabilities but having a considerable number of customer samples. In addition, decision tree models can more effectively handle the interactions between variables than credit scoring models. Even in the absence of some variables, the decision tree model can generate credit scores. The disadvantage of decision tree is that in some of the bottom "cells", there may be very little data, which cannot meet the requirements for sample size statistics.

- **Ensemble Learning.** Ensemble learning completes learning tasks by building and combining multiple learners. It usually can obtain significantly better generalization performance than a single learner. There are two ideas for ensemble, serial and parallel. In serial idea, the base learner is generated in turn. The output of the former learner is used as the input of the latter. If one of the base learners is wrong, it will affect the subsequent learner. Boosting is a typical serial ensemble learning method. However, in parallel idea, each base learner is built using a certain classification algorithm and works independently. Therefore, the error of a certain base learner will not affect the classification results of other base learner and has little impact on the overall decision-making. Bagging is a typical parallel ensemble learning method. Yao [11] used bagging and boosting techniques and selected CART method to build an ensemble model. The results show that Ensemble Learning methods can improve the accuracy of model classification. Wang [12] constructed base learners with logistic regression, decision tree, neural network and SVM models via bagging, boosting and stacking methods. The results show that the ensemble model has a significant improvement in accuracy and type-two errors.

In Sect. 4, we will select parts of these mathematical tools to train prediction model on our own overdue dataset which is constructed in Sect. 3, in order to get as good generalization performance as possible.

3 Data Preprocessing and Feature Extraction

In this part, we preform data preprocessing and feature extraction to construct our particular datasets of overdue data. During the construction process, we transform the data through multiple steps.

3.1 Sparseness of High-Dimensional Data

The origin overdue data contains more than eighty data items and plenty of sound recordings that record the call between the collectors and the customers. Plenty of redundant and invalid information exists in these data. In the process of preprocessing, sparseness of high-dimensional data aimed at selecting the key data items related to the business is used.

Data Item Filtering

The original form in collection data's original form has many data items that both contain useful and useless information. We filtered data items based on the following four rules.

Rule 1: Removing empty or substantially empty data items;
Rule 2: Removing data items that have exactly the same value;
Rule 3: Removing duplicate data items;
Rule 4: Removing data items that cannot be characterized.

Among them, Rule 4 is embodied in the filtering of non-numeric data items. Especially, these data items cannot be converted into ordered numerical features or feature codes, such as item 'name'.

Keywords Extraction

In the collection data, it also contains a large amount of sound recording information. We try to extract some words that the collectors pay more attention to from the recording information. These words may involve some very important features or certain important types of features. Data items are further filtered to transform the high dimension data into sparse data that retained most of the important features.

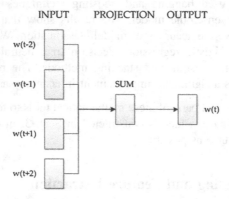

Fig. 2. CBOW Model. There are three stages: input, projection and output. The input layer is comprised of word vector of n-1 words around a word A. If n takes 5, the first two and the last two words of the word A (can be written as w(t)) are w(t−2), w(t−1), w(t + 1), w(t + 2). Correspondingly, the word vectors of the 4 words are denoted as v(w(t−2)), v(w(t−1)), v(w (t + 1)), v(w(t + 2))). The n-1 word vectors are added from the input layer to the projection layer. The output layer is a Huffman tree. Starting from the root node, the projections layer values need to be continuously classified along the tree by Logistic Regression and the intermediate vectors and word vectors are constantly modified.

Step 1: we use speech recognition applications based on hidden Markov models [13] to convert a large number of audio recordings to text information. Then we combine these audio recordings' texts into one file.

Step 2: we use the file as an input to perform word segmentation. We complete Chinese words segmentation based on mmseg4j [14] that implements the mature and efficient MMSeg algorithm [15]. As a result, we get many words that are separated by spaces.

The word segmentation idea of complex matching algorithm in MMSeg is introduced as follows. Complex algorithm takes the word length of adjacent words into consideration based on the maximum matching rules. It designs four ambiguity resolution rules to guide the word segmentation. In this realization, MMSeg considers three adjacent words as a chunk and applies the following four rules to the chunk.

- Maximum matching.
- Largest average word length.
- Smallest variance of word lengths.
- Largest sum of degree of morphemic freedom of one-character words.

Step 3: CBOW (Continuous Bag-of-Words Model) [16] is used based on the Hierarchical Softmax in word2vec to train the word-segmented texts on the network structure. The training process is shown in Fig. 2. It uses the context of the current word (w(t)), including w(t−2), w(t−1), w(t + 1), w(t + 2), to predict the current word.

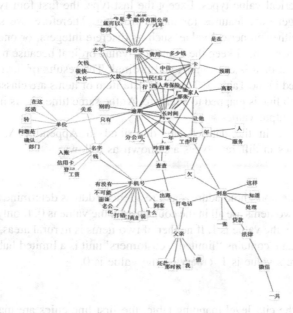

Fig. 3. Keywords Network. Each word represents a key word, which is connected with another word by a link. The closer the relationship between two words is, the darker the link's color is. (Color figure online)

Step 4: we obtain word vectors for each word. Words with similar meaning generally have similar word vectors. We import the word vector model and use word2vec to find some origin keywords' word vectors, such as 'ID', 'Company' and so on. Then we utilize these origin keywords' word vector to find similar words that might be new keywords by comparing word vectors. At the same time, all the word vectors are clustered by k-means [17]. Word clusters with higher clustering degrees are selected to construct the following keyword Network.

With word2vec, we get a series of semantically related words that might be new keywords and its corresponding similarity degrees as long as enter a keyword. So that we could build a keyword network (Fig. 3) reflecting the keywords' semantic relationship, which can help us filter data items and reduce the dimension of the data. We could further compare the differences of semantic structure between positive and negative samples' keyword network to assist data analysis.

3.2 Feature Extraction

After filtering the data items in the previous step, we can perform data feature extraction on the remaining data items. In addition, although data is divided into three parts by overdue stage in actual business, we could combine these data into one datasets because different overdue stages' data has the same data items except 'overdue stages'. Meanwhile, a bigger datasets is beneficial for us to train a more accurate model.

We can classify all data items into five types, including text, address, date, mobile number and numerical value types. Except the last type, the first four types cannot be directly used to generate features for model training. Therefore, we should convert these types into either numerical value, such as discrete integers, or one-hot encoding [18]. Meanwhile, we should keep the numerical value ordinal because many classifiers often acquiesce that data is continuous and ordinal. The results of data items' feature extraction are listed in the Table 1. The input data item or items are classified into these four types and each line is mapped to a feature. At the same time, theirs input values are transformed into output values.

Overall features in the datasets are described in Appendix A. Some typical extraction processes in different types are shown as follows.

Text Type

- Rural level. The household address and residence address determine the customers' rural level. If two items are all in the countryside, the value is 0. If only one item is in the countryside, the value is 1. If neither of two items is in rural areas, the value is 2.
- Unit. If unit name contains "limited", customers' unit is a limited liability company and the feature's value is 1, otherwise the value is 0.

Address

By constructing the city level mapping table, the first-line cities are mapped to 4, the new first-tier cities are mapped to 3, the second-tier cities are mapped to 2, the third-tier cities are mapped to 1, and the fourth-tier and fifth-tier cities are mapped to 0. We split the address text and extract the name of the prefecture-level city in the original data (or the first 6 digits from the ID card are extracted and matched with the constructed ID card city table, and the prefecture-level city name is also obtained). By joining the city-level mapping table, the address is converted to a value that represents the city level.

Table 1. Data items transformation results

Type	Data item/ items	Input value	Output value	Feature name
Text	Household address & Residence address	Two address	2, 1, 0	Rural level
	Education level	Undergraduate, specialist, vocational high school, high school,	3, 2, 1, 0	Education level
	Unit name	Unit name	1, 0	Unit
	Gender	Female, male	[1, 0], [0, 1]	Gender
	Marital status	Married, unmarried	[1, 0], [0, 1]	Marital status
	Contact' relationship with customer	Parents, family members, relatives, friends, colleagues	4, 3, 2, 1, 0	Contact relationship
	Second contact's relationship with customer	Parents, family members, relatives, friends, colleagues	4, 3, 2, 1, 0	Second contact relationship
	Overdue stage	m2, m3,..., m9	2, 3,..., 9	Overdue stage
	Credit products	Eight types of products	8 bits one-hot encoding	Product type
Address	Residence address	An address	4, 3, 2, 1, 0	Residence address
	Household address	An address	4, 3, 2, 1, 0	Household address
	Unit address	An address	4, 3, 2, 1, 0	Unit address
	Shop address	An address	4, 3, 2, 1, 0	Shop address
Date	Account opening date & First use date	Two dates	Number of days	Days between O&U
	Account opening date & First use date	Two dates	Number of months	Months between O&U
	First overdue date & First use date	Two dates	Number of days	Days between F&U
	First overdue date & First use date	Two dates	Number of months	Months between F&U
	Latest overdue date & First overdue date	Two dates	Number of days	Days between L&F
	Latest overdue date & First overdue date	Two dates	Number of months	Months between L&F
Mobile number	Mobile	A mobile number	4 bits one-hot encoding	Mobile
	Contact mobile	A mobile number	4 bits one-hot encoding	Contact mobile
	Second contact mobile	A mobile number	4 bits one-hot encoding	Second contact mobile

Date Type

We can obtain days and months by calculating difference between two different data items of the date type. For example, the account opening date can minus first use date, which will produce two kinds of results that are the difference of days and months. Similarly, pairs of first overdue date and first use date, the latest overdue date and first overdue date can be subtracted each other. Therefore, each pair of data items of date type can produce number of days and months that are always ordinal features.

Mobile Number

We find that most of mobile number can be classified by the type of tele-communications operator.

Firstly, we construct the mobile number segment mapping table, and map China Mobile's number segment to one-hot encoding [1, 0, 0, 0], China Unicom's to [0, 1, 0, 0], China Telecom's to [0, 0, 1, 0], virtual operator to [0, 0, 0, 1] respectively. Secondly, the first three digits of the mobile phone number in the original data are extracted as number segment and then match with the number segment mapping table. According to this procedure, the mobile phone numbers are converted into one-hot encoding which can be treated as features.

Transformation Rules

The transformation rules for extracting feature and assignment can be summarized as follows.

Rule 1: Using one-hot encoding to discretize feature representation for most of categorical data items;

Rule 2: If the data item's values are ordered, directly mapping to [0, N-1] (N equals to the number of its value);

Rule 3: If it satisfies Rule 2, the larger assignment value should correspond to more favorable feature value, such as higher educational level or closer relationship has larger value.

3.3 Data Normalization

Normalization of data [19] means scaling the data so that it falls into a specific range. Data normalization has many advantages, which is also used in deep learning method.

Firstly, it meets some solution's need. For example, dealing with classification problems in SVM [20] needs data to be normalizing, which has a great impact on accuracy.

Secondly, it can speed up the search for the optimal solution or convergence, as shown in Fig. 4. In Fig. 4(b), fewer search times are used to find the extremum after feature scaling, comparing with Fig. 4(a).

Thirdly, it makes data dimensionless, which removes unit restriction of data and transforms data into dimensionless numeric value. It is convenient for different units or magnitude indicators to be carried out and weighted.

We used the z-score (zero-mean) normalization known as standard deviation normalization to normalize the numerical features extracted in previous step. Normalized

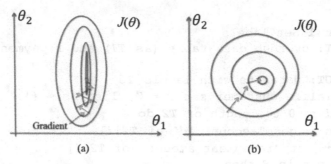

Fig. 4. Feature scaling. (a) $J(\theta)$ has elliptical contour lines if not scaled; (b) $J(\theta)$ has circular contour lines after scaling. In (a) and (b), θ_1 and θ_2 represent two features in the cost function $J(\theta)$.

data follows the normal distribution, which means its mean value is zero and its standard deviation is one. The formula is as follows:

$$x^* = (x - \mu)/\sigma \tag{1}$$

where μ is the mean of all sample data and σ is the standard deviation of all sample data. The difference between Min-max normalization and z-score normalization is that Min-max normalization is only a multiple reduction of the variance and mean deviation of the original data, while z-score normalization makes the normalized data variance one. This is more advantageous for many algorithms, but the disadvantage is that if the original data does not follow Gaussian distribution, the normalized distribution is not good.

3.4   Set Labels

After data normalization, we should set a label for each record to construct a complete datasets. We obtain all customers' repayment records from repayment table which contains each customer's repayment amount, account status and other information. The whole algorithm of setting labels is presented in Algorithm 1.

Firstly, we initialize repayment customer set and customer list as $S = \emptyset$ and Ls = {} (line 3). Then we combine repayment records that have the same 'financial account number' into one record, which adds up 'repayment amount' in these records (line 4-13).

Secondly, we merge the overdue data table with re-payment table into a new table by matching the primary key 'financial account number' in two tables (line 14). Then filling the null value in 'repayment amount' of the new table with the value of 0 (line 15).

Algorithm 1 Set labels
1: **INPUT**: overdue data table (as *T1*) and repayment table (as *T2*)
2: **OUTPUT**: datasets with labels *T3*
3: Initialize customer set *S* = ∅, list *Ls* = {}
4: **for** *i* ← 0 to length of *T2* **do**
5:      *num* ← get 'account id' of *T2*[*i*]
6:      *a* ← get 'repayment amount' of *T2*[*i*]
7:      **if** *num* **in** *S* **then**
8:        *Ls* ← get an index list in *T2* where 'account id' = *num*
9:        *pre* ← get 'repayment amount' of *T2*[*Ls*[0]]
10:       'repayment amount' of *T2*[*Ls*[0]] ← *pre* + *a*
11:       *T2* drop *T2*[*i*]
12:     **else**
13:       *S* += *num*
14: *T3* ← *T1* left join *T2* on the key 'account id'
15: Fill *null* value in 'repayment amount' of *T3* with 0
16: **for** *j* ← 0 to length of *T3* **do**
17:     *a* ← get 'repayment amount' of *T3*[*j*]
18:     *d* ← get 'total debt' of *T3*[*j*]
19:     **if** *a* ≥ 0.1 * *d* **then**
20:       'flag' of *T3*[*j*] ← 1
21:     **else**
22:       'flag' of *T3*[*j*] ← 0
23: **Return** *T3*

Thirdly, filter customers that are unwilling to pay back out of all repayment records. We set a threshold that repayment amount must be equal or greater than one-tenth of total debt. We scan the new table and compare 'repayment amount' to 'total debt' multiplied the threshold (line 16–19). Then we add a new column 'flag' that represents the label of each customer or record in the new table. If repayment amount is equal or greater than one-tenth of total debt in one record, we fill the value of 'flag' with 1 which means the record is a positive sample. Otherwise, we fill the value with 0 representing a negative sample (line 20–22). Finally, we get the complete datasets with labels (line 23).

3.5   Feature Importance Analysis

In the feature selection field, we train a model that can measure the importance of feature, such as RF (Random Forest) [21], XGBoost (Extreme Gradient Boosting) [22] (in Fig. 5) and so on. The feature importance analysis refers to the analysis of the importance relation between each feature and the target value in the datasets. The relationship can be expressed by the importance coefficient. The anomaly of features can be analyzed in order to adjust and optimize the model.

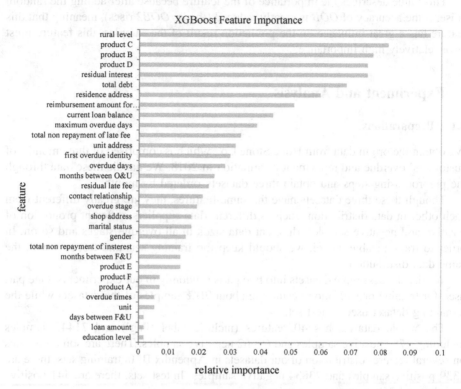

Fig. 5. XGBoost feature importance. It describes the relative importance of each feature in the model, by counting the number of occurrences of features in the middle nodes of all decision trees.

When calculating the importance of a feature X in RF, the specific steps are as follows:

1. For each decision tree, select the out of bag (OOB) [23] to calculate the out of bag error, denoted as $errOOB1$. Each time a decision tree established, training sets are obtained through bootstrap sampling. Meanwhile, one third of the data is not involved in training, called as out of bag. This part of the data is used as validation set to estimate the generalization performance of the decision tree, known as out-of-bag estimate.
2. Randomly add noise to the feature X of all samples of OOB (randomly change the value of the sample at the feature X). Then calculate the out of bag error again, which is denoted as $errOOB2$.
3. Assuming there are N trees in the forest, the importance of feature X can be calculated as follows:

$$Importance(Feature\,X) = \sum(errOOB2 - errOOB1)/N \qquad (2)$$

This value describes the importance of the feature because after adding the random noise, if the accuracy of OOB is greatly reduced (the $errOOB2$ rises), meaning that this feature has a great influence on the prediction result of the samples, this feature must have relatively high importance.

4 Experiment and Analysis

4.1 Preparations

We obtain the origin data from Hone Stone Co., which is comprised of three months of customers' overdue and repayment information in 2018. We handle these data through the preprocessing steps and obtain three datasets ordered by month.

Though these three datasets have the same features, they are greatly different from each other in data distribution, such as different data suppliers, different proportion of positive and negative samples, different data sizes in all overdue stages and so on. In order to train a stable model, we should keep the training sets and test sets with the same data distribution.

We divide these three datasets into two parts randomly in our experiments. One part used for training or validation occupying about 70% samples of the datasets while the remaining dataset used as test set.

The whole datasets has 40 features (include label 'flag') and 71442 samples including 1780 positive samples and 69662 negative samples. There are some statistics on several features distribution of our datasets in Appendix B. In training sets, there are 1339 positive samples and 43834 negative samples. In test sets, there are 441 positive samples and 25828 negative samples.

4.2 Evaluation Index

We choose AUC (Area under Curve) [24] to evaluate the prediction performance of model on training, validation and test sets. Due to its properties that reflect the sample sorting ability of classifier, AUC is not sensitive to class-imbalance. Also many machine learning and grid search methods contain the parameter like 'eval_metric' which can use AUC for validation data.

Considering that great class-imbalance make the final prediction probabilities on the low side (around 0.02) and actual business requirements do not need completely accurate classification results but need a narrowed sample spaces which contain enough positive samples. So some commonly used performance measure like precision and recall [25] cannot satisfy our need. We design a new measure method to evaluate the performance of the model generated for classification as followed.

Firstly, we sort the samples by prediction probabilities and order them from high to low.

Secondly, we choose the median of prediction probability as E and select samples whose prediction probabilities are greater than or equal to E as the final sample spaces where we regard there are rich in positive samples.

Thirdly, we calculate the accurate amount (N) of positive samples in this sample spaces and obtain a percentage $P_{50\%}$ after dividing N by total amount N_{all} of positive samples. The formula is as follows:

$$P_{50\%} = N/N_{all} \tag{3}$$

We use $P_{50\%}$ as performance measure in our prediction system and the model with higher $P_{50\%}$ has better classification performance, which is more intuitive than AUC in reflecting the fraction of coverage of positive samples.

4.3 Comparison to Existing Approaches

We use LR (Linear Regression) and other three ensemble learning methods, RF [27], Gradient Boosting Decision Tree (GBDT) [26], and XGBoost [28] train the model respectively. After optimizing the parameters of the four algorithms with cross validation, we use the test sets to evaluate the performance of the model. The results of the four algorithms are listed in Table 2. In training sets, tasks are measured by time and AUC. In test sets, tasks are measured by AUC and $P_{50\%}$.

Table 2. Evaluation of different methods

	Training		Test	
Model	Time (s)	Train-auc (%)	Test-auc (%)	$P_{50\%}$ (%)
LR	2.64	74.66	73.32	82.31
RF	7.81	74.73	73.17	83.22
GBDT	7.52	78.71	73.48	82.09
XGBoost	5.11	75.6	73.14	83.22

As we can see, the four algorithm has similar performance in test field, whose models all have around 73% test-auc and around 83% $P_{50\%}$. LR and GBDT have higher test-auc than RF and XGBoost while RF and XGBoost have higher $P_{50\%}$ than LR and GBDT.

In training field, LR has obviously faster speed than other three ensemble learning methods, which costs only 2.64 s to train a model. Then XGBoost is little slower than LR, but faster than GBDT and RF. And LR and RF has lower train-auc than GBDT and XGBoost. Comparing to test-auc, GBDT is more overfitting than LR, RF and XGBoost.

4.4 Analysis

The experimental results show that if we use the prediction model to rate and sort the unknown samples, we can get over 80% of total positive samples after filtering half of the data with lower scores. There is some information on test sets shown in Fig. 6. In Fig. 6(a), positive samples only account for 1.51% of the whole test sets. In Fig. 6(b), $P_{50\%}$ equal to 83.22% means 83.22% positive samples' prediction scores are higher than the median's.

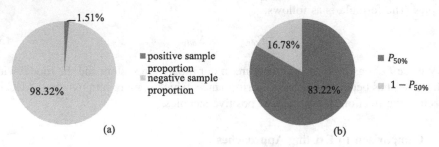

Fig. 6. Information on test sets. (a) positive/ negative sample proportion; (b) $P_{50\%}$ and 1-$P_{50\%}$. In Fig. 6(b), we choose the result of XGBoost or RF.

We propose a way to improve the repayment rate after shrinking the sample spaces by the evaluation index " $P_{50\%}$". That is sorting the samples by their prediction scores and splitting them by the median's score. Half of samples with higher scores can be reserved, on which the collectors should focus. As a result, this way can greatly reduce the sample spaces and keep as many positive samples as possible. There is some information on half of test sets with higher prediction scores shown in Fig. 7.

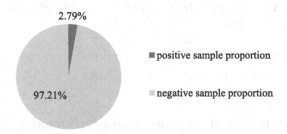

Fig. 7. Information on half of test sets with higher prediction scores.

As we can see in Fig. 7, the positive sample proportion rises by over 1.2% after filtering the test sets comparing to Fig. 6(a). Meanwhile, we can save half of time to focus on this part of data that are rich in positive samples instead of the whole data.

5   Conclusion

In this paper, our proposed data processing and feature extracting addressed to financial overdue data are been proven to be effective for predicting repayment probabilities of samples. Firstly, we complete the sparse of high-dimensional data by filtering the data items based on four principles and keywords' network extracted by Word2Vec on sound recording data. Secondly, we perform data feature extraction on the remaining data items according to their types that we classified and the transformation rules that we summarize. Thirdly, we make data normalization and then set labels of samples referring to repayment records, leading to constructing a complete datasets. Fourthly, we separate the training sets and test sets from datasets into which we combined monthly data. Fifthly, we train the model with training sets and analyze the feature importance in order to optimize the model by removing anomaly features. Finally, we use four Machine Learning methods for a comparative experiment and use test sets to evaluate the performance of the prediction model.

We also put forward a methodology to filter half of original overdue data so that the financial company can focus on the remaining data that contains most of the customers with repayment ability. Therefore, this methodology can make the financial company not only reduce manual labor and save time but also improve efficient.

Acknowledgment. This work is partially supported by the China Post-doctoral Science Foundation (Grant No. 2016M592763), the Fundamental Research Funds for the Central Universities (Grant NO. JB161006, JB161001), the National Natural Science Foundation of China (Grant NO. 61401324, 61305109), and the Natural Science Basic Research Plan in Shaanxi Province of China (Program No. 2016JQ6076).

Appendix A

Overall feature information is listed in the following table (Table 3).

Table 3. Feature information

Index	Feature name	Value	Description
1	Rural level	2, 1, 0	If household and residence is in city
2	Education level	3, 2, 1, 0	Education level
3	Unit	1, 0	Whether it is a limited company
4	Gender	[1, 0], [0, 1]	Male or Female
5	Marital status	[1, 0], [0, 1]	Married or unmarried
6	Contact relationship	4, 3, 2, 1, 0	Relation between customer and his contact
7	Second contact relationship	4, 3, 2, 1, 0	Relation between customer and his second contact
8	Overdue stage	2, 3,…, 9	Overdue stage
9	Product type	8 bits one-hot encoding	One of 8 product types
10	Residence address	4, 3, 2, 1, 0	Residence address classified by city level
11	Household address	4, 3, 2, 1, 0	Household address classified by city level

(continued)

Table 3. (*continued*)

Index	Feature name	Value	Description
12	Unit address	4, 3, 2, 1, 0	Unit address classified by city level
13	Days between O&U	Number of days	Days between account opening date and first use date
14	months between O&U	Number of months	Months between account opening date and first use date
15	Days between F&U	Number of days	Days between first overdue date and first use date
16	Months between F&U	Number of months	Months between first overdue date and first use date
17	Days between L&F	Number of days	Days between latest overdue date and first overdue date
18	Months between L&F	Number of months	Months between latest overdue date and first overdue date
19	Mobile	4 bits one-hot encoding	Telecommunications operators type of contact mobile
20	Contact mobile	4 bits one-hot encoding	Telecommunications operators type of contact mobile
21	Age	Numerical value	Age
22	Annual rate	Numerical value	Annual rate
23	Day rate	Numerical value	Day rate
24	Overdue days	Numerical value	Overdue days
25	Overdue times	Numerical value	Overdue times
26	First overdue identity	Numerical value	First overdue identity
27	Cumulative overdue days	Numerical value	Cumulative overdue days
28	Loan amount	Numerical value	Loan amount
29	Instalment periods number	Numerical value	The Number of instalment periods
30	Shop address	4, 3, 2, 1, 0	Shop address classified by city level
31	Maximum overdue days	Numerical value	Maximum overdue days
32	Total debt	Numerical value	Total debt
33	Reimbursement amount for relieving overdue	Numerical value	Reimbursement amount for relieving overdue
34	Residual interest	Numerical value	Current residual interest need to pay
35	Current loan balance	Numerical value	Current loan balance
36	Total non repayment of interest	Numerical value	Current total non repayment of interest
37	Residual late fee	Numerical value	Current residual late fee need to pay
38	Total non repayment of late fee	Numerical value	Total non repayment of late fee before the current period
39	Second contact mobile	4 bits one-hot encoding	Telecommunications operators type of second contact mobile
40	Flag	1, 0	Positive or negative sample

Appendix B

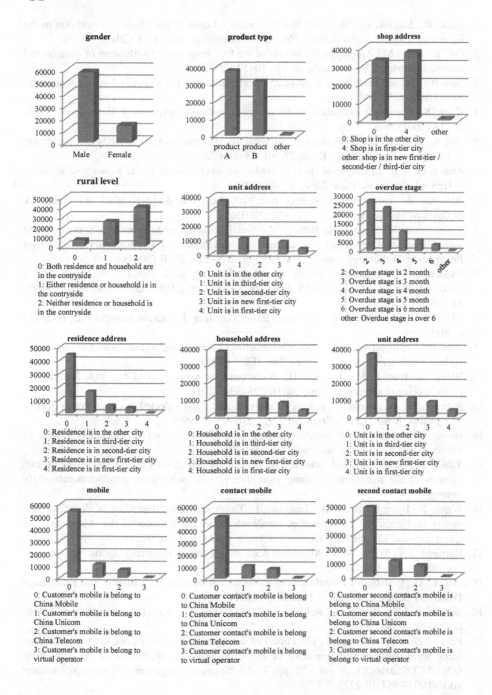

References

1. Beck, R., Jakubik, P., Piloiu, A.: Non-Performing Loans: What Matters in Addition to the Economic Cycle?. Social Science Electronic Publishing, New York (2013)
2. Gu, Y., Ding, M.: An empirical analysis of the five level loan classification of state owned commercial banks. Mod. Manag. Sci. **8**, 10–12 (2002)
3. Sun, B.: The non-performing loans of 25 listed banks panorama (2017). https://xueqiu.com/5780378715/85652443
4. Kang, S.: The credit evaluation model of small-medium enterprises. J. Hebei Univ. **32**(2), 26–33 (2007)
5. Shi, X., Zou, X.: The application of canonical discriminate analysis in credit risk evaluation of enterprise. Study Financ. Econ. **27**(10), 53–57 (2001)
6. Zhang, G., Liu, S.: Empirical study of credit risk evaluation in China's commercial banks. J. Hebei Univ. Econ. Trade **26**(4), 41–45 (2005)
7. Baesens, B.: Using neural network rule extraction and decision tables for credit-risk evaluation. Manag. Sci. **49**(3), 312–329 (2003)
8. Zekic-Susac, M., Sarlija, N., Bensic, M.: Small business credit scoring: a comparison of logistic regression, neural network, and decision tree models. In: International Conference on Information Technology Interfaces, vol. 1, pp. 265–270. IEEE (2004)
9. Dietterich, T.G.: Ensemble methods in machine learning. In: Kittler, J., Roli, F. (eds.) MCS 2000. LNCS, vol. 1857, pp. 1–15. Springer, Heidelberg (2000). https://doi.org/10.1007/3-540-45014-9_1
10. Rish, I.: An empirical study of the naive Bayes classifier. J. Univers. Comput. Sci. **1**(2), 127 (2001)
11. Yao, P.: Credit scoring using ensemble machine learning. In: International Conference on Hybrid Intelligent Systems, pp. 244–246. IEEE (2009)
12. Wang, G., Hao, J., Ma, J., Jiang, H.: A comparative assessment of ensemble learning for credit scoring. Expert Syst. Appl. **38**(1), 223–230 (2011)
13. Rabiner, L.: A tutorial on hidden Markov models and selected applications in speech recognition. Read. Speech Recognit. **77**(2), 267–296 (1990)
14. Huang, R.: rmmseg4j: R interface to the Java Chinese word segmentation system of mmseg4j. Int. J. Radiat. Oncol. **66**(1), 83–90 (2012)
15. Tsai, C.: MMSEG: a word identification system for Mandarin Chinese text based on two variants of the maximum matching algorithm (2000). http://www.geocities.com/hao510/mmseg
16. Wang, L., Dyer, C., Black, A., Trancoso, I.: Two/too simple adaptations of Word2Vec for syntax problems. In: Conference of the North American Chapter of the Association for Computational Linguistics – Human Language Technologies (2015)
17. Hartigan, J.A., Wong, M.A.: Algorithm AS 136: a k-means clustering algorithm. J. R. Stat. Soc. **28**(1), 100–108 (1979)
18. Wu, J., Coggeshall, S.: Foundations of Predictive Analytics. Data Mining and Knowledge Discovery Series. Chapman & Hall/CRC (2012)
19. Schnitzer, J.K., Rice, D.J., Robert Iii, C.F., Zajkowski, A.J.: Data Normalization. US, US20030110250 (2003)
20. Pan, J., Zhuang, Y., Fong, S.: The impact of data normalization on stock market prediction: using SVM and technical indicators. In: Berry, Michael W., Hj. Mohamed, A., Yap, B.W. (eds.) SCDS 2016. CCIS, vol. 652, pp. 72–88. Springer, Singapore (2016). https://doi.org/10.1007/978-981-10-2777-2_7

21. Menze, B.H., Kelm, B.M., Masuch, R., Himmelreich, U., Bachert, P., Petrich, W.: A comparison of random forest and its Gini importance with standard chemometric methods for the feature selection and classification of spectral data. Bmc Bioinform. **10**(1), 1–16 (2009)
22. Xia, Y., Liu, C., Li, Y., Liu, N.: A boosted decision tree approach using Bayesian hyper-parameter optimization for credit scoring. Expert Syst. Appl. **78**, 225–241 (2017)
23. Bylander, T.: Estimating generalization error on two-class datasets using out-of-bag estimates. Mach. Learn. **48**(1–3), 287–297 (2002)
24. Ling, C.X., Huang, J., Zhang. H.: AUC: a statistically consistent and more discriminating measure than accuracy. In: Proceedings of the 18th International Joint Conference on Artificial Intelligence, pp. 519–524. Morgan Kaufmann Publishers Inc. (2003)
25. Goutte, C., Gaussier, E.: A probabilistic interpretation of precision, recall and f-score, with implication for evaluation. Int. J. Radiat. Biol. Relat. Stud. Phys. Chem. Med. **51**(5), 952 (2005)
26. Ye, J., Chow, J.H., Chen, J., Zheng, Z.: Stochastic gradient boosted distributed decision trees. In: ACM Conference on Information & Knowledge Management, pp. 2061–2064 (2009)
27. Svetnik, V., Liaw, A., Tong, C., Culberson, J.C., Sheridan, R.P., Feuston, B.P.: Random forest: a classification and regression tool for compound classification and QSAR modeling. J. Chem. Inf. Comput. Sci. **43**(6), 1947 (2003)
28. Chen, T., Guestrin, C.: XGBoost: a scalable tree boosting system. In: ACM SIGKDD International Conference on Knowledge Discovery and Data Mining, pp. 785–794. ACM (2016)

The Influence of Online Community Interaction on Individual User Behavior

Xiaoyu Hu[1], Naixin Yang[1,2], and Bing Li[1(✉)]

[1] University of International, Business and Economics, Beijing, China
01630@uibe.edu.cn
[2] Beijing Jiaotong University, Beijing, China

Abstract. Social media has experienced significant growth in the past several years. As its user base expands, social media plays an increasingly important role in people's lives. However, few research discusses the impact social media communities have on their users. It is thus imperative to conduct empirical research on impacts of specific social media communities. This paper conceptually describes in a precise manner the effect that social media communities have on people who communicate and interact with each other within such communities by conducting an empirical research on a typical online social media community. We then expound the result from our analysis on data gathered and reach certain conclusions which may be useful to the general public who participate in social media communities and may of value to the regulatory agencies and commercial users of social media as well.

Keywords: Group interaction · Social interaction · Community behavior

1 Introduction

Online and offline social interactions together provide people an opportunity to establish connections and to share information and other resources—that's why we have such interactions. These interactions, in turn, affect people's everyday behavior [15].

Our research focuses on the online side of social interactions. Interactions in online communities are predominantly constituted of interactions within social groups. Social groups, as defined by Gupta and Kim [1], are formed by social media users who, without knowing each other beforehand, come together because of their common interests, goals or needs, and interact with each other in the virtual space to exchange knowledge, discuss problems, share their interests and engage in transactions. The frequent inclusion of words related to group interactions in definitions of online communities proposed by various scholars also reflects the importance of social groups to online communities [14]. Only by constant interactions within themselves and between each other can online communities generate fresh contents, draw new users, and solidify member relationships [10]. As a result, social groups are the essence and the source of value of online communities.

On the other hand, social groups influences their members as well. The prominent theory describing this influence is the peer pressure theory [18]. This theory argues that

© Springer Nature Singapore Pte Ltd. 2018
Z. Xu et al. (Eds.): Big Data 2018, CCIS 945, pp. 44–54, 2018.
https://doi.org/10.1007/978-981-13-2922-7_3

social group members feel psychologically pressured to comply with group norms and to behave in a similar manner with their peers when it comes to expressing their opinions and emotions. As a result, individuals in such groups often demonstrate behaviors similar to those of the majority of the group: they share the same decision, perception, emotion and action [13]. It is notable, however, that some individuals in social groups may be forced to change their minds or choose to be silent.

In addition, Bon [6] points out in his collective unconsciousness theory that in a crowd, homogeneity often engulfs heterogeneity. This collective unconsciousness, he argues, leads to subjective decisions at subconscious level. The convergence in behavior of crowd members is explained by social interactions [13]. Cass Sunstein, in his group polarization theory, describes changes brought to group behaviors from another perspective. He argues that "deliberation tends to move groups, and the individuals who compose them, toward a more extreme point in the direction indicated by their own preliberation judgments". Group polarization theory is another description of impacts group interaction has on the members of the group [16].

Social representation theory explains group behavior and cognition from a psychology standpoint [5]. This theory suggest that because of the difference in their life paths and experiences, members in a group do not always share the exact attitude towards everything the group encounters. In other words, although members share the same social representation, there are still discrepancies among them [17].

Researches discussed above, however, are proposed for social and group interaction in the real, physical world, I.e. interactions is the offline communities. There are insufficient examination and application of theories discussed above in online communities; as a result it is unclear whether these theories are applicable in online environments. Hence, it is imperative to conduct empirical researches of online social communities.

2 Theoretical Research and Hypothesis

Based on the socio-behavioral studies previously discussed, we conjecture that social interactions in online communities may have effects on the behaviors on the members of those communities mainly in the two following aspects.

The first aspect is that based on the crowd effect and group polarization theory, social interactions online affect members' behaviors within the same group.

The second aspect is that based on the social representation theory, online social interactions also influence individual behavior of users on the same platform.

We chose Weibo platform as our subject online platform for our research. Weibo is a Chinese microblogging (weibo) website. Launched by Sina Corporation on 14 August 2009, it is one of the most popular social media platforms in China. As of Q1 2018, Weibo has over 411 million monthly active users [19]. Weibo, compared to other platforms available, has more long term active users, greater influence, and more easily accessible information. We then selected the Marvel fans group as our subject online community. This community primarily consists of Marvel comic fans, and their interactions are mainly concentrated to that related to movie and comics related tags, major influencers in the group, and lead original content generators.

We performed in-depth research on this specific online community. By selecting sample users from Marvel fans on Weibo, crawling their posts on their Weibo timeline, as well as randomly picking accounts on Weibo as controls, we analyzed and tested our hypotheses in a quantitative approach. We were able to examine the number of Weibo each user posted and the emotions those posts have, and to eliminate noises brought by factors such as the development of the platform itself. We used methods including but not limited to web crawling, statistical analysis and time series analysis.

We measure the change in users' behavior in two metrics: the frequency in which they post Weibo, and the content of the Weibo that they post. The frequency of Weibo posting can be measured by statistical tools and time series models. An increase in such frequency can be interpreted as an increase in the users' participation in community interactions and as an overall increase in the level of activity of the community. The content of the Weibo posted, on the other hand, are relatively more difficult to describe. We focus our analysis on the emotion of each Weibo. For each Weibo, we evaluate its emotional characteristics and assign to it a numerical emotional value. We then performed time series analysis to examine whether the emotion of the Weibo posts related to the community interest, i.e., Marvel content, change over time. The change of emotions can then be attributed, at least partly, to the interaction between members of this particular community.

In order to facilitate our empirical research and explain our findings, we constructed a simple hypothesis testing model. The null hypothesis is that interactions within a community affect the behavior of community members, as reflected in the changes in the number of Weibo posted and in the emotion characteristics of such posts. The alternative hypothesis is that interactions within a community does not have a significant impact on members' behavior.

(1) **The test setup for the number of posts dimension is as follows:**

Null hypothesis: the number of Weibo posted changes significantly as the duration of community interaction increases.

Alternative hypothesis: the number of Weibo posted does not change significantly as the duration of community interaction increases.

In statistical expressions:

$$H0: \frac{\sum_{i=1}^{n} GROUP_i}{n} - \frac{\sum_{i=1}^{m} CONTROL_i}{m} \neq 0$$

$$H1: \frac{\sum_{i=1}^{n} GROUP_i}{n} - \frac{\sum_{i=1}^{m} CONTROL_i}{m} = 0$$

Where the number of users in the sample group is n, the number of users in the control group is m, $GROUP_i$ represents the number of posts of user i in the sample group, and $CONTROL_i$ represents the number of posts of user i in the control group.

(2) **The test setup for the emotional dimension of each post is as follows**

Null hypothesis: the emotion of the post changes significantly as the duration of community interaction increases.

Alternative hypothesis: the emotion of the post does not change significantly as the duration of community interaction increases.

In statistical expressions:

$$H0: Tn_i - T1_i \neq 0$$

$$H1: T3_i - T1_i = 0$$

Where Tn_i is the emotion characteristic value of community member i at time n since the initiation of group interaction. For example, $T1_i$ is the emotion characteristic value of community member i at time 1 since the initiation of group interaction.

(3) **The test setup for the individual-specific behavior is as follows**

Null hypothesis: the emotion of individual-specific behavior is similar to that of the average of the Weibo platform, i.e., community interactions does not affect individual specific behavior.

Alternative hypothesis: the emotion of individual-specific behavior is significantly different from that of the average of the Weibo platform.

In statistical expressions:

$$H0: \mu_1 = \mu_2$$

$$H1: \mu_1 \neq \mu_2$$

Where μ_1 the average emotion characteristic value of the sample group is, μ_2 is that of the control group.

3 The Approach

We used the Weibo content from the year of 2010 to 2015 of specific user group as the sample of our data analysis. We also randomly selected a control group to eliminate external noises. And we used hypothesis testing technique to validate the result of our data analysis. The research procedure is as follows (Fig. 1):

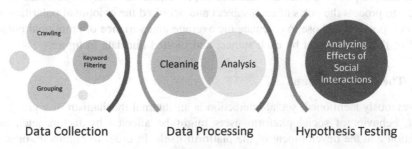

Data Collection   Data Processing   Hypothesis Testing

Fig. 1. Research procedure

Because of the nature of our research, we needed the Weibo contents of users who at least have spent a certain amount of time interacting with others in the Marvel fans community. Hence we selected users who had used Weibo for at least three years and had become Marvel fans for at least one year. The latter event can be determined by the date of the earliest Marvel-related post of each user. In addition, we needed to select more than three thousand Weibo posts for each user. In the selection process, we manually deleted side accounts that were dedicated to Marvel-related posts by users.

3.1 The Experimental Group

We first selected twenty trending Weibo's of the marvel fans community, characterized by having over one thousand comments and more than ten thousand re-posts. We then used Metaseeker to crawl all the user ids of comments posted under the twenty trending Weibo's, processed the ids in Excel, and selected the names that showed up in more than thirteen of the twenty trending posts. The users we selected commented on a considerable number of Weibo of the same theme, thus it can be recognized that they all belong to the Marvel fans group on Weibo, and that they interact with in the same group. We selected 40 users from the Marvel fans group on the Weibo platform.

We collected the information of our subjects using Metaseeker and Bazhuayu data crawler, crawling their posts from their homepages. We crawled the posts of our subjects from the beginning of time to the end of December of 2015, collected data including user id, number of followers, re-post content, original content, postdate, etc. For each subject, the number of post crawled range from 3,000 to 10,000.

Firstly, we first analyzed divide all posts into two batches, which are, group-related posts and group-unrelated posts, by recorded the high-frequency words after taking out from the account prepositions and conjunctions, then applied those high-frequency words as keywords when filtering for the group-related posts from our entire sample. After that, our data analysis focused on two aspects: the quantity and emotional polarity. The following analysis and comparison of our data focused on the same two aspects as well.

The quantity aspect analysis includes sorted the posts of each subject according to the timeline. After that we were able to acquire the monthly number of posts as well as the monthly number of each subject. We also calculated the average number of posts and group-related posts generated by all of the 40 subjects.

We performed the following when analyzing the emotional polarity characteristics of our sample posts. We utilized Semantrica, an emotional analysis software by Lexalytics, to process the posts of each subject and acquired the emotional polarities of the subjects' posts. We chose to calculate the average and variance of emotional polarities of all of the posts posted by each subject and plotted our data in figure.

3.2 The Control Group

As previously mentioned, social interaction is an internal mechanism of a group. As a result, behavior of social platform users might be affected by factors such as the operation and the development of the platform itself. In order to eliminate, or at least mitigate this effect, we randomly selected platform users to form our control group.

The requirement of the control group is relatively easy. We avoided zombie users, inactive or silent users, as well as users who existed less than three years on Weibo. We selected 50 subjects as out control group, and performed exactly the same processing and analysis compared to that we did to the sample group.

Our research is based on the following two fundamental assumptions:

First, we assumed "0" as the emotional polarity of re-posts without comments. Reason of re-posting varies greatly from user to user, thus it is difficult to identify and to quantify emotional polarities of such re-posts. The subject might agree with the original content, or unwilling to demonstrate disapproval because of certain factors. We believe that it is reasonable to assume such re-posts as emotional neutral and assign value of "0" to their emotional polarities.

Second, we assumed that by using keywords, we were able to filter out all of the community-related posts. When the originators of community-related content post their material, they usually mention related hashtags in order to facilitate the propagation of their posts. Since our keywords includes all the collectible hashtags, we believe this assumption is logical and reasonable.

4 Experiment and Result Analysis

Based on the above research ideas, we conduct corresponding experimental research and analyze the results on the basis of experiments.

4.1 The Group-Related Action Dimension

After preprocessing, we plotted the data against time, as shown in Fig. 2.

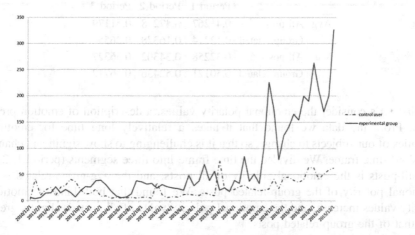

Fig. 2. Monthly variation of posts

The number of posts can be seen as an illustration of users' level of participation in the platform and community. The figure above is a line chart of the average number of posts posted by all 40 users in our sample each month from the December of 2010 to the December of 2015. We can see from the graph an increase trend in the number of Weibo posted by our subjects as time goes by.

As previously stated, the change in the number of Weibo posted can be caused by external factors, such as the platform itself. We plotted the average number of posts generated by the 50 subjects in the control group each month as well, represented by the dashed grey line in Fig. 2. The control group represents the trend in the number of post generated by an average user on the Weibo platform. It is clear that there exists a distinction in averages of the sample and the control group, after taking external factors into the account.

Using hypothesis testing techniques, we examined the previously stated hypotheses using t-test, the p-value is 0.0001343, so that we can reject the null hypotheses in favor of the alternative. And we can conclude that community interaction leads to an increase in the level of social interaction and participation.

4.2 The Emotional Polarity Dimension

We sorted the emotional polarity data and linked it to the timeline. Then we were able to examine the change in our subjects' emotions, as illustrated by the level of emotions in the speech of the subjects.

The comparison of emotional polarities can be further divided into the value of emotional polarity, which ranges from -1(most negative) to 1(most positive), with 0 representing the neutral emotion, and the variance of emotional polarity values, which represents the fluctuations of subjects' emotions. The result is shown in Table 1.

Table 1. Summary of sentiment data

		Period 1	Period 2	Period 3
Avg	All posts	0.41267	0.47238	0.51179
	Group-related	0.12214	0.16378	0.2655
Var	All-posts	0.32258	0.34302	0.36327
	Group-related	0.50121	0.52436	0.56712

First let's consider the emotional polarity values, a description of emotion preferences. From the data we found that it takes a relatively long time for emotional polarities of our subjects to change, so that it is challenging to show significant changes in a short time frame. We divided the time frame into three segments (period 1, 2, 3). Avg-all-posts is the emotional polarity of all posts, and the avg-group-related is the emotional polarity of the group-related posts. We found that both kind of emotional polarity values increase, and the emotional polarity value of all posts is slightly greater than that of the group-related posts.

We tested our hypothesis again,

$$p = 0.042199,$$

The p-value indicating a significant difference, so that we reject our null hypothesis in favor of the alternative.

Then let's consider the variance of emotional polarity values. Comparing the variances of emotional polarities of our subjects' all posts and group-related ones, we found that both variances increases with the passage of time, and variances of polarities of all posts is lower than that of the group-related ones. It can be concluded that the emotion of all posts is more stable over time, and more positive, when compared to the emotion of group-related posts. In addition, group-related posts have an average of emotional values closer to zero and a greater variance, which can be interpreted as more extremities in emotions.

To conclude, the interaction in Marvel fans community leads to an increase in the users' emotional polarity values, greater fluctuations, and more extremities in users' emotions. At least on this dataset, we can make an inference that social interactions in online communities may have effects on the behaviors on the members of those communities.

4.3   Non-community Behavior

It can be seen from the figure that the emotional mean of the user's non-community microblogging is slightly higher than the control group. This feature is verified using the method of hypothesis testing as following (Fig. 3).

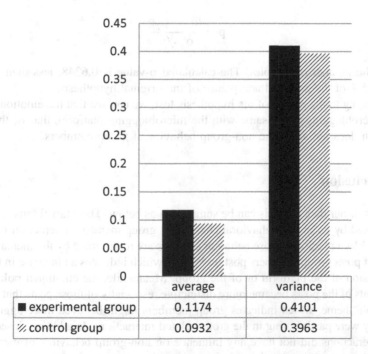

	average	variance
■ experimental group	0.1174	0.4101
▨ control group	0.0932	0.3963

Fig. 3. Average and variance of sentiment polarities

Table 2. Test parameter values

μ_1	0.0932	μ_2	0.1174
σ_1	0.3963	σ_2	0.4101

The mean and variance of the population X of interest groups are expressed as μ_1, σ_1, and the mean and variance of the control group are respectively expressed as μ_2, σ_2 (Table 2).

Put forward the hypothesis:

The original hypothesis H0: non-interest group emotion is same with the emotion of the microblogging platform, which means that interest has nothing to do with its unrelated behavior.

$$H_0 : \mu_1 = \mu_2$$

Alternative Hypothesis H1: non-interest group emotion is different from the emotion of the microblogging platform, which means that interest has nothing to do with its unrelated behavior.

$$H_1 : \mu_1 \neq \mu_2$$

Calculating the test p-value as follow:

$$p = \frac{|\mu_1 - \mu_2|}{\sigma/\sqrt{n}}$$

n is the number of samples. The calculated p-value is 0.6248, less than 1.96, so there is 95% of the accepted acceptance of the original hypothesis.

According to the results of the hypothesis test, we can see that the emotion of non-group microblogging is the same with the microblogging platform, that is, the group interaction does not affect the non-group behavior of group members.

5   Conclusion

The result of our data analysis can be summarize as below. The Marvel fans group that was selected by us shows behavioral changes as group members interact on the social platform. More precisely, those behavior changes are represented by the increase in the amount of posts group members post on Weibo, which indicates an increase in the level of immersion of the platform or of the group. Meanwhile, the emotional polarities of the contents of the posts became more distinctive, especially of those posts that relate to the group's theme, which indicates group members might behaved more aggressively when they were participating in the group-related interactions. In addition, it seems that group interactions did not have any influence on non-group behaviors of members in the specific community we chose for our research.

As shown in the result sections of this paper, the behavior of the group validates the assumption that we stated at the beginning of this paper: social interactions explain the changes in members' group-related behaviors. What's also notable here is that the emotional polarities of group-related Weibo posts are more extreme than that of the non-group-related posts, which is discussed in the social polarization theory. This characteristic of our sample community verifies the group polarization theory, and testify the applicability of this theory on online social groups. However, the insignificant of the effect of group interaction on non-group behavior shows that social representation theory is not applicable to this specific community, and further research may be necessary.

The value of this research lies in the fact that it proposed and tested hypotheses of online group interactions based on currently available theories on community interactions, and validated the applicability of those theories in an online setting. Using the data collected, it also provided a rather detailed description of group's influences on individual members' behavior. The result is applicable to fields such as public opinion monitoring and online strategic marketing.

6 Further Discussion

For this paper we conducted a well-defined empirical research on a specific group and examined the level of applicability of certain social behavior theories. Our research shows extensive details of interactions on the social media platform Weibo, and may of value to the regulatory agencies and commercial users of social media.

To eliminate the noise brought by the platform's own development, an external influence, we simulated the trend by which the behavior of an average user of Weibo evolves. This simulation process need a considerably large number of user. The number of user in the control group in this paper may not be sufficient, so sampling error may exist in our model. Besides the development of the platform itself, there are still many external factors that we did not consider, such as the development or any other aspect of selected community [11], so there are still work to be done to better this research.

Also, the community chosen in this research may be overly specific so that the duplication of this research may be of great difficulty. Researchers should continue to investigate other communities, and the same research methods, when applied to different online communities, may result in different outcomes. The conclusion reached by this paper does not have general applicability, thus it may only be used as a basis of further researches on online social interaction.

Acknowledgments. This work was supported by the China national social science foundation project (No. 16BTQ065) "Research on Multi-source Information Fusion of Emergency in Big Data Environment." & Foundation for Disciplinary Development of SIT in UIBE.

References

1. Gupta, S., Kim, H.W.: Virtual community: concepts, implications, and future research directions. In: Proceedings of the Tenth Americas Conference on Information Systems (2004)
2. Sproull, L., Faraj, S.: Atheism, sex, and databases: the net as a social technology. Public Access Internet, 62–81 (1995)
3. Romm, C., Pliskin, N., Clarke, R.: Virtual communities and society: toward an integrative three phase model. Int. J. Inf. Manag. 17(4), 261–270 (1997)
4. Hagel III, J.: Net Gain: Expanding Markets Through Virtual Communities. Hardcover. Harvard Business Pre 1997-03-01 (1997)
5. Craig, D.L., Zimring, C.: Supporting collaborative design groups as design communities. Des. Stud. 21(2), 187–204 (2000)
6. Bon, G.L.: The Crowd: A Study of the Popular Mind, vol. 34, no. 90, pp. 178–183. Dover Pub Inc. (2002)
7. Hu, C., Hu, J., Deng, S.: Research on user group interaction analysis and service development based on Web 2. J. Libr. Sci. China, 35(5) (2009). (in Chinese)
8. Li, Q.: Social psychological analysis of network communication. Theory Constr. 6, 65–68 (2007). (in Chinese)
9. Chen, Y., Fan, X.: Endogeneity in sociological quantitative analysis: a case study of social interaction effects. Acad. Annu. Meet. China Soc. Soc. Sci. (2009). (in Chinese)
10. Anthony, S.: Anxiety and rumor. J. Soc. Psychol. 89(1), 91–98 (1973)
11. Barsade, S.G.: The ripple effect: emotional contagion and its influence on group behavior. Adm. Sci. Q. 47(4), 644–675 (2002)
12. Bollen, J., Pepe, A., Mao, H.: Modeling public mood and emotion: twitter sentiment and socio-economic phenomena. In: Adamic, R., Baeza-Yates, Counts, S. (eds.) Proceedings of the Fifth International AAAI Conference on Weblogs and Social Media, pp. 450–453. AAAI Press, Palo Alto (2011)
13. Farrell, H., Drezner, D.: The power and politics of blogs. Public Choice 134(1), 15–30 (2008)
14. Suh, B., Hong, L., Pirolli, P., Chi, E.: Want to be retweeted? Large scale analytics on factors impacting retweet in twitter network. In: Pentland, A. (ed.) Proceedings of the 2nd IEEE International Conference on Social Computing, pp. 177–184. IEEE Computer Society, Los Alamitos (2010)
15. Ye, S., Wu, S.F.: Measuring message propagation and social influence on Twitter.com. In: Bolc, L., Makowski, M., Wierzbicki, A. (eds.) SocInfo 2010. LNCS, vol. 6430, pp. 216–231. Springer, Heidelberg (2010). https://doi.org/10.1007/978-3-642-16567-2_16
16. Stein, J.: The micro-foundations of international relations theory: psychology and behavioral economics. Int. Organ. 71(S1), S249–S263 (2017)
17. Conway, J.R., Bird, G.: Conceptualizing degrees of theory of mind. PNAS 115(7), 1408–1410 (2018)
18. Elhai, J.D., Tiamiyu, M., Weeks, J.: Depression and social anxiety in relation to problematic smartphone use: the prominent role of rumination. Internet Res. 28(2), 315–332 (2018)
19. https://en.wikipedia.org/wiki/Sina_Weibo

Mining Device-Specific Apps Usage Patterns from Appstore Big Data

Huoran Li[1], Xuanzhe Liu[1]([⊠]), Hong Mei[1], and Qiaozhu Mei[2]

[1] Peking University, Beijing, China
{lihuoran,liuxuanzhe,meih}@pku.edu.cn
[2] University of Michigan, Ann Arbor, USA
qmei@umich.edu

Abstract. When smartphones, applications (a.k.a, apps), and app stores have been widely adopted by the billions, an interesting debate emerges: whether and to what extent do device models influence the behaviors of their users? The answer to this question is critical to almost every stakeholder in the smartphone app ecosystem, including app store operators, developers, end-users, and network providers. To approach this question, we collect a longitudinal data set of app usage through a leading Android app store in China, called *Wandoujia*. The data set covers the detailed behavioral profiles of 0.7 million (761,262) unique users who use 500 popular types of Android devices and about 0.2 million (228,144) apps, including their app management activities, daily network access time, and network traffic of apps. We present a comprehensive study on investigating how the choices of device models affect user behaviors such as the adoption of app stores, app selection and abandonment, data plan usage, online time length, the tendency to use paid/free apps, and the preferences to choosing competing apps. Some significant correlations between device models and app usage are derived from appstore big data, leading to important findings on the various user behaviors. For example, users owning different device models have a substantial diversity of selecting competing apps, and users owning lower-end devices spend more money to purchase apps and spend more time under cellular network.

Keywords: Mobile apps · Price effect · User behavior

1 Introduction

Since Apple announced the iPhones in 2007, smartphones have been playing an indispensable role in people's daily lives. A great variety of applications (a.k.a, apps) such as Web browsers, social network apps, media players, and games make smartphones become the main access channels to Internet-based services rather than communication tools. With the ever-increasing amount of smartphone users and apps, comprehensive and insightful knowledge on what, when, where, and how the apps are used by the users is extremely important [1]. Many significant

© Springer Nature Singapore Pte Ltd. 2018
Z. Xu et al. (Eds.): Big Data 2018, CCIS 945, pp. 55–76, 2018.
https://doi.org/10.1007/978-981-13-2922-7_4

research efforts have been made in the past few years on portraying the users and understand their behaviors in term of apps.

Like all Internet users, smartphone users can be classified based on various facts, including demographics such as location, gender and age, and behaviors such as preferences to apps [4], content consumed within apps [19], and so on. Actually, in the current app ecosystem, a user is naturally identified by his/her device [23]. Many such classifications of smartphone users boil down to classifications of devices. In other words, much variance of user behaviors may be explained by the devices they use. Indeed, when users interact with their smartphones, download apps from online app stores, and use the apps for different purposes, their experiences are usually affected by various parameters of the device models they use, such as brands, hardware specifications, etc. Understanding how device models affect user behaviors can help app store operators know their users better and improve their recommender systems by considering device models. For instance, one may recommend apps with fancy graphical effects to devices that are equipped with a powerful GPU and high-resolution screen [10]. Device-specific ads is another big opportunity. For example, Facebook customizes mobile ads according to device model types since 2014 [6]. App developers are thrilled to know through which kind of device models they can gain more users and more clicks, so that they can invest their effort in customizing the ads for those models, e.g., by designing banners of proper sizes or placing videos at proper positions on the screen. Furthermore, device models may be more informative in the behavioral analyses of Android users, due to the large diversity and heavy fragmentation of Android devices [10].

Some existing efforts have been made to investigate how apps usage is affected by device models[1] [3,17,21,23]. Unfortunately, due to the lack of sufficient user behavioral data at scale, most existing studies suffer from serious selection bias, including specific user groups (e.g., in-school students) [21], fixed device models or apps [14], and limited metrics (e.g., screen size) [17]. In this paper, we present a comprehensive user study exploring whether, how, and how much the device models can really influence the user behaviors on using smartphones and apps. We collect the behavioral profiles of about 0.7 million anonymized Android users[2] by a leading app store operator in China, called *Wandoujia*[3]. Besides the largest data set to date, our study differs from existing efforts further in two aspects.

- We conduct the study from a new perspective, i.e., *the sensitivity of device's price against the app usage*. In our opinion, the price of a device model can generally reflect the level of hardware specifications when the device is

---

[1] In this paper, the term "device model" refers to the device with specific product type with hardware specifications, e.g., Samsung N7100, N9100, Xiaomi 3s, and so on.

[2] Our study has been approved by the research ethnics board of the Institute of Software, Peking University. The data is legally used without leaking any sensitive information. The details of user privacy protection are presented later in the data set description. We plan to open the data set when the manuscript is published.

[3] http://www.wandoujia.com.

released. Additionally, such a metric can imply the users' economic background, which influences user behavior at demographics level [9]. In this way, we try to categorize the users into different economic groups and explore the sensitivity of the device against the user behavior.

- Second, we explore comprehensive behavioral profiles that contain various useful information, including the apps selection, apps management activities (e.g., download, update, and uninstallation), data plan usage per app, and online time length per app, etc. In addition, we focus on only the behavioral profiles from long-term users who steadily contribute to our study. This provides solid ground for the findings from our study.

The major contributions made by this paper can be summarized as follows.

- We collect app usage from over 0.7 million users in a period of five months. Our data set covers 500 popular Android devices and over 0.2 million Android apps. The detailed user behavioral profiles include the activity log of downloading, updating, and uninstalling apps, the daily traffic and access time of every app through both Wi-Fi and cellular. Based on such a large-scale data set, we explore a comprehensive study on how the device models can impact the app usage.
- We find significant correlations between the choice of device models and app usage spanning the adoption of app stores, the selection and abandonment of apps, the online access time and data traffic of apps, the revenue of apps, and the preferences against competing apps. Some findings can be quite interesting, e.g., the users holding lower-end devices are likely to spend more money on purchasing apps and spend more time under cellular network, the selection of the apps with similar functionalities presents a substantial diversity among users, etc. The findings can be leveraged to understand the user requirements better, preferences, interests, or even the possible background such as economic or profession.
- We derive some implications that are directly helpful to several stakeholders in the app-centric ecosystem, e.g., how app store operators can improve their recommendation systems, how the app developers can identify problematic issues and gain more revenues, and how the network service providers can explore more personalized services.

The remainder of this paper is organized as follows. We first present the related work in the area of user behavior analysis of smartphone users. Next we describe the Wandoujia and the five-month data set, and present our measurement approach alongside the research questions and hypotheses. Then we conduct the correlation analysis on how the choice of device models affects the user behaviors on app usage. In addition, we propose the underlying reasons leading to such significant correlations. We also discuss about our implications for relevant stakeholders, and describe the limitation of our study and threats to the generalization of our results. Finally, we conclude the study and some outlooks to future work.

2 Related Work

Precise classification of users and understanding their behaviors of using apps are significant to every stakeholder in the app ecosystem, including app store operators, content providers, developers, advertisers, network providers, etc. Several efforts have been made in the fields.

One straightforward way to understand the user behavior is conducting field study. Usually, the field studies are conducted over some specific user groups. Rahmati et al. [15,16] made a four-month field study of the usage of smartphone apps of 14 users, and summarized the influences of long-term study and short-term study. Lim et al. [12] made a questionnaire-based study to discover the diverse usages from about 4,800 users across 15 top GDP countries. The results show that the country differences can make significant impacts on the app store adoption, app selection and abandonment, app review, and so on. Falaki et al. [7] performed a study of smartphone usage based on detailed traces from 255 volunteers, and found the diversity of users by characterizing user activities. A number of other studies have been made in similar ways [2,3,5]. To have more comprehensive behavioral data, the monitoring tools/apps are more appreciated other than questionaire. Yan et al. [24] developed an app, called *AppJoy*, and deployed such an app to collect the usage logs from over 4,000 users and find the possible patterns in selecting and using apps.

Besides the general analysis, some studies aim to investigate the diversity of user behaviors from specific perspectives. Raptis et al. [17] performed a study of how the screen size of smartphones can affect the users' perceived usability, effectiveness, and efficiency. Rahmati et al. [15] explored how users in different socio-economic status groups adopted new smartphone technologies along with how apps are installed and used. They found that users with relatively low socio-economic background are more likely to buy paid apps. Ma et al. [13] proposed an approach for conquering the sparseness of behavior pattern space and thus made it possible to discover similar mobile users with respect to their habits by leveraging behavior pattern mining. Song et al. [19] presented a log-based study on about 1 million users' search behavior from three different platforms: desktop, mobile, and tablet, and attempted to understand how and to what extent mobile and tablet searchers behave differently compared with desktop users.

Some field studies were made towards specific apps. Böhmer et al. [2,3] made a field study over three popular apps such as Angry Bird, Facebook, and Kindle. Patro et al. [14] deployed a multiplayer RPG app game and an education app, respectively, and collected diverse information to understand various factors affecting the app revenues.

Due to the difficulty of involving a large volume of users, the preceding field studies may suffer from relatively limited users and apps. At large-scale, Xu et al. [23] presented usage patterns by analyzing IP-level traces of thousands of users from a tier-1 cellular carrier in U.S. They identified traffic from apps based on HTTP signatures and present aggregate results on their spatial and temporal prevalence, locality, and correlation. Our previous work [11] was conducted over a one-month data collected by Wandoujia, and evidenced that some app usage

Table 1. Categorization of device models

Group	Price interval	# of devices	# of users	Representative devices
High-end	⩾4,000 RMB (about 600 USD)	77	265,636	Samsung N7100, Samsung S4
Middle-end	1,000–4,000 RMB (about 150–600 USD)	339	411,138	XIAOMI 3, Google NEXUS 5
Low-end	⩽1,000 RMB (about 150 USD)	84	84,488	COOLPAD 7231, LENOVO A278T

patterns in terms of app selection, management, network activity, and so on. To understand the diversity of user behaviors better, the study made in this paper is based on a more comprehensive data set that consists of the 5-month behavioral profiles from various users, and focuses on the impact made by the choice of device models.

3   The Data Set

In this section, we present the data set collected from Wandoujia, by describing the detailed information that shall be used to conduct our empirical study.

3.1   About Wandoujia

Our data is from Wandoujia[4], a free Android app marketplace in China. Wandoujia was founded in 2009 and has grown to be a leading Android app marketplace. Like other marketplaces, third-party app developers can upload their apps to Wandoujia and get them published after authenticated. Compared to other marketplaces such as Google Play, apps on Wandoujia are all free, although some apps can still support *"in-app purchase"*.

Users have two channels to access the Wandoujia marketplace, either from the Web portal, or from the Wandoujia management app. The Wandoujia management app is a native Android app, by which people can manage their apps, e.g., downloading, searching, updating, and uninstalling apps. The logs of these management activities are all automatically recorded.

Beyond these basic features, the Wandoujia management app is developed with some advanced but optional features that can monitor and optimize a device. These features include network activity statistics, permission monitoring, content recommendation, etc. All features are developed upon Android system APIs and do not require "root" privilege. Users can decide whether to enable these features. However, these advanced features are supported only in the Chinese version of Wandoujia management app.

3.2   Data Collection

Each user of Wandoujia is actually associated with a unique Android device, and each device is allocated a unique anonymous ID. In this study, we collected usage data of 5 months. The data set consists three categories of information:

---

[4] Visit its official site via http://www.wandoujia.com.

- **Device Model and Price.** The Wandoujia management app records the type of each user's device model, e.g., `Samsung Galaxy Note 2`, `Samsung S3`, `Xiaomi Note 2`, etc. In our raw data set, there are more than 10,000 different Android device models. Such a fact implies the severe fragmentation of Android devices. However, from our previous work [11], the distribution of users against device models typically complies with the *Pareto Principle*, i.e., quite small set of device models accounts for substantial percentage of users. Hence, we choose the top 500 device models according to their number of users. We then look up **the first-release price** of these device models from Jd[5]. Indeed, the price of a device model always decline over time due to the Moore's Law. However, such a price can somewhat reflect the hardware specifications and the target users at the time when the device model is released on market. Considering the large volume of users, we choose to rely on this type of price as our classification criteria.

- **App Management Activities.** The Wandoujia management app can monitor the users' management activities such as searching, downloading, updating, and uninstalling their apps. Each management action is recorded as an entry in the log file that can be uploaded to the Wandoujia servers. With the management activities, we can know which apps ever appeared on the user's device, and when these apps are downloaded, updated, and installed.

- **App Network Activities.** Besides the basic management functionalities, the Wandoujia management app provides advanced features to record the daily network activities of an app. The network statistic features are optional, and are enabled when and only when the users explicitly grant the permissions to the Wandoujia management app[6]. If the app generates network connections either from Wi-Fi or cellular (2G/3G/LTE), the network usage can be monitored and recorded at the TCP-level, including the data traffic and network connection time. Note that the network statistic feature works as a system-wide service, thereby network usage from all apps can be captured, even if the app is not installed via the Wandoujia management app. The Wandoujia management app does not record the details of every interaction session, due to the concerns of system overhead. Instead, the Wandoujia management app aggregates the total network usage of an app. In particular, the data traffic and access time generated from foreground and background are distinguished, so that we can have more fine-grained knowledge of network usage. In other words, we can use eight dimensions of the daily network usage per app, i.e., 2 metrics (access time and traffic) * 2 modes (Wi-Fi and Cellular) * 2 states (foreground and background).

To conduct a comprehensive and longitudinal measurement, we should process and filter the collected data guided by the following principles. We choose only the users who explicitly granted Wandoujia to collect their usage data from the preceding top 500 device models (denoted as the user set $\mathbb{U}$). For these

---

[5] http://shouji.jd.com, Jd is the largest e-commerce for electronic devices in China.
[6] Due to space limit, the details of how Wandoujia management app works can be referred to our previous work [11].

users, we further take into account those who continuously contribute their daily usage data for five months. Such a step assures that we can have rather complete behavioral data of these users. In this way, we obtain **761,262** users (denoted as the subset $\mathbb{U}'$) and their usage data of **228,144** apps.

Table 2. Results of U-Test among groups. The "H", "M", "L" represent high-end, middle-end, low-end, respectively.

RQ	Value	U-Test *p-value*		
		H-M	H-L	M-L
RQ1	Download & Update	0.000	0.000	0.000
	Uninstallation	0.000	0.000	0.000
RQ2	Cellular	0.000	0.000	0.000
	Wi-Fi	0.000	0.000	0.000
RQ3	Cellular	0.000	0.000	0.000
	Wi-Fi	0.000	0.000	0.000
RQ4	Expenditure	0.000	0.000	0.072

3.3   User Privacy

Certainly, the user privacy is a key issue to conduct such a measurement study based on large-scale user behavior data. Besides collecting the network activity data from only the users who explicitly have granted the permission, we take a series of steps to preserve the privacy of involved users in our data set. First, all raw data collected for this study was kept within the Wandoujia data warehouse servers (which live behind a company firewall). Second, our data collection logic and analysis pipelines were completely governed by three Wandoujia employees[7] to ensure compliance with the commitments of Wandoujia privacy stated in the *Term-of-Use* statements. Finally, the Wandoujia employees anonymized the user identifiers. The data set includes only the aggregated statistics for the users covered by our study period.

3.4   Limitation of the Data Set

The preceding data set may have some limitations. First, the management activities come from only the apps that are operated by the Wandoujia management app. The management activities of apps that are downloaded and updated in other channels such as directly from the app's websites or uninstalled via the default uninstaller of Android system, cannot be captured by our data set. Second, we can take into account only the network usage as an indicator that an

---

[7] One co-author is the head of Wandoujia product. He supervised the process of data collection and de-identification.

app is exactly used. Some apps such as calculator and book readers are often used offline, so we cannot know whether these apps have been launched and how they are used. However, our data set is the largest to date and comprehensive enough to conduct a longitudinal user study.

Table 3. The pearson correlation coefficient of each app category of RQ1–RQ4. The results are presented in form of "coefficient/p-value".

Category	RQ1		RQ2		RQ3		RQ4
	Download & Update	Uninstallation	Cellular Time	Wi-Fi Time	Cellular Traffic	Wi-Fi Traffic	Expenditure
BEAUTIFY	-0.252/0.000	-0.052/0.247	-0.162/0.000	-0.148/0.001	-0.108/0.016	0.359/0.000	-0.086/0.057
COMMUNICATION	-0.066/0.143	-0.085/0.059	0.205/0.000	0.101/0.024	-0.022/0.624	0.302/0.000	0.146/0.001
EDUCATION	-0.150/0.001	-0.200/0.000	-0.305/0.000	-0.079/0.078	-0.252/0.000	0.043/0.337	0.077/0.087
FINANCE	0.513/0.000	0.175/0.000	0.127/0.004	0.127/0.004	0.222/0.000	0.129/0.004	0.106/0.018
GAME	-0.567/0.000	-0.186/0.000	-0.093/0.038	-0.024/0.585	-0.371/0.000	-0.094/0.036	-0.221/0.000
IMAGE	0.050/0.263	0.044/0.323	0.093/0.037	0.096/0.032	0.102/0.023	0.239/0.000	0.143/0.001
LIFESTYLE	0.469/0.000	0.139/0.002	0.152/0.001	0.205/0.000	0.322/0.000	0.448/0.000	0.023/0.609
MOTHER_AND_BABY	0.080/0.073	0.030/0.505	-0.014/0.757	-0.002/0.970	-0.008/0.864	-0.014/0.764	-/-
MUSIC	-0.407/0.000	-0.231/0.000	-0.007/0.885	-0.041/0.365	-0.333/0.000	0.047/0.290	-0.019/0.674
NEWS_AND_READING	0.471/0.000	0.154/0.001	0.230/0.000	0.248/0.000	0.316/0.000	0.394/0.000	0.043/0.336
PRODUCTIVITY	0.240/0.000	0.126/0.005	0.298/0.000	0.172/0.000	0.362/0.000	0.312/0.000	0.306/0.000
SHOPPING	0.488/0.000	0.079/0.079	0.452/0.000	0.402/0.000	0.423/0.000	0.472/0.000	0.016/0.725
SOCIAL	0.071/0.115	0.143/0.001	0.203/0.000	0.228/0.000	0.188/0.000	0.280/0.000	0.085/0.057
SPORTS	0.094/0.036	0.037/0.412	0.024/0.587	0.024/0.595	0.066/0.142	-0.001/0.991	-0.131/0.003
SYSTEM_TOOL	0.042/0.352	-0.037/0.412	0.011/0.814	0.121/0.007	0.072/0.108	0.315/0.000	0.146/0.001
TOOL	0.096/0.031	0.055/0.217	-0.141/0.002	-0.058/0.198	-0.234/0.000	-0.058/0.193	0.212/0.000
TRAFFIC	0.376/0.000	0.049/0.272	0.177/0.000	0.153/0.001	0.238/0.000	0.100/0.025	-/-
TRAVEL	0.560/0.000	0.406/0.000	0.343/0.000	0.200/0.000	0.445/0.000	0.322/0.000	0.049/0.272
VIDEO	0.189/0.000	-0.091/0.042	0.066/0.140	-0.257/0.000	-0.038/0.403	-0.369/0.000	0.226/0.000
MISCs	-0.108/0.016	0.044/0.328	-0.007/0.883	-0.032/0.482	-0.035/0.433	0.216/0.000	0.010/0.819

4 Measurement Approach

In practice, it is a common fact that device models with similar price usually have similar hardware specifications and target user groups. Our measurement study then aims to evaluate whether and to what extent the choice of device models, or more specifically, the price of smartphones, can affect app usage in various metrics. According to the preceding data set of user behavioral profiles, we propose some research questions that are significant to stakeholders.

- **RQ1:** Does the choice of device models impact the usage of app stores? If such an impact exists, which users are more likely to adopt the app stores (e.g., downloading new apps and updating existing apps), and what about the various requirements of different users when they use app stores?
- **RQ2:** Does the choice of device models impact the online time spent by users? If such an impact exists, which apps do different users tend to spend their time on?
- **RQ3:** Does the choice of device models impact the traffic used by users, especially the data plan of cellular? If such an impact exists, which apps are more likely to consume the cellular data plan of different users?
- **RQ4:** Does the choice of device models impact the purchase behavior of users and thus the app revenue? If such an impact exists, which group of users are more likely to pay for apps, and which apps are more likely to be paid for by different users?

- **RQ5:** Does the choice of device models impact the choice of apps with similar functionalities or purposes? If such an impact exists, which apps are more likely to be adopted by different users?

To answer the above **RQs**, our measurement study is conducted from two aspects.

User Group Analysis. As we assume that the price of device models can possibly reflect the economic background of the users, we first study the **overall** user behaviors by categorizing the users into groups according to the first-release price. In China, the price systems of popular e-commerce web sites such as Jd, Amazon, and Taobao, the price of device models is usually segmented at every 1,000 RMB level, i.e., $\leqslant$1,000 RMB, 1,000 RMB-2,000 RMB, 2,000 RMB-3,000 RMB, 3,000 RMB-4,000 RMB, and $\geqslant$4,000 RMB. Hence, we roughly categorize the device models into three groups according to their on-sale price information that is published on the Jd, i.e., the **High-End** ($\geqslant$4,000 RMB, about 600 USD), the **Middle-End** (1,000 RMB-4,000 RMB, about 150–600 USD), and the **Low-End** ($\leqslant$1,000 RMB, about 150 USD). We list the categorization results in Table 1.

We manually check the price history evolution of the top 500 device models on Jd as well as look up some third-party data sources such as Dong-Dong[8] and Xitie[9]. Most of the device models can still fall into the above coarse-grained groups as of the start time of our data set. Very few device models cannot meet such criteria, e.g., the first-release price of Galaxy S2 was 4,399 RMB, but the price fell down to about 3,200 RMB. For this case, we still categorize the device models by the first-release price. Luckily, such is rare in our data set. In this way, the dynamics of prices can have the side effects as little as possible. For each **RQ**, we use the box-plot to report the distribution of user behaviors over the corresponding metrics in each group. To further evaluate the statistical significance among the groups, we employ the Mann-Whitney U test (U-Test for short) [8] among the three groups. The U-Test is widely adopted in large data set to test whether the two groups have significant difference. In particular, the U-Test can be applied onto unknown distributions, and fit our data set very well.

App Category Analysis. We then study whether the choice of device models can impact the user behavior on specific apps. To this end, we run the *regression analysis* between the price of device model and the corresponding metrics, by organizing the $\mathbb{U}'$ according to the first-release price of each device model. Investing each **RQ** essentially relates to the user's preferences and requirements of specific apps. Hence, we categorize the apps according to the classification system of Wandoujia, e.g., *NEWS_AND_READER*, *GAME*, *VIDEO*, and so on.

---

[8] https://itunes.apple.com/us/app/dong-dong-gou-wu-zhu-shou/id868597002? mt=8, is an app for inquiring history price of products on Jd.

[9] http://www.xitie.com, is a website for inquiring price history of products on popular e-commerce sites.

For each **RQ**, we summarize the related metrics over each app category for every single user, and thus compute the Pearson correlation coefficient between the metrics and the price of the device model. In this way, we can explore whether the choice of device models can be statistically significant to the app usage.

In the next section, we conduct our analysis in the following workflow. We first present the motivations of each research question, respectively. Then we try to examine the impact caused by the choice of device models, by synthesizing the correlation analysis results at the granularity of each device model and the groups. Finally, we summarize the findings and try to explore the underlying reasons leading to the diverse user behaviors on app usage.

5   Analysis and Results

In this section, we explore the research questions and validate all the hypotheses, respectively. We show the results of user group U-Test in Table 2 and the results of app-specific regression analysis in Table 3.

Generally, most of the pairwise U-Test results in Table 2 at $p < 0.0001$ level. The only exception is for **RQ4**, indicating that the difference on expenditure of paid apps is not quite significant between middle-end group and low-end group. Such an observation indicates that the **price of device model has a statistically significant correlation of the behaviors of users from different groups**. Hence, we can focus more on the distribution variance among groups (in the box-plots) and the results reported in Table 3.

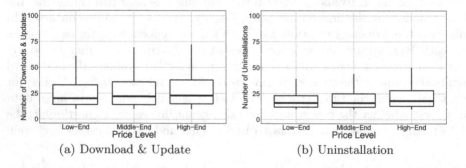

(a) Download & Update   (b) Uninstallation

Fig. 1. Distributions of the number of downloads, updates, and uninstallations among device model

5.1   Effect on App Management Activities

First, we focus on **RQ1**, i.e., whether users holding different device models perform differently in terms of app downloads, updates, and uninstallations. This research question is motivated by three folds. First, the behaviors of app downloads can indicate the adoption of app stores by users, i.e., whether the users prefer seeking apps from app stores and which users are more active. Second, we can identify the most popular/unpopular apps for a given device model, the app stores operators can accurately recommend the apps. Third, we can explore which apps are more likely to be abandoned by the users holding a specific device model, and such knowledge can help app developers identify the possible problems such as device-specific bugs.

Adoption of App Stores. In Fig. 1, we first report the number of 5-month management activities in terms of download and update. On average, we can see that most users do not frequently access the app stores. However, in each group, the standard deviation of management activities is quite significant. Such an observation indicates that users can perform quite variously in terms of management. We can observe that the users holding higher-level devices are more likely to access the app stores.

App Selection. We next investigate whether the choice of device models can impact the app selection. From our previous study of the global distribution of apps [11], we find that users can have quite high overlap in selecting the popular apps, such as WeChat, QQ, Map. Hence, we first explore the **similarity** of apps selection. We aggregate the apps that are downloaded and updated by at least 1,00 unique users of each device model, and compute the pairwise cosine similarity between the three groups. The cosine similarity values are 0.81 (high-end and middle-end), 0.86 (high-end and low-end), and 0.81 (middle-end and low-end), respectively. Such an observation evidences our preliminary findings. Then we explore the **diversity** of app selection. For simplicity, we cluster each app according to its category information provided by Wandoujia, e.g., *Game, NEWS_AND_READING*, etc. We compute the contributions of downloads and updates from every single device model against a specific app category. For example, if there are 1,000,000 downloads of *GAME* apps and 50,000 of these downloads and updates come from the device model Samsung S4, we assign the contributions made by this device model is 5%. Then we make the correlation analysis of app selection and the price of device, by means of Pearson correlation coefficient. We find that as the price of device models increases, the users are more likely to choose apps from the categories of *TRAFFIC* ($r = 0.376, p = 0.000$), *LIFESTYLE* ($r = 0.469, p = 0.000$), *NEWS_AND_READING* ($r = 0.471, p = 0.000$), *SHOPPING* ($r = 0.488, p = 0.000$), *FINANCE* ($r = 0.513, p = 0.000$), and *TRAVEL* ($r = 0.560, p = 0.000$). In contrast, the correlation analysis show that as the prices of device models increases, the users are less likely to choose the apps from *GAME* ($r = -0.567, p = 0.000$) and *MUSIC* ($r = -0.407, p = 0.000$). Such observations imply that the choice of device models can significantly impact the app selections, and infer the characteristics and requirements of the users. For example, users with high-end smartphones are more likely to have higher positions and better economic background, so they care about the apps from *NEWS AND READING, FINANCE, TRAVEL*, and *SHOPPING*. Users holding low-end device models care more about the entertainment such as *Game* and *Music*.

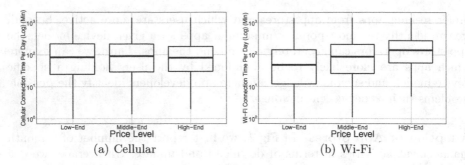

Fig. 2. Distribution of daily online time among device models.

App Abandonment. The uninstallation can indicate the users' negative attitudes towards an app, i.e., the user does not like or require the app any longer. From Fig. 1, it is observed that the median of uninstallation activities differ quite marginally among the three groups. We then perform the correlation analysis in a similar way of downloads and updates. In most categories, the Pearson coefficient is not quite significant. However, we find that the correlation in *TRAVEL* ($r = 0.406, p = 0.000$) is significant. Although the travel apps are most likely to be downloaded by higher-end users, these apps are also very possible to be uninstalled by these users.

Although the uninstallation does not take significant correlations to device models at the level of app category, investigating the individual apps that are possibly abandoned by a specific device model is still meaningful. To this end, we explore the apps which have been uninstalled for more than 100 times in our data set, and get 3,123 apps. We then draw the distribution of uninstallations according to the device model. An interesting finding is that, the manufacturer-customized or preloaded apps are more possibly uninstalled on their own lower-end device models. For example, the app `Huawei Cloud Disk` (the package `com.huawei.hidisk`) is a preloaded app on almost all device models produced by Huawei. This app has 20,985 uninstallations, while 17,641 uninstallations come from the lower-end devices. The similar findings can be found in other device models produced by `Samsung, Lenovo,` and `ZTE.` Such an observation implies that the lower-end users are less likely to adopt these customized or preloaded apps. Besides the preloaded apps, some apps are also more likely to be uninstalled by a specific device model. For example, the `Samsung Galaxy Note 2` accounts for more than 80% uninstallations of two camera apps. Such a finding implies that these apps can suffer from device-specific incompatibility or bugs.

5.2   Effect on Online Time

We next intend to validate **RQ2**, i.e., the choice of device models impacts on how long the users spend their time on the Internet. Such a research question is motivated by understanding whether the choice of device models can lead to various preferences toward cellular and Wi-Fi usage. If the app developers can know that some users from a specific device model spend more cellular time rather than Wi-Fi, they can provide customized features to these models. For example, developers can optimize the data plan usage by providing pre-downloading contents when these users are in Wi-Fi network. In addition, if users holding specific device models spend much time on a few categories of apps, the developers, web content providers, and advertisers can leverage such knowledge to customize more accurate advertisements to audience.

The Wandoujia management app can record the daily foreground/ background connection time under both cellular and Wi-Fi. Since the foreground time is computed only when the users interact with the app (by checking the stack of active apps in Android system), we exclude the background time in the online time analysis.

Figure 2 describes the distribution of online time. **For online time, we are surprised to find that users rely less on the cellular network as the price of device model increases.** In other words, the higher-end users typically spend less time under cellular network. For the average daily online time, the low-end users ($\leqslant$1,000-RMB device models) spend about 10 min more than the high-end users ($\geqslant$4,000-RMB device models) under cellular, while the high-end users spend 1 hour more than the low-end users under Wi-Fi. Immediately, we can infer that the network conditions vary a lot among different users, i.e., the lower-end users are less likely to stay in the places where Wi-Fi are covered. In contrast, the higher-end users tend to have better Wi-Fi connections. Such a difference can further imply the possible locations where different users may stay, e.g., the high-end users are more likely to stay in offices.

We then investigate whether the choice of device models affect the usage of "network-intensive" apps. Similar to the preceding analysis in the management activities, we compare the online time distribution of device models over each app category, under cellular and Wi-Fi, respectively. For cellular, the online time of apps have no significant correlation with the price of device models, except the categories of $TRAVEL$ ($r = 0.3427, p = 0.000$) $SHOPPING$ ($r = 0.452, p = 0.000$), and $EDUCATION$ ($r = -0.305, p = 0.000$).

This result can support the findings of **RQ1**. On the other hand, it is interesting to see that users holding lower-end smartphones are more likely to use $EDUCATION$ ($r = -0.305, p = 0.000$) apps under the cellular. Such a finding suggests that a considerable proportion of lower-end users may be students. The correlation between the choice of device model and online time under Wi-Fi seems not to be quite significant, either. Only in the category of $SHOPPING$ ($r = -0.304, p = 0.000$), the choice of device models seems to take significant correlation with the price of device. Such an observation is not surprising, as higher-end users are supposed to have better economic background and more likely to spend more on shopping. We can infer that users share quite similar preferences in app usage under Wi-Fi.

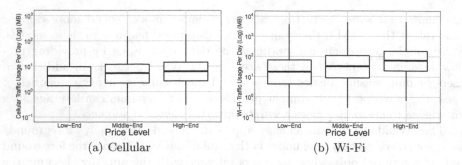

Fig. 3. Distribution of daily traffic consumption among device models.

5.3   Effect on Traffic Consumption

Another important network-related issue for smartphone users is the traffic. We next focus on **RQ3**, i.e., whether users holding different device models differ in cellular and Wi-Fi traffic usage. Generally, users do not care about the traffic from Wi-Fi, but do concern the traffic from cellular that they need to pay for, especially for low-end users who usually have relatively limited budgets. However, exploring the traffic generated from Wi-Fi can be meaningful. Intuitively, we can explore which apps are more "bandwidth-sensitive" on specific device models, so the app developers and network service provider can consider techniques to compensate for bandwidth variability.

The distribution of traffic consumption among device models is shown in Fig. 3. Interestingly, although the higher-end users are observed to spend the least time under the cellular network in **RQ2**, they spend the most traffic. In other words, we can infer that the higher-end users are more likely to use those "traffic-heavy" apps. On average, a high-end user can spend more 100 MB cellular data plan than a low-end user. In China, such a difference of data plan does matter very much.

Since the gaps between different device models are substantial, identifying which apps consume most traffic on specific device models can be quite meaningful. Similar to preceding analysis, we compute the Pearson correlation coefficients between the choice of device model from every single user and the apps on which the traffic is consumed. The cellular data plan consumed over the apps from *SHOPPING* ($r = 0.423, p = 0.000$) and *TRAVEL* ($r = 0.445, p = 0.000$) presents a quite significant positive correlation to the price of device models. In contrast, the correlations seem to be significantly negative in *GAME* ($r = -0.371, p = 0.000$) and *MUSIC* ($r = -0.333, p = 0.000$) apps. In these app categories, users with lower-end smartphones tend to spend more cellular traffic. These findings are consistent with the app download and update preferences in **RQ1**.

The traffic generated under Wi-Fi presents significant correlations with the device models in some categories. The lower-end users tend to spend a large number of Wi-Fi traffic on the *VIDEO* apps, In contrast, the higher-end users are more likely to rely on the apps of *COMMUNICATION, PRODUCTIVITY, SYS-TEM_TOOL, TRAVEL, BEAUTIFY, NEWS_AND_READING, LIFESTYLE,* and *SHOPPING* under Wi-Fi. Such a difference in the Wi-Fi traffic usage can indicate the requirements and preferences of users holding different device models, and thus implies the possibly different background of the users.

5.4 Effect on App Revenue

The above three research questions have revealed the general correlations between the choice of device models and app usage. To further explore the underlying reasons that can be more relevant to other stakeholders such as app developers, we try to focus on the **RQ4**, i.e., whether the choice of device models can affect the app revenue. It should be mentioned that all apps directly downloaded from Wandoujia are free, despite the in-app purchase of some apps. However, since the Wandoujia management app act as a system-wide service to monitor all apps that are installed on a device, we can still identify which apps should be paid ones that are downloaded from other channels. To this end, we write a crawler program to retrieve the package names of paid Android apps from other app stores including Google Play, 360safe App Store, and Baidu App Store. Then we extract the logs that are related to these paid apps from our data set. In addition, we record the fee of each paid app. By such a step, we finally obtained 27,375 users that have used paid apps. As the price of a device model can serve as an indicator of its owner's economic background, it is not uncommon to suppose users with expensive devices to pay more money in purchasing paid apps. We are interested in three facets, (1) **the possibility to buy paid apps**, i.e., are users with high-end devices more likely to use paid apps? (2) **the expenditure of apps**, i.e., do the high-end users pay more money on buying apps? (3) **the diversity of paid apps**, i.e., do users with different models pay for different purposes?

First, we investigate the possibility of buying apps. From Table 4, we can observe that the higher-end users account for about 47% of all users who have ever purchased paid apps, and 45% of purchased records. We further perform a *Chi-square test* to confirm the significant correlations between the proportion of buyers and group ($\chi^2 = 2875.05, p = 0.000$).

When moving to the expense of apps, we perform the pairwise T-test of average expenditure of users. The T-Test can identify the positive or negative correlation of the tendency of buying apps against the groups. The T-test values are -4.281 (high-end v.s. low-end) and -5.740 (middle-end v.s. low-end), respectively, given the $p = 0.000$. To our surprise, buyers with higher-end device models are likely to spend even less money than lower-end users. Interestingly, **although the lower-end users account for quite small proportion of all buyers and the purchase orders, they are likely to spend more money on average.** Such a finding indicates that the choice of device models impact the wish to buy apps, which affects the revenue of apps accordingly.

Finally, the choice of device models can impact the preferences of paid apps. The lower-end users are more likely to pay for *GAME* ($r = -0.221, p = 0.000$),

BEAUTIFY ($r = -0.086, p = 0.000$ such as themes), and *SPORTS* ($r = -0.131, p = 0.000$). In contrast, the higher-end users tend to buy *PRODUCTIVITY* ($r = 0.306, p = 0.000$, such as office suite). Such a difference can reflect the different requirements of apps which the users would like to pay for.

5.5   Effect on Choices of Competing Apps

We finally move to the **RQ5**, i.e., whether the choice of device models can affect selecting the apps of the same or similar functionalities (we name such apps as "competing apps" in the follows) Such a research question is motivated by the existence of a number of "competing apps" on the app stores. For example, there are a number of competing browsers such as `Chrome`, `FireFox`, `Opera Mini`, `Safari`, maps such as `Google Maps`, `Baidu Maps`, and `Yahoo! Maps`, and so on. Although these apps can provide very similar or even the same functionalities, they can perform quite variously such as the data traffic and energy drain, given the same user requests [18]. In addition, besides the common functionalities, the competing apps are likely to provide some differentiated features such as adjustable color and light for display. End-users often feel confusing to select the apps that are more adequate to their own preferences and requirements.

Unlike the correlation analysis of **RQ1–RQ4**, we do not directly conduct correlation analysis to all app categories. Instead, we choose three typical apps: *News reader*, *Video player*, and *Browser*, as they are observed to be commonly used in daily life. For each app, we select the top-5 apps according to the online time that the users spend on them. The reason why we employ the online time instead of the number of downloads is that the online time can be computed only when the users interact with the app. The selected competing apps are as follows. The **News** contains `Phoenix News`, `Sohu News`, `Netease News`, `Today's Top News`, and `Tencent News`; the **Video Player** contains `QVOD`, `Lenovo Video`, `Baidu Video`, `Sohu Video`, and `iQiyi Video`; the **Browser** contains `Chrome`, `UC Web`, `Jinshan`, `Baidu`, and `360Safe`.

First, we want to figure out the distribution of the user preferences against the app according to the device model. We employ the cumulative distribution function (CDF) to demonstrate such distributions, as shown in Fig. 4. For each app, the X-Axis represents the price of device models that are sorted in ascending order, and the Y-Axis refers to the percentage of the app's users holding such a device model. An app tends to have be used by more higher-end users if the curve draws near the bottom. Obviously, we can observe that the choice of

Table 4. The number of paid apps, buyers, and purchase records

Device group	# of paid apps	# of buyers	# of purchases
High-end	639	13,788	14,977
Middle-end	628	13,700	14,591
Low-end	190	1,590	1,640

device models significantly impact the selection of competing apps. For the news readers, we can see that the Phoenix News and Netease News are more likely to be adopted by higher-end users. In contrast, the Sohu News tend to be more preferred by the lower-end users.

The difference among device models is even more significant for the *Video* players. The Lenono Video takes a very significant difference compared to other 4 apps, indicating most of its users are lower-end. One possible reason is the Lenovo Video is a preloaded app that is used mainly on smartphones manufactured by Lenovo, and most of these smartphones are categorized into middle-end and low-end groups.

Finally, in the browser group, the similar findings can be observed. The most preferred browser of higher-end users is Chrome, followed by the 360Safe browser, Jinshan browser, and the UC Web browser. The Baidu browser are the most likely to be adopted by the low-end users.

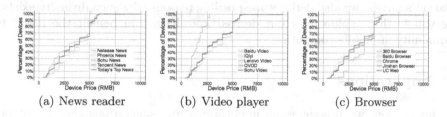

(a) News reader   (b) Video player   (c) Browser

Fig. 4. Distribution of preferences of competing apps

We can confirm that the device models can affect the choice of competing apps. In addition, we are interested in the possible underlying reasons. Referencing the app similarity model proposed by Chen *et al.* [4], we explore the profile of these competing apps, including the textual descriptions, vendors, and features. Some immediate findings are derived.

First, we can infer that the lower-end users prefer the "local" apps more than the "international apps". For example, the Baidu browser and the 360Safe browser are both developed by local app vendors in China. Second, some special features can be more attractive to different users. For example, from the profile the Baidu and *UC Web* claim that they are more "traffic-friendly" by compressing the Web page on a front-end proxy before the Web page is delivered to the users. Hence, it is not surprising that these two browsers are more appreciated by the lower-end users, given that these users have limited budgets. Third, the content providers can be an possible impact factor. For example, Sohu is famous for the fast channel of entertainment news in China, and thus the lower-end users, who are possibly in-school students and low-position employees, are more likely to take the Sohu News as the favored app. In contrast, the Phoenix News is provided by the Phoenix New Media[10], which is famous for the in-depth, objective

---

[10] http://ir.ifeng.com/.

reviews of economic, finance, and politics. As a result, the users holding higher-end device models the `Phoenix News` as a preference, since they may care more about the related topics.

5.6 Summary of Findings

So far we have answered the 5 research questions proposed previously. Although users holding different device models share similarities in some categories of apps such as *Communication*, significant diversities are observed. The correlation analysis and hypotheses testing confirm that the choice of device models exactly impacts the user behaviors in terms of adoption of app stores, app selection, app abandonment, online time, data plan, app revenue, and selection of competing apps. In summary, we can conclude that the choice of device models can significantly impact the app usage, which in turn reflects the classification of the user requirements, preferences, and possible background. Besides the correlation analysis, we also derive some possible reasons why such diversity exists. We will further explore how our derived knowledge can benefit the research community of mobile computing.

6 Implications

In this section, we present some implications and suggestions that might be helpful to relevant stakeholders in the app-centric ecosystem.

6.1 App Store Operators

App stores play the central role in the ecosystem. Intuitively, the recommender systems deployed on the app stores should accurately suggest the proper apps of users. From our findings in **RQ1–RQ5**, the choice of device models can significantly impact the user preferences of app selection, especially in some competing apps. It is reported that most app stores mainly rely on the similarity-based recommendations such as the apps frequently downloaded by users, the apps developed by the vendors, or apps with similar purposes [22]. Some advanced recommendation techniques can further improve the recommendation quality, such as those based on the similarity between apps from various aspects such as app profile, category, permission, images, and updates [4]. However, synthesizing the user requirements, preferences, and even economic background inferred from our study can be further helpful. For example, the app stores can recommend a lower-end user with the browsers such as `UCWeb` and `Baidu` instead of `Chrome`, if the user cares about the data plan. To the best of our knowledge, very few app stores take into account the impacts of device model as an influential factor, including Wandoujia. In practice, we plan to integrate the device model as a dimension to improve the recommendation quality.

From **RQ1**, we can find that a large number of users (at least in China) do not frequently download and update their apps from app stores. Although the

Android users can use other third-party app stores (e.g., those provided by the device manufacturers) other than Wandoujia, it is still worth reporting that the lower-end users are less likely to access the app stores. In this way, the app stores have to carefully consider how to expand the desires from these users.

6.2   App Developers

Developers can also learn some lessons from our study when designing and publishing their apps. From **RQ1**, we can find that some apps are more frequently uninstalled on some device models. It implies that there can be some problems such as compliance with hardware or the device-specific APIs. As reported on the StackOverFlow [20], some camera related bugs have been found on Samsung Galaxy Note2. Our finding can validate such problematic issues. When having the distribution of uninstallations according device models, the developers, OS-vendors, and device manufacturers can draw attentions and prevent problematic issues.

As we presented in **RQ2**, the choice of device models makes quite significant impacts on the online time of apps. For example, the higher-end user can spend more time on *NEWS_AND_READING* and *TRAVEL*. Due to the fragmentation of Android devices [10], currently more and more in-app ads networks allow the developers to customize the title, banner, and content of ads according to the device model [6]. Hence, the developers of these apps should consider customizing some device-specific ads networks to these "heavy" users and gain the potential revenue of ads clicking.

From **RQ4**, we observe that lower-end users are likely to pay more for apps than higher-end users. Such a finding can assist developers to locate target audiences from whom the revenue can be gained. To attract the lower-end users who are less likely to buy their apps but probably with less budgets, the developers can consider the *"try-out-and-buy"* model to increase the user interests. In addition, they can further explore some new features or in-app ads to increase more revenue.

From **RQ5**, we can find that the selection of competing apps can vary a lot among users holding different levels of device models. Developers can further explore why their apps are less adopted by users from some specific device models, and fix possible bugs, optimize the design, or provide advanced features. For example, it is said that the UC Web can save traffics for users by compressing images and refactoring the Web page layout on the sever before the page is downloaded by the device. Such optimized features can be leveraged to improve the user experiences of apps.

6.3   Network Service Providers

The network service providers or carriers such as T-Mobile, China Mobile, and China UniCom, play an important role in providing the communication infrastructure and service delivery. From **RQ2** and **RQ3**, we can derive the network usage patterns of users choosing different device models. From **RQ2**, the lower-end are more likely to connect to cellular network than the higher-end users, especially in using some specific apps. However, the findings of **RQ3** suggest that lower-end users typically spend less data plan in cellular network since they may have relatively limited budgets. Such an observation indicates that the lower-end users are less possible to use the "traffic-heavy" apps such as online music and video players under cellular network. To increase the data plan usage of these lower-end users, the network service providers do need to concern some special business models by case. For example, the network service providers can negotiate with the vendors and provide special packages of data plan, such as the "*buy-out*" ones that are customized to specific apps, e.g., music or video players. Such packages provide the users the unlimited cellular traffic that can be used but only to access the specific apps. As the customized packages are independent from the regular data plan, they are possibly appreciated by users for specific purposes. The network service providers can exploit such a business model to the device manufacturers and app store operators. For example, the device manufacturers commonly preload some apps in their devices. The network service providers can bind these apps with the customized data package, and share the commissions from the revenue of device manufacturers.

In addition, from **RQ2**, we can also infer that the lower-end users are less likely to be covered by Wi-Fi, and tend to spend more network access time under cellular. By synthesizing another findings of **RQ2**, i.e., the lower-end users spend more time on *EDUCATION* apps under cellular, we can infer that these users are possibly in-school students. As the network service providers can easily obtain the device model information and location distribution of the connected devices, e.g., from the tier-1 cellular carriers, they can estimate and allocate the radio resources around the places where lower-end users are more likely to stay.

7   Threats to Validity

Considerable care and attention should be addressed to ensure the rigor of our study. As with any chosen research methodology, it is hardly without limitations. In this section, we will present the potential threats for validating our results.

One potential limitation of our work is that the data set is collected from a single app marketplace in China. The users under study are mainly Chinese, and the region differences should be considered. In addition, the investigated apps are only Android apps. Hence, some results may not be fully generalized to other app stores, platforms, or countries. Such an limitation can hardly be addressed due to the difficulty of acquiring the similar behavioral profile from large-scale users. However, the measurement approach itself can be generalized to other similar data sets from other app stores. Additionally, China has become the biggest market of mobile devices and apps all over the world, so our findings derived from over 0.7 million Android users can provide some comprehensive and representative knowledge to the research communities of modern Internet.

Choosing the price as indicator can introduce threat since the price of devices can change quite frequently. To address such a threat, we make a coarse-grained categorization of device models. We manually checked the first-to-market price and the latest price as of the end point of our data set of each model. The price of most device models can still fall into our category. In addition, it would be interesting that if we categorize the device models based on other levels of price. Although there can be some bias caused by the varying price of device models, we believe that our measurement approach and findings can be generalized to any other data sets like ours.

8   Conclusion and Future Work

In this paper, we have presented the correlation analysis of choice of device models against the user behaviors of using Android apps. Our study was conducted over the largest to date set collected from over 0.7 million users of Wandoujia. We reported how the choice of device models can impact the adoption of app stores, app selection and abandonment, online time, data plan usage, revenues gained from apps, and comparisons of competing apps. The results revealed the significance of device models against app usage. We summarized our findings and presented implications for relevant stakeholders in the app-centric ecosystem.

Currently, we plan to take into account the device models as an important impact factor in the recommendation systems of Wandoujia. The analytical techniques shall be developed as an offline learning kernel and improve more personalized recommendation of apps. We are now developing features to collect much finer-grained information in the Wandoujia management apps, such as the traffic/access time per session in apps and the click-through logs. We believe that such detailed information can explore more diversity among users and thus improve our study.

Acknowledgement. This work was supported by the National Key R&D Program under the grant number 2018YFB1004800, the National Natural Science Foundation of China under grant numbers 61725201, 61528201, 61529201, and a Google Faculty Award. We thank Wenlong Mou for assistance with statistical analysis in this project, and for comments that greatly improved the manuscript.

References

1. Alharbi, K., Yeh, T.: Collect, decompile, extract, stats, and diff: mining design pattern changes in android apps. In: Proceedings of the MobileHCI 2015, pp. 515–524 (2015)
2. Böhmer, M., Hecht, B., Schöning, J., Krüger, A., Bauer, G.: Falling asleep with angry birds, Facebook and Kindle: a large scale study on mobile application usage. In: Proceedings of the MobileHCI 2011, pp. 47–56 (2011)
3. Böhmer, M., Krüger, A.: A study on icon arrangement by smartphone users. In: Proceedings of the CHI 2013, pp. 2137–2146 (2013)
4. Chen, N., Hoi, S.C.H., Li, S., Xiao, X.: SimApp: a framework for detecting similar mobile applications by online kernel learning. In: Proceedings of the WSDM 2015, pp. 305–314 (2015)

5. Church, K., Ferreira, D., Banovic, N., Lyons, K.: Understanding the challenges of mobile phone usage data. In: Proc. MobileHCI 2015, pp. 504–514 (2015)
6. Facebook: Facebook: how do i run ads only on specific types of phones? (2014). https://www.facebook.com/business/help/607254282620194
7. Falaki, H., Mahajan, R., Kandula, S., Lymberopoulos, D., Govindan, R., Estrin, D.: Diversity in smartphone usage. In: Proceedings of the MobiSys 2010, pp. 179–194 (2010)
8. Freund, R.J., Mohr, D.: Statistical Methods, 3rd edn. Academic Press, Cambridge (2010)
9. Kelly, D., Smyth, B., Caulfield, B.: Uncovering measurements of social and demographic behavior from smartphone location data. IEEE Trans. Hum.-Mach. Syst. 43(2), 188–198 (2013)
10. Khalid, H., Nagappan, M., Shihab, E., Hassan, A.E.: Prioritizing the devices to test your app on: a case study of android game apps. In: Proceedings of the FSE 2014, pp. 610–620 (2014)
11. Li, H., et al.: Characterizing smartphone usage patterns from millions of android users. In: Proceedings of the IMC 2015, pp. 459–472 (2015)
12. Lim, S.L., Bentley, P.J., Kanakam, N., Ishikawa, F., Honiden, S.: Investigating country differences in mobile app user behavior and challenges for software engineering. IEEE Trans. Softw. Eng. 41(1), 40–64 (2015)
13. Ma, H., Cao, H., Yang, Q., Chen, E., Tian, J.: A habit mining approach for discovering similar mobile users. In: Proceedings of the WWW 2012, pp. 231–240 (2012)
14. Patro, A., Rayanchu, S.K., Griepentrog, M., Ma, Y., Banerjee, S.: Capturing mobile experience in the wild: a tale of two apps. In: Proceedings of the CoNEXT 2013, pp. 199–210 (2013)
15. Rahmati, A., Tossell, C., Shepard, C., Kortum, P.T., Zhong, L.: Exploring iPhone usage: the influence of socioeconomic differences on smartphone adoption, usage and usability. In: Proceedings of the MobileHCI 2015, pp. 11–20 (2012)
16. Rahmati, A., Zhong, L.: Studying smartphone usage: lessons from a four-month field study. IEEE Trans. Mob. Comput. 12(7), 1417–1427 (2013)
17. Raptis, D., Tselios, N.K., Kjeldskov, J., Skov, M.B.: Does size matter?: investigating the impact of mobile phone screen size on users' perceived usability, effectiveness and efficiency. In: Proceedings of the MobileHCI 2013, pp. 127–136 (2013)
18. Sani, A.A., Tan, Z., Washington, P., Chen, M., Agarwal, S., Zhong, L., Zhang, M.: The wireless data drain of users, apps, & platforms. Mob. Comput. Commun. Rev. 17(4), 15–28 (2013)
19. Song, Y., Ma, H., Wang, H., Wang, K.: Exploring and exploiting user search behavior on mobile and tablet devices to improve search relevance. In: Proceedings of the WWW 2013, pp. 1201–1212 (2013)
20. StackOverflow: Android camera fails (2014). http://stackoverflow.com/search?q=android+camera+samsung+fail
21. Tossell, C., Kortum, P.T., Rahmati, A., Shepard, C., Zhong, L.: Characterizing web use on smartphones. In: Proceedings of the CHI 2012, pp. 2769–2778 (2012)
22. Viennot, N., Garcia, E., Nieh, J.: A measurement study of Google play. In: Proceedings of the SIGMETRICS 2014, pp. 221–233 (2014)
23. Xu, Q., Erman, J., Gerber, A., Mao, Z.M., Pang, J., Venkataraman, S.: Identifying diverse usage behaviors of smartphone apps. In: Proceedings of the IMC 2011, pp. 329–344 (2011)
24. Yan, B., Chen, G.: AppJoy: personalized mobile application discovery. In: Proceedings of the MobiSys 2011, pp. 113–126 (2011)

An Optimized Artificial Bee Colony Based Parameter Training Method for Belief Rule-Base

Man-Na Su, Zhi-Jian Fang, Shao-Zhen Ye, Ying-Jie Wu,
and Yang-Geng Fu[✉]

College of Mathematics and Computer Science,
Fuzhou University, Fuzhou, China
ygfu@qq.com

Abstract. In order to solve the problems of poor portability, complex implementation, and low efficiency in the traditional parameter training of the Belief rule-base, an artificial bee colony algorithm combined with Gaussian disturbance optimization was introduced, and a novel Belief rule-base parameter training method was proposed. By the light of the algorithm principle of the artificial bee colony, the honey bee colony search formula and the cross-border processing method were improved, and the Gaussian disturbance was employed to prevent the search from falling into a local optimum. The parameter training was implemented in combination with the constraint conditions of the Belief rule-base. By fitting the multi-peak function and the leakage detection experiment of oil pipelines, the experimental error were compared with the traditional and existing parameter training methods to verify its effectiveness.

Keywords: Belief rule-base (BRB) · Artificial bee colony algorithm
Swarm intelligence · Parameter optimization model

1 Introduction

Currently, D-S evidence theory [1, 2], decision theory [3], fuzzy theory [4] and traditional IF-THEN rule-base [5] have provided a good theoretical basis for solving multi-attribute decision making problems. However, in view of the large amount of uncertain information existing in the data, a decision method that can effectively handle uncertain information is still urgently needed. Therefore, on the basis of D-S evidence theory, decision theory, fuzzy theory and traditional IF-THEN rule-base, Yang et al. [6] proposed a Belief rule-base system (BRB) based on the Belief rule-base inference methodology using the evidential reasoning Approach (RIMER) in 2006. Compared with the traditional IF-THEN rules, BRB has proposed a new way to deal with fuzzy information on the basis of the traditional processing of uncertain information by constructing the rule with confidence distribution result and calculating the activation weights by matching the candidate attributes of the antecedents in reasoning. This method has been successfully applied to the engineering fields of oil pipeline leak detection [7], graphite component detection [6], smart traffic lights [8], etc.

© Springer Nature Singapore Pte Ltd. 2018
Z. Xu et al. (Eds.): Big Data 2018, CCIS 945, pp. 77–93, 2018.
https://doi.org/10.1007/978-981-13-2922-7_5

In the Belief rule-base system, rules deal with fuzzy information by candidate attributes of the antecedents, and the uncertain information is processed by using the premise attribute weight, rule weight and result confidence level distribution. Therefore, how to select many parameters in the set rules has become the decisive factor that directly affects the performance of the system. Rules in traditional BRB system are mostly given by experts based on historical experience information. But, when dealing with complex decision problems, it is difficult for experts to accurately determine the values of parameters. Therefore, how to build the parameter training framework model of BRB has become an urgent problem. At first, Yang et al. [9] proposed to use FMINCON function in MATLAB to train relevant parameters, although it greatly improved the performance of the system, but the MATLAB code was compiling, and the portability was low. Then, on the basis of the Chen's [12] global optimization parameter training model, Chang [10] and Wu [11] respectively propose the gradient descent method and the algorithm for improving the efficiency of gradient descent convergence using binary search. However, it involves the cumbersome calculation of the derivation formula in the process. Su et al. [13] first put forward the combination of particle swarm optimization and parameter training to introduce group intelligence into the training model of Belief rule-base, but it is not ideal in convergence precision, and there is still optimization space.

In this paper, we use the new global optimization parameter training model proposed in Wu, Su [11, 13], and introduce the artificial bee colony algorithm [14] combined with Gaussian disturbance optimization to train the parameters in the Belief rule-base system. Artificial bee colony algorithm essentially simulates biological behavior of bees collecting honey. The feasible solution can be expressed by nectar source, and the parameters can be trained through honey bees approaching the optimal nectar source step by step. Based on the research of artificial bee colony algorithm, we optimize the bee colony honey search formula and cross-border processing method to improved the precision and performance of the system, and the Gauss disturbance is added to prevent the search from falling into the local optimal. In the experimental analysis, the validity and superiority of the algorithm are verified by comparing the experimental accuracy of the traditional method and the existing parameter training method.

The organizational structure of this paper is as follows: Sect. 2 introduces the relevant knowledge of the Belief rule-base theory and problems in the training of system parameters. The third section introduces the algorithm principle of artificial bee colony, and the optimization method is proposed for parameter training of Belief rule-base. The fourth section proves the effectiveness of the method by fitting multi-peak functions and leak detection experiments of oil pipelines. Finally, summarize the article and propose the direction for further research.

2   Belief Rule-Base System

The Belief rule-base system proposed by Yang et al. [6] contains two aspects: the Belief rule- base representation and the Belief rule-base inference methodology using the evidential reasoning Approach. The following brief introduction to the relevant theoretical knowledge of the Belief rule-base.

2.1 Representation of a Belief Rule-Base (BRB)

In order to express uncertain information, Yang et al. [6] introduced a confidence distribution in the results section based on the traditional IF-THEN rules. And adding the weight of the premise attribute and the weight of the rule to weight the importance of the attribute and the rule. The k th rule in the rule-base can be expressed as follows:

$$R_k: \quad \text{if } x_1 \text{ is } A_1^k \wedge x_2 \text{ is } A_2^k \wedge \ldots \wedge x_{T_k} \text{ is } A_{T_k}^k$$
$$\text{Then } \{(D_1, \bar{\beta}_{1,k}), (D_2, \bar{\beta}_{2,k}), \ldots, (D_N, \bar{\beta}_{N,k})\}$$

With a rule weight θ_k and attribute weight $\bar{\delta}_1, \bar{\delta}_2, \ldots \bar{\delta}_{T_k},$.

Where $A_i^k (i = 1, \ldots T_k; k = 1, \ldots, L)$ represents the first i premise attribute candidate value of the k th rule; T_k indicates the number of premise attribute included in the k th rule; L represents the number of rules in the system; $\bar{\beta}_{j,k}$ denotes confidence in the j th result evaluation level D_j of the k th rule; N is the number of evaluation grades; θ_k and $\bar{\delta}_i$ denote the weight of the rule of the k th rule and the weight of the premise attribute of the i th premise attribute, respectively.

2.2 Belief Rule-Base Inference Methodology Using the Evidential Reasoning Approach (RIMER)

The Belief rule-base inference methodology using the evidential reasoning Approach is the core of the Belief rule-base system. It mainly consists of two parts, rule activation and activation rule synthesis.

Activation Weight Calculation and Confidence Correction. For a system input sample, its activation weight for the k th rule can be expressed as follows:

$$\omega_k = \frac{\theta_k \prod_{i=1}^{T_k} \left(\alpha_i^k\right)^{\bar{\delta}_{k,i}}}{\sum_{l=1}^{L} \left[\theta_l * \prod_{i=1}^{T_l} \left(\alpha_i^l\right)^{\bar{\delta}_{l,i}}\right]} \tag{1}$$

Among them, the weight of the premise attribute is normalized $\bar{\delta}_{k,j} = \frac{\delta_{k,i}}{\max_{i=1,\ldots,T_k}\{\delta_{k,i}\}}$.

α_i^k represents the individual matching degree of the input sample $X(x_1, x_2, \ldots, x_{T_k})$ for the i th premise attribute, and the calculation formula is as follows:

$$\alpha_i^j = \begin{cases} \frac{A_i^{k+1} - x_i}{A_i^{k+1} - A_i^k}, j = k, A_i^k \leq x_i \leq A_i^{k+1} \\ \frac{x_i - A_i^k}{A_i^{k+1} - A_i^k}, j = k+1, A_i^k \leq x_i \leq A_i^{k+1} \\ 0, \quad otherwise \end{cases} \tag{2}$$

When confidence results $\sum_{j=1}^{N} \bar{\beta}_{j,k} = 1$, the rule is called complete. Conversely, when the confidence result $\sum_{j=1}^{N} \bar{\beta}_{j,k} < 1$, we need to correct the confidence result:

$$\beta_{i,k} = \bar{\beta}_{i,k} \frac{\sum_{t=1}^{T_k} \left[\tau(t,k) \sum_{j=1}^{|A_t|} \alpha_{t,j} \right]}{\sum_{t=1}^{T_k} \tau(t,k)} \tag{3}$$

among them, $\tau(t,k) = \begin{cases} 1, U_t \in R_k, t = 1, \ldots, T_k \\ \quad 0, \quad otherwise \end{cases}$

Rule Synthesis. For the activated rules, the application of Evidence Reasoning (ER) algorithm [6] is used for synthesis.

First. Convert the resulting confidence distribution to the probabilistic quality for ER synthesis:

$$m_{j,k} = \omega_k \beta_{j,k} \tag{4}$$

$$\tilde{m}_{D,k} = \omega_k \left(1 - \sum_{j=1}^{N} \beta_{j,k} \right) \tag{5}$$

$$\bar{m}_{D,k} = 1 - \omega_k \tag{6}$$

$m_{j,k}$ represents the credibility of the j th result of the k th rule; $\tilde{m}_{D,k}$ denotes the credibility of the k th rule that is not assigned to any result evaluation level due to incomplete confidence distribution; $\bar{m}_{D,k}$ denotes the credibility of the k th rule that is not assigned to any result evaluation level due to activation weights;

Second. For probability quality, apply the ER analytical formula [15] to synthesize the activation rules:

$$C_j = k \left[\prod_{l=1}^{L} \left(m_{j,l} + \tilde{m}_{D,l} + \bar{m}_{D,l} \right) - \prod_{l=1}^{L} \left(\tilde{m}_{D,l} + \bar{m}_{D,l} \right) \right] \tag{7}$$

$$\tilde{C}_D = k \left[\prod_{l=1}^{L} \left(\tilde{m}_{D,l} + \bar{m}_{D,l} \right) - \prod_{l=1}^{L} \bar{m}_{D,l} \right] \tag{8}$$

$$\bar{C}_D = k \prod_{l=1}^{L} \bar{m}_{D,l} \tag{9}$$

$$k^{-1} = \sum_{j=1}^{N} \prod_{l=1}^{L} \left(m_{j,l} + \tilde{m}_{D,l} + \bar{m}_{D,l} \right) - (N-1) \prod_{l=1}^{L} \left(\tilde{m}_{D,l} + \bar{m}_{D,l} \right) \tag{10}$$

According to the rule synthesis results, the confidence distribution of each result level is obtained:

$$\beta_j = \frac{C_j}{1 - \bar{C}_D}, n = 1, \ldots, N \tag{11}$$

$$\beta_D = \frac{\tilde{C}_D}{1 - \bar{C}_D} \tag{12}$$

Where β_j represents the confidence of the system inference for the j th result evaluation level D_j; β_D indicates the confidence level not assigned to any result evaluation level D_j;

According to the result confidence distribution and grade utility value $\mu = \{\mu_1, \mu_2, \ldots, \mu_N\}$ we can get result utility:

$$\mu_{min} = (\beta_1 + \beta_D)\mu_1 + \sum_{j=2}^{N} \beta_j \mu_j \tag{13}$$

$$\mu_{max} = (\beta_N + \beta_D)\mu_N + \sum_{j=1}^{N-1} \beta_j \mu_j \tag{14}$$

$$\mu_{avg} = \frac{\mu_{min} + \mu_{max}}{2} \tag{15}$$

2.3 Belief Rule-Base Parameter Training Model

When solving complex attribute decision problems, it is difficult for experts to give accurate system parameter values based on historical experience. Therefore, Yang et al. [9] proposed to use prediction results of the BRB system compare with the actual values to correct the system parameters.

Traditional Belief Rule-Base Parameter Training Model. Yang et al. [9] construction parameter training model as shown in Fig. 1:

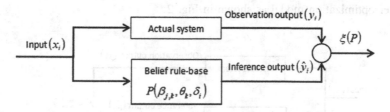

Fig. 1. Traditional parameter optimization model

This model trains the confidence distribution $\beta_{j,k}$, rule weight θ_k, and premise attribute weights δ_i in the BRB system. Based on the comparison inference output and the observation output, the model objective function can be expressed as follows:

$$\min\{\xi(P)\}$$

s.t.$A(P) = 0, B(P) \geq 0$

$\xi(P)$ is the objective function, which represents the error between the system's inference output and the observation output. When $\xi(P)$ is smaller, the output of the system can be fitted to the true result, that is, the system performance is better.

Where $A(P)$ is an equality condition and $B(P)$ is an inequality constraint condition is as follows:

First. The rule weight indicates the degree of importance of the rule, and is normalized to a value between 0 and 1, that is:

$$0 \leq \theta_k \leq 1, k = 1, \ldots, L$$

Second. The premise attribute weight represents the importance of the premise attribute, and is normalized to a value between 0 and 1, that is:

$$0 \leq \bar{\delta}_i \leq 1, i = 1, \ldots, M$$

Third. The confidence distribution of the result indicates the credibility of the rule at the result evaluation level D_j, which belongs to a value between 0 and 1, that is:

$$0 \leq \beta_{j,k} \leq 1, j = 1, \ldots, N, k = 1, \ldots, L$$

Fourth. If the result confidence is distributed, then $\sum_{j=1}^{N} \bar{\beta}_{j,k} = 1$, otherwise $\sum_{j=1}^{N} \bar{\beta}_{j,k} < 1$.

Global Parameter Optimization Training Model. Chen [12] proposed a global parameter optimization model by considering the candidate values of the premise attribute as a parameter on the basis of Yang [9]. Wu [11], Su [13] based on Chen's parameter optimization model also included the premise attribute weights, rule weight and result evaluation utility grades etc. into parameter training, and constructed a new parameter optimization model as shown in Fig. 2:

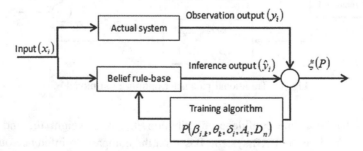

Fig. 2. Global parameter optimization training model

A new parameter constraint inequality is added to the model:

Fifth. It is required that the difference between the adjacent values of the same premise attribute is not greater than the infinitesimal $\varepsilon_{i,j}$:

$$A_{i,j} - A_{i,j+1} \leq \varepsilon_{i,j}, i = 1, \ldots, M, j = 1, \ldots, |A_i| - 1$$

Sixth. The level of evaluation of result indicates the degree of preference. The higher the degree of preference is, the higher the utility expectation is:

$$D_j - D_{j+1} \leq 0, j = 1, \ldots, N - 1$$

2.4 Question Putting Forward

In the existing parameter training method, The FMINCON function [16] in MATLAB which is applied by Zhou et al. has weak portability, and for complex systems, there are problems such as low training efficiency and low accuracy. The methods introduced by Chang [10], Wu [11] introducing gradient training and dichotomize to improve the convergence speed have significant improvements system performance in the convergence time and accuracy, but the formula derivation process is complex and difficult to implement. Wang et al. [17] proposed the method of expert intervention increased the rationality of the parameters. Therefore, on the basis of this, Su [13] and others proposed the combination of group intelligence and parameter training, but there is still room for improvement in the accuracy of decision making. However, none of the above methods has the space to further convergence in terms of accuracy. Therefore, this paper proposes an optimized artificial bee colony algorithm to further approximate the parameters in parameter training to achieve better system performance.

3 Parameter Training Method Base on Optimization of Artificial Bee Colony

This paper proposes a parameter training method on the basis of the global optimization parameter model of the Belief rule-base and the optimized artificial bee colony algorithm [14].

3.1 Artificial Bee Colony Algorithm

The algorithm simulates the biological behavior of bee colony gathering honey and divides the population into collection bee, observation bee and reconnaissance bee. The collection bees randomly searched for the location of the nectar source and collect honey one by one. Observe bees select nectar source for honey collection based on the probability of the degree of nectar source feedbacked by collection bees. When there are no better results for finding the nectar many times, the collection bee is changed to reconnaissance bee with the location of random to collect the honey.

Simulating the biological process, the artificial bee colony algorithm is described as follows:

Step 1. Generate a bee with a population of $2N$, where the number of collection bees and the number of observation bees are both N. And initialize the nectar source, that is, collection bee location. Randomly initialize the nectar source location based on the upper and lower bounds of the parameters, and set the nectar source as the vector of the d dimension $x_i = \{x_{i,1}, x_{i,2}, \ldots, x_{i,d}\}$:

$$x_{i,j} = low_j + \phi_{i,j}(up_j - low_j), \quad i = 1, \ldots, N, j = 1, \ldots, d \tag{16}$$

Among them, up_j and low_j are the upper and lower bounds of the j dimension parameter variable respectively. $\phi_{i,j}$ is a random number in $[0,1]$ interval.

Step 2. Collection bees collect honey, and Collection bees randomly search for new nectar according to formula (17)

$$\tilde{x}_{i,j} = x_{i,j} + \varphi_{i,j}(x_{i,j} - x_{k,j}), i = 1, \ldots, N, k = 1, \ldots, N, k \neq i, j = 1, \ldots, d \tag{17}$$

Where $\varphi_{i,j}$ is the random number in $[-1,1]$ interval, and k denotes the random collection bee individual number which is not equal to i.

Step 3. Observation bees according to the nectar source fitness feedbacked by collection bees, use the roulette probability to choose the nectar source to collect honey. The probability that the i th position of collection bee is located in the honey position is:

$$P_i = \frac{fit(x_i)}{\sum_{i=1}^{N} fit(x_i)} \tag{18}$$

The fitness value formula:

$$fit(x_i) = \begin{cases} \frac{1}{[1+f(x_i)]}, & f(x_i) \geq 0 \\ 1 + |f(x_i)|, & else \end{cases} \tag{19}$$

$f(x_i)$ represents the error of individual x_i in the BRB system.

Step 4. Check the number of iterations of collection bee. If collection bee does not update the position when it exceeds a certain number of thresholds, the collection bee is converted to reconnaissance bee to collect honey at the global random honeypot location.

Step 5. Repeat steps 2 to 4 until the iteration reaches a set number of times or the error is less than a certain threshold.

3.2   Algorithm Improvement

It is explored that the convergence speed of the artificial bee colony algorithm is slow and it is easy to fall into the local optimal solution. This paper proposes the following improvements to the artificial bee colony.

Honey sources searching formula

Aiming at the problem that the artificial bee colony algorithm converges too slowly, this paper will modify the moving formula (17) of bee randomly searching for honey source to:

$$\tilde{x}_{i,j} = x_{i,j} + \varphi_{i,j}(x_{i,j} - best_j), i = 1, \ldots, N, k = 1, \ldots, N, k \neq i, j = 1, \ldots, d \qquad (20)$$

Among them, *best* is the current optimal individual position.

By moving to the current optimal nectar source location directly, the number of non-meaningful searches is reduced, thereby reducing the convergence time and improving the convergence efficiency.

Cross-Border Processing

For problems where random values exceed the upper and lower limits of the parameters, the artificial bee colony algorithm bound the bees' behavior path by forcibly giving the upper and lower bounds, ignoring the individual information that the bees originally had, slowing the convergence rate. So this paper will modify the cross-border treatment of artificial bee colonies as follows:

$$x_{i,j} = \begin{cases} best_j, & \gamma > threshold \\ low_j + \gamma(up_j - low_j), & else \end{cases} \qquad (21)$$

Where, $\gamma \in [0, 1]$ a random number.

When the cross-border occurs, the algorithm gives a certain probability to the current optimal honey position or random position to re-apply collection bee individual information to quickly approach the current optimal honey position, further accelerating the convergence speed.

Gaussian Disturbance

In order to prevent the bee colony search from falling into the local optimal solution prematurely, the Gaussian disturbance is added in each update of the population to make the bee get out of the local optimal solution.

According to Gaussian disturbance function $Disturb = \frac{\lambda}{G_m} Gauss()$ new a location of nectar source:

$$Disturb\ x_j = best_j + Disturb_j, \quad j = 1, \ldots, d \qquad (22)$$

Among them, G_m is the algebra of the current population and λ is the empirical constant.

With the iteration of the population, the step size of the disturbance is gradually shortened, and the approximation of the optimal solution is realized.

If the new nectar source location is better than the current best nectar source, it will be added to the population to iterate:

$$x_{k,j} = Disturb \; x_j, \quad j = 1, \ldots, d \tag{23}$$

Where k represents the randomly selected k th bee individuals.

3.3   Algorithmic Description

Combining the optimized artificial bee colony algorithm and the Belief rule-base parameter training model, the parameter training algorithm in this paper can be represented as the following flow chart in Fig. 3:

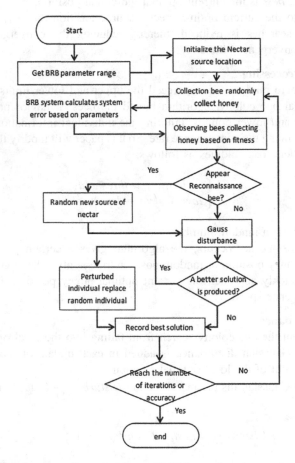

Fig. 3. Algorithm flowchart for optimizing artificial bee colony parameter training method

4 Experimental Analysis

The experiment runs in Intel(R) Core(TM) i5-4570@3.20 GHz CPU,8 GB RAM, Windows7 operating system environment. It is coded with Visual C++ 6.0 and analyzed with MATLAB drawings.

Through the experiments of multi-peak function fitting and oil pipeline leak detection experiments, the effectiveness of the experiment is verified by MSE (Mean Squared Error), MAE (Mean Absolute Error) respectively, and shows its superiority through comparison with other algorithms.

4.1 Multimodal Function Fitting Experiment

A common multimodal function is adopted in the experiment [12]:

$$g(x) = x\sin(x^2), 0 \le x \le 3$$

The experiment evenly takes the 1000 sample points selected in the domain of function definition as the system test data. According to the [12] expert estimate, the initial BRB rule is shown in Table 1 as follows:

Table 1. Initial BRB system rules.

Rule number	Rule weight	Alternative value	Result confidence distribution $\{D_1, D_2, D_3, D_4, D_5\} = \{-2.5, -1, 1, 2, 3\}$
1	1	0	$\{(D_1, 0), (D_2, 0.5), (D_3, 0.5), (D_4, 0), (D_5, 0)\}$
2	1	0.5	$\{(D_1, 0), (D_2, 0.44), (D_3, 0.56), (D_4, 0), (D_5, 0)\}$
3	1	1	$\{(D_1, 0), (D_2, 0.08), (D_3, 0.92), (D_4, 0), (D_5, 0)\}$
4	1	1.5	$\{(D_1, 0), (D_2, 0), (D_3, 0.84), (D_4, 0.16), (D_5, 0)\}$
5	1	2	$\{(D_1, 0.34), (D_2, 0.66), (D_3, 0), (D_4, 0), (D_5, 0)\}$
6	1	2.5	$\{(D_1, 0), (D_2, 0.53), (D_3, 0.47), (D_4, 0), (D_5, 0)\}$
7	1	3	$\{(D_1, 0), (D_2, 0), (D_3, 0.8), (D_4, 0.2), (D_5, 0)\}$

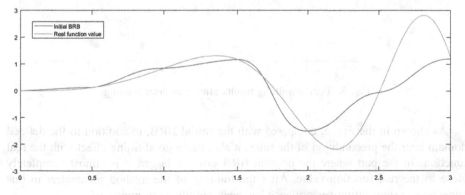

Fig. 4. Initial system function fitting results

It is easy to see that by fitting the image, the system has good fitting effect near the domain defined by the premise attribute in the rules given by the initial BRB system. Because the premise attribute of the initial BRB rule is averagely selected within the function definition domain, it is not defined near the local maximum and minimum values such as peaks and troughs of the function. Therefore, the error of the system prediction output near the extreme value is more obvious, the overall prediction result is more average, and the overall error is larger (Fig. 4).

According to the algorithm in this paper, the number of populations is set to 300 iterations and the mean square error MSE is used as the objective function. After the training, the BRB rules are shown in Table 2:

Table 2. System rules after parameter training.

Rule number	Rule weight	Alternative value	Result confidence distribution $\{D_1, D_2, D_3, D_4, D_5\} = \{-2.5, -1, 1, 2, 3\}$
1	0.876	0	$\{(D_1, 0.052), (D_2, 0.330), (D_3, 0.217), (D_4, 0.328), (D_5, 0.073)\}$
2	0.708	0.586	$\{(D_1, 0.004), (D_2, 0.234), (D_3, 0.348), (D_4, 0.194), (D_5, 0.220)\}$
3	0.732	0.911	$\{(D_1, 0032), (D_2, 0.034), (D_3, 0.435), (D_4, 0.116), (D_5, 0.383)\}$
4	0.987	1.330	$\{(D_1, 0.004), (D_2, 0.224), (D_3, 0.117), (D_4, 0.394), (D_5, 0.261)\}$
5	0.631	2.183	$\{(D_1, 0.250), (D_2, 0.490), (D_3, 0.252), (D_4, 0.007), (D_5, 0.001)\}$
6	0.578	2.846	$\{(D_1, 002), (D_2, 0.041), (D_3, 0.008), (D_4, 0.531), (D_5, 0.418)\}$
7	0.506	3	$\{(D_1, 0.103), (D_2, 0), (D_3, 0.222), (D_4, 0.385), (D_5, 0.290)\}$

The BRB fitting effect after training is shown in Fig. 5:

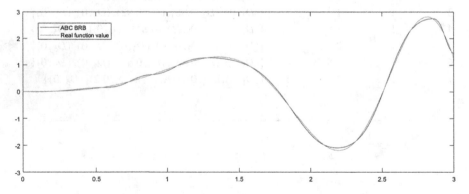

Fig. 5. Function fitting results after parameter training

As shown in the Fig. 5, compared with the initial BRB, in addition to the defined domain near the precondition of the rules, it also has a good fitting effect with the real function. In the part where the original BRB error is larger, it is almost completely close to the real function value. After the training of the method parameters in this paper, the system fitting performance has been significantly improved.

In the following Table 3, the method in this paper and the Chen method in the literature [12] and the Wu in the literature [11] are compared to the MSE as the standard for the precision performance in follow Table 3:

Table 3. Comparison of function fitting results with different parameter training methods

BRB system	MSE
Initial BRB	0.159
Chen-BRB	0.0126
Wu-BRB	0.0223
ABC-BRB	**0.0025**

From the above table comparison, we find that the accuracy of this paper have achieved better system performance compared with other methods.

4.2   Leak Detection Experiment of Oil Pipeline

In order to further verify the effectiveness of the method, an example of leak detection in oil pipeline [18] is introduced. A 100-km-long oil pipeline installed in the United Kingdom was used as the observation object in the experiment. The 2008 data from the normal to the leakage fault of the oil pipeline was selected as experimental data. Data selection records the difference between input and output traffic as FlowDiff, the pressure difference between oil and pipeline generated by PresssureDiff and the size of leakage as LeakSize. Data obtained from actual observation data have been tested by Zhou [18], Xu [7] and Yang et al. [19] for BRB experiments.

In literature [18], FlowDiff and PresssureDiff are used as the premise attributes to build the Belief rule-base to predict the size of the leakage, the candidate values and the result utility levels given by the expert experience:

$$\text{FlowDiff} = \{-10, -5, -3, -1, 0, 1, 2, 3\}$$
$$\text{PressureDiff} = \{-0.042, -0.025, -0.01, 0, 0.01, 0.025, 0.042\}$$
$$\text{LeakSize} = \{0, 2, 4, 6, 8\}$$

Based on the premise attribute candidate values, 56 rules were constructed, 25% of the data from each period of time was uniformly selected as training data, and 2008 data was used as test data, and the average absolute error MAE was used as the objective function measurement standard. Compare the prediction results of the initial BRB in Fig. 6 with the prediction of the ABC-BRB system in Fig. 7 after training in this method.

From Fig. 6, it can be clearly seen that when the initial BRB system has a low PressuresureDiff value and FlowDiff value, there is a significant error between the BRB system prediction results and the actual observation values of a large number of samples. When the PresssureDiff and FlowDiff values are high, some samples are lower than the actual observations, and most of the predictions are concentrated at a more uniform level with no significant difference. There is a high level of error and parameter training is needed to further improve system performance.

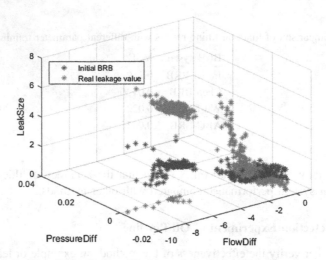

Fig. 6. Prediction of leakage of the initial BRB system

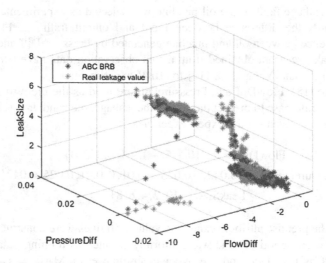

Fig. 7. Prediction of leakage of BRB system after parameter training

In Fig. 7, the system ABC-BRB trained by the method in this paper basically coincides with the actual observation value when the PresssureDiff value and FlowDiff value are low. At the same time, the partial sample with higher PresssureDiff value and FlowDiff value also have further error reduction. The performance of the prediction accuracy of the system is improved more comprehensively and the overall error is reduced.

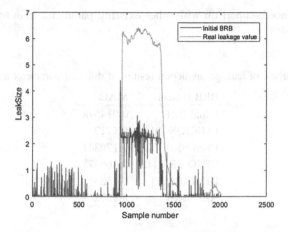

Fig. 8. Comparison between the predicted of the initial BRB system and actual values

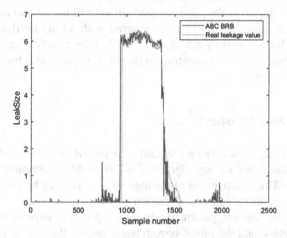

Fig. 9. Comparison between the actual values and the predicted of the BRB system after parameter training

Observing the system output with parameter before and after training and leakage values, we can see that at the initial stage of initial BRB leakage prediction as shown in Fig. 8, that is, sample 0-1000, the initial BRB prediction results have a significant difference from the actual observations of 0 actual observations. And in the peak period of leakage, that is, the sample 1000-1500, the prediction is significantly lower than the actual observation value, the overall prediction value is more average and the error is larger. As shown in Fig. 9, the ABC-BRB after training in this paper has significantly improved the initial BRB prediction and fitting effect, not only in the early and late leaks, the predicted value is stable, less minor fluctuations, and the predicted value in the peak period of leakage is also relatively high, the system predictions are more in line with actual observations.

The performance comparison with other existing parameter training methods is as follows in Table 4:

Table 4. Comparison of leakage prediction results of different parameter training methods

BRB system	MAE
Initial BRB	0.914578
FMICION [9]	0.1717
Chen-BRB [12]	0.170341
VPSO-BRB [13]	0.166478
DE-BRB [17]	0.1653
ABC-BRB	**0.1572**

From the above table, it can be seen that after the training of parameters in this paper, compared with the initial BRB system, the MAE has decreased by 82.81%, which significantly improves the precision performance of the system and it verifies the necessity of system parameter training. Compared with VPSO-BRB system and DE-BRB system, MAE decreased by 5.57% and 4.90%, respectively. Experimental results show that the proposed method is superior to the existing parameter training methods in accuracy and performance.

5   Summary and Prospect

Aiming at the accuracy performance problem of parameter training in Belief rule-base, an optimized artificial bee colony algorithm is proposed to optimize the training of global parameters. The effectiveness of the algorithm is verified by two experiments of multimodal function fitting and leak detection of oil pipeline.

However, for the current parametric training methods, experts need to provide initial rule parameters, and the initial system has a great influence on the final training results. Therefore, how to obtain representative rules from data instead of experts' rules based on experience is the direction we will study next. In addition, the BRB system not only exhibits quantitative processing ability under uncertain information, but also has the ability to process qualitative information [20]. Therefore, how to better construct the BRB system to deal with qualitative information model based on parameter training is also an important research direction.

Acknowledgements. This research was supported by the National Natural Science Foundation of China (Nos. 71501047 and 61773123).

References

1. Dempster, A.P.: A generalization of Bayesian inference. J. R. Stat. Soc. **30**(2), 205–247 (1968)
2. Shafer, G.: A Mathematical Theory of Evidence. Princeton University Press, Princeton (1976)
3. Huang, C.L., Yong, K.: Multiple Attribute Decision Making Methods and Applications a State-of Art Survey. Springer, Berlin (1981). https://doi.org/10.1007/978-3-642-48318-9
4. Zadeh, L.A.: Fuzzy sets. Inf. Control **8**(3), 338–353 (1965)
5. Sun, R.: Robust reasoning: integrating rule-based and similarity-based reasoning. Artif. Intell. **75**(2), 241–295 (1995)
6. Yang, J.B., Liu, J., Wang, J., et al.: Belief rule-base inference methodology using the evidential reasoning approach-RIMER. IEEE Trans. Syst. Man Cybern. Part A Syst. Hum. **36**(2), 266–285 (2006)
7. Xu, D.L., Liu, J., Yang, J.B., et al.: Inference and learning methodology of belief-rule-based expert system for pipeline leak detection. Expert Syst. Appl. **32**(1), 103–113 (2007)
8. Lin, Y.Q., Li, M., Chen, X.C., et al.: A belief rule base approach for smart traffic lights. In: International Symposium on Computational Intelligence and Design, pp. 460–463. IEEE (2017)
9. Yang, J.B., Liu, J., Xu, D.L., et al.: Optimization models for training belief-rule-based systems. IEEE Trans. Syst. Man Cybern. - Part A: Syst. Hum. **37**(4), 569–585 (2007)
10. Chang, R., Wang, H.W., Yang, J.B.: Parameter training method of belief rule base based on gradient and dichotomy. Syst. Eng. **25**(S), 287–291 (2007)
11. Wu, W.K., Yang, R.H., Fu, Y.G., et al.: Parameter training method of belief rule base based on accelerating gradient method. Comput. Sci. Explor. **8**(8), 989–1001 (2014)
12. Chen, Y.W., Yang, J.B., Xu, D.L., et al.: Inference analysis and adaptive training for belief rule based systems. Expert Syst. Appl. **38**(10), 12845–12860 (2011)
13. Su, Q., Yang, R.H., Fu, Y.G., et al.: Parameter training method of belief rule base based on variable speed particle swarm optimization. Comput. Appl. **34**(8), 2161–2165 (2014)
14. Karaboga, D.: An idea based on honey bee swarm for numerical optimization (2005). https://www.researchgate.net/publication/255638348_An_Idea_Based_on_Honey_Bee_Swarm_for_Numerical_Optimization_Technical_Report_-_TR06
15. Wang, Y.M., Yang, J.B., Xu, D.L., et al.: The evidential reasoning approach for multiple attribute decision analysis using interval belief degrees. Eur. J. Oper. Res. **175**(1), 35–66 (2006)
16. Jiang, J., Li, X., Zhou, Z.J., et al.: Weapon system capability assessment under uncertainty based on the evidential reasoning approach. Expert Syst. Appl. **38**(11), 13773–13784 (2011)
17. Wang, H.J., Yang, R.H., Fu, Y.G., et al.: Differential evolution algorithm for parameter training of belief rule base under expert intervention. Comput. Sci. **42**(5), 88–93 (2015)
18. Zhou, Z.J.: Belief Rule Base Expert System and Complex System Modeling. Science Press, Berlin (2011)
19. Zhou, Z.J., Hu, C.H., Yang, J.B., et al.: Online updating belief rule based system for pipeline leak detection under expert intervention. Expert Syst. Appl. **36**(4), 7700–7709 (2009)
20. Chang, L., Zhou, Z.J., You, Y., et al.: Belief rule based expert system for classification problems with new rule activation and weight calculation procedures. Inf. Sci. **336**(C), 75–91 (2016)

Search of *Center-Core* Community in Large Graphs

Linlin Ding, Yu Xie, Xiaohuan Shan, and Baoyan Song[(⊠)]

School of Information, Liaoning University, Shenyang 110036, Liaoning, China
bysong@lnu.edu.cn

Abstract. Community search plays an important role in complex network analysis. It aims to find a densely connected subgraph containing the query node in a graph. However, the most existing community search methods do not consider the influence of nodes and can not perfectly support the search in large graphs, making them have limitations in practical applications. In this paper, we introduce a community model called *center-core* community based on k-core decomposition, which can both capture the influence of nodes and guarantee the cohesiveness of community. Then we devise a *center-core* community graph index (CCG-Index), and online search algorithms (SingleQuery and MultiQuery) which support efficient search of *center-core* community in optimal time. Extensive experiments on four real-world large networks demonstrate the efficiency and effectiveness of our methods.

1 Introduction

Many real-world complex networks, such as the Internet, social networks, and biological neural networks, contain community structures. Because community structures can highly reflect the correlation of the complex networks, finding community structures of real-life networks is an important problem. Community search aims to find the most likely community structure that a node belongs to. Usually, a good community structure is described by the closeness of the nodes and defined as a densely connected subgraph in most previous studies. However, in many application domains, they need stricter requirements which consider the potential influence (or importance) of nodes and ask for the influence of nodes within a community is not lower than query node. For example, in military intelligence analysis domain, consider an ordinary soldier getting the confidential information in an army interaction network. He will try his best to pass the information to the soldiers with higher influence who usually take important positions as soon as possible in the detachment containing him. Finding a core subgraph made up of these people helps to analyze military intelligence. More, finding a subgraph containing the terrorists with important roles in communication network and finding a subgraph containing the researchers with higher influence and in the same research filed in research collaborator network are other examples of our applications.

© Springer Nature Singapore Pte Ltd. 2018
Z. Xu et al. (Eds.): Big Data 2018, CCIS 945, pp. 94–107, 2018.
https://doi.org/10.1007/978-981-13-2922-7_6

The most existing methods only find the well connected community containing query node based on various dense subgraph structures, such as degree [1–3], distance [4,5], triangle [6,7], etc. Among them, a k-core is the largest subgraph of graph G such that each node in it has at least a degree k. However, the existing community search methods based on k-core have limitations applying to the above alike applications since they do not consider the influence of nodes, such as global search (GS) proposed in [8], an algorithm based on CoreStruct named GrCon [9], a local search (LS) proposed in [10], etc.

In order to solve the above problems, we use k-core as a qualifying dense structure for modeling a densely connected community. To further consider the node influence, we use coreness to evaluate the node influence by k-core decomposition algorithm [11]. Coreness of a node is the highest order of a core that contains it. The node influence increases as coreness increases. Thus we propose a new community model called *center-core* community based on k-core decomposition. It requires the community is a connected component of the maximal k-core containing the query node q and the coreness of nodes in community is no less than q. In addition, for the nodes whose corenesses are equal to q in the graph, the community needn't contain those nodes which are only reached from q via nodes with larger corenesses. The intuition behind this definition can be easily understood through the example in the army interaction network mentioned above. Assume Fig. 1 is the army interaction network, then the coreness of v_2, v_9, v_{10} is 2 respectively, the coreness of v_3 is 3 and the coreness of the rest nodes is 4 respectively. So the influence of all nodes in Fig. 1 is no less than v_2. Certainly, $A \cup B$ can be seen as a community for v_2. But in fact, its not good to make too many soldiers know the information according to the intelligence confidentiality requirements. For instance, though the influence of v_9 is equal to v_2, v_9 is insignificant to v_2 because the information has already pass to v_7 with higher influence before reaching v_9. Thus all the nodes in B are insignificant to v_2 without the link between v_7 and v_9 due to the connectivity requirement. So A is the *center-core* community containing v_2 by our definition.

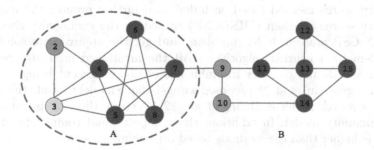

Fig. 1. $A \cup B$ is a connected 2-core of v_2; A is the *center-core* community of v_2

In this paper, we study the problem of *center-core* community search in large and dynamic graphs. First, we introduce a community model called *center-core* community based on the concept of k-core decomposition. This model can reflect the cohesiveness of the community, and also as much as possible contain the important nodes to query nodes by considering the potential influence of nodes and be more meaningful in practical applications. Then, we devise a linear-space index structure, *center-core* community graph index, called CCG-Index according to the characteristics of *center-core* community. The index only takes up a linear space and can be efficiently constructed. We also propose online search algorithms based on the index to find the *center-core* community for given query node(s) in large networks. Finally, we conduct extensive experiments on four real-world large networks, and the results demonstrate the efficiency and effectiveness of the proposed methods.

In summary, our contributions are as follows:

- We propose a community search model called *center-core* community based on k-core decomposition and motivate the problem of finding *center-core* community containing the given query nodes;
- We design a space-efficient index structure called CCG-Index which can well preserve the information about *center-core* community;
- We propose a method to effectively search *center-core* community in large graphs based on CCG-Index;
- Extensive experiments over four real-world graphs demonstrate the efficiency and effectiveness of the proposed algorithms.

2   Related Work

2.1   Community Search

Sozio and Gionis study the community search problem (CSP) based on minimum degree in social networks that aims to find the maximal connected k-core with largest k containing the query nodes [8]. Cui et al. [10] propose a more efficient local search algorithm for a query node. Barbieri et al. [9] propose a very efficient community search method based on index, and further propose the minimum community search problem (MINCCSP) to reduce the community size. They show MIN-CSP problem is NP problem, and get the approximate solution of the problem by a heuristic algorithm. All the mentioned methods are static algorithm and do not consider the influence of nodes. Except being based on minimum degree, Cui et al. [2] proposed a quasi-clique model to study the overlap community search problem. Huang et al. [12,13] study the CSP based on a k-truss community model. In addition, the computational complexity of those methods is higher than the methods based on minimum degree.

2.2   *K*-core Decomposition

Seidman introduces the concept of k-core for measuring the group cohesion in a network. The cohesion of the k-core increases as k increases [1]. Recently, the

k-core decomposition in graph has been successfully used in identifying the influential spreader in complex network [14,15]. Batagelj and Zaversnik [11] propose a $O(n+m)$ algorithm for k-core decomposition in general graphs. This algorithm may be inefficient for the disk-resident graphs. Cheng et al. [16] propose an efficient k-core decomposition algorithm for the disk-resident graphs. Their algorithm works in a top-to-down manner that calculates the k-cores from higher order to lower order. To make the k-core decomposition more scalable, Montresor et al. propose a distributed algorithm for k-core decomposition by exploiting the locality property of k-core. Li et al. [17] propose an efficient core maintenance algorithm in dynamic graphs.

3   Problem Statement

Let $G(V, E)$ denote an undirected graph with node set V and edge set E. For any subset $H \subseteq V$, the subgraph induced by H, denoted as $G[H]$, is the graph whose node set is H and whose edge set is $(H \times H) \cap E$. Table 1 summarizes the notations used in this paper.

Table 1. Notations

Notation	Description
$G = (V, E), |V| = n, |E| = m$	An undirected graph G and n, m is the number of nodes and edges respectively
$G(H)$	The subgraph induced by H
C_k	The node set of the k-core
$d(v, G)$	The degree of v in G
$\mu(G)$	$\mu(G) = min_{v \in G} d(v, G)$
S_k	$S_k = C_k / C_{k+1}$
$c(v)$	The coreness of v
$c(Q)$	$c(Q) = min_{q \in Q} c(q)$
$Con_{k\text{-}core}(v/Q)$	The k-core connected components containing node v or node set Q
$Con_{k\text{-}shell}(v/Q)$	The k-shell connected components containing node v or node set Q
$indexV(v)$	The index vertex containing node v
X_v	$X_v = |u : u \in N(v), c(u) \geq c(v)|$
$N(v)$	The set of neighbors of node v

Definition 1 (*k-core*). *Given a graph* $G = (V, E)$, *a* k-*core of* G *is a maximal subgraph of* G *such that the degree of its each node is at least* k, *that is,* $d(v, G[C_k]) \geq k, v \in C_k$.

Denote the node set of a core as C_k to represent and identify a k-core. It is easy to see that the order of a core corresponds to the minimum degree of a node in that core i.e., $\mu(C_k) \geq k$. As shown in Fig. 2, the entire graph is 1-core i.e. $C_1 = \{v_1, v_2, v_3, ..., v_{20}\}$, $\mu(G[C_1]) = 1, C_2 = \{v_2, ..., v_{15}, v_{17}, v_{18}, v_{19}, v_{20}\}$, $\mu(G[C_2]) = 2, C_3 = \{v_3, ..., v_8, v_{11}, ..., v_{15}\}$, $\mu(G[C_3]) = 3$, $C_4 = \{v_4, v_5, v_6, v_7, v_8\}$, $\mu(G[C_4]) = 4$.

Property 1. Each k-core with different k value is unique and may not be connected.

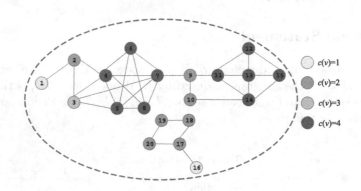

Fig. 2. An example graph G

As shown in Fig. 2, the k-core where the k is from value 1 to 4 are unique. Among them, 1-core, 2-core, 3-core have many connected components which are not connected.

Property 2. The cores are nested. The k-core with smaller k contains the k-core with larger k i.e., $C_j \subseteq C_i, i < j$.

As shown in Fig. 2, $C_4 \subseteq C_3 \subseteq C_2 \subseteq C_1 = V$.

Definition 2 (*k-shell*). *A k-shell is the induced subgraph by the set of nodes that only belongs to the k-core but not to the $(k+1)$-core i.e., the subgraph induced by the set of all nodes whose coreness is k.*

Denote the node set of a shell as S_k to represent and identify a k-shell, then $S_k = C_k/C_{k+1}$. Specifically, $S_{k_{max}} = C_{k_{max}}$, where the k_{max} is the highest order in all cores. As shown in Fig. 2:

$S_1 = C_1/C_2 = \{v_1, v_2, v_3, ..., v_{20}\}/\{v_2, ..., v_{15}, v_{17}, ..., v_{20}\} = \{v_1, v_{16}\}$.
$S_2 = C_2/C_3 = \{v_2, ..., v_{15}, v_{17}, ..., v_{20}\}/\{v_3, ..., v_8, v_{11}, ..., v_{15}\} = \{v_2, v_9, v_{10}, v_{17}, ..., v_{20}\}$.
$S_3 = C_3/C_4 = \{v_3, ..., v_8, v_{11}, ..., v_{15}\}/\{v_4, ..., v_8\} = \{v_3, v_{11}, ..., v_{15}\}$.
$S_4 = S_{k_{max}} = C_4/C_4 = \{v_4, v_5, v_6, v_7, v_8\}$.

Definition 3 (*coreness*). *The coreness of a node $v \in V$ is the highest order of a core that contains v, that is, the nodes with coreness k belong to k-core, but not belong to $(k+1)$-core.*

Denote the coreness of a node v as then $c(v)$, then $c(v) = max\{k | v \in C_k, k \in [0, 1, ..., k_{max}]\}$. In Fig. 2, take node v_6 as an example. v_6 belongs to C_1, C_2, C_3 and C_4 respectively, the highest order of the core containing v_6 is 4, so $C(v_6) = 4$.

Definition 4 (*connected k-core*). *A connected k-core is one of the connected components of the subgraph induced by k-core, denoting as $Con_{k-core}(v/Q)$.*

As shown in Fig. 2, 2-core is the induced graph of C_2, which has two connected components. Each one is a connected 2-core, $\{v_2, v_3, ..., v_{15}\}, \{v_{17}, v_{18}, v_{19}, v_{20}\}$. $Con_{k-core}(v)$ stands for the connected k-core of v, i.e. $Con_{2-core}(v_2) = \{v_2, v_3, ..., v_{15}\}$. According to Property 2, node v may belong to many connected k-core, i.e. v_2 belongs to one connected 2-core, and also belongs to one connected 1-core.

Definition 5 (*connected k-shell*). *A connected k-shell is one of the connected components of the subgraph induced by k-shell, denoting as $Con_{k-core}(v/Q)$.*

As shown in Fig. 2, the 2-shell that is the induce subgraph of S_2 has three connected components:$\{v_2\}, \{v_9, v_{10}\}, \{v_{17}, v_{18}, v_{19}, v_{20}\}$. According to Definition 2, the connected k-shell of v is unique, $Con_{shell}(v)$, i.e. $Con_{shell}(v_2) = \{v_2\}$.

Definition 6 (*center-core*). *Given a graph $G = (V, E)$ and a query node set $Q = \{q_1, q_2, ..., q_r\}, |Q| = r, r \geq 1, Q \subseteq V$. Set the k value of the maximum connected k-core containing Q as c. H is a center-core, if H satisfies the following conditions:*

1. *H is a connected c-core containing Q with the minimum degree c, that is, $H \subseteq Con_{c-core}(Q), \mu(H) = c$;*
2. *$\forall v \in H, c(v) \geq c(q)$;*
3. *The connected k-shell of node w, which coreness is equal to c in H, contains any query q or connects with any two query nodes, that is, $\exists q \in Q, q \in Con_{c-shell}(w)$ or $\exists q_1 \in Q, q_2 \in Q, q_1$ reaches q_2 with nodes in $Con_{c-shell}(w)$.*

Condition 1 ensures that *center-core* community containing q is densely connected since it has the largest minimum degree based on the concept of k-core. Condition 2 makes sure the influence of each node in *center-core* community is no less than q. And condition 3 ensures the nodes whose influence are equal to q needn't contact with q via nodes whose influence are larger than q. Because those nodes are certainly in the same connected shell with q, so that they can reach each other without the other nodes that have different coreness.

Problem Definition (*center-core* community search). *Given a graph $G = (V, E)$ and a query node set $Q = \{q_1, q_2, ..., q_r\}, |Q| = r, r \geq 1, Q \subseteq V$, find a node set $R \subseteq V$ where the subgraph induced by R, $G[R]$, is the center-core community containing Q.*

Example 1. In Fig. 2, suppose the query node $q = v_2$, then by the problem definition, the subgraph induced by node set $\{v_2, v_3, v_4, v_5, v_6, v_7, v_8\}$ is the *center-core* community containing v_2.

4   Querying *Center-core* Community

4.1   The Novel CCG-Index

4.1.1   CCG-Index Structure

We propose the CCG-Index structure, which can reflect the hierarchical structure of the graph. It's a hierarchical structure according to coreness. As shown in Fig. 3, the top level is the first level (level $= 1$). The number of level increases by the level increasing, where the level number corresponds to coreness. Each index item vertex (we call the node element in index as index item vertex to distinguish from the nodes in graph) in index is a connected k-shell with different k. The nodes in the same connected k-shell must be connected, so it can compress storage space since the edges between those nodes needn't be preserved. Besides, each vertex keeps the children and parents information for query and update later. Let the vertices in S_k point to the connected vertices in S_{k+1}, and each directed edge can both indicate the hierarchical and connected relationship between vertices. The level k index item vertex is the parent index item vertex of the connected $k + 1$ level, that is, the level $k + 1$ index item vertex is the child index item vertex of the connected k level. We can directly output the *center-core* community containing a query node according to the direction relationship in index. It can avoid repetitively visiting nodes if we find the nodes higher than query node from top to bottom level by level. Figure 4 is the CCG-Index of G in Fig. 2. The index vertices are A, B, C, D, E, F, G, H.

We can see that the index is composed of connected components of all S_k. If query node is v_2, which coreness is 2. We first find the index item vertex $indexV(v_2) = D = \{v_2\}$, then find the nodes that have the same coreness and connected directly, and then find nodes with higher coreness level by level. As shown in Fig. 4, first find the index item vertex $B = \{v_3\}$ in the next level, then continue find the vertex $C = \{v_4, v_5, v_6, v_7, v_8\}$ then continue find the vertex

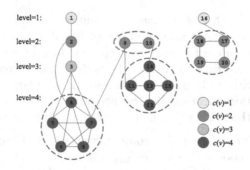

Fig. 3. The hierarchical division graph of G

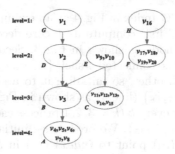

Fig. 4. The CCG-index of G

in the next level. So the induced subgraph of all the nods in these vertexes $\{D, B, A\}$ is the *center-core* community containing v_2 : $\{v_2, v_3, v_4, v_5, v_6, v_7, v_8\}$ as shown in the subgraph A in Fig. 1.

4.1.2 CCG-Index Construction

The construct course of CCG-Index is shown in Algorithm 1. First, calculate the coreness of each node by k-core decomposition algorithm in [11]. Second, compute each connected component in each connected k_{max}-shell where k is the largest coreness in graph and initialize it to a single index item vertex. The coreness of nodes in the bottom level index vertex is the maximum without any child vertex. Each vertex in S_k can be created by seeking the neighbors of each node in S_k. Next, recursively find and let the vertices in S_k point to the connected vertices in S_{k+1} from $k = k_{max-1}$ to $k = 1$ according to the neighbors of nodes in S_k. That is to say, the S_k vertex is the parent vertex of S_{k+1} connected with itself, and the S_{k+1} vertex is the child vertex of S_k connected with itself, $k = k_{max}$-1, ..., 1.

Algorithm 1. Construct CCG-Index

Input: $G = (V, E)$
Output: The CCG-Index
1: Compute the k-core decomposition for G and keep the core index;
2: $k_{max} = max\{c(v) | v \in V\}$;
3: CCG-Index=$\emptyset$;
4: Create a vertex for each connected k_{max}-shell in CCG-Index;
5: **for** $k = k_{max} - 1$ to 1
6:    **for** all nodes $v \in S_k$
7:       Create a vertex containing v if v is not visited in CCG-Index;
8:     **for** $w \in N(v)$
9:       **if** $c(w) == c(v)$ **then**
10:         Merge $indexV(v)$ and $indexV(w)$;
11:      **if** $c(w) > c(v)$ **then**
12:         Let $indexV(v)$ points to the vertex in S_{k+1} which is connected to $indexV(w)$;
13: return The CCG-Index;

Example 2. Take the CCG-Index in Fig. 4 as an example to illustrate the construction course in Fig. 5. First, compute its k-core decomposition (line 1) and save the coreness of each node as shown in Fig. 3. Second, compute $k_{max} = 4$ (line 2), and there is only one connected 4-shell $\{v_4, v_5, v_6, v_7, v_8\}$ since the nodes in S_4 are adjacent to each other, so initialize it to a single index item vertex $indexV(v_4) = \{v_4, v_5, v_6, v_7, v_8\}$ (line 4) as shown in Fig. 5(a). Then recursively process each node in S_k for every k ($k = 3, 2, 1$ process each node in) (lines 5–12). For $S_3 = \{v_3, v_{11}, v_{12}, v_{13}, v_{14}, v_{15}\}$, We process v_3: initialize $indexV(v_3) = \{v_3\}$ (line 7); let $B = indexV(v_3)$ point to $indexV(v_4)$ in S_4 because $v_4 \in N(v_3)$ and the coreness of v_4 is larger than v_3 (lines 11–12). Process v_{11}: initialize $indexV(v_{11}) = C = \{v_{11}\}$; add neighbors of v_{11} in S_3 into $C = indexV(v_{11})$ so $indexV(v_{11}) = \{v_{11}, v_{12}, v_{13}, v_{14}\}$. Because no neighbors of $v_{11}, v_{12}, v_{13}, v_{14}$ have larger coreness respectively, further process v_{15}: $indexV(v_{15}) = \{v_{15}\}$. Now, v_{15} is adjacent to v_{12}, so merge $indexV(v_{15})$ and $indexV(v_{12})$ and the vertex becomes $C = \{v_{11}, v_{12}, v_{13}, v_{14}, v_{15}\}$ (lines 9–10) as shown in Fig. 5(b). The operations for S_3 are completed. We can finally get the CCG-Index of G as shown in Fig. 5(c) and (d) by recursively processing S_2 and S_1.

In Algorithm 1, the calculation of k-core decomposition require $O(n + m)$ time [11]. The main cycle in lines 5–12 which process each nodes need $O(n'+m')$ time, n' and m' is the number of vertexes and edges in index respectively, and $n' \ll n, m' \ll m$. The other operation can be processed in constant time, so the time complexity of Algorithm 1 is $O(n+m)$. In addition, it only needs $O(n)$ space since each node is only storaged once in the index.

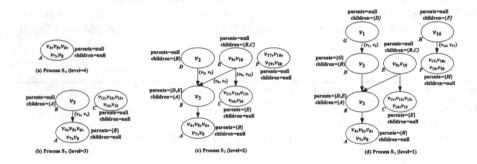

Fig. 5. The construction of CCG-index of G

4.2   *Center-core* Community Query Processing

4.2.1   Single Query Node Processing

For single query node, the *center-core* community query processing is straight-foward based on the CCG-Index. Treat the index structure as a tree, then the larger index level, the larger number of coreness of index vertex. The nodes in the subtree taking $indexV(q)$ as root is the result of *center-core* community search for q. Because the $indexV(q)$ includes all the nodes whose coreness equals q and in the same connected k-shell with q. The coreness of the nodes in its children vertices in larger levels is all larger than q. Thus such subtree is the *center-core* community for q. The query algorithm is shown in Algorithm 2.

Algorithm 2. SingleQuery

Input: The CCG-Index, a given node q
Output: The center-core containing q
1: $R = indexV(q)$;
2: **if** children($indexV(q)$)==null **then**
3:     return R;
4: Push all children of $indexV(q)$ into the stack s;
5: **while** s is not empty **do**
6:     Pop the vertex in top of s;
       add the nodes in it into R;
       push its children into s;
7: return R;

Example 3. Consider Fig. 2, given a query node $q = \{v_2\}$, we query the *center-core* community R containing q by Algorithm 2. By retrieving the CCG-Index as shown in Fig. 4, first initialize $R = indexV(v_2) = D = \{v_2\}$ (line 1). We use a stack s to process this traversal of children of each vertex that be visited (lines 3–6). Because D only has one child vertex B, push B to stack s, and then pull, $R = \{v_2, v_3\}$. Continue the iterated operation, then push vertex A, which is the child of vertex B, and then pull, $R = \{v_2, v_3, v_4, v_5, v_6, v_7, v_8\}$, utill the stack s is null. The query course is over, without any node to be visited. So the search result is $R = \{v_2, v_3, v_4, v_5, v_6, v_7, v_8\}$, the subgraph induced by R is the subgraph A in Fig. 1.

Algorithm 2 just needs to traverse the vertices related to query node in CCG-Index, so its time complexity is no more than $O(n')$. Due to this algorithm need not to consider the branches connected to the connected k-shell without query node, it narrows the search scope and accelerates the retrieval speed.

4.2.2 Multiple Query Nodes Processing

In CCG-Index, for the multiple query nodes processing, that is the query set $Q = \{q_1, q_2, ..., q_r\}$, we find a substree which root node is the least common ancestors (LCA) of index vertex of all the query nodes. The subtree is the *center-core* dense subgraph containing Q. Because the LCA of quey nodes is the nodes with maximum coreness making the nodes connected of Q, where the minimum coreness of subgraph can be maximized. Specially, if two query nodes with the same coreness locating in different index item vertices and they are connected, the union of these two index item vertices is the root node.

Algorithm 3. MultiQuery

Input: The CCG-Index; a set of query nodes $Q = \{q_1, q_2, ..., q_r\}$
Output: The *center-core* containing Q
1: $R = \emptyset$;
2: $Root = indexV(q_1)$;
3: **for** $i=2$ to r
4:     **for** $v \in Root$;
5:         $Root = findLCA(v, indexV(q_1))$;
6:         $R = R \cup subtree(Root)$;
7:     **if** $(Root == \emptyset)$ **then** break;
8: return R;

Example 4. As shown in Fig. 2, suppose the query set is $Q = \{v_3, v_6, v_{11}\}$. According to Algorithm 3, search the CCG-Index in Fig. 4. First, initialize $R = \emptyset$ (line 1); $Root = indexV(v_3) = B$ (line 2); $indexV(v_6) = A$, find the LCA of B and A, $C = min\{c(B), c(A)\} = min\{3, 4\} = 4$. The LCA of forth level of B and A is B, so R=$\{substree(B)\}$. Then, $indexV(v_{11} = C)$, also continue to find the LCA of B and C, that is E, so the final R=$\{subtree(B), subtree(e)\} = \{B, A, C, E\} = \{v_3, v_4, ..., v_{15}\}$ (lines 3–8).

Algorithm 3 shows the multiple query nodes *center-core* dense subgraph algorithm. We use a traversing algorithm to traverse the index vertices corresponding to R from top to bottom with the time complex $O(n'')$. n'' is the number of nodes in R. So, the time complex of Algorithm 3 is $O((|Q| - 1)n'')$. Because $n'' \ll n$, the time complex is still small in actual applications.

5   Experiments

We conduct extensive experiments on four real-world large networks to evaluate the efficiency and effectiveness of the proposed algorithms in this paper. We implemented all of the algorithms in Java and ran the experiments on a PC with Intel quad core at 3.2 GHz, 8G memory. The experimental datasets are from four real-world networks named Twitter, DBLP, Amazon and Youtube. Twitter is a social network, where each node represents a user and each edge represents the friendships of users. DBLP is an author collaboration network, where each node represents an author and each edge represents a coauthor relationship. Amazon is a e-commerce network, where each node stands for a product and each edge stands for purchasing this product. Youtube is a user-to-user link network. The statistics of these graphs are reported in Table 2, containing the number of nodes n, the number of edges m and the maximum coreness h gaining from k-core decomposition.

For index construction, we compare the index construction time and index size of CCG-Index (Algorithm 1) and CoreStruct [9] respectively. For query processing, we implement three algorithms, CCG (*center-core* community search, Algorithm 2 and Algorithm 3), GS [8], GrCon [9] and compare their execution time with different query nodes on different datasets.

Table 2. Datasets

Dataset	n	m	h	Description
Twitter	81,306	1,768,149	38	Social network
DBLP	317,080	1,049,866	40	Collaboration network
Amazon	410,236	3,356,824	41	Product network
Youtube	1,134,890	2,987,624	51	Social network

5.1   Index Construction

Figure 6 shows the index construction time on different datasets of CCG-Index and CoreStruct. The time of CCG-Index is slightly less than the CoreStruct. They are almost the same because they all need to traverse the each node and its incident edges in G. However, the construction time of CoreStruct depends on the maximum coreness of different k-core which needs to traverse a part of nodes repeatedly. So, the larger of value h, the more time of construction index.

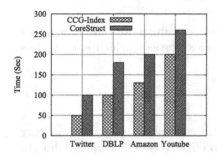

Fig. 6. Index construct time   **Fig. 7.** Index storage space

Figure 7 is the index size in different datasets. It can be seen that the storage space of CCG-Index is obviously lower than CoreStruct because there is nested feature of k-core. CoreStruct repeatedly stores many nodes. However, CCG-Index only stores the node information once by fully using the structure feature of *center-core*.

5.2   Query Processing

The query node set Q is designed by randomly choosing 1, 4, 8, 16 and 32 nodes from four real datasets respectively. So, for single query processing, $|Q| = 1$, CCG is the algorithm SingleQuery, Algorithm 2. For the multiple nodes query processing, $|Q| > 1$, CCQ is the algorithm MultiQuery, Algorithm 3.

Figure 8 shows the performance of query processing of three algorithms in four real datasets. In each dataset, with the increasing of $|Q|$, the query time of CCG is much better than the other two algorithms. Because the more nodes of

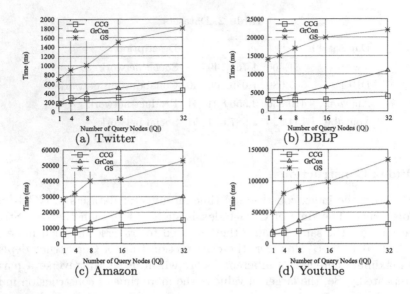

Fig. 8. Query processing time

query processing, the more nodes to be traversed, causing more time. The query time of CCG algorithm is mainly about the index vertices, which is much less than the whole nodes.

6 Conclusions

In this paper, we mainly study such meaningful community search for a query node, called *center-core* community search. To effectively reduce query time to apply to large and dynamic graphs, we further propose the index called CCG-Index to preserve information about the level classification of node influence and connected relationship of nodes. Thus we can quickly query *center-core* community by retrieving index. Extensive experiments over four real-world graphs demonstrate the efficiency and effectiveness of the proposed algorithms.

Acknowledgement. This work is supported by National Natural Science Foundation of China (No. 61472169, 61502215), Science Research Normal Fund of Liaoning Province Education Department (No. L2015193), Doctoral Scientific Research Start Foundation of Liaoning Province (No. 201501127), National Key Research and Development Program of China (No. 2016YFC0801406).

References

1. Seidman, S.B.: Network structure and minimum degree. Soc. Netw. **5**(3), 269–287 (1983)
2. Cui, W., Xiao, Y., Wang, H., Lu, Y., Wang, W.: Online search of overlapping communities. In: ACM SIGMOD International Conference on Management of Data, pp. 277–288 (2013)

3. Tsourakakis, C., Bonchi, F., Gionis, A., Gullo, F., Tsiarli, M.: Denser than the densest subgraph: extracting optimal quasi-cliques with quality guarantees. In: ACM SIGKDD International Conference on Knowledge Discovery and Data Mining, pp. 104–112 (2013)

4. Bta, A., Krsz, M.: A high resolution clique-based overlapping community detection algorithm for small-world networks. Informatica **39**(2), 177–187 (2015)

5. Koujaku, S., Takigawa, I., Kudo, M., Imai, H.: Dense core model for cohesive subgraph discovery. Soc. Netw. **44**, 143–152 (2016)

6. Wang, N., Zhang, J., Tan, K.L., Tung, A.K.H.: On triangulation-based dense neighborhood graph discovery. Proc. VLDB Endow. **4**(2), 58–68 (2010)

7. Li, R.H., Yu, J.X.: Triangle minimization in large networks. Knowl. Inf. Syst. **45**, 617–643 (2015)

8. Sozio, M., Gionis, A.: The community-search problem and how to plan a successful cocktail party. In: ACM SIGKDD International Conference on Knowledge Discovery and Data Mining, Washington, D.C., USA, pp. 939–948, July 2010

9. Barbieri, N., Bonchi, F., Galimberti, E., Gullo, F.: Efficient and effective community search. Data Min. Knowl. Discov. **29**(5), 1406–1433 (2015)

10. Cui, W., Xiao, Y., Wang, H., Wang, W.: Local search of communities in large graphs. In: ACM SIGMOD International Conference on Management of Data, pp. 991–1002 (2014)

11. Batagelj, V., Zaversnik, M.: Fast algorithms for determining (generalized) core groups in social networks. Adv. Data Anal. Classif. **5**(2), 129–145 (2011)

12. Huang, X., Cheng, H., Qin, L., Tian, W., Yu, J.X.: Querying k-truss community in large and dynamic graphs. In: ACM SIGMOD International Conference on Management of Data, pp. 1311–1322 (2014)

13. Huang, X., Lakshmanan, L.V.S., Yu, J.X., Cheng, H.: Approximate closest community search in networks. Proc. VLDB Endow. **9**(4), 276–287 (2015)

14. Miorandi, D., Pellegrini, F.D.: K-shell decomposition for dynamic complex networks. In: Proceedings of the International Symposium on Modeling and Optimization in Mobile, Ad Hoc and Wireless Networks, pp. 488–496 (2010)

15. Zhao, Q., Lu, H., Gan, Z., Ma, X.: A *K*-shell decomposition based algorithm for influence maximization. In: Cimiano, P., Frasincar, F., Houben, G.-J., Schwabe, D. (eds.) ICWE 2015. LNCS, vol. 9114, pp. 269–283. Springer, Cham (2015). https://doi.org/10.1007/978-3-319-19890-3_18

16. Cheng, J., Ke, Y., Chu, S., Ozsu, M.T.: Efficient core decomposition in massive networks. In: IEEE International Conference on Data Engineering, pp. 51–62 (2011)

17. Li, R.H., Yu, J.X., Mao, R.: Efficient core maintenance in large dynamic graphs. IEEE Trans. Knowl. Data Eng. **26**(10), 2453–2465 (2014)

Real-Time Scientific Impact Prediction in Twitter

Zhunchen Luo[1](✉), Jun Chen[1], and Xiao Liu[2]

[1] Information Research Center of Military Science,
PLA Academy of Military Science, 100142 Beijing, China
`zhunchenluo@gamil.com`, `13501239808@126.com`
[2] School of Computer Science and Technology,
Beijing Institute of Technology, 100081 Beijing, China
`xiaoliu@bit.edu.cn`

Abstract. As the number of scientific publication is getting larger and larger, scientific impact prediction has become an urgent need. However, traditional scientific impact prediction, which is mainly based on longtime accumulated citation networks, metadata and the whole text of papers, is relatively hysteretic and can hardly fit the rapid development of technology. Moreover, Twitter has become one of the most import channels to spread latest technique information because of its fast information spread speed. The advantage of publishing new messages in real-time can compensate the imperfections of traditional scientific impact prediction methods. Therefore, we propose a new approach to predict scientific impact in Twitter in real time before publishing the paper content. After filtering scholarly tweets (*ST tweets*), and extracting Tweet Scholar Blocks (*TSBs*) indicating metadata of papers to help predict scientific impact in real time, author social features, venue popularity features, and title features are exploited to predict whether the article will increase h-index of its first author after five years. Our model achieves an outstanding result that its best accuracy is 80.95%. The best feature conjunction consists of the sum of friends and followers of all the co-authors, followers count of the first author and title embeddings. And the amount of followers of all the co-authors is the most critical feature. Our finding reveals that Twitter has the potential to predict scientific impact in real time. We hope that real-time scientific impact prediction in Twitter can help researchers to expand their influences and more conveniently "stand on the shoulders of giants".

Keywords: Twitter · Scientific impact prediction · Real-time

1 Introduction

Scientific impact prediction has become an urgent need since the number of scientific publication is getting larger and larger. As an instance, the number

Z. Xu et al. (Eds.): Big Data 2018, CCIS 945, pp. 108–123, 2018.
https://doi.org/10.1007/978-981-13-2922-7_7

of e-print publications in $arXiv^1$ has exceeded 1,354,091. Scientific publications are spread in various channels and platforms, such as Twitter, Mendeley, printed journals.

Twitter is now one of the biggest social networks, and the vast volume of tweets posted on Twitter per day is highly attractive for information retrieval purpose. There not only is a tremendous amount of unrevealed information about scientific papers in Twitter but also are lots of scholars post tweets to express their excitement when their papers got accepted [12,22]. We call the tweets that imply accepted papers scholarly tweets (*ST tweets*) and the rest non-scholarly tweets (*NST tweets*).

The volume of information about scientific papers is enormous on Twitter, and most data is real-time, even before the paper content is published and shortly after the notifications of acceptance. However, previous work shows that most scientific impact prediction works are based on citation networks [16], metadata of papers [15], or text content of articles [32], and those methods are quite time-consuming, as the analysis requires the publication content of the paper. The wish to predict the scientific impact of a newly published paper may be delayed to a great degree. On the contrary, for example, the *ST tweet*[2] illustrated in Fig. 1 published on May 25, 2016, implies that the paper: "Domain Adaptation for Authorship Attribution: Improved Structural Correspondence Learning" co-authored by Manuel Montes is accepted by the Association for Computational Linguistics 2016 (*ACL 2016*). Notifications of acceptance[3] of long papers were delivered on May 24, 2016. And the conference was held from August 7 to August 12, 2016. Apparently, the *ST tweet* was posted before the date of publication which shows content. If we can predict the scientific impact of the paper once it has been accepted even before it comes to publication, we can use the real-time prediction to boost later information analyzation in much more advanced.

Fig. 1. An example of a *ST Tweet*

Toward this end, we propose a new approach to predict scientific impact in Twitter, so that the impact can be calculated in real time, before the publication of the related paper. At first, we trace a data stream by tracking "paper accepted" in Twitter, but there are some *NST tweets* in the data stream, so we build a classification model to filter *ST tweets*. To predict scientific impact in Twitter in real time, we want to make use of the metadata of papers. It is investigated that most *ST tweets* consist of text blocks called *Twitter Scholar Blocks* (*TSBs*) indicating meta data. According to the investigation, we then build a sequence tagger to extract *TSBs* to gather metadata information. Finally, we build a binary classification model by combining *TSBs* with information in social networks in Twitter to predict whether the paper implied in *ST tweet* will increase the h-index [9] of its first author after five years. The best accuracy of our model is 80.95%, which outperforms the baseline based on citation networks. We find the best feature conjunction consists of the sum of friends and followers of all the co-authors, followers count of the first author and title embeddings, besides, amount of followers count of all the co-authors is the most critical feature.

Contributions: The main contributions of this paper are three-fold:

(1) We show that social networks like Twitter have the potential to predict scientific impact in real time.
(2) We propose *TSBs* in *ST tweets* and a novel approach utilizing *TSBs* to predict scientific impact in real time with 80.95% accuracy.
(3) We discover that the best feature conjunction consists of the sum friends and followers count of all the co-authors, followers count of the first author and title embeddings. And sum followers count of all the co-authors is the most critical feature.

2 Related Work

Our research is related to two aspects of work; the one is traditional scientific impact prediction that is regarded as a regression problem on citation numbers, the other is scientific analysis in social media.

2.1 Regression Scientific Impact Prediction

Scientific impact prediction often seems like a regression problem on citations. Information extracted from citation networks is widely used. [5] use temporal elements and topological elements to predict future citation. [16] use encoding method based on citation network of Scopus database. [26] investigated the factors determining the capability of academic articles to be cited in the future using topological analysis of citation networks.

Text information seems to be popular in recent years. [32] consider predicting measurable responses to scientific articles primarily based on their text content. [15] analyze the usefulness of rich information derived from the full text of the articles through a variety of approaches, including rhetorical sentence analysis, information extraction, and time-series analysis and they combine metadata and whole text to achieve a better result.

There are also works that combine these two types of information. [31] adapt a discriminative approach that can make use of any text or metadata and show that lexical knowledge offers substantial power in modeling out-of-sample response and forecasting response for future articles. They show that various social factors influence written scientific communication and they can uncover these factors by measuring language similarity between articles.

Although approaches mentioned above perform efficiently, they do not utilize the information on Twitter. Furthermore, the content of most implied papers is not public when *ST tweets* are posted, so content is not a good factor to help predict scientific impact in real time.

2.2 Social Media Scientific Analysis

While most of the previous work focuses on structured data sources, there is some work focus on tweets. [28] explored the feasibility of measuring social impact and public attention to scholarly articles by analyzing buzz in social media. They explored the dynamics, content, and timing of tweets relative to the publication of a scholarly article, as well as whether these metrics are sensitive and specific enough to predict highly cited articles.

[29] studied approaches for defining and measuring information flows within tweets during scientific conferences. They suggest refinements of analyzing datasets based on tweets collected during scientific conferences and present our results from applying novel forms of intellectual tweet content analyses.

Many papers have only zero or one tweet mentioned, how to restrict the impact analysis on only those journals producing a considerable Twitter impact is a problem. [2] defined the Twitter Index (TI) containing journals with at least 80% of the papers with at least one tweet each. For all papers in each TI journal, they calculated normalized Twitter percentiles (TP) which range from 0 (no impact) to 100 (highest impact).

The approaches mentioned above are not appropriate for real-time prediction in Twitter because the formation of citation networks is time-consuming. In this paper, we use h-index instead of citation number as metric to evaluate the scientific impact and convert the traditional regression problem on citation number to a classification problem on h-index.

3 Overview

We look deeply into the tweets from our "paper accepted" data stream and find that some tweets are *NST tweets*. For example, the tweet: "*can the bank accept the toilet paper issued as by @UKenyatta as collateral??*", is a *NST tweet*, because the word "*paper*" means anything but a thesis in that tweet. Thus we need to build a classification model to filter *ST tweets*.

To predict scientific impact in Twitter in real time, we want to make use of the metadata of papers. It is surprising to find that *ST tweets* are always consisted of text blocks indicating metadata, such as authors and titles of papers and names, time and places of venues. We call these text blocks *Twitter Scholar Blocks* (*TSBs*) and build a sequence tagger to extract them.

We build a binary classification model to predict scientific impact in Twitter in real time. We use the model to judge whether the paper implied in *ST tweet* will increase the *h*-index [9] of its first author after five years. The *ST tweets* that imply accepted papers that will increase the *h*-indices of the first authors after five years are called High Impact Scholarly Tweets (*HIST tweets*). *TSBs* and information in Twitter social networks are combined in our model to predict *HIST tweets*. The framework of our approach is shown in Fig. 2.

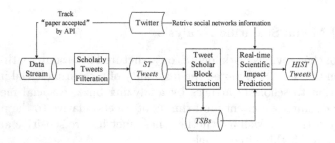

Fig. 2. Framework of our approach

4   Scholarly Tweets Filtration

We take filtering *ST tweets* from the data stream as a classification problem. To solve this problem, we propose an approach called Scholarly Tweets Filtration (*STF*) and build a classification model based on support vector machine (*SVM*) [3,4]. It is not easy to resolve the problem since there are two types of *ST tweets*. Some of them have specific paper titles, while others do not. For example, the tweet: "*Our paper 'Latent Space Model for Multi-Modal Social Data' accepted at @www2016ca #www2016!*", has an explicit title "*Latent Space Model for Multi-Modal Social Data*" surrounded by a pair of double quotation marks, while the tweet: "*New paper accepted!*", does not. The features we exploit are listed in Table 1.

To capture the information in social networks, the following features related to the author of the tweet are designed: **user's scholarly membership of academic institutions**. Obviously and empirically, the members of academic institutions have the higher probabilities to mention accepted papers. For simplicity, we collect names of academic institutions from the Internet and make a list containing the top sixty high-frequency words, such as "university", "institute", "research". Then we examine whether user descriptions contain words in the list to judge the existence of scholarly memberships.

Table 1. Features exploited in scholarly tweets filtration

Feature	Description
Scholar	Is the user a scholar
Bag-of-words	The bag of words of the tweet
Symbols	Words starting with symbols
Length	Text content length
Sentiment	The sentiment of tweet

To capture the information in tweet text, the following features are designed: **bag of words, words with trending symbols (e.g., "#", "*"), length of the tweet and the sentiment label of the tweet**. Words with trending symbols are commonly used to express topics on Twitter. In *ST tweets*, topics are often abbreviations of conferences, journals and research fields. We think the sentiment label is significant because our intuition is that no one would hide her happiness if her paper were accepted. Previous work shows sentiment analysis in citation context helps achieve better result [27]. The result of sentiment analysis is one of the three labels: positive, neutral and negative, according to our statistics, few of *ST tweets* is negative. In the experiment we used a free and open source tweet-specified sentiment analysis API[4] to generate sentiment labels for tweets.

To evaluate our *STF*, we manually labeled 5,400 tweets from our "paper accepted" data stream, nearly 45% are *ST tweets* and the ratio between *ST tweets* with and without explicit title is 10:7. Five-fold cross-validation was used in this experiment. FastText [10] was chosen as our baseline. The architecture of fastText is similar to the CBOW model [17], and it utilizes hierarchical softmax to reduce time expenditure. By training with SVM, the accuracies of *STF* and baseline are listed in Table 2. The performance of *STF* is 5.96% higher than the performance of the baseline. Although fastText model uses bag of n-gram as additional features to capture some partial information about the local word order, it only focuses on the text content. Thus the assistance of social features might improve the performance. The performance on *ST tweets* with explicit titles is 35.62% higher than the performance on *ST tweets* without explicit titles, which confirms the difficulties to filter *ST tweets* without explicit titles.

Table 2. Results of scholarly tweets filtration

Tweets	Number	*STF*	Baseline
All	5400	86.26%	81.37%
With titles	1429	98.04%	97.27%
No titles	1001	72.22%	70.43%

---

[4] https://dev.exploreyourdata.com/index.html.

5   Tweet Scholar Block Extraction

To help predict scientific impact in Twitter in real time, we want to extract metadata from the implied papers. [13] found that a series of conventions allow users to tweet in structural ways using the combination of different blocks of texts which are combinations of plain text, hashtags, links, mentions and so on. We investigate that researchers post *ST tweets* also in structural ways using combinations of different Tweet Scholar Blocks (*TSBs*). Each *TSB* carries a part of meta data. Furthermore, the combinations of *TSBs* encode scholarly information about papers, venues, and authors. Six types of *TSB* are proposed by us. A *ST tweet* consisted of different types of *TSBs* is illustrated in Fig. 3 and every underlined sequence of tokens shows a type of *TSB*.

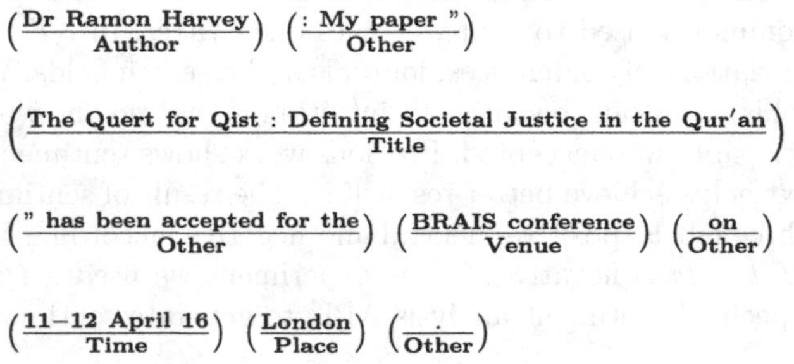

$$\left(\frac{\text{Dr Ramon Harvey}}{\text{Author}}\right) \left(\frac{\text{: My paper ”}}{\text{Other}}\right)$$

$$\left(\frac{\text{The Quert for Qist : Defining Societal Justice in the Qur'an}}{\text{Title}}\right)$$

$$\left(\frac{\text{” has been accepted for the}}{\text{Other}}\right) \left(\frac{\text{BRAIS conference}}{\text{Venue}}\right) \left(\frac{\text{on}}{\text{Other}}\right)$$

$$\left(\frac{\text{11–12 April 16}}{\text{Time}}\right) \left(\frac{\text{London}}{\text{Place}}\right) \left(\frac{.}{\text{Other}}\right)$$

Fig. 3. An example of tweet scholar blocks

Author: The names of authors (e.g., Dr. Ramon Harvey).
Title: The title of the paper (e.g., The Quert for Qist: Defining Societal Justice in the Qur'an).
Venue: The short name of the venue (e.g., BRAIS conference) or its entire name (e.g., The British Association for Islamic Studies conference).
Time: The time expression when the venue will be held (e.g., 11–12 April 16).
Place: The place where the venue will be held (e.g., London).
Other: The rest part of tweet text that does not belong to the above five types.

In order to test the validity that *ST tweets* are constructed by different combinations of *TSBs*, we randomly chose 1,400 *ST tweets*. Firstly, we used an annotator[5] [8,20] to tokenize those tweets. Secondly, we manually labeled *TSBs* in BIO schema [23]. The *ST tweets* that are consisted of only *Other* type of *TSBs* are 17.73%. It means most *ST tweets* contain at least one non-*Other* type of *TSBs*. Therefore, we can extract some metadata from papers by extracting *TSBs*.

---

[5] http://www.cs.cmu.edu/~ark/TweetNLP/.

We regard extracting *TSBs* from *ST tweets* as a sequence labeling problem. To solve this problem, we propose an approach to extract *TSBs* called Tweet Scholar Block Extraction (*TSBE*) and build a sequence tagger based on conditional random fields (*CRFs*) [11]. Due to the informal and short nature of the tweets, we apply a tweet-specified POS tagger which is the same annotator we use to tokenize to produce POS labels. We use the tweet NER tagger [24,25] to extract features for *Time* and *Place* types of *TSBs*.

By analyzing *ST tweets*, we discover that most of the Twitter account mentioned in *ST tweets* are co-authors. So we use tokens starting with "@" (e.g., @LiuQunMTtoDeath) in *ST tweets* to help extract *Author* type of *TSBs*.

It is also investigated that the titles of papers usually occupy up to 40% of the text content which are often surrounded by pairwise symbols (e.g., "" and '') or all capitalized to show their differences. So to extract *Title* type of *TSBs*, we judge whether a token is capitalized or surrounded by pairwise symbols.

In *ST tweets*, nearly 87% words with trending symbols indicate venues (e.g., #SPAFACON2016). Some Twitter accounts mentioned (e.g., @acl2016) and preposition phrases (e.g., *by the Intl Jrl of Osteopathic Medicine, in Chem. Sci.*) also mean venues. Therefore, we use words with trending symbols and prepositions to help extract *Venue* type of *TSBs*.

Table 3. Statistics of the five none-*other* types of *TSBs*

Type	Number	Precision	Recall	F1 Value
Title	304	93.55%	74.36%	82.86%
Author	732	82.81%	72.60%	77.37%
Place	217	79.23%	72.94%	75.96%
Venue	973	83.12%	72.73%	77.58%
Time	235	77.40%	74.55%	75.95%

To evaluate our *TSBE*, we used the 1,400 *ST tweets* described above to train and test. Five-fold cross-validation was used in this experiment. The precision rates, recall rates and F1 values of the five none-*Other* types of *TSBs* are shown in Table 3. The *Other* type of *TSBs* is neglected because the label of this kind is "O" in BIO schema and we do not care in this paper. The performance of *Title* type of *TSBs* is the highest. Judging whether a token is capitalized or surrounded by pairwise symbols might be useful features. The performances of *Place* and *Time* types of *TSBs* are close. And the performances of *Author* and *Venue* types of *TSBs* are also close. It might be that they are strongly correlated to the representation of the leading symbols and prepositions, so similar features take effects. We analyze some blocks with wrong predictions and find that some *Time* and *Place* types of *TSBs* were affected by the errors produced in the tweet NER tagger. To improve our performances, we need to decrease the errors produced in the tweet NER tagger and enhance the representation of leading symbols and prepositions.

6 Real-Time Scientific Impact Prediction

We regard predicting scientific impact in Twitter in real time as a classification problem of judging whether the paper implied in *ST tweet* will increase the *h*-index of its first author after five years. To solve this problem, we propose an approach called Real-time Scientific Impact Prediction (*RSIP*) and build a classification model based on support vector machine (*SVM*). Previous work shows that the scientific citation process acts relatively independently of the social dynamics on Twitter [30], so we take both social networks information in Twitter and text information generated from *TSBs* into account. As the paper implied in *ST tweets* may be not public, we can not use its content. Thus we can only think of metadata of articles. We categorize the features we exploit into three categories: author social features, venue popularity features, and title features. The features we exploit is listed in Table 4.

Table 4. Features exploited in real-time scientific impact prediction

Feature	Description
Sum Friends Count	Sum friends number of all the authors
Sum Followers Count	Sum followers number of all the authors
Sum Statuses Count	Sum statuses number of all the authors
Maximum Friends Count	Maximum friends number of all the authors
Maximum Followers Count	Maximum followers number of all the authors
Maximum Statuses Count	Maximum statuses number of all the authors
Minimum Friends Count	Minimum friends number of all the authors
Minimum Followers Count	Minimum followers number of all the authors
Minimum Statuses Count	Minimum statuses number of all the authors
Average Friends Count	Average friends number of all the authors
Average Followers Count	Average followers number of all the authors
Average Statuses Count	Average statuses number of all the authors
Friends Count	Friends number of the first author
Followers Count	Followers number of the first author
Statuses Count	Statuses number of the first author
Individual Verification Status	Is the first author verified
Group Verification Status	Is anyone among all the authors verified
Retweets Count	Retweets number of the *ST tweet*
Replies Count	Replies number of the *ST tweet*
Liked Count	Liked number of the *ST tweet*
Current Tweets Count	Tweets number in the venue
History Tweets Count	Average history tweets number in the venue
Title Embedding	Sum of word embeddings in title

6.1   Author Social Features

To capture the author social information, we try to find reasonable representations of influences of authors. It is investigated that the first author usually leads the collaboration. Besides, previous work shows that the overall impact of all co-authors should have the potential to influence a paper's quality and popularity, which will further affect a researcher's h-index [6]. We use the *Author* type of *TSBs* extracted to find the authors in *ST tweets*. And the first author is defined as follows. If the *ST tweet* is original, its author is the first author of the paper. Otherwise, the first *Author* type of *TSB* indicates the first author of the paper. We think the influence of an individual is related to her friends number, followers number, statuses number. To show the influence of a group, we calculate the sum, maximum value, minimum value and average value of influences of the participants in that group. In spite of these, we take statuses of user verification into account. Verification is used by Twitter mostly to confirm the authenticity of celebrity accounts. Previous work found that 91% of tweets written by verified users are retweeted, compared with 6% of tweets where the author is not verified [21], which means that tweets posted by celebrities are more popular. Additionally, we calculate retweets count, replies count and liked count of the *ST tweet*.

6.2   Venue Popularity Features

Google Scholar metrics[6] shows that different venues have large differences in their $h5$-indices (the h-index when only considering articles published within the last 5 complete years). Since the well-respected venues are better platforms for researchers to publish their work or results, our intuition is that top sites help scholars spread their scientific impact. And increasing the citation counts of their papers further offers an enormous potential to increase their h-indices.

Scholars often use Twitter as a note-taking tool [14] during venues, so the tweets number in the venue topic may reflect the popularity and influence of the site. We use the quantity of statuses in the venue topic to represent the popularity of the venue. Considering the developments and the trends of the venues, we also take the historical total quantity of statuses into account.

6.3   Title Features

Every scientific paper has its specified topics, while the popularity of topics may influence the speed of the appearance of scientific impact [1]. We think the title is the most direct and attractive way to declare research topic.

To capture the influence of topics, we attempt to extract information from learning representations of the titles of scientific papers. To learn a good representation of the titles of scientific papers, we first use word2vec

---

[6] https://scholar.google.com/citations?view_op=top_venues.

[17–19] and pre-trained 300-dimensional word embedding *GoogleNews-vectors-negative300.bin.gz*[7] which is trained from the Google News corpus to obtain the representations of words, and then we sum up all the corresponding embeddings of the words in title split with whitespaces. If there is no explicit title in the *ST tweets*, we set the Title Embedding all zeros to make this feature not work.

7 Experiments

To evaluate our approaches, firstly, we manually labeled tweets from our "paper accepted" data stream and set experiments to compare the performances between *RSIP* and the baseline. Then we did feature selection experiment to find the best feature conjunction of *RSIP* to improve performance. At last, we did feature analysis experiment to find the effectiveness of each feature in the best feature conjunction of *RSIP* and which features, in particular, are highly valued. Five-fold cross-validation was used in our experiments. Accuracy was used as the evaluation metric.

7.1 Data Set

We randomly chose 273 *ST tweets* posted in 2011 from our "paper accepted" data stream and found the true names of authors, titles of papers, names, time and places of venues by a scholarly search engine such as Google Scholar[8]. There are no *NST tweets* in the data set, since *NST tweets* do not imply accepted papers and it is meaningless to feed them into the baseline we chose. In these *ST tweets*, 142 *ST tweets* of them are without specific titles, while others are with explicit titles. According to the information we found, we then use Google Scholar to gather *h*-index of every first author and citation number of every corresponding paper in 2016. Now we can know whether the papers accepted in 2011 will increase *h*-indices of their first authors after five years in 2016. If the paper's citation number in 2016 is more substantial than its author's *h*-index in 2016, it means the article accepted in 2011 increased its primary author's *h*-index after five years in 2016. In such way, we manually labeled the data.

7.2 Real-Time Scientific Impact Prediction Evaluation

Since there are few related works about scientific impact prediction in real time, we took the approach of [5], which is the state-of-the-art method to predict the scientific impact on citation networks, as our baseline to simulate real-time prediction. In the baseline, temporal and topological features derived from citation networks are used to predict a paper's future impact (e.g., number of citations). The baseline uses a behavioral modeling approach in which the temporal change

---

[7] https://drive.google.com/file/d/0B7XkCwpI5KDYNlNUTTlSS21pQmM/edit?usp=sharing.

[8] https://scholar.google.com.

in the number of citations a paper gets is clustered, and new papers are evaluated accordingly. Then, within each cluster, the impact prediction is modeled as a regression problem where the objective is to predict the number of citations that a paper will get in the near or far future, given the early citation performance of the paper. The baseline produced the citation number of each paper in our dataset after five years. And we compared the citation numbers to the real first authors' h-indices in 2016 to judge whether the papers increased its first author's h-index in 2016.

Table 5. Comparing results between baseline and *RSIP*

Approach	Accuracy
Baseline	63.00%
TSBE+RSIP	73.99%
RSIP	78.02%

We compared the result of using *TSBE* and *RSIP* (*TSBE+RSIP*) with the result of using *RSIP* on manually labeled *TSBs* and the result of the baseline. Results are shown in Table 5. Overall, it is feasible to predict scientific impact in Twitter in real time. The performance of *RSIP* is higher than the performance of the baseline. The reason might be that the baseline is not appropriate for predicting scientific impact in real time. And the performance of *TSBE+RSIP* is lower than the performance of *RSIP* on manually labeled TSBs. The errors produced in *TSBE* might affect the performance of *RSIP*.

7.3 Feature Selection

To find the best feature conjunction of the features to improve the performance of our real-time scientific impact prediction model, we used an advanced greedy feature selection method the same as [7] used. Figure 4 shows the feature selection approach mentioned above.

Since greedy feature selection approach suffers from data sparseness, it is always blocked by a local optimum feature set. To find a global optimum feature set, this approach uses random techniques to generate several feature sets first and then run greedy feature selection on the best one among them. Finally, we find that the best feature conjunction consisted of *Sum Friends Count, Sum Followers Count, Followers Count* and *Title Embedding*. We call it *RSIP_Best*.

Results in Table 6 illustrate that the best feature conjunction (*RSIP_Best*) outperforms *RSIP* by about 3.76% on our manually labeled dataset. The three kinds of features, namely maximum, minimum, average counts are not in the best feature conjunction. It might be that these three kinds of counts do not reflect the entire influences of groups and are often limited by the variances of individual authorities. Additionally, the performance of *TSBE+RSIP_Best* is lower than the performance of *RSIP_Best* and is 4.96% higher than the performance of *TSBE+RSIP*. It might be that the errors produced in *TSBE* effect *RSIP_Best*.

> *An advanced greedy feature selection algorithm.*
> **Input:** All features we extracted.
> **Output:** the best feature conjunction *BFC*
> **Procedure:**
> Step1: Randomly generate 80 feature set *F*.
> Step 2: Evaluate every feature set in *F* and select the best one denoted by *RBF*.
> Features excluded those in *RBF* are denoted as *EX_RBF*
> Step 3: t = 0,*BFC*(t)=*RBF*;
>   Repeat
>   Foreach feature in *EX_RBF*
>   If   Evaluation(*BFC*)
>   < Evaluation(*BFC*, feature)
>   *BFC*(t+1) = {*BFC*(t), feature}
>   *EX_RBF*(t+1) = *EX_RBF*(t) − {feature}
>   While *BFC*(t+1) ≠ *BFC*(t)
> Note: Evaluation(*BFC*) refers to the performance of ranking function trained from features in *BFC* on validation data.

Fig. 4. Advanced greedy feature selection algorithm used in feature selection

Table 6. Comparing results with best feature conjunction

Approach	Accuracy
RSIP	78.02%
RSIP_Best	80.95%
TSBE+RSIP	73.99%
TSBE+RSIP_Best	77.66%

7.4   Ablation Study

To find the effectiveness of each feature and which features, in particular, are highly valued by *RSIP_Best*, we also removed each feature from *RSIP_Best* and *TSBE+RSIP_Best* respectively to evaluate the effectiveness of each feature by the decrement of accuracy.

By comparing the results shown in Table 7, we can see that *Sum Followers Count* is very effective to our *RSIP_Best*. The reason might be that *Sum Followers Count* is more suitable to stand for the influence of the authors' group.

Meanwhile, *Title Embedding* is not so efficient in our data. Perhaps the reason is that 52.01% of the *ST tweets* do not have specific titles. So the feature only works on the rest *ST tweets*.

Table 7. Comparing results by decaying every feature one by one

Approach	Accuracy
RSIP_Best	80.95%
RSIP_Best-Sum Friends Count	75.09%
RSIP_Best-Sum Followers Count	73.99%
RSIP_Best-Followers Count	76.56%
RSIP_Best-Title Embedding	79.85%
TSBE+RSIP_Best	77.66%
TSBE+RSIP_Best-Sum Friends Count	69.60%
TSBE+RSIP_Best-Sum Followers Count	65.93%
TSBE+RSIP_Best-Followers Count	68.86%
TSBE+RSIP_Best-Title Embedding	73.63%

8 Conclusion

In this paper, we propose *STF*, *TSBE* and *RSIP* to predict scientific impact in real time. The accuracy of *RSIP_Best* is 80.95%, which outperforms the baseline based on citation networks.

The best feature conjunction consists of the sum friends and followers count of all the co-authors, followers count of the first author and title embeddings. And sum followers count of all the co-authors is the most critical feature. The results show that Twitter has the potential to predict scientific impact in real time and our novel approach can achieve comparable performance. Hope real-time scientific impact prediction in Twitter can help researchers to expand their influences and more conveniently "stand on the shoulders of giants".

Acknowledgments. We very appreciate the comments from anonymous reviewers which will help further improve our work. This work is supported by National Natural Science Foundation of China (No. 61602490).

References

1. Bethard, S., Jurafsky, D.: Who should I cite: learning literature search models from citation behavior. In: CIKM, pp. 609–618 (2010)
2. Bornmann, L., Haunschild, R.: How to normalize Twitter counts? A first attempt based on journals in the Twitter index. Scientometrics **107**, 1405–1422 (2016)
3. Boser, B.E., Guyon, I., Vapnik, V.: A training algorithm for optimal margin classifiers, pp. 144–152 (1992)
4. Cortes, C., Vapnik, V.: Support-vector networks. Mach. Learn. **20**(3), 273–297 (1995)
5. Davletov, F., Aydin, A.S., Cakmak, A.: High impact academic paper prediction using temporal and topological features. In: CIKM, pp. 491–498 (2014)

6. Dong, Y., Johnson, R.A., Chawla, N.V.: Will this paper increase your h-index?: Scientific impact prediction. In: WSDM, pp. 149–158 (2015)
7. Duan, Y., Jiang, L., Qin, T., Zhou, M., Shum, H.: An empirical study on learning to rank of tweets. In: COLING, pp. 295–303 (2010)
8. Gimpel, K., et al.: Part-of-speech tagging for twitter: annotation, features, and experiments. In: Meeting of the Association for Computational Linguistics: Human Language Technologies: Short Papers, pp. 42–47 (2011)
9. Hirsch, J.E.: An index to quantify an individual's scientific research output. Proc. Natl. Acad. Sci. U. S. A. **102**(46), 16569–16572 (2005)
10. Joulin, A., Grave, E., Bojanowski, P., Mikolov, T.: Bag of tricks for efficient text classification. In: EACL, pp. 427–431 (2017)
11. Lafferty, J., Mccallum, A., Pereira, F.: Conditional random fields: probabilistic models for segmenting and labeling sequence data, pp. 282–289 (2001)
12. Letierce, J., Passant, A., Breslin, J.G., Decker, S.: Using Twitter during an academic conference: the iswc2009 use-case. In: ICWSM, pp. 279–282 (2010)
13. Luo, Z., Osborne, M., Petrovic, S., Wang, T.: Improving Twitter retrieval by exploiting structural information. In: AAAI, pp. 648–654 (2012)
14. Mapes, K.: A qualitative content analysis of 19,000 medieval studies conference tweets. In: ACM International Conference on the Design of Communication, p. 48 (2016)
15. Mckeown, K., et al.: Predicting the impact of scientific concepts using full text features. J. Assoc. Inf. Sci. Technol. **67**, 2684–2696 (2015)
16. McNamara, D., Wong, P., Christen, P., Ng, K.S.: Predicting high impact academic papers using citation network features. In: Li, J., et al. (eds.) PAKDD 2013. LNCS (LNAI), vol. 7867, pp. 14–25. Springer, Heidelberg (2013). https://doi.org/10.1007/978-3-642-40319-4_2
17. Mikolov, T., Chen, K., Corrado, G.S., Dean, J.: Efficient estimation of word representations in vector space. CoRR abs/1301.3781
18. Mikolov, T., Sutskever, I., Chen, K., Corrado, G.S., Dean, J.: Distributed representations of words and phrases and their compositionality. In: NIPS, pp. 3111–3119 (2013)
19. Mikolov, T., tau Yih, W., Zweig, G.: Linguistic regularities in continuous space word representations. In: HLT-NAACL, pp. 746–751 (2013)
20. Owoputi, O., O'Connor, B.T., Dyer, C., Gimpel, K., Schneider, N., Smith, N.A.: Improved part-of-speech tagging for online conversational text with word clusters. In: HLT-NAACL, pp. 380–390 (2013)
21. Petrovic, S., Osborne, M., Lavrenko, V.: RT to win! predicting message propagation in Twitter. In: ICWSM, pp. 586–589 (2011)
22. Priem, J., Costello, K.L.: How and why scholars cite on Twitter. Proc. Asist Ann. Meet. **47**(1), 1–4 (2010)
23. Ratinov, L.A., Roth, D.: Design challenges and misconceptions in named entity recognition. In: CoNLL, pp. 147–155 (2009)
24. Ritter, A., Clark, S., Mausam, Etzioni, O.: Named entity recognition in tweets: an experimental study. In: EMNLP, pp. 1524–1534 (2011)
25. Ritter, A., Mausam, Etzioni, O., Clark, S.: Open domain event extraction from Twitter. In: KDD, pp. 1104–1112 (2012)
26. Shibata, N., Kajikawa, Y., Matsushima, K.: Topological analysis of citation networks to discover the future core articles. JASIST **58**, 872–882 (2007)
27. Small, H.G.: Interpreting maps of science using citation context sentiments: a preliminary investigation. Scientometrics **87**, 373–388 (2011)

28. Thelwall, M., Priem, J., Eysenbach, G.: Can tweets predict citations? Metrics of social impact based on twitter and correlation with traditional metrics of scientific impact. J. Med. Internet Res. **13**, e123 (2011)
29. Weller, K., Dröge, E., Puschmann, C.: Citation analysis in Twitter: approaches for defining and measuring information flows within tweets during scientific conferences. In: Proceedings of the ESWC2011 Workshop on 'Making Sense of Microposts': Big Things Come in Small Packages, Heraklion, Crete, Greece, 30 May 2011, pp. 1–12 (2011)
30. de Winter, J.C.F.: The relationship between tweets, citations, and article views for plos one articles. Scientometrics **102**, 1773–1779 (2014)
31. Yogatama, D., Heilman, M., 'connor, B.O., Dyer, C., Routledge, B.R., Smith, N.A.: Predicting responses and discovering social factors in scientific literature predicting responses and discovering social factors in scientific literature (2011)
32. Yogatama, D., Heilman, M., O'Connor, B.T., Dyer, C., Routledge, B.R., Smith, N.A.: Predicting a scientific community's response to an article. In: EMNLP, pp. 594–604 (2011)

Online Matrix Factorization Hashing
for Large-Scale Image Retrieval

Lei Wang, Quan Wang, Di Wang$^{(\boxtimes)}$, Bo Wan, and Bin Shang

Xidian University, No. 2 South Taibai Road, Xi'an, Shaanxi 710071, China
wangdi@xidian.edu.cn

Abstract. In recent years, approximate nearest neighbor search methods based on hashing have received considerable attention in large-scale data. There are plenty of new algorithms have been created and applied to different applications successfully. However, Due to the coming of big-data era, the data increasing rapidly and constantly. The batch-mode methods cannot process data efficiently. To solve the problem, online hashing has attracted more attention. Online methods can reduce storage and increase speed of computing. But existing online hashing algorithms also have some problems. The first one is the label information often cannot be got. Because of that, supervised approaches are not practicable. Another problem is online hashing methods process data as a stream, so the relations between old data and new arriving data is taken into account. It is the reason why a novel approach is proposed in this paper which combines matrix factorization with the idea of online hashing. This method considers the relationship between the previous data and newly arriving data. In addition, it updates the hashing learning model by the matrix factorization when the new data is arrived. The experimental results demonstrate superiority of the proposed approach. It outperforms most state-of-the-art online hashing methods and batch-mode methods.

Keywords: Online hashing · Matrix factorization · Large-scala data
Retrieval

1 Introduction

With the explosive growth of data, it's becoming difficult to search for information with high speed and accuracy. As a critical component of big data, unstructured data is abundant in the real world and image data is one of them. Traditional nearest neighbor search methods based on linear scanning always have bad performance because the datasets are huge. To handle the problem, many experts came up with Approximate Nearest Neighbor, and plenty of efforts have been devoted to research. The most successful search techniques can be roughly split into two categories: tree-based approaches and hashing-based approaches. K-D tree is one of the most representative tree-based methods. It's a promotion of a binary search tree designed and proposed by Finkel and Bentley.

We can split the dataset which consists of k-dimensional data by using K-D tree. Although these methods can make search extremely fast, the performances of tree

© Springer Nature Singapore Pte Ltd. 2018
Z. Xu et al. (Eds.): Big Data 2018, CCIS 945, pp. 124–134, 2018.
https://doi.org/10.1007/978-981-13-2922-7_8

models are decreasing when the data become high-dimensional. In addition, because the data structure based on tree model is bigger than the original data itself, it suffers from a storage problem. To handle this problem, hashing technique is developed by many experts for high-dimensional data.

The goal of the hashing-based methods is to learn a set of hashing functions, which can preserve similarity of the original data. Since the hamming distances computing by XOR, the hashing-based nearest neighbor search is extremely fast. Owing to the speed of it, it's possible to be an online algorithm. Early work, such as Locality Sensitive Hashing (LSH) [1], construct hash functions by random projection. This is a typical data-independent method and always requires long length to achieve satisfactory search accuracy. And most of popular hashing algorithms are batch-based methods. Representative works include Iterative Quantization (ITQ) [2], Anchor Graph Hashing (AGH) [3], Latent Semantic Minimal Hashing (LSMH) [7] etc. These methods can obtain good performances, but they all get training labelled data at the beginning. However, in the real application, the data is not available at the beginning. Furthermore, there are two limitations of the batch-mode methods as following.

First, the real data in the applications, they are always coming in streaming fashion. To clarify this situation, there will have an example. When the search engine will search huge data from the data centers, it cannot get the whole data at one time in general. The data, such as images, used to arrive continuously. In batch-mode methods, the algorithm has to wait all the data are accumulated completely and re-train the hashing functions. Obviously, this is an inefficient approach to learn the hash code. In this time, an online algorithm is needed to improve the search efficiency. In the paper, our method can learn the hash code from the newly arrived data chunks.

Second, when the data is coming continuously, we always need a large disk to store it. But for huge datasets, it is so difficult to read the data into memory. Apparently, the reading speed will affect the speed of retrieval. So, an algorithm which can solve the storage problem and also can search quickly is needed. In this paper, our proposed method can just store little information for the old data and update the hashing function by the newly data.

To solve these problems, online hashing method has attracted more attention. New online hashing learning algorithms are proposed. Online hashing (OKH) [4, 6, 8] is paid more attention. It's a supervised learning method. The method can accept a pair of examples and calculates loss function based on the label information. But in this method, we cannot deal with the data chunk and need label information to decide whether it needs to update the hashing model. In real applications, the labels always unavailable and the data chunks need to be solved rather than the pair of examples. Then, online sketching hashing was proposed [5]. It can get data sketching from the data chunk. It not only can solve the storage problem but also can deal with the data chunk. But it does not consider the relations between the old data and the newly arrived data. The proposed method in this paper can solve all the problems. The details of this method will be introduced in the next section.

In this paper, we put forward a novel online hashing algorithm to solve the above problems. The proposed method, called as Online Matrix Factorization Hashing for Large-Scale Retrieval (OMFH). In our method, we do not need to store all the data in the disk, and we just need to store some information in the memory. Figure 1 shows the

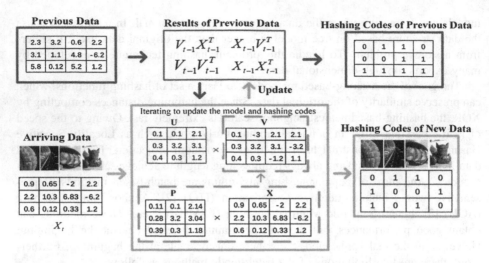

Fig. 1. The framework of OMFH, illustrated with toy data. Data in blue square represent the previous data and in the orange square represent the arriving data. When the new data is coming, it will be decomposed and then the result will be used to update the previous results. In every round, the previous data results will be updated based on the new model and the binary codes of new data will be get.

process of the OMFH. Our work is largely inspired by the collective matrix factorization (CMF) [9]. The algorithm achieves the optimized procedure by gradient descent and iterations. Our paper has main contributions as the following:

1. We put forward an online matrix factorization method preserving the semantic similarity to facilitate image retrieval process. In addition, we can update the hashing function by the new data with a high efficiency.
2. To solve the storage problem, we just to record the calculation result to deliver it to the next update of the hashing function.

The paper is organized as follows: in Sect. 2, we will give the details of the proposed method. Including the integrated model and the specific optimized process. In Sect. 3, the experimental results will be presented. And in Sect. 4 conclusions are displayed.

2 The Proposed Method

In this section, details about the OMFH are expounded. The key of this algorithm is flexible use of additive model, a variant of the additive model. There are three parts below. First, notations will be introduced for easy reading of the rest content in the paper. Second, we will introduce the whole framework of OMFH. Then, the details of the algorithm and the optimization method will be displayed in the last part.

2.1 Notation

In real world, because the data always coming rapidly and constantly, the whole dataset always cannot be get in one time. To simulate the real situation, we suppose the data coming as data chunk, such as $X_t = \left[x_t^1, x_t^2, \ldots, x_t^N \right] \in R^{d \times N}$ where each column $x_t^i \in R^{d \times 1}$ is a sample of the chunk, and t presents this chunk is coming at round t. The aim of hashing-based method tries to learn a series of hashing function $h_k(x) \in H(k = 1, 2, \ldots, K)$ to map each feature to a k-dimensional binary code, where $h_k(x)$ belongs to $\{-1, 1\}$. The original data X_t in the Euclidean space is encoded to binary codes as Y_t in Hamming space.

2.2 Online Framework and Matrix Factorization

The key problem for online algorithm is to adapting to the stream-fashion data. In order to improve the practicability of the algorithm, our method is an unsupervised method and we use the "new additive model" to handle the data. The primary idea is that we need a modifiable model which we just need to add an item when the new sample is coming. Then, the model framework is proposed as:

$$F_t = F_{t-1} + f_t \tag{1}$$

where F_t is the model of the t round and f_t is new data of t round. In this Framework, the function just needs to add the new variate at each round. The details of the F_t will be presented in the next section.

Matrix factorization has been applied to retrieve algorithms successfully. There are lots of approaches to matrix decomposing, such as SVD, QR etc. Given a matrix X, the latent semantic feature can be learned by matrix factorization:

$$X \approx UV \tag{2}$$

where $U \in R^{d \times k}$, $V \in R^{k \times N}$, and k is the number of hashing code. The hash codes can be learned from the formula, such as y = sign(v).

2.3 Online Matrix Factorization Hashing

To learn the hash codes, we need a cost function to represent the problem. The cost function can measure the similarity of the original matrix X between the two matrices by the square of the Euclidean distance. Based on the last section, the first part of the cost function as:

$$E(U, V) = \min_{U,V} \|X - UV\|_F^2 + \gamma(\|U\|_F^2 + \|V\|_F^2) \tag{3}$$

where γ is the tradeoff parameter of the regularization term. Now, we do not need to think more about the online structure, in the next part we will give you the details about it. The formulation in (3), the matrix U is treated as a basis which is the mapping relation between the original matrix X and the compressed matrix V, and each column of V can be regarded as the code of the X. From this formulation, we can learn the hashing code, but it is not fast and accurate enough.

We are inspired by the LSMH, ITQ, CMFH, the data encoding process can be regarded as assigning each data point to the closest vertex of the hypercube. Then, a projection matrix is needed to map each column of the X to the binary code.

$$f(X) = PX \tag{4}$$

where $P \in R^{k \times d}$ is the projection matrix, and this is the second part of the whole cost function.

Now, we need to think how to combine the above with the online framework. The overall objective function combines the matrix factorization part given in Eq. (3), the linear embedding part in Eq. (4) and regularization term:

$$F = min_{U,V,P} \|X - UV\|_F^2 + \mu \|V - PX\|_F^2 + \gamma \left(\|U\|_F^2 + \|V\|_F^2 + \|P\|_F^2 \right) \tag{5}$$

where μ and γ are tradeoff parameters, and the regularization term is defined to avoid overfitting. $\|\cdot\|_F^2$ is the Frobenius norm. To learn the hash code, the Eq. (5) has to be optimized. But it is a non-convex problem with three matrix variables U, V, P. Inspired by the existing hashing approaches, the one of the variables can be learned by fixing the other two. Then, the problem above can be optimized.

The key of our algorithm is updated the model online, so we consider F_t represent the t round of the cost function. Based on the last section, the online formulation can be represented as:

$$\begin{aligned} F_t &= F_{t-1} + f_t \\ &= \|X_{t-1} - U_t V_{t-1}\|_F^2 + \mu \|V_{t-1} - P_t X_{t-1}\|_F^2 + \|X_t - U_t V_t\|_F^2 + \mu \|V_t - P_t X_t\|_F^2 \\ &\quad + \gamma (\|U\|_F^2 + \|P\|_F^2 + \|V\|_F^2) \end{aligned} \tag{6}$$

The t in the Eq. (6) represent the round number. Because the Eq. (6) is not a convex problem, it can be optimized by iteration.

Fix P, V, take the derivative of objective function with respect to U:

$$U_t = \left(X_{t-1} V_{t-1}^T + X_t V_t^T \right) \left(V_{t-1} V_{t-1}^T + V_t V_t^T + \gamma I \right)^{-1} \tag{7}$$

Fix U, V, take the derivative of objective function with respect to P:

$$P_t = \left(V_{t-1} X_{t-1}^T + V_t X_t^T \right) \left(X_{t-1} X_{t-1}^T + X_t X_t^T + \frac{\gamma}{\mu} I \right)^{-1} \tag{8}$$

Fix P, U, take the derivative of objective function with respect to V:

$$V_t = \left(U_t^T U_t + (\mu + \gamma) I \right)^{-1} \left(U_t^T X_t + \mu P_t X_t \right) \tag{9}$$

From the optimized results, we can see the item as $X_{t-1} V_{t-1}^T$, $V_{t-1} V_{t-1}^T$ etc. They all the product of the last round results, in other words, they have been gotten in last round.

Because of that, they just need to be stored in the memory, and in a new round we can use it immediately. Because of this, the model's updating just need addition and multiplication in every round which are very fast in computer calculation. Not only we improve the computation speed, but also we can solve the storage problem in this way. Because the computer does not need to store all the data in the past, because they just need the four products of last round. In addition, the proposed method is an unsupervised setting algorithm which is more practicable (Table 1).

Table 1. The summary of the algorithm.

Algorithm 1 Online Matrix Factorization Hashing
Input:
Data chunk X_t coming as a stream, parameter μ, γ, k.
Output:
Hashing code sign(V), the matrix factorization result U, the projected matrix P.
When the data chunk X_t is arrived at present:
1. Initialize U, P, V at the first round by random matrices, and centering X_t by means.
2. While not converge do:
Fix P, V, update U by Equation (7)
Fix U, V, update P by Equation (8)
Fix U, P, update V by Equation (9)
End
3. $Y = sign(V)$

For out-of-sample data, we minimize the Eq. (3), then we got V as:

$$V = (U^T U + \gamma I)^{-1} U^T X \tag{10}$$

Then, we can get the hash code as Y = sign(V).

3   Experiments

In this section, three issues have been presented to examine the performance of the proposed method. The first experiment is to verify the changes of the accuracy. With the data coming constantly, we can see the result is changing. Then, the next experiment is a comparison between OMFH (our method) and other existing online hashing algorithms. The last experiment is going to compare OMFH between batch-mode methods to verify our online model's practicability. In these experiments, the parameters $\mu = 2 \times 10^{-5}$ and $\gamma = 10^{-3}$ respectively.

3.1   Datasets

The MNIST is a database of handwritten digits has a training set of 69,000 examples, and a test set of 1000 examples. The digits have been size-normalized and centered in a fixed-size image. It is one of the most popular datasets to test hashing algorithms. Each image in this dataset is matched to a label from 0 to 9. In this paper, we used 784-dimensional greyscale vector as the feature representation.

CIFAR-10 is another popular database which is collected by Alex Krizhevsky, Vinod Nair and Geoffrey Hinton, to test hashing methods. This dataset consists of 60,000 images in 10 classes, with 6000 images per class. The database is divided into two parts, including 50,000 training examples and 10,000 test examples. In this paper, we extract 512-dimensional GIST descriptor to represent images.

3.2   The Evaluation Criterion

To verify the algorithm result, we choose the common criterion to evaluate. The retrieval search performance is always evaluated with mean average precision (MAP). It's one of the most comprehensive criterion in retrieval. MAP is calculated with the area under precision-recall curve. The proposed method in this paper is an unsupervised algorithm. And all the results in the following are averaged over 5 times independent runs.

3.3   The MAP at Different Rounds

The first experiment, we focus on the proposed method OMFH. Because we want to know changes of the accuracy. Because of the online setting, the training data is separated into plenty of data chunks to make the data is available as a stream. By this method, we can simulate the real conditions in the real world. The source codes of them are offered by authors. In addition, the training data is divided into 100 chunks, to simulate 100 rounds as a data stream, for the two datasets. And we calculate the MAP scores after each round. In other words, we need to run training process and test process 100 times and get a MAP result after every round.

Figure 2 shows MAP scores at every round on both of datasets with 16, 32 and 64 bits codes. It is obvious that scores are increased continuously with the round accumulating. Our proposed OMFH achieves stable improvement when the data chunk coming as a stream. Furthermore, we can see the MAP scores has increased significantly with the length of hashing code. This is because the longer length of hashing codes, the less information will be reduced. And, the longer length will decrease the possibility of the hash collision. The figure (a) and (b) show the MAP scores in different datasets.

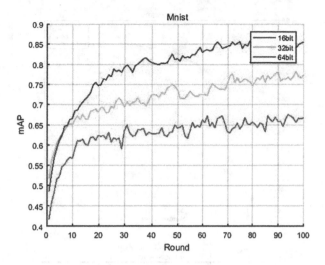

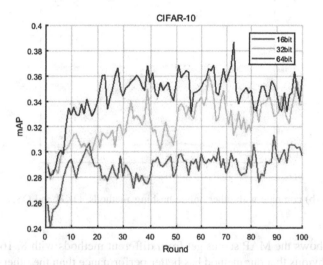

Fig. 2. (a) and (b) show the performances of MAP scores at every round in Mnist dataset and CIFAR-10 dataset.

3.4 Comparison to Online Hashing

In the second experiment, we have comparison to the other online hashing algorithms. The first one is online hashing (OKH), and another one is online sketching hashing (OSH). They all have good performances in practice. Because we have the same online setting as the Sect. 3.2, the only difference is the size of the data chunk. We give the OMFH 10,000 examples at one time. In this experiment, we simulate the data like a stream. The source codes of OKH and OSH are provided by the authors.

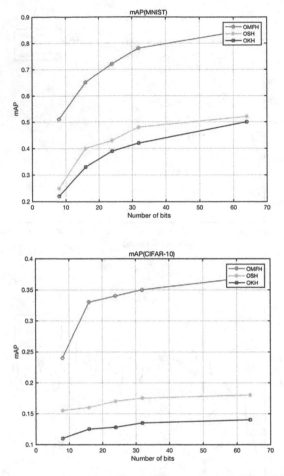

Fig. 3. (a) and (b) compare with other online hashing methods, OKH and OSH, with different code length

Figure 3 shows the MAP scores get from different methods with 8, 16, 24, 32 and 64 bits. It is obvious that our method has better performance than the others. Moreover, the growing rate of OMFH is increasing rapidly than other methods.

3.5   Comparison to Batch Based Methods

In the last experiment, we want to confirm that our online hashing function also can have good performance with the comparison to batch-mode methods. We compare our method with the batch-mode hashing methods such as ITQ, AGH, SH, LSMH. The source codes of AGH, SH and ITQ are provided by the authors. The LSMH is coded by myself based on the paper. Before the experiment, we assume all the training data is available and zero-centered. For our method, we use the same size dataset as batch-mode methods. The only difference lies in data form. We give the OMFH data as a stream, but give others the whole dataset.

Figure 4 shows the MAP scores of all algorithm using Hamming ranking on both of databases with different hashing code length. In the picture, we can observe that the proposed OMFH achieves comparable accuracy with other methods. In addition, the accuracy of OMFH exceeded some of the batch based hashing function. The performance of OMFH is slightly inferior to the best competitor LSMH but still acceptable. In the experiment, the training datasets have the same size. And in the test process, we use the final model which learned from the hashing function. All the result indicate that the online algorithm does not lose accuracy than batch-mode hash function. And in this picture, we can observe obviously that MAP score of the OMFH can reach almost 0.9.

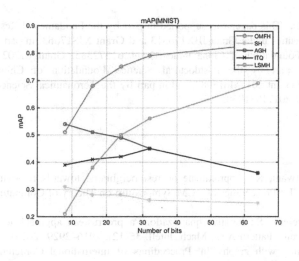

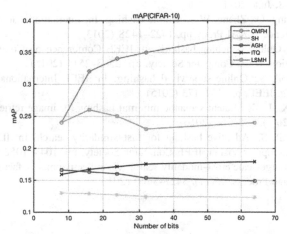

Fig. 4. (a) and (b) are different methods in Mnist dataset and CIFAR-10 dataset with 8, 16, 24, 32 and 64 bits.

4 Conclusion

In this paper, we put forward a novel online hashing learning method called online matrix factorization hashing, which updates the hashing model when the new data is coming continuously. Our proposed method applies the matrix factorization to the online hashing to preserve the semantic information. In addition, the newly arriving data has contract with the previous data in this hashing learning model. The three aspects of the experiment show that our approach achieves satisfactory results, though it is an online method which cannot get all training data available in advance. It is promising that OMFH can be a serviceable method for real applications with streaming data for online searching.

Acknowledgments. This paper was supported in part by the Fundamental Research Funds for the Central Universities under Grant JBX170313 and Grant XJS17063, in part by the National Natural Science Foundation of China under Grant 61572385, Grant 61702394, and Grant 61711530248, in part by the Postdoctoral Science Foundation of China under Grant 2018T111021 and Grant 2017M613082, and in part by the Aeronautical Science Foundation of China under Grant 20171981008.

References

1. Indyk, P., Motwani, R.: Approximate nearest neighbors: towards removing the curse of dimensionality. In: Proceedings of ACM Symposium on Theory of Computing, pp. 604–613 (1998)
2. Gong, Y., Lazebnik, S.: Iterative quantization: a procrustean approach to learning binary codes. IEEE Trans. Pattern Anal. Mach. Intell. **35**(12), 2916–2929 (2013)
3. Liu, W.: Hashing with graphs. In: Proceedings of International Conference on Machine Learning, pp. 1–8, June 2011
4. Huang, L.K., Yang, Q., Zheng, W.S.: Online hashing. In: International Joint Conference on Artificial Intelligence. AAAI Press, pp. 1422–1428 (2013)
5. Leng, C., et al.: Online sketching hashing. In: IEEE Conference on Computer Vision and Pattern Recognition. IEEE Computer Society, pp. 2503–2511 (2015)
6. Cakir, F., Sclaroff, S.: Online supervised hashing. In: IEEE International Conference on Image Processing. IEEE, pp. 162–173 (2015)
7. Lu, X., Zheng, X., Li, X.: Latent semantic minimal hashing for image retrieval. IEEE Trans. Image Process. **26**(1), 355–368 (2016)
8. Cakir, F., Sclaroff, S.: Adaptive hashing for fast similarity search. In: IEEE International Conference on Computer Vision. IEEE Computer Society, pp. 1044–1052 (2015)
9. Ding, G., et al.: Large-scale cross-modality search via collective matrix factorization hashing. IEEE Trans. Image Process. **25**(11), 5427–5440 (2016)

Multi-constrained Dominate Route Queries in Time-Dependent Road Networks

Fangjun Luan, Qi Li$^{(\boxtimes)}$, Keyan Cao, and Liwei Wang

Department of Information and Control Engineering, Shenyang Jianzhu University,
110168 Shenyang, China
luanfangjun@163.com, liqiarya@gmail.com, caokeyan@gmail.com,
wangliwei@163.com

Abstract. With the rapid development of location-based services, there
is more and more personalized demand for route planning. The existing
studies on route queries on time-dependent network to find the optimal
path, for example: shortest route, highest scoring route, etc. However, in
practical application, users will want to be satisfied with the constraint
and evaluate the good routes to make a choice, for example, users want
to look for the well-evaluation route through bank, café shop, restaurant
in ordered within 3 h for reference. Based on this requirements, a new
location-based service is proposed in this paper, which is called time-
dependent and dominated route query (MTDDR). In order to solve the
MTDDR problem, three algorithms are designed in this paper, which are
respectively the precise algorithm BSL algorithm, the time-dependent
estimation algorithm TDER algorithm, a heuristic FTDR algorithm,
while the design further reduces the pruning strategy of the search space.
Using the real network data set on OpenStreetMap, to test the validity
of three algorithms under different parameters.

Keywords: Time-dependent network · Route query · Dominate route

1   Introduction

Route search has been an important problem that application in online map
services and location based services for route planning. The optimal sequenced
route (OSR) query was originally introduced in [1]. It aims at finding the optimal route from an origin location passing through a number of points of interest
(POIs), each belonging to a specific sequence of categories of interest (COIs).
This query has several applications within location-based services or car nav-
igation systems. Some variations and solutions to the OSR query have been

Supported by National Natural Science Foundation of China (No. 611602323); China
Postdoctoral Science Foundation (2016M591455); and Liaoning Province Doctor
Startup Fund (201601209).

proposed in [4–7]. In the keyword-aware optimal route (KOR) query [3], it aims to find an optimal route such that it covers a set of user-specified keywords. To improve KOR query, the keyword-aware dominant route (KDR) [8] query is proposed, which returns a subset of dominant routes from all the feasible routes such that each returned dominant route is not dominated by any other feasible route. The Skyline algorithm in [11–13] is combined with route queries. On a typical road network, the time that the user needs to traverse the edge depends on the departure time, that is, the road network is a time-dependent graph. In [10] and [14–16] considering the time-dependent graph in route query.

We use the feature: time-dependent as an inherent part of the problem, and as our main contribution, we propose effective algorithms to solve the new multi-constrained time-dependent and dominated route query (MTDDR) query, which takes as input a time-dependent graph TDG and returns routes for a user departing at time t from s towards d, visiting exactly one POI from each COI in the ordered list K, such that the query will return all possible paths that are not dominated by other paths, under a specified budget constraint.

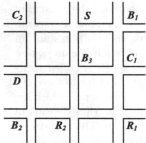

Fig. 1. A partial road network.

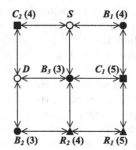

Fig. 2. A graph representing the road network in Fig. 1.

The small TDG instance with three different COIs: banks $K_B = \{b_1$ and $b_2\}$, restaurants $K_R = \{r_1$ and $r_2\}$ and café shops $K_C = \{c_1$ and $c_2\}$, a starting node s and a destination node d. The broken lines shown in Fig. 3 represent the travel time costs of some edges of the network. Let us consider an MTDDR query where $t = 8$ and $t = 15$, respectively. The order of K is $K = [K_C, K_R]$ and the time constraint is 35 min. For the sake of simplicity, but without loss of generality, we assume that the time spent at each COI depends only on the COI itself, i.e., it is equal for all POIs in the same category, e.g., the time spend at any café shop in K_C and any restaurant in K_R is 15 min and 60 min respectively. In the time-dependent network, if $t = 8$, the query returns two routes: (i) $P_1 = \langle s, C_1, R_1, d \rangle$, with budge score is 34 min and object value is 10. (ii) $P_2 = \langle s, C_2, R_2, d \rangle$, with budget score is 26 min and object value is 9. Similarly, if $t = 15$, the query returns only one routes: $P_1 = \langle s, C_1, R_1, d \rangle$, with budget score is 33 min and object value is 10, since the budget score of $P_2 = \langle s, C_2, R_2, d \rangle$ is 37 min which exceeds the time limit 35 min.

We also make the practically reasonable assumption that the (time-dependent) travel cost of each edge in the TDG satisfies the FIFO property, i.e., an object that starts traversing an edge first has to finish traversing this edge first as well. The general time-dependent shortest path problem in TDGs is NP-hard [9], but it has a polynomial time solution in FIFO networks. In the context of our problem, the FIFO property guarantees that there is no improvement in travel time if one "waits" at a vertex for the "best time" to traverse it, thus in what follows such "waiting" is not allowed.

The graph shown in Fig. 1 where a partial network is shown. Figure 2 shows a graph representing TDG.

The main contributions of this paper are as follow:

-We propose a new time-dependent route query problem, called MTDDR query, to find a route in a road network which covers a set of user-specified keywords, satisfied a budget constraint, and is not dominated by any other feasible route in terms of the ratings of places.

-We propose an exact basic solution BSL algorithm, a time-dependent estimation algorithm TDER and a heuristic algorithm FTDR.

-We evaluate the effectiveness and efficiency of our approach with extensive experiments on real and synthetic datasets, showing that our algorithm is more efficient than the one using the basic solution.

The paper is organized as follows. The problem is formally defined in Sect. 2. In Sect. 3 the basic algorithm and the heuristic algorithm TDER are presented. The experimental results are presented in Sect. 4. Section 5 concludes this paper.

2   Preliminaries

In this section, we formalize the concept of time-dependent graph and explain the concept of dominance relation between feasible routes. We also provide other basic definitions, such as travel time and arrival time, which is useful to illustrate the issues we are discussing.

Definition 1 (A time-dependent graph). *Given a graph TDG $G = (V, E, W)$ is a graph, where $V = \{v_1, ..., v_n\}$ is a set of vertices, where each node in V is associated with a set of keywords represented by K, and each node is also associated with a rating score $o(v)$, which indicate the popularity of the location; $E = \{(v_i, v_j) \mid v_i, v_j \in V, i \neq j\}$ is a set of edges associated with a cost value $w(v_i, v_j)$ which can be travel duration, travel distance or travel cost. $w = \{(v_i, v_j) \mid v_i, v_j \in E\}$, where $w(v_i, v_j): [0, T] \rightarrow R^+$ is a function which attributes a positive weight for $w(v_i, v_j)$ depending on a time instant $t \in [0, T]$ and where T is a domain-dependent time length.*

A TDG is a graph where the edge costs vary with time. For each edge (u, v), a function $w(u, v)(t)$ gives the cost of traversing (u, v) at departure time t and $[0, T]$ is the domain of the functions in w. For example, $T = 24\,h$ means that $w(u, v)(t)$ is defined for a full day.

Definition 2 (Budget score). *Route P_i is a path that goes through v_1 to v_n sequentially, following the relevant edges in TDG. The budget score (BS) of a route P_i is defined as sum of the cost of its edges, i.e. $BS(P_i) = \sum_{i=1}^{n} w(v_i, v_j)$.*

Definition 3 (Object value). *Route P_i is a path that goes through v_1 to v_n sequentially, following the relevant edges in TDG, and each node is also associated with a rating score $o(v_1)$. The object value (OV) of a route P_i is $OV(P_i) = \sum_{i=1}^{n} o(v_1)$.*

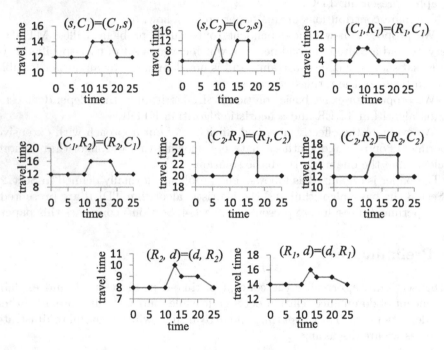

Fig. 3. Basic algorithm for sequenced keywords.

Figures 2 and 3 show an example of the two definitions above. Given a route $P_1 = \langle s, C_1, R_1, d \rangle$ and the departure time $t = 8$. The budget score $BS(P_1) = 12 + 8 + 14 = 34$ and the object value $OV(P_1) = 5 + 5 = 10$.

Definition 4 (Arrival time). *Given a TDG $G = (V, E, W)$, the arrival time to traverse an edge $(v_i, v_j) \in E$ at departure time $t \in [0, T]$ is given by $AT(v_i, v_j, t) = t + w(v_i, v_j)(t)/60 \bmod T$.*

Definition 5 (Travel time). *Given a TDG $G = (V, E, W)$, a path $P_1 = \langle s, v_1, \ldots, v_n \rangle$ in G and a departure time $t \in [0, T]$, the travel time of P is the time-dependent cost to traverse this path, $TT(P, t) = \sum_{i=1}^{n-1} w(v_i, v_j)(t_i)$, where $t_1 = t$ and $t_{i+1} = AT(v_i, v_j, t)$.*

3   Solutions to the MTDDR Query

We propose an exact algorithm in Sect. 3.1, an algorithm based on pre-processing and pruning strategy in Sect. 3.2 and a heuristic algorithm in Sect. 3.3.

3.1   Baseline Algorithm

A basic and exact method is proposed to solve the MTDDR query in the time-dependent graph TDG, according to the user's query $Q = (TDG, s, d, t, K, \theta)$, traversal graph under the time constraints in all feasible routes, through comparing the BS and OV of the routes, determined finally returned dominant routes to the users for personalized selection (Fig. 4).

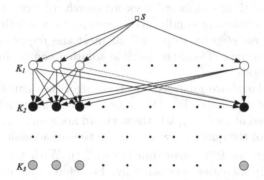

Fig. 4. Basic Algorithm for sequenced keywords.

Definition 6 (Multi-constrained Dominate Route). *Given a MTDDR query $Q = (TDG, s, d, t, K, \theta)$, takes as input a time-dependent graph TDG $G = (V, E, W)$ and POI $\in V$ be a set of points of interest in G, which returns a set of dominant routes DR for a user departing at time t from s towards d, visiting exactly one POI from each COI in the ordered list K, such that the query returns all possible dominant routes $P_i = \langle s, v_{K_1}, \ldots, v_{K_n}, d \rangle$, under a specified budget constraint θ.*

Definition 7 (Dominance relation between feasible routes). *Given a MTDDR query $Q = (TDG, s, d, t, K, \theta)$ the objective value of feasible route P_i and $P_j (i \neq j)$ are $OV(P_i)$ and $OV(P_j)$, and budget score are $BS(P_i)$ and $BS(P_j)$, respectively. We say that P_i is dominated by P_j, denoted by $P_i \prec P_j$ iff (i) $OV(P_i) \leq OV(P_j)$ and $BS(P_i) > BS(P_j)$; (ii) $OV(P_i) < OV(P_j)$ and $BS(P_i) = BS(P_j)$.*

In the graphs of Figs. 2 and 3, given a MTDDR query $Q = (TDG, s, d, 8, \{K_C, K_R\}, 35)$. If departure time $t = 8$, there are three feasible routes in $P_1 = \langle s, C_1, R_1, d \rangle$ with $BS(P_1) = 34$ and $OV(P_1) = 10$, $P_2 = \langle s, C_1, R_2, d \rangle$ with $BS(P_2) = 32$ and $OV(P_2) = 9$, $P_3 = \langle s, C_2, R_2, d \rangle$ with

$BS(P_3)$= 26 and $OV(P_3)$=9. Considering the dominance relation between feasible routes, route 2 is dominated by route 3, denoted by $P_2 \prec P_3$, according to $OV(P_2) = OV(P_3)$ and $BS(P_3) > BS(P_2)$.

Definition 8 (Potential route descendant). *For routes P and P', we say route P' is the descendant of P, if $P' = P \oplus v_i$ and under a specified budget constraint θ.*

There are many ways to traverse all the feasible routes under the time constraints when we know the start node. In this paper, we use the breadth first search strategy to find all feasible solutions to meet the time constraints.

According to the query $Q = (TDG, s, d, t, K, \theta)$ by users, the basic algorithm is based on the breadth first search strategy. Search all nodes to meet the requirements of the current keyword, whether conform to the time constraints, it will from one node to continue downward keyword search conforms to the next node, if do not meet the constraint conditions, exceeding to constraint threshold, the search stops, traverse step by step until all meet the requirements of feasible solution time constraints. Then, the feasible solution can be further optimized in all feasible solutions.

According to the above analysis, the BSL algorithm can finally return all routes that are not dominated by other routes by traversing all feasible routes. A vertex is expanded at most k_m, k is the keyword node set size, $k_m = max\{k_n\}$. $|K|$ is the number of the query keywords. If the time of accessing a node is $O(1)$, the time complexity of BSL algorithm is $O(k_m^{|K|})$. With the increase of $|K|$, the time complexity increases exponentially. Therefore, this method is not only inefficient, but also not scalable.

3.2   TDER Algorithm

To measure the potential of a vertex v_i, a function $L_{v_i} = TT_{v_i} + h(v_i)$ is used, where TT_{v_i} is the current cost from the starting point s to v_i and $h(v_i)$ is a estimation, which in our case is a lower bound estimate of the time to pass through all the categories of POIs and to reach the destination d. The smaller the value given by the sum of the current cost to a vertex plus the function value for it, the greater is potential. If the estimate $h(v_i)$ is a lower bound to the actual cost from v_i to d, L_{v_i} is said to be admissible.

Pre-processing Step. Aiming at reducing the execution time of the query, a pre-processing step is performed in our solutions, in order to calculate bounds to guide the expansion of vertices.

In order to calculate the heuristic function value for vertices in TDG, we construct the lower bound graph $\underline{G}$ and the upper bound graph $\overline{G}$. The lower bound graph of a TDG $G = (V, E, W)$ is a graph $\underline{G} = (V, E, \underline{W})$ where V and E are the same set of vertices and edges in G and $\underline{W}$ is the set of costs $\underline{w}(v_{i-1}, v_i)$=$\min_{t \in [0,T]}\{w(v_{i-1}, v_i)(t)\}$. The upper bound graph $\overline{G} = (V, E, \overline{W})$ has the same set of vertices and edge in G and $\overline{W}$ is the set of edge costs

$\overline{w}(v_{i-1}, v_i) = \max_{t \in [0,T]} \{w(v_{i-1}, v_i)(t)\}$, for all $w(v_{i-1}, v_i) \in W$. For example, considering the TDG shown in Fig. 3, the cost of the edge(s, C_1) in $\underline{G}$ would equal to 12 min and in $\overline{G}$ would equal to 14 min.

For each vertex v_i, the distance between v_i and other nodes which match one keyword, a non-time-dependent (i.e. static) breadth first traversal is executed in graph $\underline{G}$ and $\overline{G}$. For example, consider vertex s in Fig. 2. The distance between s to all the café shops in the $\underline{G}$ is recorded as $h(v_i)$ with the café shops object value in descending order. The distance between s to all the café shops in the $\overline{G}$ is recorded as $dist$ in descending order.

Table 1. Value of preprocessing in $\underline{G}$ and $\overline{G}$

v_{li}	$o(v_i)$	$h(v_i)$	v_{ui}	$dist$
C_1	5	12	C_8	50
C_2	4	3	C_5	35
C_7	4	10	C_4	30
C_4	4	26	C_6	28
C_5	3	17	C_1	20
C_8	3	35	C_7	15
C_3	2	5	C_2	10
C_6	1	15	C_3	8

The result of our preprocessing step is shown in Table 1. The Table 1 shows the preprocessing for node s to the nodes match the keyword café shop. These values are used to prune vertices that lead to points of interest farther than a set of candidates.

Query Processing. One hard constraint of our MTDDR problem is the budget score θ, we can get some lemmas.

Lemma 1. *Given a path* $P = \langle s, v_1, \ldots, v_i \rangle$, $L_{v_i} = TT_{v_i} + h(v_i) > \theta$, $P' = P \oplus v_{i+1}$ *should not be expanded anymore.*

Proof. For path $P = \langle s, v_1, \ldots, v_i \rangle$, $L_{v_i} = TT_{v_i} + h(v_i)$, since $h(v_i)$ is the shortest path from v_i to d, for $P' = P \oplus v_{i+1}, L_{v_{i+1}} = TT_{v_{i+1}} + h(v_{i+1}) = TT_{v_i} + w(v_i, v_{i+1}) + h(v_{i+1})$. According to the Pythagorean theorem, the sum of both sides is greater than the third, $w(v_i, v_{i+1}) + h(v_{i+1}) > h(v_i)$, so we can get $L_{v_{i+1}} > L_{v_i} > \theta$. □

To reduce the calculation in the procedure of a routes refinement, we must seek for some reasonable and effective prune strategies. We can find a route in $\overline{G}$ which satisfies the specified budget constraint, and have the highest object value, then budget score of the route should become the new budget constrain, denoted by $dist_m$. We know that the $dist_m \leq \theta$. If the budget score of a potential route P violates the $dist_m$, and the descendant path should not be expanded anymore, because any of its descendants will violate the budget limit too. It can be proven as follows.

Lemma 2. *Given a path $P = \langle s, v_1, \ldots, v_n, d \rangle$, if $TT_{v_i} + h(v_i) > dist_m$, $P = P \oplus v_{i+1}$ should not be expanded anymore.*

Proof. A route $P_1 = \langle s, v_1, \ldots, v_n, d \rangle$ is the highest score route, the $dist_1 = \sum_{i=1}^{n}(v_{i-1}, v_i)$ in the $\overline{G}$ and $BS(P_1) \leq dist_m$. According to Definition 7, if a route P_2 is a potential route with $TT_{v_i} + h(v_i) > dist_m$, and the object value $OV(P_2) \leq OV(P_1)$, P_2 must be dominated by P_1, and the descendant of potential route $P_2' = P_2 \oplus v_i$ must be dominated by P_1 too (Table 2).   □

Table 2. Symbols meaning

Symbols	Meaning
AT_{v_i}	The arrival time at v_i
TT_{v_i}	The travel time at P
v_i, K_n	Node v contain uncovered keyword K
$|K|$	The number of query COIs
$P.K$	The number of the COIs in route P
$\oplus$	Insert a keyword vertex
$Fdominate$	The function to check the dominate relationship

Algorithm 1 shows the pseudo-code for the TDER algorithm. An entry in Q_1 is a tuple $\langle v_i, AT_{v_i}, TT_{v_i}, L_{v_i}, OV_{v_i} \rangle$. L_{v_i} is given by $TT_{v_i} + h(v_i)$, i.e. a sum of the travel time from s to v_i in the path in which v_i was found plus the minimum travel time in $\underline{G}$ to expect if the route satisfies the constraint time. The $L(v_i)$ value represents an optimistic expectation for the travel time of a path that leads v_i to the destination node d. First, the breadth first traversal is used to calculate the costs of the paths from v_i to each $v_j \in K_n$ and add to $Pre[]$(line 2). Next, the algorithm begins the network expansion and insert s in a priority queue Q_1, which stores the set of candidates for expansion in the next step (line 4). An entry in Q_1 is a tuple $\langle s, AT_s, TT_s, L_s, OV_s \rangle$. While Q_1 is not empty, the vertices are de-queued from Q_1(lines 5) and expanded. If the expanded route satisfies the budget constraint and have the uncovered keywords, then we expand it with an unmatched keyword node, if all the keywords are coved, we can get the route P (lines 6–13). If the expanded route P is not dominated by a feasible route $Fdominate(DR, P)$ is false, we remove the expanded route from DR (lines 16–17).

After the expansion of routes, we output the final dominant routes by computing the exact path of each remaining route P in DR. The TDER algorithm

---

Algorithm 1. TDER Algorithm($Q = (TDG, s, d, t, K, \theta), TDG = (V, E, W)$)

Input: A TDG, and a query Q;
Output: A set of dominate routes in DR ;
1: $AT_S \leftarrow t; TT_S \leftarrow 0; OV_S \leftarrow 0$;
2: $Pre[] \leftarrow TRA(\underline{G}), TRA(\overline{G})$;
3: Initialize a queue Q_l and an array DR;
4: En-queue $(s, AT_s, TT_s, L_s, OV_s)$ as P in Q_l;
5: **while** Q_l is not empty **do**
6:     $(u, AT_u, TT_u, L_u, OV_u) \leftarrow$ De-queue Q_l;
7:     Add u to P.
8:     **if** $|P.K| = |K|$ **then**
9:         $DR \leftarrow$ En-queue P;
10:        **if** $|P.K| < |K|$ **then**
11:            **if** $L_s < dist_m$ **then**
12:                **for** all the nodes $v_i.K_n$ **do**
13:                    $P \leftarrow P \oplus v_i.K_n$;
14:                **end for**
15:            **else**
16:                **if** $Fdominate(DR, P)$ is false **then**
17:                    **insert** P into DR;
18:                **end if**
19:            **end if**
20:        **end if**
21:    **end if**
22: **end while**
23: **return** DR.

---

finds the exact answer for any MTDDR query. The whole search space equals to the breadth first traversal without any pruning strategy, so the algorithm is correct.

3.3   FDER Algorithm

FTDR is inspired by the observation that people normally do not backward during their journey. FTDR just expands current routes which match part of query keywords by following a forward direction toward the target node, i.e., each expanding will make the last node of the route closer to the destination. Specifically, when we choose a node containing an unmatched keyword to expand the current route, the selected node should make the route after expanding becomes closer to d. With the search strategy, FTDR would explore few keywords nodes than does the TDER algorithm, but may miss result routes.

Given a MTDDR query $Q = (TDG, s, d, t, K, \theta)$. We say the expansion from v_i to v_{i+1} obeys the forward expansion, i.e., after the expansion, the distance from v_{i+1} to d must be smaller than or equal to the distance from v_i to d in $\underline{G}$; A positive Δ defines a weak forward expansion, which allows a constrained backward under limit Δ, i.e., $dist(v_i) - dist(v_j) \geq \Delta$, $dist(v_i) = \sqrt{(X_{v_i} - X_d)^2 + (Y_{v_i} - Y_d)^2}$, $dist(v_j) = \sqrt{(X_{v_j} - X_d)^2 + (Y_{v_j} - Y_d)^2}$.

Algorithm 2. FTDR Algorithm($Q = (TDG, s, d, t, K, \theta)$, $TDG = (V, E, W)$)

Input: A TDG, and a query Q;
Output: A set of dominate routes in DR ;
1: $AT_S \leftarrow t; TT_S \leftarrow 0; OV_S \leftarrow 0$;
2: $Pre[] \leftarrow TRA(\underline{G}), TRA(\overline{G})$;
3: Initialize a queue Q_l and an array DR;
4: En-queue $(s, AT_s, TT_s, L_s, OV_s)$ as P in Q_l;
5: **while** Q_l is not empty **do**
6:      $(u, AT_u, TT_u, L_u, OV_u) \leftarrow$ De-queue Q_l;
7:      **Foreach** node u belonging to uncovered keywords in K and
8:      $isForward(v, u)$=true **do**
9:      Add u to P.
10:      **if** $|P.K| = |K|$ **then**
11:          $DR \leftarrow$ En-queue P;
12:          **if** $|P.K| < |K|$ **then**
13:              **if** $L_s < dist_m$ **then**
14:                  **for** all the nodes $v_i.K_n$ **do**
15:                      $P \leftarrow P \oplus v_i.K_n$;
16:                  **end for**
17:              **else**
18:                  **if** $Fdominate(DR, P)$ is false **then**
19:                      **insert** P into DR;
20:                  **end if**
21:              **end if**
22:          **end if**
23:      **end if**
24: **end while**
25: **return** DR.

Algorithm 2 shows the pseudo-code for the FTDR algorithm. According to the forward expansion mechanism $isForward$, we expand the initial route with an uncovered keyword node(line 7–9). If the route is a feasible route which is from s to d under the time constrained covering all the keywords, we use function $Fdominate(DR, P)$ to check whether route P is dominated by a route in DR, if false, insert route P into DR (lines 18–19). The worst time complexity of FTDR is the same as that of TDER algorithm, but in practice, FTDR checks much fewer nodes.

4 Experiments

To evaluate the performance of the proposed algorithms for MTDDR query, we use real-world road-network data extracted from the OpenStreetMap database. We focus on the city of Beijing, and we obtained 10,000 POIs with information on their latitudes and longitudes. There are nine different POI in the road network, which are: Bank, Café shop, Government, Hospital, Hotel, Public, Restaurant, School, Supermarket. The Bank, Café shop, Hospital, Hotel, Public, Restaurant and Supermarket are evenly distributed, and the School and Government are normally distributed too. Three datasets (denoted by Node2000, Node6000, Node10000) are generated from the data of Beijing, which contain 2,000 POI nodes, 6,000 POI nodes and 10,000 POI nodes, respectively. The distance serves as the budget score of the edge. To assign rating values for different POIs, we randomly generate it from a uniform distribution over {1, 2, 3, 4, 5}. Those rating values are used as the objective scores, which will be maximized in the MTDDR query. According to the change rule of morning and evening peak time in real life, the whole day is divided into 24 periods with the hour as a single digit, and the time weight function of each period is obtained.

Table 3. Experimental parameters and values

Parameters	Values
Network size	2000,6000,10000
Number of query keyword	(2,5)
Number of keyword node	(100,500)
Budget limit	(10000,50000)

We study the performance of our algorithms by varying network size, the number of query keywords, the number of keyword nodes, and budget limit, as shown in Table 3. The default values of some parameters are as follows: $\theta = 20,000$, $k = 200$ and $|K| = 3$. The default dataset is Node6000. Each set comprises 50 queries. The starting and ending locations are selected randomly.

All the evaluations are implemented in C++, carried out in Windows 7, running on Intel Core(TM) i3-3.3 running 8 GB RAM.

4.1 Efficiency

This set of experiments is to study the efficiency of the proposed algorithms with variation of network size, the number of query keywords, the number of keyword nodes, and budget limit.

Effect of Network Size. From the runtime of queries on different datasets, we do not see a clear tendency of the effect of the different datasets. A possible reason is that the time is mainly affected by factors such as query keywords, θ, start and end nodes, etc.

Effect of the Number of Query Keywords. The runtime of BSL, TDER and FTDR increases because more permutations are needed when the number of nodes increases and the keyword covering size remains the same.

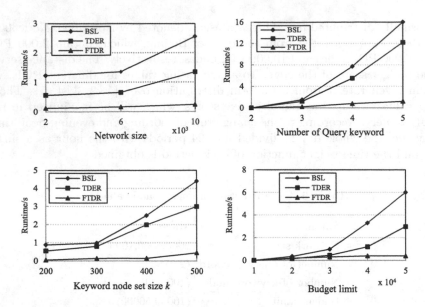

Fig. 5. Efficiency of different parameters.

Effect of the Size of Keyword Node Set. With the increase of keyword nodes, TDER slows down and FTDR runs slightly slower. This is because the complexity of TDER is exponential to the size of cover size of each keyword while FTDR prunes more routes (Fig.5).

Effect of the Budget Limit θ. At each Δ, the average runtime is reported over 10,000, 20,000, 30,000, 40,000, and 50,000 with a number of keywords being 3. The runtime of three algorithms grows when budget limit increases as a larger limit corresponds to more keywords nodes. The runtime of FTDR increases much more slightly because the pruning strategy in direction reduces the complexity of FTDR to be linear to while TDER and BSL are exponential.

4.2 Number of Returned Routes

We report the number of returned routes of the two algorithms TDER and FTDR, for the routes of BSL equal to TDER.

Varying the Budget Limit θ. The number of keywords is set to three. The size of result route set of three algorithms grows when θ increases, which permits more keyword nodes.

Varying the Number of Query Keywords. The number of returned routes of FTDR decreases significantly as the querying keyword size increases. FTDR may fail to return routes when the number of keywords is large, as shown in Fig. 6.

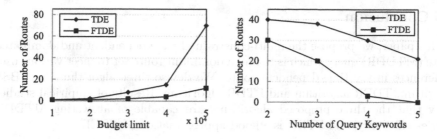

Fig. 6. Number of return routes under different parameters.

5   Related Work

Due to its practical significance, the route query problem and has attracted a lot of attentions. This section introduces some representative works related to this problem and compares them with our work.

In the Trip Plan query (TPQ) [2], the user can specify a set of keyword categories, while TPQ retrieves the shortest path between the specified source and destination that contains all the keywords. Sharifzadeh et al. [1] defines a variant problem of TPQ and dubs it as the Optimal Sequenced Route query (OSR) in both vector and metric spaces. An OSR strives to find a route of minimum length starting from a given source location s and passing through a number of typed locations in a particular order imposed on the types of the locations. To further improve the flexibility for users to specify their requirements for route search, the keyword-aware optimal route (KOR) query is proposed in [3], which aims to find an optimal route such that it covers a set of user-specified keywords, satisfies a budget constraint (e.g., time), and the total popularity (or ratings) of the route is maximized. To further improve the flexibility for users to specify their requirements for route search, The keyword-aware dominant route (KDR) [8] query is proposed, which returns a subset of dominant routes from all the feasible routes such that each returned dominant route is not dominated by any other feasible route in terms of the ratings of places, covering a set of keywords under a specified budget constraint. Different from the above queries, the MTDDR query proposed in this paper considers the time-dependent which takes as input a time-dependent graph TDG.

Cruz [16] proposed the Time Dependent Incremental Network Expansion algorithm. In this method, vertexes in a network are visited in order of travel time from the query point until K data objects are found. Time-expanded graphs allow exploit previous solutions in static networks to solve TD-kNN queries. An improved solution is proposed by [10]. In this work, the authors proposed an algorithm that is based on INE expansion and use an $A*$ search to guide this expansion. However, none of these time-dependent queries take into account the popularity as we do in this paper.

6   Conclusion

In this paper, we propose the multi-constrained time-dependent and dominated route (MTDDR) query which is to find a dominant route set to meet various user preferences in the time-dependent route. We devise three algorithms, i.e., BSL algorithm, TDER algorithm and FTDR algorithm. Results of empirical studies show that the three proposed algorithms are capable of answering MTDDR queries efficiently, and FTDR is a good approximation of TDER.

References

1. Sharifzadeh, M., Kolahdouzan, M., et al.: The optimal sequenced route query. VLDB J. **17**(4), 765–787 (2008)
2. Li, F., Cheng, D., Hadjieleftheriou, M., Kollios, G., Teng, S.-H.: On trip planning queries in spatial databases. In: Bauzer Medeiros, C., Egenhofer, M.J., Bertino, E. (eds.) SSTD 2005. LNCS, vol. 3633, pp. 273–290. Springer, Heidelberg (2005). https://doi.org/10.1007/11535331_16
3. Cao, X., Chen, L., Cong, G., Xiao, X.: Keyword-aware optimal route search. PVLDB **5**(11), 1136–1147 (2012)
4. Anwar, A., Hashem, T.: Optimal obstructed sequenced route queries in spatial databases. In: International Conference on Extending Database Technology, EDBT (2017)
5. Sasaki, Y., Ishikawa, Y., Fujiwara, Y., Onizuka, M.: Sequenced route query with semantic hierarchy. In: EDBT, pp. 37–48 (2018)
6. Ohsawa, Y., Htoo, H.: Top-k sequenced route queries. In: IEEE International Conference on Mobile Data Management IEEE, pp. 320–323 (2017)
7. Dai, J., Liu, C., Jiajie, X., Ding, Z.: On personalized and sequenced route planning. World Wide Web **19**(4), 679–705 (2016)
8. Li, Y., Yang, W., Dan, W., Xie, Z.: Keyword-aware dominant route search for various user preferences. In: Renz, M., Shahabi, C., Zhou, X., Cheema, M.A. (eds.) DASFAA 2015. LNCS, vol. 9050, pp. 207–222. Springer, Cham (2015). https://doi.org/10.1007/978-3-319-18123-3_13
9. Orda, A., Rom, R.: Shortest-path and minimum-delay algorithms in networks with time-dependent edge-length. J. ACM **37**(3), 607–625 (1990)
10. Costa, C.F., et al.: Optimal time-dependent sequenced route queries in road networks. In: Sigspatial International Conference on Advances in Geographic Information Systems. ACM, pp. 1–4 (2015)
11. Kriegel, H.P., Renz, M., Schubert, M.: Route skyline queries: a multi-preference path planning approach. In: IEEE International Conference on Data Engineering IEEE, pp. 261–272 (2010)
12. Sharifzadeh, M., Shahabi, C.: The spatial skyline queries. In: International Conference on Very Large Data Bases, Seoul, Korea. VLDB, pp. 751–762 (2006)
13. Awasthi, A., et al.: K-dominant skyline join queries: extending the join paradigm to k-dominant skylines. In: 2017 IEEE International Conference on Data Engineering. IEEE, pp. 99–102 (2017)
14. Ahmadi, M.H., Haghighatdoost, V.: General time-dependent sequenced route queries in road networks. In: DASC/PiCom, pp. 949–956 (2017)
15. Li, L., et al.: Minimal on-road time route scheduling on time-dependent graphs. Proc. VLDB Endow. **10**(11), 1274–1285 (2017)
16. Cruz, L.A., Nascimento, M.A., et al.: K-nearest neighbors queries in time-dependent road networks. J. Inf. Data Manag. **3**(3), 221–226 (2012)

Multi-class Weather Classification Fusing Weather Dataset and Image Features

Shan Wang, Yidong Li[✉], and Wenhua Liu

Beijing Jiaotong University, Beijing 110000, China
ydli@bjtu.edu.cn

Abstract. Weather classification is getting more and more attractive because it has many potential applications, such as visual systems and intelligent transportation, especially in transportation. However, the researches about it were based weather dataset are designed to be executed in clear weather conditions. And it is challenging tasks due to the diversity of weather cues and lack of discriminate feature. Therefore, in addition to the image features, real-time weather factors is needed to provide additional information. Most existing weather classification methods involve many pre-processing techniques, for instance sky detection and boundary detection, which makes the model highly rely on the performance of the aforementioned steps. Therefore, in this paper, we propose a novel method for multi-class weather classification. We newly constructed multi-class weather dataset contains 5K images which has more adverse weather conditions. To improve the discrimination of image representation and decrease the influence of pre-processing, our approach combines the real-time weather data with the image feature as the final feature vector to identify different weather. With experiments on our own dataset, we demonstrate that the proposed framework achieves superior performance compared to the state-of-the-art methods.

Keywords: Multi-class weather classification · Dark channel
Real-time weather dataset

1 Introduction

The weather conditions affects our daily lives in many ways, from outdoor sporting events to the clothes we wear. What's more, it not only strongly influences us in our daily lives and outdoor sporting events, but also affects the functionality of many visual systems including outdoor video surveillance and vehicle assistant driving systems. In different weather, the visibility is different. Especially in rainy days and hazy days, visibility will be greatly reduced, which will have a great impact on the picture and video clarity. In these conditions, it will be difficult to lock the target, which greatly affected the progress of law enforcement agencies. In addition, if we are driving on the road, it is important that vehicle assistant driving systems can give drivers some suggestions according to the real-time weather situation. Such as appropriate speed, distance and so on, otherwise it is easy to cause safety problems such as traffic

© Springer Nature Singapore Pte Ltd. 2018
Z. Xu et al. (Eds.): Big Data 2018, CCIS 945, pp. 149–159, 2018.
https://doi.org/10.1007/978-981-13-2922-7_10

accidents. Therefore, there is no doubt that weather classification plays an important role in many visual systems and weather systems.

The approaches of typical image or scene classification rely on structural information for categorizing the scene into different classes, such as [1, 2]. The structural information is based on illumination-invariant features such as SIFT or HOG. Compared to the typical image classification task, weather classification from images is affected by various factors, for instance illumination and reflection which are more complicated and not scene specific, making conventional scene classification methods inapplicable. Outdoor images that shot at different times of the day under different weather conditions may be different. What's more, many researches regards the rain streaks and haze as noise, and try to remove it [3, 4]. Most of the existing methods in the field of computer vision are based on the assumption that the weather in outdoor images is clear. However, different weather conditions, for example rain or haze will decrease the quality of images or videos. Such effects may significantly decrease the performances of outdoor vision systems which relies on image or video feature extraction or visual attention modeling. Hence, the applications of weather classification are getting more and more attention in the observation of weather conditions and the reliability improvement of video surveillance systems.

Multi-class weather classification has not been thoroughly studied. Previous researches on weather recognition just analyzed weather images or weather dataset provided by the Weather Bureau e.g. [5, 6]. Recently, the authors of [7, 8] proposed a collaborative learning framework via analyzing multiple weather factors for weather recognition. However, this method involves many pre-processing techniques, for instance sky detection and boundary detection, which makes the model highly rely on the performance of the aforementioned steps. Dark channel prior has been well studied in single image haze removal [3] and from [7, 8] we can see that dark channel can be well used to estimate the thickness of the haze.

The rest of the paper is organized as follows. Section 2 gives a brief introduction about the related work of weather classification. Section 3 describes the proposed approach, including our own dataset and the features of experimental datasets. Section 4 represents the experimental results and provides some discussion. Section 5 draws the conclusions.

2   Related Work

A large body of research has been focusing on weather classification. At first, most of it just analyzed weather dataset provided by the Weather Bureau, e.g. [5, 6]. Recently, many works have been proposed to investigate the weather classification from image [9–11]. In this section, we give a brief review about the related work of weather classification.

Most of the existing methods in the weather classification has been focusing on weather recognition, including sunny, cloudy and proposed a collaborative learning framework via analyzing multiple weather cues, for instance [7]. The authors of [12] proposed a method to label images of the similar scene with three weather conditions including sunny, cloudy, and overcast. However, it just analyzed only 1K weather images for experiment. In addition, there are many proposed approaches for multi-class weather classification just can be used for fixed scene, for example the authors of [13–15] proposed an approach for multi-class weather classification, which could be used for the traffic scene only. As far as we know, only very few works have been proposed to investigate multi-class weather classification which used handmade features.

In general, most of the aforementioned methods can be divided into three basic steps. The first step is to extract the specific Regions of Interest (ROIs) in a weather image, including sky region [16], shadow region [17] and so on, then use several illumination-invariant features such as SIFT [18], GIST or HOG to represent the different regions and finally to solve classification problem by some classifiers, such as Support Vector Machine (SVM) and Adaboost. However these approaches involves many pre-processing techniques which makes the model highly relies on the performance of the aforementioned techniques.

Recently, deep learning is used in weather classification as well. [19] considered two-class weather classification by CNNs. [5] analyzed weather data to predict PM2.5. We can see that the previous works provide many solutions for weather classification, but the performances of these approaches are not well and the steps of these approaches are complicated.

Weather classification based on weather dataset and weather images are often considered independently and most efforts have been made to tackle them separately. Due to weather characteristics change over time in a day. So, it is challenging tasks due to the diversity of weather cues. Therefore, in addition to the image features, real-time weather factors is needed to provide additional information. As for the status of weather classification topic, we have summarized the following two challenges:

(1) Experiment Data. The images of existing weather picture database are clear and discriminate. However, there is no obvious distinction between actual images, as shown in Fig. 1.
(2) Weather features. For each type of weather has to learn a variety of features, which makes the model highly rely on the performance of the aforementioned steps.

Fig. 1. The actual weather pictures. The actual weather pictures has blurred boundaries, there is no obvious distinction between images.

Therefore, the main contributions made in this paper can be summarized as follows:

(1) As for experiment data. We collect an outdoor image set which has more adverse weather conditions. It provides an extensive tested for the evaluation of our approach. In addition, we recorded the real-time weather factors as well. Both of them are detailed in Sect. 3.
(2) In order to decrease the influence of pre-processing and complicated of afore-mentioned methods, our approach combines the real-time dataset which can provide additional information with the image features as the final feature vector to identify different weather conditions.

3 Our Approach

3.1 Dataset and Features

There are a large body of weather dataset, such as Flicker, Picasa, Fengniao. But all the pictures of these dataset are clear and we cannot obtain the related weather factors from it. Therefore, in order to achieve the actual situation of weather classification, we collect an outdoor image set under unlimited shooting conditions and evaluate our approach on our own dataset. The weather images includes 4 weather categories, i.e., sunny, rainy, hazy and snowy. In addition, we recorded the real-time weather factors as well, e.g. temperature and humidity. Then, we label all the images according to MojiWeather. If the information provided by the MojiWeather does not match the actual situation, we will subject to the actual situation. We randomly selected 4K images from the dataset to evaluate our method. As shown in Fig. 2, most of the images have totally different backgrounds with different weather condition. The distribution statistics on our experiment dataset is listed in Table 1.

Table 1. The distribution statistics on our experiment dataset.

Label	Sunny	Rainy	Hazy	Snowy
Number	2858	1142	1586	750

In this paper, the weather factors selected to train a model are as follows:

- Wind Speed (WS)
- Temperature (TEMP)
- Humidity (HUM)
- Time(TM)
- Dark Channel (DC)

First three of these (WS, TEMP, HUM, TM, DC) were provided by the Moji-Weather in real time, as shown in Table 2. From [7, 8] we can see that dark channel can be well used to estimate the thickness of the haze. So in addition to our existing structural data, we combine it with the dark channel value which are detailed in Sect. 3.2 as the final feature vector to identify different weather.

From [7, 8] we can see that dark channel can be well used to estimate the thickness of the haze. The humidity of different weathers is different, so the clarity of different pictures is also different. In summary, we decided to use dark channel values for image features. In weather data, we select the wind speed, temperature and humidity. Because in different weather, the temperature and humidity gaps are large. For example, the temperature of sunny days is higher than the temperature of snow days. The humidity of the air during rainy days and snow days is greater than the humidity of other weather conditions.

Fig. 2. Weather images in experiment. The images are collected by us. (a) Sunny images. (b) Rainy images. (c) Haze images. (d) Snowy images.

Table 2. Features and weather condition provided by the MojiWeather.

ID	TEMP	HUM	WS	TM	Weather Condition
1	13	15%	2	2017/02/28 11:21:00	Sunny/normal
2	14	79%	3	2017/03/21 12:07:55	Rainy/hazy
3	5	80%	1	018/03/17 10:28:19	Snowy/hazy
...	...	...	...	...	...
4000	8	80%	2	018/03/17 17:43:00	Snowy/hazy

3.2   Proposed Algorithm

The details of weather classification is described in Algorithm 1. The main idea is to calculate the dark channel value of each image, and then to combine it with the features provided by MojiWeather. Different weather conditions have different thickness of the haze. Therefore, using dark channel prior with the haze imaging model which proposed in [3], we can directly estimate the thickness of the haze and identify different weather conditions.

Step (3) calculate dark channel value matrix M according to paper [3]. J^{dark} is the dark channel of J. Different from the [7] and [8] divided each image into 80 and 100 patches respectively and take the median of dark channel value of each patch. To improve the discrimination of image representation, we divided each image into 1000 patches and take the average of dark channel value of each patch. Normally, the sky is at the top of the images and buildings are at the bottom. In order to better obtain the features of these two parts, for M we make the size of the row smaller than the size of the column. Therefore, we divide M into 10×100 non-overlapping patches. Then we calculate the average value of each patch to form a 1000 dimensional feature vector to indicate hazy, as described in step (5) and (6). Finally, to cope with multi-class weather classification task using AdaBoost. The architecture of the proposed method is shown in Fig. 3.

Algorithm 1 Pseudocode of the weather classification algorithm

---

Input: weather image file I and features provided by the weather dataset D

Output: the accuracy P of classification

(1) for each image J

(2) resize J into 300×500

(3) calculate dark channel value matrix M

$$J^{dark}(x) = \min_{c \in \{r,g,b\}} \left(\min_{y \in \Omega(x)} (J^c(y)) \right)$$

where J^c is a color channel of J and $\Omega(x)$ is a local patch centered at x.

(4) set 30×5 as the cell size.

(5) divide M into 10×100 non-overlapping patches

(6) calculate average value of each patches to form the 1000-D vector DC

(7) end for

(8) read D, and combine it with DC as the final feature vector F

(9) solve classification problems by AdaBoost

(10) calculate P

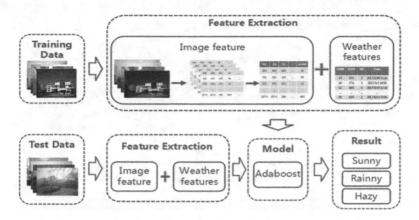

Fig. 3. The architecture of the proposed method

4 Experiment

In this section, we conduct experiments to evaluate the effectiveness of our method. After introducing experimental settings, we compare our method with the state-of-the-art methods of weather classification on our own weather datasets.

4.1 Experiment Setting

In our experiment, we randomly selected 4K images from the dataset to evaluate our method. And each image has four labels in experiment. Finally, we use AdaBoost to cope with multi-class weather classification task.

Because the proposed method uses real-time recording of relevant weather factors, and the existing weather images standard database does not have corresponding real-time weather factors. Therefore, our experiments only selected some related image classification methods which have better performance in the previous experiment to complete comparative tests on our data set.

4.2 The Result of Classification

To evaluate the features of our approach, we use different feature extraction methods, e.g. HSV, LBP. Figure 4 shows the classification accuracy of different features. For each feature, we calculate the classification accuracy under different proportion of training set. And from it, we can see that the best accuracy of our features is better than other method. Especially at 80%, so our follow-up experiment used 80% of the dataset as training set.

Table 3 shows the accuracy of each weather condition classification via our method. We list some successful cases and some failure cases in Fig. 5. As for haze weather, the classification accuracy only about 75%. In order to validate our method, we choose the method of other papers which have better performance in the previous experiment to compare the haze weather classification performance. Table 4 shows the

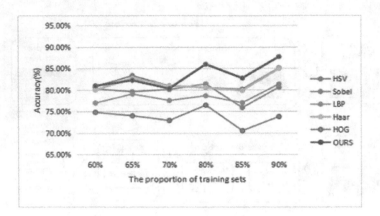

Fig. 4. Classification accuracy with different features. Classification accuracy under the different proportion of training sets with different features.

classification results of related methods on our dataset. We can see that all the haze weather classification methods are performs poorly on our weather dataset. It shows that the haze classification method in this paper has improved compared with the previous papers, especially with the real-time recorded weather factors.

Table 3. The accuracy of each weather classification.

Class	Sunny	Rainy	Hazy	Snowy
Accuracy (%)	88.38%	87.50%	74.25%	82.24%

Table 4. Haze weather classification results of related methods on our dataset

Method	[7]	[8]	[12]	DC	Ours
Accuracy (%)	55.50%	56.50%	63.5%	69.13%	74.25%

As for weather classification, we also compared our method with some related image classification methods as shown in Table 5. We choose the method of [12], SIFT and HOG based on traditional scene classification feature classification methods. A method based on collaborative learning framework for single natural images to two-class weather classification in [7]. And method in [8] which extended the similar framework four-class weather classification. They designed features for sunny, rainy, snowy and haze weather. Figure 6 illustrates the classification performance on each weather category by related methods. The result demonstrate that the proposed method achieves superior performance compared to the state-of-the-art methods.

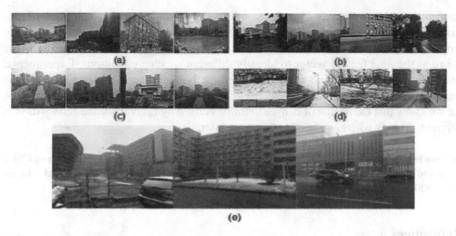

Fig. 5. Classification results. (a) Sunny images. (b) Rainy images. (c) Haze images. (d) Snowy images. (e) Three snowy images mis-detected as rainy images.

Table 5. Weather classification results of related methods on our dataset

Method	[7]	[8]	[12]	DC	Ours
Accuracy (%)	26.91%	76.42%	74.51%	79.65%	85.94%

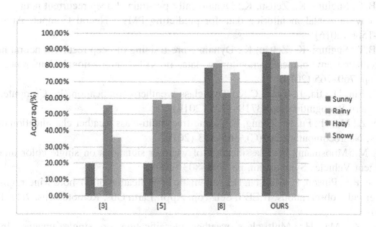

Fig. 6. Classification accuracy on weather category via related methods

5 Conclusion

Most of vision based weather dataset are designed to be executed in clear weather conditions. However, limited visibility in rain or cloudy strongly affects the accuracy of vision systems. In this paper, we newly constructed a multi-class weather dataset contains 4 labels which has more adverse weather conditions. In order to capture a discriminate feature for every weather condition and avoid involving complicated pre-processing techniques. Our approach combines the real-time weather dataset which can provide additional information with the dark channel value which is calculated from the image as the final feature vector to identify different weather conditions. Experimental results show the effectiveness of the proposed method and the importance of real-time weather factors. Our method mis-detected snowy days with low visibility as rainy days, but we know that the degree of danger to traffic caused by rainy days and snowy days is different. This is where we need to improve later.

Acknowledgment. This work is supported in part by the National Science Foundation of China Grant (#61672088 and #61370070), and Fundamental Research Funds for the Central Universities (#2016JBM022 and #2015ZBJ007).

References

1. Li, L.J., Su, H., Lim, Y., Li, F.F.: Object bank: an object-level image representation for high-level visual recognition. Int. J. Comput. Vis. **107**(1), 20–39 (2014)
2. Zou, J., Li, W., Chen, C., Du, Q.: Scene classification using local and global features with collaborative representation fusion. Inf. Sci. **348**, 209–226 (2016)
3. He, K., Sun, J., Tang, X.: Single image haze removal using dark channel prior. IEEE Trans. Pattern Anal. Mach. Intell. **33**(12), 2341–2353 (2011)
4. Chen, D.Y., Chen, C.C., Kang, L.W.: Visual depth guided color image rain streaks removal using sparse coding. IEEE Trans. Circ. Syst. Video Technol. **24**(24), 1430–1455 (2014)
5. Ong, B.T., Sugiura, K., Zettsu, K.: Dynamically pre-trained deep recurrent neural networks using environmental monitoring data for predicting $PM_{2.5}$. Neural Comput. Appl. **27**(6), 1553–1566 (2016)
6. Ong, B.T., Sugiura, K., Zettsu, K.: Dynamic pre-training of deep recurrent neural networks for predicting environmental monitoring data. In: IEEE International Conference on Big Data, pp. 760–765 (2014)
7. Lu, C., Lin, D., Jia, J., Tang, C.K.: Two-class weather classification. In: Computer Vision and Pattern Recognition, pp. 3718–3725 (2014)
8. Zhang, Z., Ma, H., Fu, H., Zhang, C.: Scene-free multi-class weather classification on single images. Neurocomputing **207**(C), 365–373 (2016)
9. Roser, M., Moosmann, F.: Classification of weather situations on single color images. In: Intelligent Vehicles Symposium, pp. 798–803 (2008)
10. Trombe, P.J., Pinson, P., Madsen, H.: Automatic classification of offshore wind regimes with weather radar observations. IEEE J. Sel. Top. Appl. Earth Obs. Remote Sens. **7**(1), 116–125 (2014)
11. Zhang, Z., Ma, H.: Multi-class weather classification on single images. In: IEEE International Conference on Image Processing, pp. 4396–4400 (2015)
12. Chen, Z., Yang, F., Lindner, A., Barrenetxea, G., Vetterli, M.: How is the weather: automatic inference from images. In: IEEE International Conference on Image Processing, pp. 1853–1856 (2012)
13. Song, H., Chen, Y., Gao, Y.: Weather condition recognition based on feature extraction and K-NN. In: Sun, F., Hu, D., Liu, H. (eds.) Foundations and Practical Applications of Cognitive Systems and Information Processing. AISC, vol. 215, pp. 199–210. Springer, Heidelberg (2014). https://doi.org/10.1007/978-3-642-37835-5_18

14. Kuehnle, A.: Method and system for weather condition detection with image-based road characterization (2013)
15. Yan, X., Luo, Y., Zheng, X.: Weather recognition based on images captured by vision system in vehicle. In: Yu, W., He, H., Zhang, N. (eds.) ISNN 2009. LNCS, vol. 5553, pp. 390–398. Springer, Heidelberg (2009). https://doi.org/10.1007/978-3-642-01513-7_42
16. Tao, L., Yuan, L., Sun, J.: Skyfinder: attribute-based sky image search. In: ACM SIGGRAPH, p. 68 (2009)
17. Lalonde, J.-F., Efros, A.A., Narasimhan, S.G.: Detecting ground shadows in outdoor consumer photographs. In: Daniilidis, K., Maragos, P., Paragios, N. (eds.) ECCV 2010. LNCS, vol. 6312, pp. 322–335. Springer, Heidelberg (2010). https://doi.org/10.1007/978-3-642-15552-9_24
18. Lowe, D.G.: Distinctive Image Features from Scale-Invariant Keypoints. Kluwer Academic Publishers, Dordrecht (2004)
19. Elhoseiny, M., Huang, S., Elgammal, A.: Weather classification with deep convolutional neural networks. In: IEEE International Conference on Image Processing (2015)

Multiple Meta Paths Combined for Vertex Embedding in Heterogeneous Networks

Tong Wu[1(✉)], Chaofeng Sha[2], and Xiaoling Wang[1]

[1] Shanghai Key Laboratory of Trustworthy Computing,
East China Normal University, Shanghai, China
51164500254@stu.ecnu.edu.cn, xlwang@sei.ecnu.edu.cn
[2] Shanghai Key Laboratory of Intelligent Information Processing, Fudan University,
Shanghai 200433, China
cfsha@fudan.edu.cn

Abstract. In the real-world many complex systems exist in the form of heterogeneous networks. As we all know, heterogeneous networks consist of various types of vertices and relations, so it is difficult to deal directly with data mining. At present, although many state-of-the-art methods of network representation learning have been developed, these methods can only deal with homogeneous networks or lose information when handling heterogeneous networks. In order to compensate for the weakness of the previous methods, we propose a multiple meta paths combined embedding (MMPCE) model to represent the heterogeneous networks. This method can automatically obtain the low-dimensional vector representation of vertices and preserve the rich semantic and structural information in the network. We conduct experiments on two real world datasets. The experimental results demonstrate the efficacy and efficiency of the proposed method in heterogeneous network mining tasks. Compare to the previous method, our model can cover a wider range of semantic information and be more flexible and scalable.

Keywords: Network embedding · Vertices embedding
Heterogeneous representation learning
Heterogeneous information embedding

1 Introduction

Network is a ubiquitous data structure. A large number of real-world applications store and present data in the network. Such as social networks in Facebook and Twitter, paper citation network in DBLP, biological protein-protein networks and so on. In order to exploit the hidden value of the network, we transform the network into a form that can be handled by machine learning. Therefore, network representation learning grows up to be a crucial step.

Network representation learning is a way of obtaining low-dimensional vector representation of vertices in the network. It can effectively represent and manage

© Springer Nature Singapore Pte Ltd. 2018
Z. Xu et al. (Eds.): Big Data 2018, CCIS 945, pp. 160–177, 2018.
https://doi.org/10.1007/978-981-13-2922-7_11

large-scale networks, and provides the support for the downstream data mining work including node classification [1,2], clustering [3], link prediction [4] and information retrieval [5]. In recent years, network representation learning has attracted many research interests and produced many state-of-the-art methods including DeepWalk [11], LINE [13] and Node2vec [12]. These efforts achieve outstanding performance in various network analysis tasks and are highly efficient when dealing with large-scale network. However, they are limited to processing homogeneous networks and cannot represent heterogeneous networks in reality.

Heterogeneous network often contains multiple types of vertices and relationships, as opposed to homogeneous network that only have a singular type of nodes and relationships. For example, for a movie network, vertices can be movie, user, genre, keyword, year and so on. In the heterogeneous network, if all vertices are treated as the same type, many important semantic information will be lost. In addition, there is just one kind of relation between vertices in a homogeneous network, and their semantics are also expressed in the same way. However, in the heterogeneous network, the relationships between vertices are different, and the meaning of each relationship is unique, such as user-movie liking relationships, movie-keyword describing relationships and so on. It can be observed that the structure of the heterogeneous network is more complicated than the homogeneous network and is more in line with the form of the data existing in the real world. Moreover, the heterogeneous network contains more kinds of vertex objects and relation types and covers more information. Therefore, the challenge of completing the representation of heterogeneous network is more realistic and significant.

At present, some representation learning methods for heterogeneous network have been proposed, for instance PTE [14] and ICE [15], but these methods can handle only a limited number of vertices and relationships, and cannot make full use of rich semantic and structural information in the network.

In order to compensate for the weakness of the foregoing methods to represent heterogeneous information networks, we study the structure of the heterogeneous network and propose a multiple meta paths combined embedding (MMPCE) method to get the low-dimensional vector representation of the vertices. More specifically, we formalize the heterogeneous network representation learning problem. The objective of this problem is to map multiple types of vertices into a same low-dimensional latent space, so that the learned vertices embeddings can preserve the structural features and semantics in the original network. In the MMPCE model, we first construct multiple meta paths according to the semantic information between vertices. Then find vertices neighbors on each meta path and get the embeddings of vertices depending on the idea of similar neighbors. Finally, all meta paths are grouped together to jointly train the vector representation of all types of vertices.

We conduct experiments on two real-world datasets and compare the MMPCE model with other methods by two different kinds of task including classification tasks and retrieval tasks. The experimental results show that compared with other baselines, the proposed model can not only significantly improve the accuracy of node classification, but also greatly increase the effective proportion of the similarity search. It demonstrates that MMPCE can learn the vertices embedding by integrating multiple meta paths and take full advantage of the rich semantic and structure information in the heterogeneous network, so that the learned vertices embedding can fully reflect the characteristics and properties of the original network.

In summary, the contributions of this paper are listed as follows:

- We propose a Multiple Meta Path Combined Embedding method, namely MMPCE. The method can automatically obtain the low-dimensional vector representation of various types of vertices and preserve the rich semantic and structural information in the heterogeneous network.
- Through extensive experiments, we demonstrate the efficacy and efficiency of the proposed method in a variety of heterogeneous network mining tasks, such as multi-class classification task (achieving 5–66% improvements over counterparts) and multi-label classification task (achieving gains of 7–66% over baselines).
- Compare to the previous method which can only select a fixed relationship, the method we proposed can cover a wider range of semantic information, and the method is more flexible and scalable.

The rest part of this paper is organized as follows: Sect. 2 summarizes the current related works. Section 3 defines the problem of heterogeneous information network embedding. In Sect. 4, we introduce the proposed MMPCE model in detail. In Sect. 5, we present and analyse the experimental result. Finally we conclude in Sect. 6.

2   Related Work

Network representation learning can map network vertices into low-dimensional vectors by analyzing the structure of a complex network in latent vector space. The vertices embeddings maintain the structural characteristics of the original network. Network representation learning facilitates machine learning techniques for mining existing complex network and is widely used in areas, such as node classification [1,2], clustering [3], link prediction [4] and information retrieval [5].

In the early days of network analysis and graph mining, some well-known works, such as IsoMap [7] and LLE [6], are all based on the spectral decomposition. They first construct the affinity graph [8] based on the feature vectors and then embed the affinity graph into a low dimensional space. However, the computational complexity of this kind of method is so expensive that it is very inefficient to handle large-scale network. With the rapid development of deep learning techniques, the neural network-based representation learning models have been extensively used. Especially in the field of NLP, the word2vec model [9,10] proposed by Mikolov et al. has significantly improved the performance of text data representation. Guided by this model, Perozzi et al. create the Deep-Walk [11], a new way of learning network representation. This method defines the co-occurrence vertices in random walk as the "context" of a vertex, and then puts multiple vertex sequences into the word2vec model and gets the vector representation of vertices. Subsequently, Grover and Leskovec [12] extend Deep-Walk by proposing a biased random walk method with two hyper-parameters, which integrates the homophily and structural equivalence of the network into the vertices representation. Tang et al. proposed LINE [13] based on the idea of similar neighbors. This method preserves the first-order proximity and the second-order proximity between vertices in the network by defining the loss function. Although these methods can deal with large-scale network efficiently, they are restricted to dealing with homogeneity network and present a huge bottle-neck to the representation learning of heterogeneous network. At present, there are some representational learning methods for heterogeneous networks, such as PTE [14] first combines three bipartite graphs, and then uses the second-order proximity between nodes to learn the vector representation, and ICE [15] is constructed by combining entity-text network and text-text network, and then the embedding representation of vertices is obtained according to the idea of neighborhood proximity, they complete the vertex representation by integrating multiple heterogeneous bipartite graphs. However, these methods can only handle a limited number of vertex types and relationships, and cannot make full use of rich semantic and structural information in the network.

In order to better complete the representation of heterogeneous network, we propose the MMPCE model. This method integrates multiple meta paths to capture the rich structural and semantic information in the network, so that the learned vertices embedding can more fully preserve the structural characteristics of the original network. Compared with the previous methods, the MMPCE model can cover more information about network and have a more obvious effect in data mining applications.

3   Problem Definition

In this section, we first introduce the definition of heterogeneous information network and abstract the heterogeneous information network using network schema, and then introduce the concept of meta path and explain the meta path instance and path concatenation. Finally, a formal definition of heterogeneous information network embedding problem is given.

3.1   Heterogeneous Information Network

Definition 1. A heterogeneous information network (HIN) [16] is defined as a directed graph $G = (V, E, O, R)$, where V is the set of vertices, E is the set of edges, O is the set of vertex types and R is the set of relation types. Each vertex v in the graph is associated with a vertex type mapping function $\phi(v) : V \rightarrow O$ and only belongs to one vertex type. Each edge e is associated with a relation type mapping function $\psi(e) : E \rightarrow R$ and only belongs to one relation type. In addition, the HIN requires $|O| + |R| > 2$ (the types of vertices $|O| > 1$ and the types of relations $|R| > 1$).

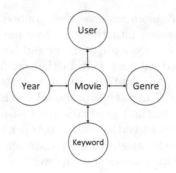

Fig. 1. Schema of a movie network

In the heterogeneous information network, each edge is accompanied with a weight w, which indicates the strength of a particular relation between vertices. The weights of edges can be either binary or any positive real values. Note that in this study we only consider non-negative weights.

Usually, in the face of complex heterogeneous network, we construct the network schema $T_G(O, R)$ to better understand the structural relationship between multiple vertex types. For example, for a movie network, the vertex types O contain five entities of vertex type: $Movie(M)$, $User(U)$, $Genre(G)$, $Keyword(K)$, and $Year(Y)$, the set of relation types R include *comment* between user and movie, *belong to* between movie and genre, *contain* between movie and keyword, and *post* between movie and year. The movie network schema is shown in Fig. 1.

3.2　Meta Path

Definition 2. Meta path [17] is a composite relation path consisting of multiple vertex types and relation types defined on the network schema T_G. Meta path is denoted in the form of $O_1 \xrightarrow{R_1} O_2 \xrightarrow{R_2} \cdots \xrightarrow{R_n} O_{n+1}$, short as $L = (O_1 O_2 \cdots O_{n+1})$. For example, in the movie network, a meta path made up of users who like movies in the same genre can be expressed as: $user \xrightarrow{comment} movie \xrightarrow{belong\ to} genre \xrightarrow{contain} movie \xrightarrow{commented} user$, short as $UMGMU$.

The meta path can distinguish the semantics among paths connecting two verties. For example, the relationship between users may include users who like the same genre and users who have evaluated the same movie. Different meta paths can contain different semantics. In addition, the meta path contains the different neighbors of the verties. The relationships between neighboring verties can reflect the structural characteristics of the verties in the network. So the mate paths can the rich semantic and structural information in the heterogeneous network.

If a specific path $l = (o_1 o_2 \cdots o_{n+1})$ between o_1 and o_{n+1} in heterogeneous network G follows the meta path L, and the order of vertex type and relationship type in the path is same as L. We call this specific path l as a meta path instance of L, i.e. $l \in L$. For example, $Jack \xrightarrow{comment} ToyStory \xrightarrow{belong\ to} Animation$ is the instance of UMG.

For any two meta paths, $L1 = (O_1 O_2 \cdots O_{n+1})$ and $L2 = (O'_1 O'_2 \cdots O'_{m+1})$, they can be concatenated if O_{n+1} and O'_1 are the same type and written as $L = (L1L2) = \left(O_1 O_2 \cdots O_n O'_1 O'_2 \cdots O'_{m+1} \right)$. For example, the path UM and the path MG may be concatenated into UMG, which represents the user favorite movie genres.

3.3   Heterogeneous Information Network Embedding

Different from homogeneous network with only a singular type of vertex and relation, a heterogeneous network contains a variety of vertex and relation types. The main challenge of the problem that comes from heterogeneous information network is that those state-of-the-art methods that perform well in homogenous graphs can not be applied directly to a heterogeneous network. Therefore, the representation learning of the HIN needs to be redefined.

Problem 1. Heterogeneous information network embedding: Given a heterogeneous network G, the problem of heterogeneous information network embedding aims to map multiple type vertex $v \in V$ into the same low-dimensional latent space, and then learn a matrix $U \in \mathbb{R}^{|V| \times d}$ to represent the embedding of all vertices. d is the dimension of low-dimensional latent space. Each row of U represents feature vectors of vertices in the network. The feature vectors are distributed and can preserve the structural features and semantic information of vertices in the original heterogeneous information network.

By embedding heterogeneous information network, the low-dimensional feature vectors of vertices can not only directly measure the similar characteristics of vertices, but also have great benefits to the mining of heterogeneous information network data, such as node classification, clustering, visualization, similarity search, link prediction and so on.

4   Methodology

In this section, we will provide a detailed description of the proposed model. First of all, we introduce the adjacency matrix representation of one-hop meta path and k-hop meta path. And then we present the design process of multiple meta path combined embedding (MMPCE) in detail and the optimization of the objective function. The goal of MMPCE is to maximize the probability of occurrence of heterogeneous network by taking into account various types of vertices and relationships on the meta paths.

Table 1. Adjacency matrix M^L of meta path $L(UM)$

	Jack	Adam	Bill	David
Swan princess	2	0	0	0
Toy story	3	0	0	3
Heat	0	1	1	0
Guardian angle	0	1	1	0
Dragonheart	0	0	2	3

4.1 Matrix Representation of Meta Path

In the original heterogenous network, we can directly extract one-hop meta path. Such as $movie - user$, $movie - genre$, $movie - year$ and so on. The k-hop meta path can be obtained by concatenating multiple one-hop meta paths. For a meta path usually contains multiple path instances, different path instance contains different vertices in the heterogeneous network. In order to better demonstrate the connectivity between vertices in the meta path, we use the adjacency matrix to represent the meta path.

Table 1 shows the adjacency matrix M^L of one-hop meta path $L(UM)$, which demonstrates the connection between vertex type $User$ and $Movie$ in the network. Any element value m_{ij} of M^L indicates the number of occurrences of the path instance l_{ij}. for instance, $m_{(Jack, Toy Story)} = 3$ means $Jack$ viewed $Toy Story$ three times or rated it three times. Here, we take the number of occurrences of the path instance as the weight value of this path instance. If no path instance is observed between path endpoint vertices in L, set the path instance weight value to zero.

The matrix representation of one-hop meta path is very simple. But in practice meta path is usually k-hop meta path with more semantic information. In order to obtain the k-hop meta path, we can use the path concatenating method to make one-hop meta path connect end to end according to the order of vertices in the given meta path. The corresponding adjacency matrix can be calculated by matrix multiplication between one-hop adjacency matrices. For example, given a meta path $L = (O_1O_2O_3 \cdots O_n)$, its corresponding adjacency matrix $M_L = M_{O_1O_2} \times M_{O_2O_3} \times \cdots \times M_{O_{n-1}O_n}$.

4.2 Multiple Meta Path Combined Embedding

In this subsection, we will implement vertices embedding using the proposed MMPCE model to explore the rich structural and semantic information in heterogeneous network.

Most of the existing state-of-the-art techniques [11–13] firstly define the neighbors of vertices, and then encode vertices into low-dimensional latent space using the co-occurrence or similarity between vertices and neighbors. In Deep-Walk and Node2vec, they obtain many sequences of vertices by the random walk.

Neighbors refer to those vertices that are located in the same sliding window as the given vertex in a vertex sequence. Their core idea is that if vertices tend to co-occur on short random walk vertex sequences. Those vertices have more similar embeddings. In LINE, neighbors are defined as those vertices that are directly connected to a given vertex in the original network (i.e. one-hop neighbor), as well as those vertices within two-hop distance of a given vertex. By adopting the idea of similar neighbors, vertices with similar neighbor structures are more similar, so as to obtain the vertices embeddings.

Similarly, our model also adopts the idea of similar neighbors, and the difference comparing with the existing method is that the definition of neighbors has changed. In MMPCE, the given vertex type is treated as the initial vertex of meta path, the vertex type at the end of meta path is used as the neighbor of a given vertex type. For example, given vertex type $User$, in meta path UMG, $Genre$ is the neighbor vertex of $User$.

In heterogeneous network, for a given vertex type V_A, there are multiple meta paths starting from V_A. Suppose there is a meta path L, where the terminal vertex type is V_B, i.e. V_B is the neighbor of V_A. The conditional probability of connecting to vertex $v_j \in V_B$ through meta path L from vertex $v_i \in V_A$ is:

$$p\left(v_j|v_i;L\right) = \frac{exp\left(\boldsymbol{u_j}^\top \cdot \boldsymbol{u_i}\right)}{\sum\limits_{s\in T(L)} exp\left(\boldsymbol{u_s}^\top \cdot \boldsymbol{u_i}\right)} \tag{1}$$

where $\boldsymbol{u_i} \in \mathbb{R}^d$ is the embedding vector of vertex v_i, $\boldsymbol{u_j} \in \mathbb{R}^d$ is the embedding vector of vertex v_j, $T\left(L\right)$ is the set of vertices of meta path terminal. Its empirical probability can be defined as:

$$\hat{p}\left(v_j|v_i;L\right) = \frac{M_{ij}^L}{W_i^L} \tag{2}$$

where M_{ij}^L is the weight of path instance l_{ij}, $W_i^L = \sum\limits_{s\in T(L)} M_{is}^L$, is the sum of path instance weight containing vertex v_i. For each vertex v_i in V_A, Eq. (1) actually defines a conditional distribution $p(\cdot|v_i;L)$ over V_B. To preserve the second-order proximity [13], we let the conditional distribution $p(\cdot|v_i;L)$ be as close as possible to its empirical distribution $\hat{p}(\cdot|v_i;L)$, which can be achieved by minimizing the following objective function:

$$O = \sum_{i\in S(L)} \lambda_i D_{KL}\left(\hat{p}\left(\cdot|v_i;L\right), p\left(\cdot|v_i;L\right)\right) \tag{3}$$

where $D_{KL}\left(\cdot,\cdot\right)$ denotes the distance between two distributes, and here we use the Kullback-Leibler divergence to measure the distribution distance. $S\left(L\right)$ is the set of meta path starting vertices. λ_i represents the importance of vertex v_i in meta path L, which can be represented by the weight of the path instance containing v_i, i.e., $\lambda_i = W_i^L$.

Equation (3) is only the objective function of one meta path. In order for all vertices to preserve rich semantic and structural information in the heterogeneous network, we jointly train the objective functions of all paths in the meta path set $\Gamma(L)$. By neglecting some constants, the objective function becomes:

$$O = -\sum_{L \in \Gamma(L)} \left(\sum_{l_{ij} \in L} M_{ij}^L \log p\left(v_j | v_i; L\right) \right) \tag{4}$$

In the objective function (4), the conditional probability $p\left(v_j | v_i; L\right)$ is actually calculated by softmax function, which requires the summation over the entire set $T(L)$ on meta path L. In fact, in heterogeneous network, the number of path instances on meta path may be very large, and the number of vertices in $T(L)$ is also very massive, which can result in softmax computationally expensive.

In order to deal with this problem, we adopt negative sampling [9] to reduce the computational overhead of softmax by sampling negative examples of path instances according to the noise distribution of each meta path $P_n^L(v)$. Let $\log p\left(v_j | v_i; L\right)$ be approximately equal to:

$$\log \sigma\left(\boldsymbol{u_j}^{\top} \cdot \boldsymbol{u_i}\right) + \sum_{k=1}^{K} \mathbb{E}_{v_k \sim P_n^L(v)} \left[\log \sigma\left(-\boldsymbol{u_{v_k}}^{\top} \cdot \boldsymbol{u_i}\right)\right] \tag{5}$$

where $\sigma(x) = \frac{1}{1+exp(-x)}$ is the sigmoid function, K is the number of negative samples, P_n^L is the noise distribution of L, set $P_n^L(v) \propto W_v^{L\frac{3}{4}}$, W_v^L the sum of path instance weight containing v in meta path L.

For the purpose of mapping each vertex contained in $\Gamma(L)$ into a low-dimensional vector space, we need to minimize the objective function (4). Here we adopt stochastic gradient descent algorithm (SGD) with the techniques of edge sampling [13] and negative sampling to perform the objective function optimization.

In our model, meta path set $\Gamma(L)$ includes multiple types of meta paths and vertices. Even if the vertex types at both ends of meta path are the same, the semantic information and the number of path instances are different, e.g., UMU and $UMGMU$. Therefore, in negative sampling, we need to construct a negative sampling table for each vertex type at the end of the meta path.

In the process of objective function optimization, the edge sampling is first performed, we treat all path instances of $\Gamma(L)$ as edges, and the edges are sampled according to the weights of the edges by using the alias table method [18]. Then we select the corresponding negative sampling table according to the type of meta path to which the sampled path instance belongs. Finally, negative path instances sampled from the negative sampling table participate in gradient computation and update model parameters. We summarize the detailed training algorithm in Algorithm 1.

Algorithm 1. Training Algorithm for MMPCE

Input: meta path set $\Gamma(L)$, number of samples T, number of negative samples K
Output: matrix of vertices embeddings $U \in \mathbb{R}^{|V| \times d}$
  1: Initialize negative sampling tables according to $\Gamma(L)$
  2: Random initialization matrix of vertex embeddings U'
  3: **repeat**
  4:    Sample a meta path instance p from Alias Table
  5:    Draw K negative path instances from $NegativeSample(p)$
  6:    Update parameters U' according to graditents
  7: **until** converge
  8: Obtain the vertex embeddings $U = U'$

Table 2. Statistics of the two real-world datasets

	DBLP	MovieLens
Papers/movies	184057	4058
Authors/users	111108	114994
Keywords	68924	15502
Venues/genres	22	19

5 Experiment

In this section, we measure the performance of the MMPCE model on two real-world datasets. We have created two different types of tasks: classification tasks and retrieval tasks. In both of these tasks, we compare the proposed model with other approaches and verify the validity of vertices embeddings learned from our model. Then we analyze and assess the experimental results. Finally, we also demonstrate the effectiveness of our model in terms of similarity search through a case study.

5.1 Dataset

In the experiments, two real-world datasets are used to assess the proposed model, including one paper reference dataset and one movie dataset. Both datasets are publicly available.

The paper reference dataset is extracted from the DBLP dataset (denoted as DBLP) [19]. We picked 180,457 papers from 22 venues, including 111,108 authors and 68,924 keywords. The keywords are drawn from the titles and abstracts. We used this dataset to construct a heterogeneous information network about papers, including four types of nodes: papers, authors, venues and keywords.

This movie data set is pulled from the MovieLens 20M dataset (denoted as MovieLens) [20]. We selected 4,058 movies from 19 genres and rated by 114,994 users. These movies contain a total of 15,502 keywords, which are extracted from the movie description. We have also constructed this dataset into a heterogeneous information network about movies. Table 2 shows in the detailed statistical results of these two datasets.

5.2 Experimental Setup

In the experiment, two different types of tasks were used to assess the performance of the MMPCE model, including classification task and retrieval task. The purpose of setting these two tasks is to demonstrate the efficacy and efficiency of our model for node representation in heterogeneous information network. Next, we will detail the setting of these two tasks separately.

Settings for the Classification Task. We built two different classification tasks on each of the two datasets. We constructs a multi-class classification task on DBLP and a multi-label classification task on MovieLens. The input feature vectors of classification model are respectively generated by four representation learning methods:

DeepWalk [11]: A typical NLP model. This method first obtains the vertices sequences through a short uniform random walk on the network, then puts multiple vertices of sequences into the word2vec model and gets the low-dimensional vector representation of vertices.
LINE [12]: This method preserves the first-order proximity and the second-order proximity between vertices in the network by defining the loss function, and then concatenates the first-order proximity vectors and the second-order proximity vectors to obtain the final node vector representation.
PTE [14]: The method first combines three bipartite graphs (word-word, word-document, word-label), and then uses the second-order proximity between nodes to learn the vector representation of the vertices by using the LINE model.
ICE [15]: The ICE network is constructed by combining entity-text network and text-text network, and then the embedding representation of vertices is obtained according to the idea of neighborhood proximity.

During the experiment, we randomly sample some labeled vertices as the training set T(R) and the remaining nodes for testing. This process is replicated 10 times. The size of T(R) is increased from 5% to 90%. Macro-F1 and Micro-F1 are invoked as evaluation indicators to present node classification results.

Settings for the Retrieval Task. For the retrieval tasks, two different kinds of retrieval tasks are established in the MovieLens, including one *word-to-movie* retrieval task and one *movie-to-movie* retrieval task. Since we construct retrieval tasks on heterogeneous network, the comparison methods we select are RAND, ICE, PTE and MMPCE. Deepwalk and LINE are not involved in experimental comparison here because they can only be used for homogeneous network. RAND is the baseline method, in each query task, randomly select the specified number of movies as a search result. Among three embedding methods (i.e., ICE, PTE and MMPCE), we choose the movies whose embeddings are the most similar to the specified keyword or movie in terms of the cosine similarity.

During the experiment, we select the keywords and movies according to the genres to perform the retrieval operation, and then retrieved movie results are divided according to the genres. In the *word-to-movie* retrieval task, we extracted the top 5 tf-idf words from the movie descriptions contained in each genre as query keywords. In the *movie-to-movie* retrieval task, we randomly select six movies from each genre as the movie to get searched. For the retrieval results, we adopt Precision@K as a measure to assess the performance of these methods.

Table 3. Multi-class classification result in DBLP data (%)

Metric	Method	5%	10%	20%	30%	40%	50%	60%	70%	80%	90%
Macro-F1	DeepWalk	62.54	65.76	67.64	68.40	68.86	69.06	69.52	69.30	69.51	69.40
	LINE	52.73	54.03	55.30	55.82	55.89	55.96	56.24	56.29	56.60	56.65
	ICE	47.93	50.68	52.36	52.94	53.39	53.30	53.56	53.79	53.80	53.57
	PTE	78.72	80.97	82.34	83.16	83.42	83.60	83.67	83.84	83.73	83.87
	MMPCE	**84.93**	**86.33**	**87.15**	**87.52**	**87.73**	**87.81**	**87.74**	**87.91**	**87.92**	**88.00**
Micro-F1	DeepWalk	63.86	67.02	68.88	69.51	70.01	70.19	70.69	70.42	70.61	70.56
	LINE	53.48	54.91	56.19	56.84	56.96	57.02	57.36	57.40	57.75	57.70
	ICE	49.11	52.21	53.99	54.64	55.09	55.04	55.27	55.50	55.53	55.23
	PTE	79.18	81.36	82.74	83.54	83.84	84.02	84.13	84.28	84.15	84.33
	MMPCE	**85.40**	**86.73**	**87.49**	**87.82**	**87.99**	**88.09**	**88.00**	**88.19**	**88.22**	**88.24**

Parameter Settings. For multi-class classification tasks, we adopt LIBLinear toolkit to choose L2 regularized logistic regression as a classifier to classify papers. For multi-label classification tasks, we use OneVsRestClassifier in sklearn and select linear SVM to finish multi-label classification of movies. For the network embedding approaches, on both datasets, the embedding dimension of all vertices is set to 128, the number of negative samples K is set as 5, the total number of samples $T = 1$ billion, and the learning rate is set with the starting value 0.025. For the Deepwalk, we set the window size to 5, the walk length to 20, and the walks per vertex to 20.

5.3   Classification Task Quantitative Results

Performance of the Multi-class Classification Task. In the multi-class classification experiment, since Deepwalk and LINE are only applicable to homogenous network, the input data for both methods are a paper reference network that contains only paper vertices. For PTE method, we construct three bipartite heterogeneous networks (paper-paper, paper-author, paper-keyword) as input to the model. For the proposed MMPCE model, we select five meta paths (P-P, P-A, P-K, P-K-P, P-A-P) as input data. In the DBLP dataset, since each paper belongs to only one venue, we utilise the venue as a classification label to accomplish the multi-class classification tasks for the paper.

Table 3 lists the classification results for all the papers in 20 venues. The results in comparison with the five baselines, the accuracy of the proposed MMPCE model consistently and significantly outperform the other methods in the training set from 5% to 90% in terms of both metrics. It demonstrates that our method can take full advantage of the rich semantic information and structural information in the heterogeneous network. The learned vertices vector representation can be better generalized to the classification tasks than other methods. As the scale of the training set gradually increases, MMPCE achieves an averagely 5–66% improvement in term of Macro-F1 and 5–62% gains in terms of Micro-F1 over DeepWalk, LINE, ICE and PTE.

Table 4. Multi-label classification result in MovieLens data (%)

Metric	Method	5%	10%	20%	30%	40%	50%	60%	70%	80%	90%
Macro-F1	DeepWalk	62.99	65.35	65.84	65.89	65.82	65.82	65.67	65.59	65.99	65.47
	LINE	46.70	46.70	46.76	46.75	46.76	46.82	46.84	46.88	46.84	46.71
	ICE	60.29	61.54	63.48	64.83	65.60	66.47	66.68	66.95	67.47	67.84
	PTE	65.83	68.00	70.49	72.32	73.53	74.49	74.85	75.08	75.40	75.44
	MMPCE	**70.37**	**73.61**	**77.00**	**78.45**	**79.20**	**79.70**	**80.09**	**80.36**	**80.43**	**80.99**
Micro-F1	DeepWalk	85.50	87.46	89.41	89.82	89.99	90.09	90.10	90.03	90.10	89.96
	LINE	87.81	87.80	87.83	87.82	87.79	87.78	87.89	87.98	88.00	88.10
	ICE	87.04	86.92	87.99	88.87	89.40	89.88	90.13	90.38	90.55	90.89
	PTE	86.38	86.51	88.20	89.41	90.20	90.77	91.04	91.28	91.54	91.57
	MMPCE	**88.00**	**89.61**	**91.20**	**91.94**	**92.29**	**92.52**	**92.74**	**92.96**	**93.14**	**93.48**

Performance of the Multi-label Classification Task. In the MovieLens dataset, we get to make a connection between the movies with the same tags, and construct a network of relationships between movies as input to Deepwalk and LINE. For PTE, we constructed three bipartite graphs of movie-movie, movie-user and movie-keyword as input to this model. In addition, we select five meta paths (M-M, M-U, M-K, M-M-U and M-M-K) as the input of the MMPCE model to learn the vector representation of the movie. In the movie dataset, each movie usually belongs to one or more genres, so the genre is invoked as a movie category label to achieve multi-label classification tasks.

Table 4 lists the classification results for all movies in 19 genres. Due to the unbalanced distribution of movies in each genre, there is a definite gap in the accuracy of classification between Macro-F1 and Micro-F1. For example, there are greater than 1,400 movies in both the Drama and Comedy genres in the dataset, while the Western and IMAX genres include less than 100 movies. From the experimental results, the proposed MMPCE model has a higher classification accuracy than other baseline methods in each training set proportion, which proves that it is helpful for us to integrate multiple meta paths for joint training to learn vertices embedding. In a gradual increase in training set, the constant

Table 5. Word-to-movie retrieval task

	Method	Drama	Romance	Action	Crime	Horror	Animation	Average
P@50	Rand	0.408	0.176	0.212	0.108	0.132	0.060	0.183
	ICE	0.428	**0.328**	**0.444**	**0.220**	0.496	0.708	0.437
	MMPCE	**0.444**	0.292	0.384	0.200	**0.556**	**0.824**	**0.450**
P@100	Rand	0.374	0.162	0.256	0.132	0.126	0.066	0.186
	ICE	0.420	**0.342**	**0.414**	**0.190**	0.512	0.586	0.411
	MMPCE	**0.468**	0.324	0.398	0.180	**0.672**	**0.850**	**0.482**

Table 6. Movie-to-movie retrieval task

	Method	Drama	Romance	Action	Crime	Horror	Animation	Average
P@50	Rand	0.397	0.170	0.227	0.120	0.130	0.063	0.184
	ICE	0.500	0.257	0.287	0.080	0.323	0.163	0.268
	PTE	0.212	0.393	0.390	0.143	0.657	0.353	0.358
	MMPCE	**0.744**	**0.490**	**0.863**	**0.467**	**0.727**	**0.627**	**0.653**
P@100	Rand	0.400	0.157	0.207	0.117	0.123	0.060	0.177
	ICE	0.534	0.238	0.275	0.100	0.251	0.120	0.253
	PTE	0.245	0.343	0.326	0.147	**0.610**	0.243	0.319
	MMPCE	**0.694**	**0.380**	**0.698**	**0.330**	0.608	**0.475**	**0.531**

gain achieved by the proposed method is around 2–4% in Micri-F1 score and 7–66% in Macro-F1 score over other benchmark algorithms.

Based on the experimental results of two classification tasks, it can be seen that the vertices embedding learned by Deepwalk and LINE are not effective in classification tasks. It indicates that the representation learning methods of homogeneous network is flawed and limited. These methods can only learn the same type of vertices, and do not possess the ability to handle complex heterogeneous network in reality. Although ICE can learn the representation of heterogeneous node embedding, it is restricted to deal with the relationship between item-word and word-word, and cannot include more relationship types and node types, so the ICE also performs poorly in the end result. Compared with ICE, PTE can accommodate more network information, so that the learned vertices embedding performs well in classification tasks, but its ability is limited to the joint training of three heterogeneous bipartite graphs, and the relationship is the link that is directly observable in the original network. However, MMPCE can learn the vertices embedding by integrating multiple meta paths and make full use of the rich semantic information and structure information in the heterogeneous network, so that the learned vertices embedding can fully reflect the characteristics and properties of the source network graph.

5.4 Retrieval Task Quantitative Results

In the MovieLens, two different kinds of retrieval tasks are established, including one *word-to-movie* retrieval task and one *movie-to-movie* retrieval task. The goal

Table 7. Case study of similarity search in DBLP data

Rank	AAAI	NIPS	ACL	VLDB	SIGGRAPH	SIGCOMM
0	AAAI	NIPS	ACL	VLDB	SIGGRAPH	SIGCOMM
1	IJCAI	ICML	EMNLP	SIGMOD	VR	CoNEXT
2	ICML	AISTATS	COLING	ICDE	EuroVIS	INFOCOM
3	ECAI	CVPR	EACL	PVLDB	VIS	MOBICOM
4	ACL	COLT	CoNLL	CIKM	ISMAR	ICNP
5	COLING	ICCV	SIGIR	SIGIR	CVPR	SenSys
6	EMNLP	ECML/PKDD	LREC	KDD	LREC	NIPS
7	KDD	MOBICOM	IJCNLP	MOBICOM	EUROCRYPT	PVLDB
8	ECML/PKDD	EMNLP	IJCAI	LREC	ICCV	LREC
9	SIGIR	KDD	CIKM	COLING	ACISP	ICICS
10	NIPS	ACL	AAAI	NIPS	IPSN	ESORICS

of this retrieval task is to determine a movie that belongs to the same genre as the keyword or the specified movie.

In the *word-to-movie* task, we extract the top 5 tf-idf words from the description of the movie contained in each genre as the query keyword, and then obtain the cosine similarity between the learned keyword embedding and all the movies. We take out the top 50 and top 100 movies that have the greatest similar keywords as query results.

Since the DeepWalk and LINE do not involve the representation of keywords, the results generated by the PTE method contain only the vector representation of the movies, so these three methods are not listed in Table 5. As shown in the table, the proposed MMPCE method is the best performing among all baseline methods in terms of average precision@50 and precision@100. Although the retrieval performance of MMPCE in some genres is slightly lower than ICE, for drama, horror and animation movies, the proposed MMPCE model significantly outperform compared with the other baseline algorithms. Note that in the word-to-movie retrieval task, ICE retrieval results are relatively poor compared with the proposed MMPCE method. This result is due to ICE only covering the relationship between movie-keyword and ignoring the information association of movie-movie, resulting in a part of similar movies is not retrieved.

In the *movie-to-movie* retrieval task, we randomly select six movies from each genre as the query movie, then calculate the cosine similarity between the query movie and all the movies, and extract the top 50 and the top 100 movies that have the greatest similar query movie as results.

Since we constructed a heterogeneous retrieval task, Deepwalk and LINE are not involved in experimental comparison. Table 6 displays the comparison of the four methods in terms of precision@50 and precision@100. It can be seen from the table that the MMPCE method has the best average query performance among all the counterparts. For some genres, MMPCE generally has a substantial improvement in the search similarity movies. Compared with ICE and PTE, the proposed MMPCE approach outperforms with a considerable amount. This is because it not only considers the link that is directly observable in the source network, but also adds multiple meta paths according to the semantic information, so the information preserved in the vertices embedding is more abundant, and the effective proportion of the query result is higher.

5.5   Case Study: Similarity Search

We have constructed a case that search similarity venues to prove the efficiency of our method. In the DBLP dataset, we selected a total of 48 academic conferences from 6 research areas as the total query dataset. The cosine similarity is utilized to calculate the distance between the specified venue and the remaining venues, and finally the top 10 venues closest to the specified venues are returned as the query result.

Table 7 lists the top 10 similar results that are the same as the specified venues in the six research areas. It can be seen from the table that the proposed MMPCE model can well retrieve the venues in the same domain that have the closest match with the specified venue. The closer the venue of the specified venue theme, the higher the ranking. For example, in the field of NLP, we choose ACL as the query node. Among the corresponding query results, EMNLP, COLING, ECAL and CoNLLC are most relevant to the NLP in the top 4 positions of the search result. The similarity search results can also serve as observed in other five fields.

6   Conclusion

In this work, we propose a Multiple Meta Path Combined Embedding method, namely MMPCE. The method can automatically obtain the low-dimensional vector representation of various types of vertices and capture both the semantic and structural information of a heterogeneous network. Compared with the previous methods, the proposed model can cover a wider range of semantic information, and the method is more flexible and scalable. Extensive experiments demonstrate the efficacy and efficiency of the proposed method in a variety of heterogeneous network mining tasks.

Our future work will be improved as follows. In the heterogeneous network, not all meta paths are meaningful and reliable for learning. Some meta paths fail to provide valuable information for representation of vertices. For example KMY, it does nothing to get the embedding of movie. We need to select reliable meta paths from all meta path sets for representation learning. In addition, we will attempt to extend the model to represent the dynamic heterogeneous network.

Acknowledgment. The authors acknowledge the financial support from the following foundations: National Key R&D Program of China (No. 2017YFC0803700), National Natural Science Foundation of China (No. 61532021 and 61472141) and Shanghai Knowledge Service Platform Project (No. ZF1213).

References

1. Tang, L., Liu, H.: Relational learning via latent social dimensions. In: Proceedings of the 15th ACM SIGKDD International Conference on Knowledge Discovery and Data Mining, pp. 817–826. ACM (2009)
2. Krizhevsky, A., Sutskever, I., Hinton, G.E.: ImageNet classification with deep convolutional neural networks. In: Advances in Neural Information Processing Systems, pp. 1097–1105 (2012)
3. Ng, A.Y., Jordan, M.I., Weiss, Y.: On spectral clustering: analysis and an algorithm. In: Advances in Neural Information Processing Systems, pp. 849–856 (2002)
4. Ou, M., Cui, P., Pei, J., et al.: Asymmetric transitivity preserving graph embedding. In: Proceedings of the 22nd ACM SIGKDD International Conference on Knowledge Discovery and Data Mining, pp. 1105–1114. ACM (2016)
5. Weiss, Y., Torralba, A., Fergus, R.: Spectral hashing. In: Advances in Neural Information Processing Systems, pp. 1753–1760 (2009)
6. Roweis, S.T., Saul, L.K.: Nonlinear dimensionality reduction by locally linear embedding. Science **290**(5500), 2323–2326 (2000)
7. Tenenbaum, J.B., De, S.V., Langford, J.C.: A global geometric framework for nonlinear dimensionality reduction. Science **290**(5500), 2319–2323 (2000)
8. Yan, S., Xu, D., Zhang, B., et al.: Graph embedding and extensions: a general framework for dimensionality reduction. IEEE Trans. Pattern Anal. Mach. Intell. **29**(1), 40–51 (2007)
9. Mikolov, T., Sutskever, I., Chen, K., et al.: Distributed representations of words and phrases and their compositionality. In: Advances in Neural Information Processing Systems, pp. 3111–3119 (2013)
10. Mikolov, T., Chen, K., Corrado, G., et al.: Efficient estimation of word representations in vector space. arXiv preprint arXiv:1301.3781 (2013)
11. Perozzi, B., Al-Rfou, R., Skiena, S.: DeepWalk: online learning of social representations. In: Proceedings of the 20th ACM SIGKDD International Conference on Knowledge Discovery and Data Mining, pp. 701–710. ACM (2014)
12. Grover, A., Leskovec, J.: node2vec: scalable feature learning for networks. In: Proceedings of the 22nd ACM SIGKDD International Conference on Knowledge Discovery and Data Mining, pp. 855–864. ACM (2016)
13. Tang, J., Qu, M., Wang, M., et al.: LINE: large-scale information network embedding. In: Proceedings of the 24th International Conference on World Wide Web. International World Wide Web Conferences Steering Committee, pp. 1067–1077 (2015)
14. Tang, J., Qu, M., Mei, Q.: PTE: predictive text embedding through large-scale heterogeneous text networks. In: Proceedings of the 21th ACM SIGKDD International Conference on Knowledge Discovery and Data Mining, pp. 1165–1174. ACM (2015)
15. Wang, C.J., Wang, T.H., Yang, H.W., et al.: ICE: item concept embedding via textual information. In: Proceedings of the 40th International ACM SIGIR Conference on Research and Development in Information Retrieval, pp. 85–94. ACM (2017)
16. Sun, Y., Han, J.: Mining Heterogeneous Information Networks: Principles and Methodologies, pp. 20–28. Morgan & Claypool Publishers, San Rafael (2012)
17. Sun, Y., Han, J., Yan, X., et al.: PathSim: meta path-based top-K similarity search in heterogeneous information networks. Proc. VLDB Endow. **4**(11), 992–1003 (2011)

18. Li, A.Q., Ahmed, A., Ravi, S., et al.: Reducing the sampling complexity of topic models. In: Proceedings of the 20th ACM SIGKDD International Conference on Knowledge Discovery and Data Mining, pp. 891–900. ACM (2014)
19. Tang, J., Zhang, J., Yao, L., et al.: ArnetMiner: extraction and mining of academic social networks. In: Proceedings of the 14th ACM SIGKDD International Conference on Knowledge Discovery and Data Mining, pp. 990–998. ACM (2008)
20. Harper, F.M., Konstan, J.A.: The movielens datasets: history and context. ACM Trans. Interact. Intell. Syst. (TiiS) **5**(4), 19 (2016)

A Review of Recent Advances in Identity Identification Technology Based on Biological Features

Jianan Tang[1], Pengfei Xu[1], Weike Nie[1(✉)], Yi Zhang[2], and Ruyi Liu[3]

[1] School of Information Science and Technology, Northwest University,
Xi'an, China
weikenie@163.com
[2] School of Electronic and Information Engineering, Xi'an Jiaotong University,
Xi'an, China
[3] School of Computer Science and Technology, Xidian University, Xi'an, China

Abstract. With the development of social informatization technology, the problems of individual information security are becoming serious. Nowadays identity identification has been required essentially in government and business field. In this paper, we summarize and analyze the identification principles and identification methods based on biometrics, including the present researches fingerprint, palmprint, iris, human face, vein, gait and signature, and make comparative analysis of the differences of the error recognition rate, stability, acquisition difficulty and counterfeiting degree. Finally, the prospects of biometric recognition technologies are discussed additionally.

Keywords: Biometric recognition · Identity authentication · Fingerprint
Palmprint · Face · Vein

1 Introduction

Identity is the basis of individual life in the society. It is not only a sign of the difference among individuals, but also it gives individuals a lot of social attributes. The generalized identity identification refers to the individual identity by means of ID card and fingerprint. In the Internet field, identity authentication refers to the process of confirming the identity of the operator in the computer network system. The problems of personal identification can be divided into two categories: Verification and Recognition. Verification refers to verifying that the user is the identity he/she claims. Recognition refers to determining the specific identity of the user.

Personal verification or identification is an important part of the information security field, which greatly affects the privacy of individuals. With the rapid development of all kinds of information technology, modern life has put forward higher requirements for the practicability and reliability of user identification. For traditional authentication methods, such as passwords and passwords, authentication is theoretically relatively safe as long as the password strength is high enough. However, because people's memory is limited, the chosen password is usually not so complicated, so that

Z. Xu et al. (Eds.): Big Data 2018, CCIS 945, pp. 178–195, 2018.
https://doi.org/10.1007/978-981-13-2922-7_12

people can log in again. Although this method is convenient, it is easily stolen by others. Once the intruder breaks the security line, it is easy to get the user's information.

An actual biometric system should: (1) Achieve acceptable recognition accuracy and speed with reasonable resource requirements; (2) Be harmless to people and be accepted by people; (3) For various types of fraud, the method has sufficient robustness. Biometrics is the most convenient and secure authentication technology. It does not need to remember complex passwords, and does not need to carry keys, smart cards and the like. Biometrics recognizes people themselves, which directly determines that this type of certification is safer and more convenient. Because each person's biometrics have uniqueness different from others and constant stability within a certain period of time, it is not easy to forge and counterfeit. Therefore, biometric identification technology is used for identity identification, which is safe, reliable and accurate. In addition, biometric technology products are realized by means of modern computer technology, and it is easy to integrate with computers and security, monitoring, and management systems to achieve automated management.

Biometrics provide a more reliable and efficient way to authenticate. Specifically, each person has a unique physical feature or behavior that can be automatically measured or identified, called a biological feature, which can be divided into physiological features (such as fingerprints, facial images, irises, palm prints, etc.) and behavioral features (such as gait, sound, handwriting, etc.). Biological characteristics are relatively stable over a period of time, so they do not disappear and are difficult to forge. So far, various biometric technologies [1] have been developed for many years.

At present, biometrics has three main application directions in life: 1. As an important means of criminal investigation and identification; 2. To meet the needs of enterprise security and management. For example, physical access control, logical access control, attendance, patrol and other systems have fully introduced biometric technology; 3. self-service government services, immigration management, financial services, e-commerce, information security (personal privacy protection).

The development potential of biometric applications is mainly reflected in the following aspects: First, there are many sources of human identity information and a large amount of data. Whether it is static management or dynamic control, identity is the primary factor. Second, with the development of economic globalization, large factories can use automatic identity authentication instead of manual labor to manage. In addition, economic globalization has a more direct impact on the need for frequent personal identification. The prevalence of biometric authentication in e-commerce and e-government is also the best solution at this stage for the foreseeable future.

In this paper, the authors conducted a comprehensive analysis of current biometric-based recognition methods. Compared to existing articles on biometric identification methods, our differences lie in the detailed description of various biometric methods and the summary of some improved algorithms in recent years. In addition, the results of 3D technology and the rapid development of deep learning networks in the field of identity identification are briefly introduced (Fig. 1).

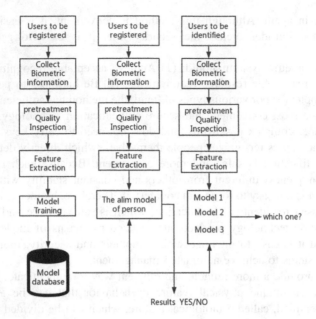

Fig. 1. The process of verification/recognition

2   Biometric Feature Recognition Technology

2.1   Fingerprint Recognition

Fingerprint recognition is one of the most widely used biometric technologies. Fingerprints have two important characteristics: uniqueness and stability [2, 3]. Almost no two fingerprints are identical. Galton first estimated the unique level of fingerprints as 1.45×10^{-11}, which provides a theoretical basis for a series of fingerprint identification studies [4]. At the same time, the characteristics of each fingerprint from birth, its shape may remain unchanged for life, unless it is seriously affected or diseased.

The whole process of fingerprint recognition technology is divided into four steps: fingerprint acquisition, fingerprint preprocessing, feature extraction and fingerprint matching. The development of fingerprint collection has gone through the original ink printing stage. Most existing fingerprint identification systems are equipped with contact-based two-dimensional sensors, such as China's second-generation ID card and optical fingerprint collector for company attendance. Solid-state fingerprint sensors installed on mobile electronic devices, including fingerprint smart phones and other mobile electronic devices. With the rapid development of 3D scanning and reconstruction technology, non-contact 3D fingerprint reconstruction and recognition technology is used to solve the shortcomings of contact-based 2D fingerprint recognition system and improve recognition accuracy [5].

Fingerprint preprocessing and feature extraction are important components of fingerprint recognition. After processing the fingerprint image by denoising and refining, the original fingerprint image is converted into a refinement map of the pixel range. The task of feature extraction is to extract the main feature points of the fingerprint in the refinement graph, including two singular points (center point and triangle point) and two points (bifurcation point and end point) with higher probability of occurrence. In addition, we need to match the directional field distribution around each point [6] and the type of feature points. This information will be built into the feature vector and stored in the data (Fig. 2).

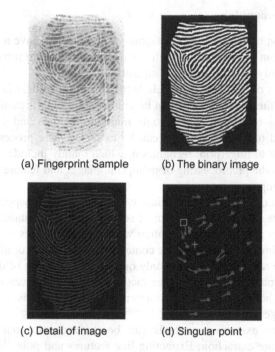

(a) Fingerprint Sample   (b) The binary image

(c) Detail of image   (d) Singular point

Fig. 2. The process of fingerprint recognition

Fingerprint matching is a pattern recognition process based on fingerprint pattern classification. At present, there are three different types of recognition methods according to the selected feature types: 1. Detailed feature recognition method [7], in which the details of the fingerprint information are extracted and detailed feature vectors are constructed to match. This method requires high quality images and a certain amount of feature information. Finally, the threshold is used to determine whether the collected fingerprints are from the same finger; 2. The structural pattern recognition method [8], which pays more attention to the overall distribution of fingerprints, including the pattern. This method can better resist noise, but the description of the fingerprint is rough and cannot be unique. 3. The method of combining the structure mode and the detail mode first narrows the matching range according to the classification of the fingerprint, thereby reducing the matching time and improving the matching efficiency. However, since the current fingerprint classification method is not highly accurate, the failure rate of this method is also high.

At present, in view of the great success of the CNN method, Lin et al. [9] proposed to develop a non-contact 3D fingerprint identification method to solve the limitations of the existing method by integrating the 3D fingerprint information of the side view. By integrating the 3D fingerprint information of the side view and the top view with the corresponding 3D contour map, the depth feature representation of the global shape and texture information can be enriched.

2.2   Palmprint Recognition

Palmprint recognition has many special advantages. Palm prints have a larger area and richer information than fingerprints. A high-performance palmprint recognition system can be built using only lower resolution acquisition devices.

The main features of the palm print include the main line, folds, triangle points and detail points. All of the above features can be extracted in a high-resolution palmprint image, while for low-resolution images, only mainline features and wrinkle features (collectively referred to as line features) can be extracted. The process of palmprint recognition is similar to fingerprint recognition, which mainly includes the acquisition of palmprint images, the preprocessing of palmprint images and feature extraction, and the matching of palmprint images.

The pretreatment of palm prints is mainly carried out in three aspects. First, due to the uncertainty of the position of the palm, it is necessary to maintain all images in a substantially uniform position and orientation with varying degrees of translation and rotation. Second, palmprint images mostly contain images of some or all of the fingers, but palmprint image recognition focuses only on the central region of the palm, so we need to segment the palmprint image and extract and normalize the central region of the palmprint image. In addition, due to the different sizes of the palms, it is necessary to standardize the images.

Palmprint feature extraction methods can be divided into four categories: 1. Structure-based feature extraction: Extracting line features and point features; 2. Feature extraction based on time-space domain transform: Transforming the original space palmprint image into the frequency domain and defining several feature vectors in the frequency domain, which can reflect the texture in the palmprint density and depth. 3. Statistical feature extraction: Redefine the original image with a stochastic model composed of statistical features; 4. Subspace-based feature extraction: Convert the image into a low-dimensional vector or matrix. The palm print is represented and matched in a low dimensional space.

The basic idea of Do Gcode [10] proposed by WU.XQ is to use the reciprocal of Gaussian function as the horizontal and vertical filter of the image. Then we use the Do G code as the extracted feature. When matching, the approximate degree of the image is determined by calculating the distance between the Hamming images. Compared with 2D image feature matching, 3D palm print is more favorable. In [11], the l1 norm or l2 norm regularized 3D palmprint recognition based on the cooperative representation (CR) framework is proposed.

2.3 Face Recognition

Face recognition technology, as a technique based on physiological feature recognition in the field of biometric recognition, extracts facial features by computer and performs identity verification based on these features.

At present, face recognition research mainly includes five aspects: (1) face detection: finding the coordinates of the face and the area occupied by the face in different situations; (2) face representation: extracting the facial features of the person Determine the detected face and the existing face description in the database. (3) Face matching. The object to be tested is compared with the existing face image of the database, and the result is obtained. The key to this step is to select the appropriate face expression method and matching algorithm. (4) Facial table analysis. The computer analyzes and understands the meaning of the person's emotions by recognizing the expression of the face and face. (5) Physiological classification: A careful analysis of the physiological characteristics of the human face, and information about the gender, age, race, occupation, etc. of the person.

Common algorithms for face recognition include: (1) Face recognition based on geometric features first proposed by Bledsoe [12]. The facial feature vectors (such as eye, nose, mouth position coordinates, and width distance) are obtained by recognizing the relative distance between the face feature points of the face. Find the best matching faces by performing vector comparisons. (2) Identification based on feature faces. The subspace pattern is formed by extracting statistical features of the face. The principal component analysis method is mainly used for orthogonal transformation. (3) Face recognition based on template matching requires a face template of some different standard samples to be given in advance. First, a global range search is performed on the face to be measured, and the similarity between the faces of the target person is compared according to the templates. It is judged whether the image window contains the target face by using the size of the image window. (4) The face recognition method based on Markov model is represented by the corresponding set of characteristic values for the five most important areas of the face (forehead, eyes, nose, mouth, and chin). (5) Face recognition based on elastic matching method is an algorithm based on dynamic link structure. It uses the sparse degree of the lattice to represent the facial image. The feature vector mark is obtained by the representation node in the sparse graph by performing Gabor wavelet decomposition on the image position, and the edge of the sparse graph is marked by the distance vector of the connected node. (6) The face recognition method based on neural network is currently popular, mainly using BP algorithm for face facial feature extraction. Through the learning process, we can get the invisible expression of the rules and rules of face recognition, which is more adaptable and the processing speed is very fast. With the further development of deep learning, it provides a new idea for the solution of the face recognition problem.

In 2014, Facebook proposed DeepFace [13], which uses convolutional neural networks and large-scale face images for face recognition. It achieves 97.35% accuracy on LFW, and performance is comparable to manual recognition. The VGG [14] network achieved a depth of 98.95% with a deep network topology and a large input image. The DeepId [15–18] network proposed by the Chinese University of Hong

Kong has made a series of improvements to the convolutional neural network. The accuracy is increased to 99% by combining local and global features, using joint Bayesian processing convolution features, and training using both identification and authentication information. With the expansion of the face data set, the face recognition accuracy has also been improved accordingly. In 2015, Google's FaceNet [19, 20] achieved an accuracy of 99.63%. It uses TripletLoss as its supervisory information and uses 200Mllion images for training. After that, Baidu even got an accuracy of 99.77% [21]. In addition, DCNN's network structure is getting bigger and deeper: VGGFace has 16 layers, FaceNet has 22 layers, 2015 ResNet [22] has reached 152 layers.

In view of the large amount of data required in face training, the literature [24] proposed the idea of artificially synthesizing a large amount of training data, and the face synthesis from three directions: perspective, shape, expression. Recently, different kinds of generation tasks can be completed by using Generative Adversarial Networks (GAN), which can generate real-life pictures. Google's new BEGAN model is used for face data sets, and the author focuses on image generation tasks. Even at even higher resolutions, new milestones in visual quality have been established. Deep learning can also achieve very good results on smaller data sets. Lightened CNN [23] uses a new activation function MFM function and uses a small network structure to achieve an accuracy of 98.13%. Caffe-Face achieved an accuracy of 99.28% using the Center Loss function.

With the development of science and technology, face automatic recognition technology has made great achievements, but it still faces difficulties in practical application. It not only requires accurate and rapid detection and segmentation of face parts, but also effective compensation, feature description and accurate classification effect. The following aspects need to be emphasized and improved: 1. The combination of local and global information can effectively describe the characteristics of human face, and the method based on hybrid model is worthy of further study. 2. Multi-feature fusion and multiple classifier fusion methods are also a means to improve recognition performance. 3. Because the human face is not rigid, the similarity between human faces and the influence of various factors lead to the difficulty of accurate face recognition. In order to meet the real-time requirements of automatic recognition, it is necessary to study the fusion method of face and fingerprint, iris and speech recognition technology. 4.3D deformation model can deal with many kinds of change factors and has a good development prospect. The method of simulation or compensation is adopted to achieve a better recognition effect. Three-dimensional face recognition algorithm is still being explored, which needs to be improved and innovated on the basis of traditional recognition algorithm. 5. Surface texture recognition algorithm is a kind of newest algorithm, which waits for us to continue to study.

In a word, face recognition is a challenging subject. It is difficult to achieve good recognition effect by using only one existing method. How to combine with other technologies, improve the recognition rate and speed, reduce the calculation amount, improve the robustness, adopt the embedded and hardware realization and practical is worth studying in the future.

2.4   Iris Recognition

The iris recognition performs the authentication or recognition of the user identity by extracting the texture features in the test iris image and comparing the feature with the feature template pre-stored by the user.

The pre-processing of the image includes positioning of the inner and outer boundaries of the iris, normalization and image enhancement, and denoising operations. Iris feature extraction uses texture extraction algorithm, such as the classical Gaobor filter method, to extract iris texture features from the uniformized normalized iris image, and performs binary coding to obtain a feature template composed of 0 and 1. Pattern matching can use a normalized Hamming distance to align feature templates of two iris images for identification (Fig. 3).

Fig. 3. CASIA iris database sample

The matching is mainly to separate and encode the texture features in the iris by performing pre-analysis, and then compare the obtained data with known samples in the database. According to the analysis similarity, it is diagnosed whether the two irises match, and thus the identity detection is realized. At present, iris-based identity checking methods are developing rapidly. The more classical algorithms are as follows: (1) Texture-based feature recognition algorithm. The most classic is the iris image of four different resolutions based on the Gauss-Laplace pyramid structure proposed by Wildes [25]. The iris was first analyzed and clustered using the Fisher linear criterion, and then the correlation between the iris images was compared; (2) Iris recognition method based on zero-crossing detection. The most representative method of iris recognition based on zero-crossing detection is the algorithm proposed by Boles [26] et al. The core idea uses four-scale wavelet to decompose the one-dimensional signal of the iris image, and then selects the zero-crossing point as the iris recognition standard. According to different two functions, the iris is further classified and matched to achieve identity detection; (3) Phase-based identification method. Daugman [27] 's iris recognition method is representative of the phase-based recognition method. Its core idea is to use a multi-scale Gabor filter to analyze the iris image. The features are separated and processed by phase encoding. Finally, iris matching is realized by Hamming distance. (4) Multi-channel Gabor filter bank and Daubechies-4 wavelet identification algorithm. The method is an iris recognition system independently developed by the Chinese Academy of Sciences. The core idea is to use multi-channel

Gabor filter filtering and two-dimensional wavelet transform for the iris image after normalization. In the feature matching stage, the variance reciprocal weighted Euclidean distance algorithm is used. The experiment proves that the method can obtain better identification results.

2.5 Retinal Identification

Retinal recognition is accomplished by extracting retinal vascular features and comparing them to pre-stored retinal data to confirm identity. The retina is composed of complex blood vessels distributed around the tiny nervous system at the back of the eyeball. It can remain unchanged for life without damage or disease. At the same time, it is also the most difficult system to be deceived, so retina recognition can be said to be one of the most reliable biometric technologies.

Retinal image acquisition is the first step in achieving retinal recognition. In a real environment, a professional fundus camera is generally used to scan a human retina to obtain a retinal image. The quality of the acquired image will affect the later recognition.

At present, methods for feature extraction of retinal blood vessels fall into two broad categories. One type is to extract the gray scale information of the retinal blood vessels and the characteristic information around the optic disc from the whole image, and the other is to extract the feature points from the shape of the blood vessel by the blood vessel segmentation and refinement.

With the research and development of retinal recognition technology, three matching methods are gradually formed: 1. Image-based matching method. Two typical methods: the American professor Rahman [28] proposed to use the overall gray information of the retinal blood vessel image and the statistical features around the optic disc to identify; Norwegian professor Ashokkumar [29] uses the visible spectrum for retinal recognition. The main disadvantage is that the structural characteristics of the blood vessels are neglected, resulting in insufficient matching accuracy. And the positioning of the disc itself becomes a problem. There is also a literature [30] that proposes the entire retinal vessel tree as a matching feature. 2. Node-based matching method. First, the blood vessels in the retinal image are segmented, and then nodes (cross points, bifurcation points) are selected from the segmented blood vessel network as feature points for one-to-one comparison. The advantage of this method is that the accuracy of the matching is higher, but the amount of calculation is relatively large. 3. Retinal vascular morphology recognition method based on structural features. Based on the fact that the mutual position between the nodes and the nodes is fixed, the matching feature of the retinal vascular structure is extracted by using the nearest neighbor triangle, which can effectively reduce the influence of image rotation and scaling.

2.6 Vein Recognition

Vein recognition uses the veins of the venous vessels to achieve identity authentication. Images are typically acquired using near infrared imaging.

Vein recognition mainly includes four stages: image acquisition, preprocessing, feature extraction and matching. For the acquisition of vein images with insufficient quality, the corresponding pre-processing techniques are needed for image localization and normalization. The extracted features mainly include vein grain features, texture features, minutiae features, and features obtained through learning. The grain features are a good representation of the topology of the entire vein network. Texture features are often expressed using local two-code values (gray value comparison), which are the bifurcation points, endpoints, etc. of the blood vessels. The features obtained through learning are the component features obtained by dimensional reduction after principal component analysis. The last step is matching. Matching refers to one of the basic technical aspects of vein recognition. In the case that a user registers multiple templates, a multi-template matching method is generally used to improve the matching precision, including a fuzzy fusion matching method of a three-valued template; matching method based on pixel matching rate, Hamming distance matching, and pixel non-matching rate fusion; a hierarchical strategy based on multiple templates.

In view of the problem that the finger vein image pixels are low and easy to extract errors during the extraction process, the literature [31] proposed the local binary pattern (LBP) and a local derivative pattern (LDP) algorithm for vein recognition. Rosdi et al. proposed using a new LBP algorithm local a line binary pattern (LLBP) for identification.

In 2016, Matsuda [32] evaluated a new method of finger vein authentication based on feature point matching. He proposed matching the feature points of irregular shadow and vein deformation, and extracting features using the curvature of the intensity image section. In order to increase the number of feature points, the points extracted from any position are non-linear. In addition, a finger shape model and a non-rigid matching method are proposed, which have a higher recognition success rate.

2.7 Gait Recognition

Gait characteristics are an emerging behavioral feature that mainly refers to people's habitual walking style. Gait characteristics are unique to everyone. Since each person has a different physiological structure and walking habits, it is difficult to find two identical gaits. Gait is the only feature of all current biometrics that is suitable for long-range observations at present.

A complete gait recognition system usually consists of three parts: gait contour extraction, gait feature extraction, and gait classification. The gait contour is extracted using motion detection algorithms to extract moving objects from gait video images. If there is more than one moving object in the video, you need to classify the moving targets. After the motion object is extracted, it needs to be filtered by noise and scaled to a standard size to obtain a gait contour. Gait features contain two important components: structured components and dynamic components. The structured component captures the individual's body shape, height, limb length, step size, etc. The dynamic component captures the motion characteristics of the human body during walking, such as the swinging of the arms, hips, and legs.

The current gait recognition methods can be roughly divided into two categories: one is a model-based approach. This type of method identifies individuals by modeling the anatomy and extracting image features to map the gait features into structural components of the model, or to derive motion trajectories from the human body. The other type is the non-model approach. This type of method acquires a compact representation of the motion characteristics by characterizing the entire motion pattern of the human body, so there is no need to consider the underlying structure of the human body. After the gait features are extracted, the classification algorithm classifies and identifies its features. The classification algorithms currently used in the field of gait recognition include neighborhood classification, support vector machine, hidden Markov model, multivariate discriminant analysis, linear discriminant analysis, tensor analysis, etc. In addition, principal component analysis is in gait. Identification is often used to preprocess feature matrices.

In the existing literature, there are many researches on the gait recognition algorithm of human body parameters. The literature [33] is based on the vibration model commonly used in physics to build a model of human body posture and shape, and uses self-learning algorithm to update in real time. In literature [34], based on the particle filter algorithm, the gait shape tracker is designed to realize the gait recognition and tracking algorithm based on the overall contour data of the step. The literature [33, 34] is a more representative gait recognition algorithm based on human contours.

2.8   Signature Recognition

Signature recognition is a technique for biometric recognition based on human behavior characteristics. It is a widely used biometric recognition technology. It can be divided into two modes according to the timing of the signature: offline mode (off-line) and online mode (on-line). Online signatures retain more valuable information. The movement speed, acceleration, pressure and other information of the nib make the recognition rate higher. For offline signature recognition, the written dynamic information is almost completely lost. It can only rely on the static information (handwriting features) of the signature image to reflect the signature writing style and habits, so identification is more difficult (Fig. 4).

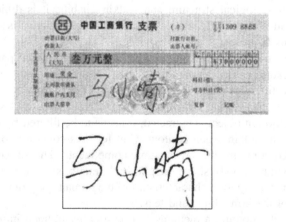

Fig. 4. Signature sample

Domestic and foreign scholars have gradually deepened and developed research on the field of signature recognition, and many new theories and methods have appeared in this field. Ferrer [35] proposed three kinds of gray feature fusions: Local binary pattern, Local directory pattern and Local derivative pattern. The LS-SVM was used for classification and identification, and verified on GPDS and MCYT databases. Kaur et al. [36] used the SURF algorithm on the pre-processed pictures and then classified them using HMM. Abdoli et al. [37] proposed a signature recognition system for feature extraction of Geodesic derivative pattern (GDP), which uses KNN as a classifier to compare feature vectors with each other. Shekar [38] applied Discrete Cosine Transform (DCT) to the binary image of the signature image, achieving a recognition rate of 92.74%. Vargas [39] uses the combination of gray level co-occurrence matrix and LBP, and uses SVM to identify. Wen [40] extracted the probability density function based on the structural distribution of the signature image and achieved good results. Azmi [41] proposed a signature verification system. The Freeman Chain Code (FCC) is used as the representation of the signature feature to extract a total of 47 dimensions, which is identified by the ANN classifier. Das [42] uses Gabor filtering technology to extract the topological features and statistical features of signature images, and forms the signature recognition system through the attribute reduction of rough set theory. Hatkar [43] extracted the geometric features of the image and used neural networks for classification and recognition.

As a deep learning method, DBN is a combination of multiple RBNs. The method solves the problem that the traditional neural network training method is not suitable for the training of the multi-layer network, and shows certain advantages in the field of handwritten digit recognition. By preprocessing the signature image and eliminating the interference, the local binary pattern (LBP) texture features of the handwritten signature image are extracted and used as the input of the DBN to construct the deep belief network model. It can automatically learn different levels of abstract features from the bottom up, and finally obtain a structural description of the signature image features, and classify at the top.

2.9   Other Biometric Feature Recognition Techniques

Biometrics technology is far more than the above mentioned, and more biometrics technologies have been proposed to meet the market needs of different fields. Each biometric identification technology has certain advantages over other biometric identification technologies in its application field, because no biometric identification technology can meet all real security requirements.

1. Non-contact recognition method

The characteristics of the human ear are unique, including shape, earlobe, skeletal structure, size, and the like. Therefore, ear shape recognition is also a non-contact biometric technology. With the popularity of computer technology, the use of keyboards has become the basic skill of work. Keyboard built-in sensors capture keystroke dynamics for authentication. When he or she frequently uses the keyboard, the user's identity can be authenticated by detecting the speed, pressure, time, and time the user clicks on each keystroke.

Voiceprint recognition converts a sound signal into an electrical signal that is then recognized by a computer. The voice spectrum of any two people is different. The acoustic characteristics of each person's voice are relatively stable but easy to change. This change may come from physical, pathological, psychological, etc., but it is also related to environmental disturbances. However, because each person's pronunciation is different, people can still distinguish the voice of different people or judge whether it is the voice of the same person.

2. The identification based on medical technology

Gene identification identifies a particular human identity by using a gene fragment having biological characteristics on the DNA sequence. The identification device is costly, the recognition algorithm is complex, and the identification method is difficult to implement quickly. It is now only used for a few specific applications.

In addition, the researchers found that everyone's heart is not exactly the same. It can be identified by ECG differences. Each value in the ECG wave has a profound meaning. These values change when a heart attack changes, but the sub-mode in which the value is shared with the value does not change. Because different hearts correspond to different ECGs, we can use this feature for identity authentication. However, the ECG acquisition process is more complicated than other methods, and the equipment requirements are higher. A non-contact heartbeat method has been successfully developed that uses radar remote scanning to analyze the shape and size of the heart and its geometric features, such as fluctuations in heart recognition.

There is also a method of identifying through the skin. The identity of a person is identified by illuminating infrared light into a small piece of skin to measure the wavelength of the reflected light. The theoretical basis is that each human skin with different skin thickness and subcutaneous layer has its own special mark. Because of the personality and specificity of the skin, cortex and different structures, these reflect light of different wavelengths. This method requires less computing ability.

3 The Comprehensive Comparison and Analysis of Several Biometric Identification Techniques

Each biometric technology has its own characteristics. In general, recognition techniques based on physiological logic features require near-end acquisition. Its stability is generally safer and more reliable than behavior-based identification techniques. But because it requires near-end collection, it is often inconvenient to use in remote authentication situations, and there is a risk of being easily imitated. However, behavioral characteristics are not easily imitated. We compared the advantages and disadvantages of each of the above biological features. These advantages and disadvantages determine their application in different fields, and also put forward new requirements for future research and development.

Fingerprint functionality is usually unique. The greater the number of features, the higher the accuracy of fingerprint recognition. On the one hand, current fingerprint readers are not expensive and can be scanned very quickly. However, because of the need to artificially limit fingerprint acquisition data, the location of each constraint is not exactly the same. Differences in focus can also lead to varying degrees of distortion, which makes high-precision fingerprint recognition not meet the original requirements; on the other hand, fingerprints are often used for criminal records, which makes some people afraid to "record" fingerprints.

The advantage of face recognition lies in its nature and the characteristics that are not easy to be detected by the participants. Humans usually confirm each other's identities according to their faces. The human face has the similarity of structure. Even for different individuals, the structure is still very similar. This characteristic is advantageous to the localization human face, but is disadvantageous to the human face distinction. The shape of the human face is also very unstable. The change of facial expression is very easy to change the character of face, and the image difference of face is larger in different observation angle. In addition, face recognition is also susceptible to illumination conditions, human face cover and other factors.

Iris recognition is probably the most reliable biometric feature recognition technology, but the iris image acquisition equipment needs expensive camera equipment and the size of the equipment is very difficult to be miniaturized. In practical applications, the camera may produce image distortion. The iris images obtained in the darker environment are often of poor quality and cannot complete the recognition task.

Voice-print recognition does not involve user privacy issues, so the user acceptance is higher. On the one hand, the acquisition of voice print is the most convenient. A microphone or phone and mobile phone can collect user voice feature information and complete identity authentication. The complexity of the voice-print recognition algorithm is low. With some other measures, such as content identification through speech recognition, the accuracy rate can be improved. These advantages make the application of voice recognition more and more favored by system developers and users. The Voice-print recognition technology is based on the speaker's invariance, but the same person's sound is susceptible to age, mood, physical condition and so on, resulting in lower recognition performance. In addition, different microphones and channels have varying degrees of influence on recognition performance. The environment noise and the mixed speaker situation also have great influence.

The signature recognition technology also has the characteristic which is easy to be accepted by the populace. As people's experiences and habits change, so does the way they sign up. The electronic signature board used for signature recognition is complex and expensive, and it is very different from the touchpad on the notebook, so it is difficult to be a remote identity authentication technology on the Internet. At the same time, the current physical and material technology is also difficult to the electronic signature board miniaturization.

The 2 key metrics for measuring biometric performance are False Rejection Rate (FRR) and false Accept Rate (FAR). FRR refers to mistaking a test sample from a real person for a rate of rejection by the impostor, while far refers to the ratio of mistaking a test sample from a pseudo person to a real man (Table 1).

Table 1. The comprehensive comparison and analysis of several biometric identification techniques

The methods of identity recognition	False Accept Rate (FAR)	False Rejection Rate (FRR)	Stability	The degree of counterfeiting
Fingerprint recognition	Very low	Very low	General	General
Palmprint recognition	Low	5%	General	General
Face recognition	0.5%	<0.2%	General	General
Iris recognition	Very low	10%	All the same for life	Extremely difficult
Retinal recognition	<0.0001%	<0.1%	All the same for life	Extremely difficult
Vein recognition	<0.0001%	<0.01%	All the same for life	Extremely difficult
Gait recognition	Unknown	Unknown	Unstable	General
Voice recognition	Low	Low	Unstable	Difficult
Signature recognition	Low	10%	Unstable	General

4 Summary and Prospect

This article reviews current research and applications of biometrics such as facial, voiceprint, fingerprint, palm print, iris, retina, gait, vein and signature. The characteristics and safety of biometrics are compared and discussed. The application of biometric technology in different fields is summarized.

Due to its unique advantages, biometric technology has shown great application prospects. At present, fingerprint recognition has been basically popularized. In the near future, biometrics will move toward high security, portability, non-contact and low cost. Face recognition and vein recognition have shown great advantages. With the development of science and technology, the cost of 3D face recognition and vein recognition will become smaller and smaller, gradually replacing other biometric methods.

Identity technology is an important technology in the government and business sectors. Biometric technology has the advantages of traditional certificate and key authentication technologies. However, a single biometric system is not reliable in the application field. A powerful biometric fusion recognition system is the research direction. The maturity of this technology will bring about great changes in the economy and society.

In the future, there is no doubt that the application of this technology will have a profound impact on our future economic and social life.

References

1. Glossary of Key Information Security Terms. Diane Publishing, Collingdale (2011)
2. Wang, Y., Hu, J.: Global ridge orientation modeling for partial fingerprint identification. IEEE Trans. Pattern Anal. Mach. Intell. **33**(1), 72–87 (2011)
3. Maltoni, D., Maio, D., Jain, A.K., Prabhakar, S.: Handbook of Fingerprint Recognition. Springer, London (2009). https://doi.org/10.1007/978-1-84882-254-2
4. Pankanti, S., Prabhakar, S., Jain, A.K.: On the individuality of fingerprints. IEEE Trans. PAMI **24**(8), 1010–1025 (2002)
5. Kumar, A.: Contactless 3D Fingerprint Identification. Springer, Heidelberg (2018). https://doi.org/10.1007/978-3-319-67681-4
6. Zhang, L.: Extraction of direction features in fingerprint image. Appl. Mech. Mater. **518**, 316–319 (2014). Trans Tech Publications
7. Maio, D., Maltoni, D.: Direct gray-scale minutiae detection in fingerprints. IEEE Trans. Pattern Anal. Mach. Intell. **19**, 27–40 (1997)
8. Hrechak, A.K., McHugh, J.A.: Automated fingerprint recognition using structural matching. Pattern Recognit. **23**(8), 893–904 (1990). 27–40
9. Lin, C., Kumar, A.: Contactless and partial 3D fingerprint recognition using multi-view deep representation. Pattern Recognit. **83**, 314–327 (2018)
10. Wu, X., Wang, K., Zhang, D.: Palmprint texture analysis using derivative of Gaussian filters. In: 2006 International Conference on Computational Intelligence and Security, vol. 1, pp. 751–754. IEEE (2006)
11. Li, W., Zhang, D., Zhang, L.: Three dimensional palmprint recognition. In: Proceedings of IEEE International Conference on Systems, Man, and Cybernetics, October 2009, pp. 4847–4852 (2009)
12. Chan, H., Bledsoe, W.W.: A man-machine facial recognition system: some preliminary results. Panoramic Research Inc., Palo Alto, CA, USA1965 (1965)
13. Taigman, Y., Yang, M., Ranzato, M.A., et al.: DeepFace: closing the gap to human-level performance in face verification. In: Proceedings of the IEEE Conference on Computer Vision and Pattern Recognition, pp. 1701–1708 (2014)
14. Parkhi, O.M., Vedaldi, A., Zisserman, A.: Deep face recognition. In: BMVC, vol. 1, no. 3, p. 6 (2015)
15. Sun, Y., Wang, X., Tang, X.: Deep learning face representation from predicting 10,000 classes. In: Proceedings of the IEEE Conference on Computer Vision and Pattern Recognition, pp. 1891–1898 (2014)
16. Sun, Y., Chen, Y., Wang, X., et al.: Deep learning face representation by joint identification-verification. In: Advances in Neural Information Processing Systems, pp. 1988–1996 (2014)
17. Sun, Y., Liang, D., Wang, X., et al.: DeepID3: face recognition with very deep neural networks. arXiv preprint arXiv:1502.00873 (2015)
18. Sun, Y., Wang, X., Tang, X.: Deeply learned face representations are sparse, selective, and robust. In: Proceedings of the IEEE Conference on Computer Vision and Pattern Recognition, pp. 2892–2900 (2015)
19. Schroff, F., Kalenichenko, D., Philbin, J.: FaceNet: a unified embedding for face recognition and clustering. In: Proceedings of the IEEE Conference on Computer Vision and Pattern Recognition, pp. 815–823 (2015)
20. Szegedy, C., Liu, W., Jia, Y., et al.: Going deeper with convolutions. In: CVPR (2015)
21. Liu, J., Deng, Y., Bai, T., et al.: Targeting ultimate accuracy: face recognition via deep embedding. arXiv preprint arXiv:1506.07310 (2015)

22. He, K., Zhang, X., Ren, S., et al.: Deep residual learning for image recognition. In: Proceedings of the IEEE Conference on Computer Vision and Pattern Recognition, pp. 770–778 (2016)
23. Wu, X., He, R., Sun, Z.: A lightened cnn for deep face representation. In: 2015 IEEE Conference on IEEE Computer Vision and Pattern Recognition (CVPR), p. 4 (2015)
24. Wen, Y., Zhang, K., Li, Z., Qiao, Y.: A discriminative feature learning approach for deep face recognition. In: Leibe, B., Matas, J., Sebe, N., Welling, M. (eds.) ECCV 2016. LNCS, vol. 9911, pp. 499–515. Springer, Cham (2016). https://doi.org/10.1007/978-3-319-46478-7_31
25. Wildes, R.: A system for automated iris recognition. In: Proceedings of the Second IEEE Workshop on Application of Computer Vision, pp. 121–128 (1994)
26. Boles, W.W., Boashash, B.: A human identification technique using images of the iris and wavelet transform. IEEE Trans. Signal Process. **46**(4), 1185–1188 (1998)
27. Daugman, J.: High confidence recognition of person by rapid video analysis of iris texture. In: Proceedings of European Convention on Security and Detection, pp. 244–251 (1995)
28. Rahman, N.A., Mohamed, A.S., Rasmy, M.E.: Retinal identification. In: Biomedical Engineering Conference. CIBEC 2008. Cairo International, pp. 1–4. IEEE (2008)
29. Borgen, H., Bours, P., Wolthusen, S.D.: Visible-spectrum biometric retina recognition. In: IIHMSP 2008 International Conference on Intelligent Information Hiding and Multimedia Signal Processing 2008, pp. 1056–1062. IEEE (2008)
30. Mariño, C., Penedo, M.G., Penas, M., et al.: Personal authentication using digital retinal images. Pattern Anal. Appl. **9**(1), 21 (2006)
31. Lee, E.C., Jung, H., Kim, D.: New finger biometric method using near infrared imaging. Sensors **11**(3), 2319–2333 (2011)
32. Matsuda, Y., Miura, N., Nagasaka, A., et al.: Finger-vein authentication based on deformation-tolerant feature-point matching. Mach. Vis. Appl. **27**(2), 237–250 (2016)
33. Hu, H.: Enhanced gabor feature based classification using a regularized locally tensor discriminant model for multiview gait recognition. IEEE Trans. Circuits Syst. Video Technol. **23**(7), 1274–1286 (2013)
34. Lee, H., Baek, J., Kim, E.: A probabilistic image-weighting scheme for robust silhouette-based gait recognition. Multimed. Tools Appl. **70**(3), 1399–1419 (2014)
35. Ferrer, M.A., Vargas, J.F., Morales, A., et al.: Robustness of offline signature verification based on gray level features. IEEE Trans. Inf. Forensics Secur. **7**(3), 966–977 (2012)
36. Kaur, R., Choudhary, P.: Offline signature verification in Punjabi based on SURF features and critical point matching using HMM. Int. J. Comput. Appl. **111**(16), 4–11 (2015)
37. Abdoli, S., Hajati, F.: Offline signature verification using geodesic derivative pattern. In: 2014 22nd Iranian Conference on Electrical Engineering (ICEE) IEEE (2014)
38. Shekar, B.H., Bharathi, R.K.: DCT-SVM-based technique for off-line signature verification. In: Sridhar, V., Sheshadri, H., Padma, M. (eds.) Emerging Research in Electronics, Computer Science and Technology. LNEE, vol. 248, pp. 843–853. Springer, New Delhi (2014). https://doi.org/10.1007/978-81-322-1157-0_85
39. Vargas, J.F., et al.: Off-line signature verification based on grey level information using texture features. Pattern Recognit. **44**(2), 375–385 (2011)
40. Wen, J., Chen, M., Ren, J.: Off-line signature verification based on local structural pattern distribution features. In: Li, S., Liu, C., Wang, Y. (eds.) CCPR 2014. CCIS, vol. 484, pp. 499–507. Springer, Heidelberg (2014). https://doi.org/10.1007/978-3-662-45643-9_53
41. Azmi, A.N., Nasien, D.: Freeman chain code (FCC) representation in signature fraud detection based on nearest neighbour and artificial neural network (ANN) classifiers. Int. J. Image Process. (IJIP) **8**(6), 434 (2014)

42. Das, S., Roy, A.: Signature verification using rough set theory based feature selection. In: Behera, H.S., Mohapatra, D.P. (eds.) Computational Intelligence in Data Mining—Volume 2. AISC, vol. 411, pp. 153–161. Springer, New Delhi (2016). https://doi.org/10.1007/978-81-322-2731-1_14
43. Hatkar, P.V., Salokhe, B.T., Malgave, A.A.: Off-line hand written signature verification using neural network. Methodology **2**(1), 1–5 (2015)

Interpretable Verification of Visually Similar Vehicle Images Using Convolutional Networks

Liugen Hong, Wenzhong Wang, Yanhui Pang, Huai Hu, and Jin Tang[✉]

School of Computer Science and Technology, Anhui University, Hefei 230601, China
ahu_hlg@foxmail.com, wenzhong@ahu.edu.cn, jtang99029@foxmail.com

Abstract. This paper presents a simple and effective method to verify similar vehicle images. In order to provide a meaningful interpretation of the verification, we propose to detect the local differences between two images. We frame this task as a saliency map regression problem, where the saliency map measures the degree of discrepancy at every pixel. To achieve this goal, we use a convolutional neural network (CNN) to map two aligned vehicle images to one saliency map. Our network design enables end-to-end training. We validate our algorithm on a vehicle image dataset. Experimental results show that our approach is accurate, fast and robust, and it achieves better performance than other methods.

Keywords: Vehicle verification · Saliency map

1 Introduction

Vehicle verification is an important task for access control in public traffic security. In vehicle access control, the license plate number is widely used as the reliable and unique ID of a vehicle image. However, in real world applications, license plates may be faked, occluded, missed, or can not be recognized. In all these cases, the license plates cannot be relied on. So there have been some research on image-based vehicle verification. Most conventional image verification methods verify images by comparing holistic image features. There are two drawbacks with these methods when dealing with vehicle images.

Firstly, in access control applications, the images of vehicles with the same attributes (such as type, maker, and color, etc.), and images shotted from nearby viewpoints are very similar to each other (Fig. 1). It is hard to distinguish these similar images using holistic image features.

Secondly, these methods can not interpret the verification (i.e. they do not provide any evidences for the verification), while interpretability is desirable in vehicle access control applications.

In order to solve these problems, we propose a method for interpretable vehicle verification. **Our goal is to not only verify whether two images are identical, but also provide evidences for the verification.** We focus

© Springer Nature Singapore Pte Ltd. 2018
Z. Xu et al. (Eds.): Big Data 2018, CCIS 945, pp. 196–209, 2018.
https://doi.org/10.1007/978-981-13-2922-7_13

Fig. 1. Five pairs of vehicle images. Each pair consists of two similar images, which have the same attributes including maker, model, year and color. The two images in the first three pairs are from different vehicles, and the images in (d) and (e) are from identical vehicles. We labeled the local differences presented in the first three image pairs with green boxes. Note that the images in the first three pairs are very similar to each other, and can only be distinguished from these local fine-grained differences. (Color figure online)

on vehicle access monitoring applications where the images are typically shot from the frontal views. We notice that there are some local details such as customized paintings, decorations, inspection marks, and so on, which can tell fine-scale differences between vehicle images (Fig. 1 shows some examples). So we propose to automatically discover their fine-grained differences. Based on these fine-grained differences, we can tell where the differences located and explain how much they differ from each other.

We develop a multi-task deep convolutional encoder-decoder network for this problem. The encoder is a Siamese network [4] composed of two identical convolutional neural nets (CNN) [3]. We use the encoder to classify two images as identical or not. The decoder is a deconvolutional network [6]. Given a pair of vehicle images, we use the decoder to predict a saliency map [11] which indicates the significance of their difference at every pixel. Our network design enables end-to-end training. We validate our method using real world vehicle images, and show that our approach achieves much better performance than the other methods.

The main innovation of our method is to combine verification task with regression task based on saliency map. In other words, our approach not only predicts whether two images are identical, but also provides the evidences for its predictions. This is achieved by predicting a score and a saliency map, where the score indicates the dissimilarity of the two images, and the saliency map indicates the subtle regions which differentiate the images.

The rest of the paper is organized as follows. Section 2 reviews the related work. Then Sect. 3 introduce algorithm in detail. In Sect. 4, we conduct various experiments. Finally, we conclude this paper in Sect. 5.

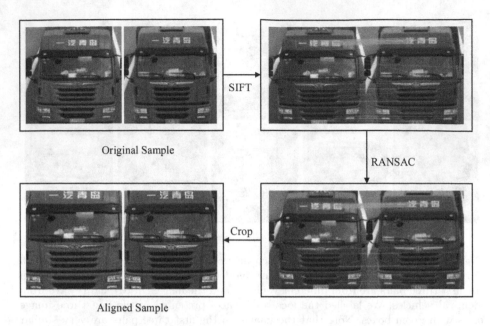

Fig. 2. Image alignment via feature matching

2   Related Work

Image verification has been researched immensely in recent years. The typical ideas of image verification are either to classify whether two images are of the same ID or to predict a similarity score by comparing their features [15]. Face verification [14] and signature verification [7] are the most successful endeavors. For example, the common outline of face verification consists of four stages: face detection [8], face alignment, feature representation and classification. In feature representation stage, feature extractor is designed to extract key features like structure of eyes and nose, but ignore other element. But in vehicle verification task, there is no such key features, for every small image patch is crucial. For example, in the same class, two vehicle images with inspection marks or without should be different. So above method cannot perform well on such fine-grained verification tasks.

Therefore, some improved methods are proposed, and have achieved good results. An intuitive improvement is to combine the classification and the similarity constraints together to form a CNN based multi-task learning framework [1]. For example, [13,17] propose to combine softmax with triplet loss together to reach a better performance. Obviously, these methods improve conventional verification method because similarity constraints may provide additional information for training the network. However, these methods still have limitation. Similarity constraints represent each image as a feature vector. Although we can verify whether samples are identical by comparing similarity of feature vector in same class, these holistic feature vectors can hardly reveal those subtle yet

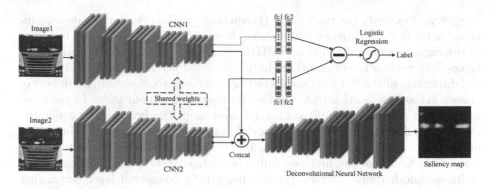

Fig. 3. Deep convolutional Encoder-Decoder network for vehicle image verification.

significant different regions. As shown in Fig. 1, the different images look almost identical except in some very small regions. In vehicle image verification, it's desirable to spot exactly the regions where two vehicles are different, so that we can interpret the verification results.

3   Approach

3.1   Image Alignment

Due to the difference in viewpoints, the misalignment between vehicle images are influential to precise location of the small different regions in the images. Therefore, the first step of our approach is to align two vehicle images. Note that we focus on the frontal vehical truck image verification which is ubiquitous in real-life transportation system. The front facade of a truck can be roughly modelled as a plane, therefore the images from different viewpoint are related via an perspective transformation. We first match the two images via SIFT [5] features, and then estimate the transformation between the images using RANSAC [2] algorithm. Two original images are transformed and cropped according to their overlapping regions. Figure 2 illustrates the alignment process.

3.2   Network Architecture

Let X_1, X_2 be two aligned images with the same size of $H \times W$. Our approach predicts a label $y \in \{0, 1\}$ and a saliency map $S \in (0, 1)^{H \times W}$ for X_1 and X_2:

$$(y, S) = f(X_1, X_2; \theta) \tag{1}$$

where y indicates whether X_1 and X_2 are identical ($y = 0$) or not ($y = 1$), $S(i, j) \in [0, 1]$ indicates the degree to which $X_1(i, j)$ differs from $X_2(i, j)$, and θ is the parameter of our model.

We represent f(Eq. (1)) as a deep convolutional encoder-decoder network [16] (Fig. 3). The encoder is a Siamese architecture [4] to extract features from two images, and to verify the two images (predicting y). The decoder is a deconvolutional network [6] to regress a saliency map from these features. The two branches of the encoder are modified from VGG16 [12]. There are thirteen convolutional layers, five pooling layers and three fully connected layers in the VGG16. Since the dimension of the last fully connected layer usually represents the numbers of classes in the classification task, but this is unnecessary in our tasks. In addition, in the CNN network, the fully connected layer maps the feature map generated by the convolutional layers into a vector of fixed length. However, the dimension of the vector in the last fully connected layer of VGG16 is lower than the dimension of the vector of the first two fully connected layers, which will cause loss of image informations. Moreover, removing a fully connected layer can reduce redundancy of the parameter in the fully connected layer. Therefore, we remove the last fully connected layer and use the first two FC layers as feature extractor for label prediction. In this way, the convolutional network is composed of thirteen convolutional layers, five pooling layers and two fully connected layers.

Vehicle Verification. Let Z_1 and Z_2 be the feature vector for X_1 and X_2, respectively, which are extracted from the output of the 2nd FC layer of the encoder. We predict the label y for X_1 and X_2 using logistic regression:

$$Pr(y = 1|X_1, X_2) = \frac{1}{1 + e^{-(W_y^T |Z_1 - Z_2| + b_y)}}$$ (2)

where $|Z_1 - Z_2|$ is a vector of the point-wise absolute difference of Z_1 and Z_2. W_y and b_y are parameters.

Saliency Prediction. Let V_1 and V_2 be the feature maps of X_1 and X_2, output from the POOL-5 layer of the convolutional network. We concatenate V_1 and V_2 into V, and regress a saliency map S from V using a deconvolutional network. The deconvnet is designed by reversing the VGG16 network. We concatenated feature maps V as input. The output of its last layer, F, the size of which is 224×224. We apply a 1×1 convolution on this feature map, followed by a sigmoid activation, to generate the saliency map S:

$$S(i, j) = \frac{1}{1 + e^{-(W_S^T F(i,j) + b_S)}}$$ (3)

where $F(i, j)$ is the feature vector of pixel (i, j) on the feature map F, W_S and b_S are parameters of the 1×1 convolution kernel.

3.3   Network Training

Given a labeled training set $\{X_1^i, X_2^i, y^i, M^i\}_{i=1}^N$. $M^i \in \{0, 1\}^{H \times W}$ is the ground truth saliency map for X_1^i and X_2^i. The parameter θ is estimated by minimizing the following multi-task loss:

$$l = l_{verify} + l_{saliency}$$ (4)

where l_{verify} is the verification loss, which is typically defined as log-loss:

$$l_{verify} = \frac{1}{N} \sum_{k=1}^{N} -y^k log(\rho^k) - (1 - y^k)log(1 - \rho^k) \tag{5}$$

where $\rho^k = Pr(y^k = 1|X_1^k, X_2^k)$ is the output of the logistic regressor (Eq. (2)).

$l_{saliency}$ represents the saliency regression loss in Eq. (4), which is defined as the mean squared error between the ground truth map and the predicted map:

$$l_{saliency} = \frac{1}{N} \sum_{k=1}^{N} \sum_{i=1}^{H} \sum_{j=1}^{W} (M^k(i, j) - S^k(i, j))^2 \tag{6}$$

where S^k is the predicted saliency map for X_1^k and X_2^k.

We use stochastic gradient descent algorithm to train our network. The hyper-parameters for training are almost the same as VGG16 [12], Except for the initial global learning rate which is 0.01 in our experiments. And we reduce the learning rate by 0.1 after 1000 iterations.

3.4   Improved Loss Function

We analyzed the verification results on the training set, and found that the average saliency of most of the false positives are less than those of the true positives. Their distributions are shown in Fig. 4. From Fig. 4, we can found that the average saliency of false positive samples is slightly lower than the average saliency of true positive samples, this is because the dissimilarity of false positive samples is lower than the dissimilarity of true positive samples. It means that the model of verification task can not distinguish highly similar vehicle image pairs (i.e. it outputs a lower ρ). This is consistent with our intuitions. Two images with less different regions (low saliency value) are more hard to discriminate. In order to reduce the false positive rates of the verification, we modified the verification loss as:

$$l'_{verify} = \frac{1}{N} \sum_{k=1}^{N} -\frac{1}{\mu_S^k} y^k log(\rho^k) - (1 - y^k)log(1 - \rho^k) \tag{7}$$

where $\rho^k = Pr(y^k = 1|X_1^k, X_2^k)$ is the output of the logisitc regressor (Eq. (2)), and $\mu_S^k \in (0, 1)$ is the average saliency of the saliency map M^k.

In this new loss function, we add a penalty factor, $\frac{1}{\mu_S^k}$, to the first term of the verification loss. The reason is that when the average saliency μ_S^k is small, the logistic regressor tends to output a small ρ, which will lead to a false positive, so we add a larger penalty factor to $-log(\rho)$. When μ_S^k is large, the output of the regressor will be large, results in a true positive, so we add a smaller penalty factor to $-log(\rho)$.

We compare the output probability ρ of the logistic regressor using the two different loss function (5) and (7). We found that probability ρ's trained with

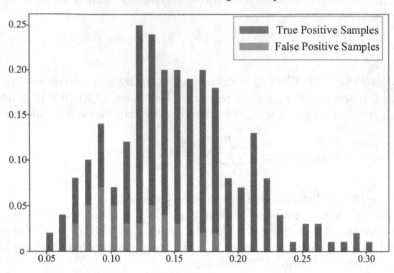

Fig. 4. The distribution of average saliency.

the modified loss (7) are larger than those trained with (5). By using the new model, the verification accuracy reached to 88.7%, and the false positive rate reduce to 0.025 (as shown in Table 1).

In addition, we analyzed the relationship between label prediction and average saliency by calculating the correlation coefficient. The formula of correlation coefficient is:

$$C = \frac{\frac{1}{N-1} \times \sum_{i=1}^{N}(\mu_S^i - \bar{S}) \times (\rho^i - \bar{\rho})}{\sqrt{\sum_{i=1}^{N}(\mu_S^i - \bar{S})^2 \times (\rho^i - \bar{\rho})^2}} \tag{8}$$

where μ_S^i is the average saliency of the saliency map, and $\bar{S}$ is the mean of average saliency for all samples. ρ^i represents the output of logistic regressor, and $\bar{\rho}$ represents the mean of output of logistic regressor for all samples.

The correlation coefficient is 0.8034 when trained with loss (5), and increased to 0.8407 when trained using loss (7). This means that using our modified loss function, the output probability of logistic regressor is more correlated with the difference saliency of the input images.

4   Experiments and Analysis

4.1   Dataset

In order to validate our method on verificating very similar vehicle images, we labeled 5K pairs of frontal view truck images. The trucks in each image pair share the same maker, type and color. The label of each pair is determined according to their license plate number. We randomly select 4K pairs as training data, of which 2.5K pairs are positive samples (different vehicle image pairs) and the other pairs are negative samples (identical image pairs). We use the remaining 1000 image pairs as test data.

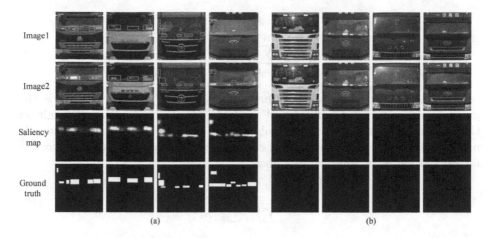

Fig. 5. Samples of qualitative results. (a) Vehicle pairs with local differences. (b) Identical vehicle pairs.

For each image pair, we have labeled every different part of two images by using a rectangle except for the crews in the cockpits. The ground truth labels are converted to a binary saliency map with the same size of the images, where every pixel in the rectangles are set to 255, and others are set to 0. Some of the data and their ground truth saliency maps are shown in Figs. 5 and 8. When training our network, we augment the training data set by randomly cropping a subimage. The augmented dataset contains 8000 image pairs.

4.2 Qualitative Results

We tested 1000 vehicle image pairs and calculated the verification accuracy. The accuracy reached to 88.7% (as shown in Table 1). Figure 5 shows ten sets of qualitative results. As we can see, the local differences such as decorations, inspection marks can be detected precisely. Besides, some interference factors can be ignored, such as the slight stains and the crews in the cockpit. Because these interference factors are labeled as background in the training data.

4.3 Evaluation of Saliency Map

In this section, we evaluated the generated saliency map from various two predicted saliency map and ground-truth map. The first step is to binarize saliency map, then we calculate MAE using the following formula:

Table 1. Comparison of two loss functions.

Evaluation criteria	Accuracy	FPR	Correlation coefficient
Loss (5)	0.830	0.096	0.8034
Loss (7)	0.887	0.025	0.8407

$$MAE = \frac{1}{W * H} \sum_{i=1}^{W} \sum_{j=1}^{H} |\bar{M}(i,j) - \bar{S}(i,j)| \tag{9}$$

where $\bar{S}$ is the binarized saliency map for X_1^i and X_2^i, and where $\bar{M}$ is the ground-truth map for X_1^i and X_2^i. W and H indicate the width and height of the map, respectively. Then, we calculated the pixel accuracy, which is a simple metric that aims to mark the ratio of correct pixels to total pixels. The formular for calculating pixel accuracy as follows:

$$PA = \frac{\sum_{i=1}^{k} P_{ii}}{\sum_{i=1}^{K} \sum_{j=1}^{K} P_{ij}} \tag{10}$$

where P_{ii} represents the numbers of pixels that are predicted correctly, and P_{ij} are the numbers of total pixels. The values are shown in Table 2.

Table 2. Quantitative results of saliency map prediction.

MAE	Pixel accuracy
0.1836	0.7615

4.4 Quantitive Comparison with Other Methods

For comprehensive experiment, we compared and analyzed our approach with other methods. The corresponding experimental details and analysis will be discussed as follows.

Object Detection. This approach uses object detection [9] method to consider local differences between two images. The main idea is to stack two images as the input and to take differences between two images as the object. The details are as follows: firstly, two RGB images are stacked into a 6-channel image [18] and fed into the network. Then we follow the pipeline of state-of-the-art object detection framework. In particular, we use Faster R-CNN [10] as the detection pipeline, i.e., "CNN feature extraction + region proposal + classification". Input layer is a 6-channel image.

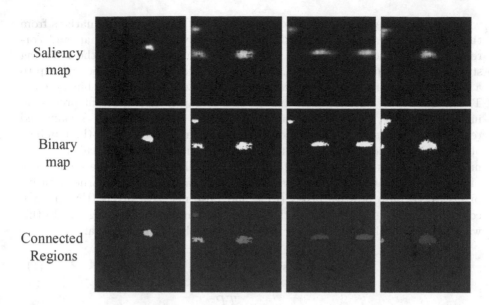

Fig. 6. The top is the saliency map. Binary map and connected regions map are in the middle and bottom, respectively.

Analysis. In this section, we analyzed the object detection and our approach. Both methods consider the local informations between vehicle image pairs. In object detection method, the feature of "difference" is learned via CNN, and regions of possibly local differences are proposed by Region Proposal Network in Faster R-CNN [10]. The advantage of this method is that features of local differences will not fade away in a whole image.

In our approach, we developed a multi-task deep convolutional encoder-decoder network. First, we used a convolutional neural network (CNN) to classify two images as identical or not. Then, we used deconvolutional networks to predict a saliency map which indicates the degree of their difference at every pixel. We not only considered the holistic feature, but also analyzed the differences of local informations. Specifically, we not only predicted the label, but also generated the saliency map to interpret the local differences. Besides, we analyzed the correlation between label prediction and average saliency. Therefore, our approach is more comprehensive than the object detection method.

In order to make the analysis more adequate, we calculated the precision and recall of two methods and compared the average precision. In detection object method, we calculated intersection-over-union (IOU $\in (0,1)$), which is the overlap ratio between the bounding box generated by the model and the ground truth. The best result is that the bounding box and ground truth overlap completely, i.e. the IOU is equal to 1. Then, we set a threshold τ, and specify that the corresponding bounding boxes are correctly detected when IOU is greater than or equal to thereshold τ. In this way, we finally calculated the precision and recall.

However, since we can hardly obtain the bounding boxes informations from the saliency map, we can not calculate IOU between saliency map and corresponding ground-truth directly in our approach. So we adopt the following strategy to evaluate our approach. We converted the predicted saliency map to a binary map at first. Then we extracted connected regions from binary map. The extraction of connected regions in binary map is an important process in image processing, which can be applied in many fields. We used 8-connected region algorithm to mark each pixel. The algorithm can calculate the numbers of connected regions during one-time image scanning, and can calculate the numbers of pixels in each connected region. Finally, according the numbers of pixels, we calculated the numbers of true positive pixels (TP), true negative pixels (TN), false positive pixels (FP), and false negative pixels (FN) in each connected region. Figure 6 shows the binary map and connected regions. In this way, we finally calculated the precision and recall by follow formula:

$$precision = \frac{TP}{TP + FP} \qquad (11)$$

$$recall = \frac{TP}{TP + FN} \qquad (12)$$

It should be noted that when the saliency map is binarized, the threshold is selected from 0 to 255. When we select each threshold, we calculate a set of corresponding precision and recall for all binary map. Then we averaged the

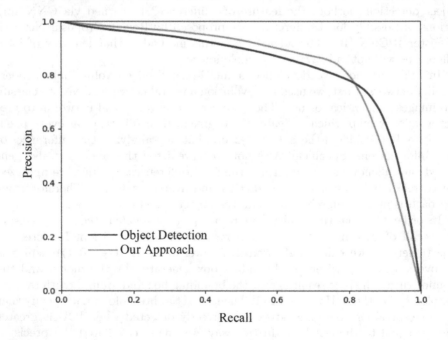

Fig. 7. The PR curves of two methods.

precision and recall values for all image pairs in this threshold. In this way, 256 pairs of precision and recall values will be obtained. Using recall as the abscissa and precision as the ordinate, we plot the precision-recall (PR) curve. The PR curves of two methods is in Fig. 7. Besides, we calculated average precision (AP) of two methods. By calculation, The AP of our approach is 0.8068, while the value of the object detection is 0.8013. The AP of two methods are fairly close, among which our method has a small advantage.

4.5   Limitation

In the actual traffic scenes, due to the illumination changes between different viewpoints, vehicle images may be captured under different illumination, which will result in specular reflection at some viewpoints. Figure 8 demonstrates some inaccurate results affected by high light.

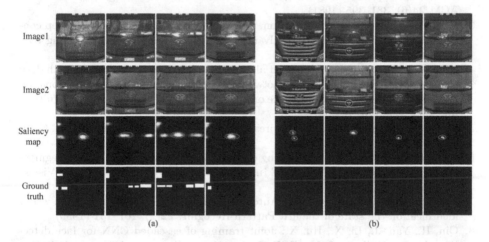

Fig. 8. Some inaccurate results. (a) There are illumination changes between viewpoints, which result in specular reflection at a specific viewpoint, so that the prediction of saliency map is inaccurate, but the label prediction is correct. (b) Specular reflection led to both the saliency map prediction and label prediction are incorrect.

5   Conclusion

We present a novel method to discriminate similar vehicle images. Our method use a convolutional neural network (CNN) to extract two vehicle images features. We first verify two vehicle whole images. We use logistic regression to measure holistic difference of features. Then we concatenate the features of two images, from which we predict a saliency map using a deconvolutional network. The saliency map shows the fine-grained differences between two vehicle images. Our network design enables end-to-end training. We validate our algorithm on a vehicle image dataset. Experimental results show that our approach is fast, effective while using very cheap annotation. For similar vehicle images, we can perform fast and efficient verification. Most importantly, we can provide evidences for the verification. In order to make the experiments more comprehensive, we have added two comparison methods and conducted comparative experiments, and experimental results show our approach achieves better performance than other methods. Furthermore, we can extend our framework into wider verification applications.

References

1. Bhattarai, B., Sharma, G., Jurie, F.: CP-mtML: coupled projection multi-task metric learning for large scale face retrieval. In: Proceedings of the IEEE Conference on Computer Vision and Pattern Recognition, pp. 4226–4235 (2016)
2. Fischler, M.A., Bolles, R.C.: Random sample consensus: a paradigm for model fitting with applications to image analysis and automated cartography. Commun. ACM **24**(6), 381–395 (1981)
3. Krizhevsky, A., Sutskever, I., Hinton, G.E.: Imagenet classification with deep convolutional neural networks. In: Advances in Neural Information Processing Systems, pp. 1097–1105 (2012)
4. Kumar, B., Carneiro, G., Reid, I., et al.: Learning local image descriptors with deep siamese and triplet convolutional networks by minimising global loss functions. In: Proceedings of the IEEE Conference on Computer Vision and Pattern Recognition, pp. 5385–5394 (2016)
5. Lowe, D.G.: Distinctive image features from scale-invariant keypoints. Int. J. Comput. Vis. **60**(2), 91–110 (2004)
6. Noh, H., Hong, S., Han, B.: Learning deconvolution network for semantic segmentation. In: Proceedings of the IEEE International Conference on Computer Vision, pp. 1520–1528 (2015)
7. Plamondon, R., Lorette, G.: Automatic signature verification and writer identification-the state of the art. Pattern Recognit. **22**(2), 107–131 (1989)
8. Qin, H., Yan, J., Li, X., Hu, X.: Joint training of cascaded CNN for face detection. In: Proceedings of the IEEE Conference on Computer Vision and Pattern Recognition, pp. 3456–3465 (2016)
9. Redmon, J., Divvala, S., Girshick, R., Farhadi, A.: You only look once: unified, real-time object detection. In: Proceedings of the IEEE Conference on Computer Vision and Pattern Recognition, pp. 779–788 (2016)
10. Ren, S., He, K., Girshick, R., Sun, J.: Faster R-CNN: towards real-time object detection with region proposal networks. In: Advances in Neural Information Processing Systems, pp. 91–99 (2015)
11. Simonyan, K., Vedaldi, A., Zisserman, A.: Deep inside convolutional networks: visualising image classification models and saliency maps. arXiv preprint arXiv:1312.6034 (2013)
12. Simonyan, K., Zisserman, A.: Very deep convolutional networks for large-scale image recognition. arXiv preprint arXiv:1409.1556 (2014)
13. Sun, Y., Chen, Y., Wang, X., Tang, X.: Deep learning face representation by joint identification-verification. In: Advances in Neural Information Processing Systems, pp. 1988–1996 (2014)

14. Sun, Y., Wang, X., Tang, X.: Deeply learned face representations are sparse, selective, and robust. In: Proceedings of the IEEE Conference on Computer Vision and Pattern Recognition, pp. 2892–2900 (2015)
15. Taigman, Y., Yang, M., Ranzato, M., Wolf, L.: DeepFace: closing the gap to human-level performance in face verification. In: Proceedings of the IEEE Conference on Computer Vision and Pattern Recognition, pp. 1701–1708 (2014)
16. Tan, Z., Liu, B., Yu, N.: PPEDNet: pyramid pooling encoder-decoder network for real-time semantic segmentation. In: Zhao, Y., Kong, X., Taubman, D. (eds.) ICIG 2017. LNCS, vol. 10666, pp. 328–339. Springer, Cham (2017). https://doi.org/10.1007/978-3-319-71607-7_29
17. Yi, D., Lei, Z., Liao, S., Li, S.Z.: Learning face representation from scratch. arXiv preprint arXiv:1411.7923 (2014)
18. Zagoruyko, S., Komodakis, N.: Learning to compare image patches via convolutional neural networks. In: Proceedings of the IEEE Conference on Computer Vision and Pattern Recognition, pp. 4353–4361 (2015)

Learning Noise-Aware Correlation Filter for Visual Tracking

Xinyan Liang[1], Xiao Wang[1], Jin Tang[1], and Chenglong Li[1,2(✉)]

[1] School of Computer Science and Technology, Anhui University, Hefei, China
lxy2626@foxmail.com, wangxiaocvpr@foxmail.com, jtang99029@foxmail.com,
lcl1314@foxmail.com
[2] Center for Research on Intelligent Perception and Computing, NLPR, CASIA,
Beijing, China

Abstract. Correlation filter has recently attracted much attention in visual tracking due to their excellent performance on both accuracy and efficiency. However, the adopted features, such as Colors, HOG and deep features, usually include noises and/or corruptions which might disturb the tracking performance. To handle this problem, we propose a novel noise-aware correlation filter method for robust visual tracking. In particular, we decompose the input feature matrix into a "clean" feature matrix and a sparse noise matrix, and then use the "clean" feature to train the correlation filter. To optimize the proposed correlation filter, we design an efficient ADMM (alternation direction of multipliers) solver. Extensive experimental results on the OTB-2013 dataset show that the proposed approach performs favorably against state-of-the-art trackers.

Keywords: Correlation filter · Feature decomposition
"Clean" feature

1 Introduction

Visual tracking is one of the most challenging and active tasks in computer vision and has drawn much attention due to its wide applications, such as video surveillance, human-computer interactions, and self-driving cars. Given the ground truth in the initial frame, the goal of visual tracking is to find all bounding boxes of the target object in subsequent frames. Despite many recent breakthroughs in visual tracking, it still remains challenge due to diverse factors, such as occlusion, object deformation, scale variation and background clutter.

Correlation Filter (CF) has recently attracted much attention in visual tracking [1,6,7,11,14,15,24,29–31] due to their excellent performance in both accuracy and efficiency. CF trackers employ the cyclic shifts to generate dense samples, and diagonalize them in the Fourier domain by using the Fast Fourier Transform (FFT). It enables trackers robust and high speed. The seminal work of CF tracking is proposed by Bolme et al. [1], which achieves hundreds of frames

© Springer Nature Singapore Pte Ltd. 2018
Z. Xu et al. (Eds.): Big Data 2018, CCIS 945, pp. 210–226, 2018.
https://doi.org/10.1007/978-981-13-2922-7_14

Fig. 1. Some examples of noises in bounding boxes.

per second and high tracking accuracy. However, MOSSE only employs the simple feature to represent objects, *i.e.*, brightness feature, without enough to be adopted in some complicated situations. To improve the tracking performance, most successful CF trackers use a discriminative object representation with either strong hand-crafted features such as HOG [7,15,24], color names [9,24], or deep features [6,8,25]. Recent work has integrated deep features [25] trained on large dataset, such as ImageNet, to represent objects. In addition, multiple type of features are also employed together to robustly represent the tracked object, such as HOG features and color name [7,24], and HOG features, color name and deep features [6,8]. Although these trackers have achieved appealing results in both accuracy and computational efficiency, they ignore that these features might be polluted by noises or corruptions. As shown in Fig. 1, the bounding box of objects usually has several background information, which caused by occlusion or irregular shape of objects. Noises in features result in model drifting by influencing the learned appearance model and filter. Figure 2 shows the tracking results of Dual Correlation Filter (DCF) [15] on sequence **Soccer**. It illustrates that DCF will lose the object with noises in bounding box, which suggests that noises influence the tracking performance.

Motivated by the robust principal component analysis (RPCA) [3], Sui [28] decompose the feature matrix into a low-rank matrix and noise matrix. But the optimization of low-rank constraints refers to singular value decomposition (SVD). And SVD has high computation complexity and extremely time consuming. It influences the efficiency of trackers while real-time is a crucial factors in visual tracking. Therefore, we propose a simple and efficient feature decompose algorithms in this paper. According to [19–22,28], we decompose the feature into the "clean" feature and noises. We do not impose the low-rank constraints on the "clean" feature and only impose the sparse constraint on the noises due to observations from Li *et al.* [17–19,21,22]. They think that the noise matrix is the sparse sample-specific corruptions, *i.e.*, a few patches are corrupted and others are clean. Motivated by these works, we suppose that the noise in features is also sparse. We aim to learn the "clean" feature through imposing the sparse constraint on noises. The noise-aware filter is optimized by the learned "clean" feature for mitigate noises effects on filter. The simple feature decomposition is incorporated into CF tracking framework. It improves the tracking accuracy by suppressing noises effects on features and filters.

This paper makes the following contributions to CF tracking and related applications as follows. First, we propose a noise-aware correlation filter tracking

Soccer Index # 55   Soccer Index #153

Fig. 2. The example represents the effectiveness of our methods. Green, red, and blue box represents the ground truth, the tracking results of DCF, DCF_{our}, respectively. (Color figure online)

algorithms based on feature decomposition. The noise-aware filter and "clean" feature can be jointly optimized in an unified framework. Second, we also design an efficient ADMM (Alternation Direction Method of Multipliers) algorithm [2] that can optimize the filter and the "clean" features in a framework. Third, extensive experiments are carried out on public benchmark datasets. The evaluation results demonstrate the effectiveness of the proposed approach against the baseline methods.

2   Related Work

Recently, CF has obtained great achievement in visual tracking due to its accuracy and computational efficiency. Bolme *et al.* [1] first introduce the CF into visual tracking, which achieves hundreds of frames per second, and high tracking accuracy. However, there is a problem about MOSSE that it only employs the simple brightness feature of image that isn't enough to adapt to some complicated situations.

More and more discriminative features are utilized in tracking to represent objects for improving the performance, such as HOG [7,15,24], Color Names [9,24] and deep features [6,8,25]. In addition, several trackers [6–8,24] employ multiple type features to represent the object for more robust tracking. To further enhance the ability to classify objects from the background, kernel tricks, which make the inseparable samples in low-dimension are mapped to the high-dimension space to achieve the purpose of classification, are used in CF tracking. Henriques *et al.* [14,15] employ the kernel trick to improve performance. To adaptively employ complementary features, Tang *et al.* [30] propose a multi-kernel learning algorithm to improve performance. However, different kernels of MKCF may restrict each other in training and updating, which limits its improvement over KCF [15]. In addition, the increased computational cost of MKCF in comparison to KCF limits the tracking speed. Therefore, Tang *et al.* [31] employ a different way [30] to introduce the MKL into KCF. The way not only adaptively exploits multiple complementary features and non-linear kernels more effectively than MKCF, but also keeps relative high speed. CF framework usually faces a problem that boundary effect which is caused by utilizing a periodic assumption of the training samples to efficiently learn a classifier. To address the boundary effect, SRDCF [7] is proposed by introducing the spatially regularized into the learning of correlation filter to penalize the filter coefficients near the boundary. In CSR-DCF [24], spatial reliability map is constructed to adjust the filter support to the part of object suitable for tracking. To adapt the size variation, several adaptive scale processing tracker [5,23] are investigated. Danelljan *et al.* [5] utilize the two correlation filters to capture the location translation and scale estimation, respectively. Li *et al.* [23] employ an effective scale adaptive scheme and integrate the HOG features and Color Name features to boost the tracking performance.

In addition, Danelljan *et al.* [8] utilize continuous convolution to integrate multi-resolution feature maps. The factorized convolution operator and the generative sample space model are introduced into tracking [6] for addressing the over-fitting and computational complexity. CFnet [32] is the first to introduce the correlation filter into a deep neural network as a differentiable layer. Sun *et al.* [29] treat the filter as the element-wise product of a base filter and a reliability term to learn the discriminative and reliable information for improving the tracker's accuracy and robustness.

3   Review of Correlation Filters Tracking

In this section, we simply introduce the classical correlation filter tracking framework. Given the training set $T = [(\mathbf{x}_1, y_1), (\mathbf{x}_2, y_2), \ldots (\mathbf{x}_n, y_n)]$, we find a linear regression function $h(\mathbf{x}) = \mathbf{w}^T \mathbf{x}$ that minimizes the squared error over samples $\mathbf{x}_i$ and their regression targets $\mathbf{y}_i$. The model can be written as follows:

$$\min_{\mathbf{w}} \sum_i (h(\mathbf{x}_i) - \mathbf{y}_i)^2 + \lambda \|\mathbf{w}\|^2 \tag{1}$$

where λ is a regularization parameter that controls overfitting. Equation (1) has a closed-form, which is given by [26]

$$\mathbf{w} = (\mathbf{X}^T \mathbf{X} + \lambda \mathbf{I})^{-1} \mathbf{X}^T \mathbf{y}. \tag{2}$$

where $\mathbf{X}$ is circulant matrix generated by the base sample $\mathbf{x}$, $\mathbf{I}$ is identity matrix. The per row of circulant matrix $\mathbf{X}$ is one virtual sample $\mathbf{x}_i$ obtained through the cyclic shift of the base sample $\mathbf{x}$. Let $\mathbf{y} = [\mathbf{y}_1, \mathbf{y}_2, ..., \mathbf{y}_n]^T$ is a regression target of samples $\mathbf{X}$. Each element $\mathbf{y}_i$ is a regression target of $\mathbf{x}_i$.

To calculate in the Fourier domain, the solution (2) is transformed into the complex version as follows:

$$\mathbf{w} = (\mathbf{X}^H \mathbf{X} + \lambda \mathbf{I})^{-1} \mathbf{X}^H \mathbf{y}. \tag{3}$$

where $\mathbf{X}^H$ is the Hermitian transpose, that is the transpose of the complex-conjugate of $\mathbf{X}$, $\mathbf{X}^H = (\mathbf{X}^*)^T$. The circulant matrix $\mathbf{X}$ can be expressed as diagonal of $\mathbf{x}$ by the Discrete Fourier Transform (DFT) [12]:

$$\mathbf{X} = F \; diag(\hat{\mathbf{x}})F^H \tag{4}$$

where $\hat{\mathbf{x}}$ denotes the DFT of $\mathbf{x}$, that is $\hat{\mathbf{x}} = \mathcal{F}(\mathbf{x})$, $diag(\mathbf{x})$ denotes the diagonal matrix of a vector $\mathbf{x}$. F is a constant matrix that does not depend on the generating vector $\mathbf{x}$, as $\mathcal{F}(\mathbf{z}) = \sqrt{n}F\mathbf{z}$. The notation n is the size of the generating vector $\mathbf{x}$. From now on, we will alway use a hat $\hat{\mathbf{a}}$ as shorthand for the DFT of vector $\mathbf{a}$.

The property (4) of circulant matrix can be applied to the solution (3), which is expressed as follows:

$$\hat{\mathbf{w}} = \frac{\hat{\mathbf{x}}^* \odot \hat{\mathbf{y}}}{\hat{\mathbf{x}}^* \odot \hat{\mathbf{x}} + \lambda} \tag{5}$$

where $\odot$ and the fraction denote element-wise product and division, respectively. And $\mathbf{x}^*$ represents the complex-conjugate of $\mathbf{x}$.

The Eq. (1) can be transformed into dual domain. The dual objective function can be written as follows:

$$\min_{\alpha} \frac{1}{4\lambda}\alpha^T \mathbf{X}\mathbf{X}^T\alpha + \frac{1}{4}\alpha^T\alpha - \alpha^T\mathbf{y} \tag{6}$$

where α is the dual variable. The two solutions from the objective function (1) and (6) are related by $\mathbf{w} = \frac{\mathbf{X}^T\alpha}{2\lambda}$. Here, for clarity and avoiding the calculation of cyclic matrix, the dual form is rewritten as:

$$\min_{\alpha} \frac{1}{4\lambda}\alpha^T \; C(\mathbf{x})\,C(\mathbf{x})^T\alpha + \frac{1}{4}\alpha^T\alpha - \alpha^T\mathbf{y} \tag{7}$$

where $C(\mathbf{x})$ denotes the cyclic matrix generated by the base sample $\mathbf{x}$. The variable α can be optimized efficiently in the Fourier domain:

$$\hat{\alpha} = \frac{\hat{\mathbf{y}}}{\frac{1}{2\lambda}\hat{\mathbf{x}}^* \odot \hat{\mathbf{x}} + \frac{1}{2}} \tag{8}$$

where the fraction denote element-wise division.

The final response map ϕ is calculated through the following equation:

$$\phi = \mathcal{F}^{-1}(\hat{\mathbf{w}} \odot \hat{\mathbf{x}}) \; = \mathcal{F}^{-1}(\hat{\mathbf{k}}^{\hat{\mathbf{x}}\hat{\mathbf{z}}} \odot \hat{\alpha}) \tag{9}$$

where $\mathcal{F}^{-1}$ denotes the inverse operation of DFT, $\hat{\mathbf{k}}^{\hat{\mathbf{x}}\hat{\mathbf{z}}}$ is the kernel of training samples $\mathbf{x}$ and candidate patches $\mathbf{z}$. Then we can localize the target in current frame through the response map.

CF trackers usually employ several hand-craft features (e.g. HOG features, Color Name features) or deep learning features to represent the object for robust tracking. However, these features might be polluted by noises or corruptions. According to Eqs. (8) and (9), noises in features will influence the learned appearance model and filter and thus limit the tracking performance. As shown in Fig. 2, DCF and DCF$_{our}$ can both work well when the tracked object has not noises.

However, DCF loses the tracked object but the DCF$_{our}$ can successfully track the object when the object includes the noises. It illustrates that noises influence the learned filters and then limit the trackers performance. Therefore, we propose a noise-aware correlation filter tracking algorithm to mitigate the noises effect on filters. In next section, we will mainly introduce how the algorithms works in correlation filter tracking framework.

4   Methods

In this section, we give a detailed description about how to learn jointly noise-aware filter and "clean" feature through feature decomposition for suppressing the influence of noises.

4.1   Noise-Aware Correlation Filter

Objects can be represented using hand-craft features (e.g. HOG features [10], Color Name features [33]). Besides the hand-craft features, the deep features extracted from the VGG model [27] also are applied to represent the object. However, whether hand-craft features or deep features, they may be polluted by noises or corruptions. RPCA [3] has a powerful capability to suppress noises or corruptions. But the optimization of RPCA refers to singular value decomposition (SVD). And SVD has high computation complexity and extremely time consuming. It influences the efficiency of trackers. Therefore, we propose a simple and efficient feature decompose algorithms, which decomposes the feature $\mathbf{x}$ into the "clean" feature $\mathbf{z}$ and noises $\mathbf{e}$. And we impose the sparse constraint on noises and do not impose the low-rank constraint on the "clean" feature. The learned "clean" feature is used to optimize the noise-aware filter for mitigate noises effects.

$$\mathbf{x} = \mathbf{z} + \mathbf{e} \tag{10}$$

The Eq. (10) is incorporated into CF tracking framework to jointly optimize variables. The "clean" feature $\mathbf{z}$ is used to optimize the noise-aware filter and make the filter more robust. The dual form model (7) can be reformulated as Eq. (11), which can learn the "clean" features and the noise-aware filter to improve the accuracy and robustness.

$$\min_{\alpha, \mathbf{z}, \mathbf{e}} \frac{1}{4\lambda} \alpha^T C(\mathbf{z}) C(\mathbf{z})^T \alpha + \frac{1}{4} \alpha^T \alpha - \alpha^T \mathbf{y} + \beta \|\mathbf{e}\|_0$$
$$s.t.\ \mathbf{x} = \mathbf{z} + \mathbf{e} \tag{11}$$

where λ is a regularization parameters that controls overfitting, and β is balanced parameter.

Because the l_0 norm is non-convexity, it is difficult to directly optimize the Eq. (11). To overcome the obstacles, we will use convex relaxation to relax the non-convex sparsity terms into the convex sparsity terms. We replace the l_0 norm with l_1 norm using the convex relaxation. Thus, the Eq. (11) can be relaxed as:

$$\min_{\alpha, \mathbf{z}, \mathbf{e}} \frac{1}{4\lambda} \alpha^T \boldsymbol{C}(\mathbf{z}) \boldsymbol{C}(\mathbf{z})^T \alpha + \frac{1}{4} \alpha^T \alpha - \alpha^T \mathbf{y} + \beta \|\mathbf{e}\|_1$$

$$s.t. \ \mathbf{x} = \mathbf{z} + \mathbf{e}$$

(12)

Although the Eq. (12) seems complex and is not joint convex, subproblem of each variable is convex by fixing other variables and has a closed-form solution. Therefore, the model can be optimized by the ADMM (alternating direction method of multipliers) algorithm [2]. As demonstrated in the experiments, the parameters of Eq. (12) are easy to adjust, and the tracking performance is insensitive to parameter variations.

4.2   Optimization

In this section, we mainly introduce how to solve the objective function (12). Although the variables of the (12) are not joint convex, the subproblem of each variable with others fixed is convex and has a closed-form solution. The ADMM is a effective solver of the problems like (12). By introducing augmented Lagrange multipliers, the optimization function (12) can be written as the following augmented Lagrange function:

$$\mathcal{L}_{(\alpha, \mathbf{z}, \mathbf{e})} = \frac{1}{4\lambda} \alpha^T \boldsymbol{C}(\mathbf{z}) \boldsymbol{C}(\mathbf{z})^T \alpha + \frac{1}{4} \alpha^T \alpha - \alpha^T \mathbf{y} + \beta \|\mathbf{e}\|_1$$
$$+ \frac{\mu}{2} \|\mathbf{x} - \mathbf{z} - \mathbf{e} + \frac{\mathbf{p}}{\mu}\|_2^2 - \frac{1}{2\mu} \|\mathbf{p}\|_2^2$$

(13)

where $\mu > 0$ is the penalty parameter and $\mathbf{p}$ is the Lagrangian multipliers. The ADMM method updates one of the variables by minimizing $\mathcal{L}$ with other variables fixed. By updating these variables iteratively, the convergence can be guaranteed [2]. Besides the Lagrangian multipliers $\mathbf{p}$, there are three variables that need to be updated, including $\alpha, \mathbf{z}, \mathbf{e}$. The closed form solution of each subproblems are as follows.

Update α (with others fixed): The optimization (13) with respect to the variable α can be formulated as follows:

$$\alpha = \operatorname*{argmin}_{\alpha} \frac{1}{4\lambda} \alpha^T \boldsymbol{C}(\mathbf{z}) \boldsymbol{C}(\mathbf{z})^T \alpha + \frac{1}{4} \alpha^T \alpha - \alpha^T \mathbf{y}$$

(14)

For calculating the variable α, we take the derivative of the α-subproblem (14) and set it to 0. The variable α has the closed-form solution. With some algebra, the closed-form solution of the variable α can be formulated as follows:

$$\alpha = (\frac{1}{2\lambda} \boldsymbol{C}(\mathbf{z}) \boldsymbol{C}(\mathbf{z})^T + \frac{1}{2} \mathbf{I})^{-1} \mathbf{y}$$

(15)

where $\mathbf{I}$ is the identity matrix. The amount of computation cost of Eq. (15) is large, mainly from matrix inverse and multiplication in spatial domain. For the fast operation in Fourier domain, the property (4) of cyclic matrix is introduced into the solution (15). The variable α is updated with only the base sample as follows:

$$\hat{\alpha}^{k+1} = \frac{\hat{y}}{\frac{1}{2\lambda}(\hat{z}^*)^k \odot \hat{z}^k + \frac{1}{2}} \tag{16}$$

where the fraction denotes the element-wise division. Finally, the α can be obtained via $\alpha = \mathcal{F}^{-1}(\hat{\alpha})$.

Update z (with others fixed): The z is updated through solving the subproblem (17) corresponding to z with the closed-form solution.

$$z = \underset{z}{\operatorname{argmin}} \frac{1}{4\lambda} \alpha^T C(z) C(z)^T \alpha + \frac{\mu}{2}\|x - z - e + \frac{P}{\mu}\|_2^2 \tag{17}$$

To solve and calculate efficiently in Fourier domain, the Eq. (17) is formulated as follows by Parsevaal's theorem:

$$\hat{z} = \underset{\hat{z}}{\operatorname{argmin}} \frac{1}{4\lambda} \hat{\alpha}^H diag(\hat{z}) diag(\hat{z}^*)\hat{\alpha} + \frac{\mu}{2}\|\hat{x} - \hat{z} - \hat{e} + \frac{\hat{P}}{\mu}\|_2^2 \tag{18}$$

where $diag(z)$ denotes the diagonal matrix of a vector z. With some algebra, the first term of Eq. (18) can be transformed as the following forms:

$$\hat{z} = \underset{\hat{z}}{\operatorname{argmin}} \frac{1}{4\lambda} (\hat{\alpha} \odot \hat{z})^H (\hat{\alpha} \odot \hat{z}) + \frac{\mu}{2}\|\hat{x} - \hat{z} - \hat{e} + \frac{\hat{P}}{\mu}\|_F^2 \tag{19}$$

where the derivative of $(\hat{\alpha} \odot \hat{z})^H (\hat{\alpha} \odot \hat{z})$ is $2(\hat{\alpha}^* \odot \hat{\alpha} \odot \hat{z})$ with some algebra. The solution of $\hat{z}$ is obtained by setting the derivative of (19) to 0.

$$\hat{z}^{k+1} = \frac{\mu(\hat{x} - \hat{e}^k) + \hat{p}^k}{\frac{1}{2\lambda}(\hat{\alpha}^*)^{k+1} \odot \hat{\alpha}^{k+1} + \mu^k} \tag{20}$$

where fraction denotes the element-wise division. Finally, the "clean" feature z can be obtained using the formula: $z = \mathcal{F}^{-1}(\hat{z})$.

Update e (with others fixed): The optimization (13) with respect to the variable e is formulated as follows:

$$e = \underset{e}{\operatorname{argmin}} \beta\|e\|_1 + \frac{\mu}{2}\|x - z - e + \frac{P}{\mu}\|_2^2 \tag{21}$$

The noise e is obtained by the soft-thresholding (or shrinkage) method [4] with closed-form solution:

$$e^{k+1} = \mathcal{S}_{\frac{\beta}{\mu^k}}(x - z^{k+1} + \frac{p^k}{\mu^k}) \tag{22}$$

where $\mathcal{S}_{\frac{\beta}{\mu}}(x)$ is the soft-thresholding operator for a vector x with parameter $\frac{\beta}{\mu}$. Here, the $\mathcal{S}_{\frac{\beta}{\mu}}(x)$ can be calculated through the Eq. (23).

$$\mathcal{S}_{\frac{\beta}{\mu}}(x) = (sign(x)) \odot (max(0, |x| - \frac{\beta}{\mu})) \tag{23}$$

Algorithm 1. Optimization Procedure to Equation(13)

Require:
   The object feature matrix $\mathbf{x}$, and the parameter β, λ, and μ;
   Set $\mathbf{z}^0 = \mathbf{x}$, $\mathbf{p}^0 = \mathbf{e}^0 = \mathbf{0}$, $\alpha^0 = 1$, $\mu_0 = 5$, $\mu_{max} = 20$, $\rho = 3$, $\tau = 10^{-10}$,
   $maxIter = 3$, and $k = 0$.
Ensure: $\alpha, \mathbf{z}$, and $\mathbf{e}$.
 1: **while** not converged **do**
 2:    Update $\mathbf{z}^{k+1}$ by Equation(20);
 3:    Update $\mathbf{e}^{k+1}$ by Equation(22);
 4:    Update α^{k+1} by Equation(16);
 5:    Update Lagrange multipliers as followings:
 6:    $\mathbf{p}^{k+1} = \mathbf{p}^k + \mu(\mathbf{x}^{k+1} - \mathbf{z}^{k+1} - \mathbf{e}^{k+1})$;
 7:    Update μ_{k+1} by $\mu_{k+1} = \min(\mu_{max}, \rho\mu_k)$;
 8:    Update k by $k = k + 1$;
 9:    Check the convergence condition: the maximum element changes of $\mathbf{z}, \mathbf{e}$, and α
        between two consecutive iterations are less than τ or the maximum number of
        iterations reaches $maxIter$.
10: **end while**

Besides the above variables, the Lagrange multiplier $\mathbf{y}$ is updated by following the Eq. (24):

$$\mathbf{p}^{k+1} = \mathbf{p}^k + \mu^k(\mathbf{x} - \mathbf{z}^{k+1} - \mathbf{e}^{k+1}) \tag{24}$$

Since each subproblem of (13) is convex, we can guarantee that the limit point by our algorithm satisfies the Nash equilibrium conditions [35]. The details of optimization procedure are shown in Algorithm 1.

4.3   Tracking

In this section, we briefly introduce the process of the localization and update steps of the proposed algorithm.

Localization. Features are extracted from the searching area in current frame. We employ the learned variables $\mathbf{z}, \alpha$ in previous frame to locate the object in current frame. The response map can be obtained as follows:

$$\phi = \mathcal{F}^{-1}(\hat{\mathbf{k}}^{\hat{\mathbf{x}}\hat{\mathbf{z}}} \odot \hat{\alpha}) \tag{25}$$

where $\hat{\mathbf{k}}^{\hat{\mathbf{x}}\hat{\mathbf{z}}}$ denotes the kernel between $\hat{\mathbf{z}}$ and $\hat{\mathbf{z}}$, and the $\hat{\mathbf{z}}$ represents the learned target appearance model using the "clean" feature, the $\hat{\mathbf{x}}$ denotes the feature in current frame. We locate the tracked object using the response map.

Learning-Update. The searching area is extracted in current frame. The extracted feature from searching area are feed into the Eq. (13) to optimize the dual variable and the "clean" feature. For capturing the variation of tracked target's appearance, the dual variable and the target appearance model are updated by an autoregressive model with learning rate. The updated strategy of dual variable α and appearance model $\hat{\mathbf{z}}$ follows the following formulation.

$$\hat{\alpha}^t = (1 - \eta)\hat{\alpha}^{t-1} + \eta\hat{\alpha}$$
$$\hat{\mathbf{z}}^t = (1 - \eta)\hat{\mathbf{z}}^{t-1} + \eta\hat{\mathbf{z}} \tag{26}$$

where η is learning rate. According to the above solution procedure, it does not refers to the large amount of computation operations like matrix inverse.

5   Experiments

In this section, we mainly introduce the implementation details of our model, evaluation datasets and evaluation metrics, and the analysis of experimental results compared with baseline and other state-of-the-art trackers in public benchmark.

5.1   Experimental Setup

Implementation Details and Parameters: To demonstrate the effectiveness of our model, we select the two different baseline trackers to implement the model. They are dual correlation filter methods (DCF) [15] based on hand-craft features and HCF [25] based on the deep features. The two baseline models both employ the dual model to optimize the problem. Firstly, the implementation details of embedding our model in DCF (DCF$_{our}$) are introduced. Standard HOG [10] descriptors are used to represent the object in DCF. We only embedded the feature decomposition model in DCF to jointly learn noise-aware filters and the "clean" feature for improving tracking performance. The β and λ in Eq. (13) is set to 0.005 and 0.1, respectively.

Next, we mainly introduce the implementation details of embedding our model in HCF. Ma *et al.* [25] extract three layers deep features from VGGNet [27], that are conv4,conv3-4 and conv5-4. The three convolution features independently learn the filter and the appearance model. The final response map is fused by three different response maps obtained by the three convolution features. We only implement noise-aware correlation filter model based on different convolution layers. The β and λ in Eq. (13) is set to 0.01 and 1e-4, respectively. The experiments are carried out on a PC with an Intel i7 4.2 GHz CPU and 32G RAM.

As seen from the above section, the parameter settings for DCF$_{our}$ and HCF$_{our}$ have great discrepancy, but this discrepancy is considered reasonable because DCF$_{our}$ and HCF$_{our}$ are different models. All parameters are optimal by varying them on a certain scope. Moreover, when we slightly adjust the parameters, tracking performance only change a little and Table 1 shows the results of the proposed method DCF$_{our}$ and HCF$_{our}$ with different parameters.

Datasets and Evaluation Metrics: Our method is evaluated on benchmark dataset: OTB-2013 [34] with 50 sequences. The images are annotated with

Table 1. The precision rate (PR) of the proposed method DCF$_{our}$ and HCF$_{our}$ with different parameters.

	Param	Setting	PR	Param	Setting	PR		Param	Setting	PR	Param	Setting	PR
HCF$_{our}$	λ	1e-2	0.888	β	0.005	0.893	DCF$_{our}$	λ	0.05	0.724	β	0.001	0.728
		1e-3	0.890		0.01	0.895			0.1	0.739		0.005	0.739
		1e-4	0.895		0.02	0.892			0.2	0.735		0.01	0.728

ground truth bounding boxes and various visual attributes. For the OTB-2013 dataset, we employ the one-pass evaluation (OPE) and use two metrics: precision rate (PR) and success rate (SR). PR is the percentage of frames whose output location is within the given threshold distance of ground truth. That is to say, it computes the average Euclidean distance between the center locations of the tracked target and the manually labeled ground-truth positions of all the frames. SR is the ratio of the number of successful frames whose overlap between predicted and ground truth bounding box is larger than a threshold. In the legend, we report the area under curve (AUC) of success plot and precision score at 20 pixels threshold corresponding to the one-pass evaluation for each tracking method.

Compared Trackers: To identify the effectiveness of our model, we evaluate the proposed model with comparisons to several state-of-the-art methods for evaluations. Several trackers based on correlation filter are selected to evaluate the performance, including DSST [5], KCF [15], CSR-DCF [24], SRDCF [7] and SAMF [23]. In addition, we also select several representative trackers to compare with our methods, Struck [13], TLD [16], SCM [36].

Table 2. The Mean FPS compared with DCF, KCF, DSST, SAMF, CSR-DCF trackers.

	DCF$_{our}$	DCF	KCF	DSST	SAMF	SRDCF	CSR-DCF
Mean FPS	77.52	564.64	374.16	60.30	11.49	7.81	18.18

5.2   Tracking Speed

The tracking speed is crucial in many realistic tracking applications. We therefore generalize the tracking speed about DCF and DCF$_{our}$ in Table 2. In addition, Table 2 enumerates other trackers based on correlation filter, including DCF, KCF, DSST, SAMF, SRDCF, CSR-DCF. It is verified that DCF$_{our}$ performs at about 77.52 FPS (frames per second) to achieve real-time tracking (equivalent to approximately 20 FPS) although it achieves the lower tracking speed than DCF.

RPCA [3] has a powerful capability to suppress noises or corruptions. RPCA is a NP-hard problem because it simultaneously involves $rank$ and ℓ_0. To overcome these obstacles, a convex relaxation of the problem is proposed, we use $\|\cdot\|_*$ and ℓ_1 to replace $rank$ and ℓ_0, respectively. It also should be note that although the convex relaxations are leveraged, the obtained problem is still non-convex. Fortunately, this problem is convex with respect to each of them when others are fixed. ADMM algorithm [2] has proven to be an efficient and effective solver of RPCA. However, it involves SVD in the process of iterative solution using ADMM. SVD has high computational complexity and extremely time consuming. Although the proposed method is also optimized by ADMM, SVD is not involved in the optimization.

Table 3. The precision rate (PR%) and success rate (SR%) on OTB-2013 over DCF, DCF_{our}, HCF, and HCF_{our}, DSST, SAMF, SRDCF, CSR-DCF, Struck, TLD, and SCM

	DCF	DCF_{our}	HCF	HCF_{our}	DSST	KCF	SAMF	SRDCF	CSR-DCF	Struct	TLD	SCM
PR	72.8	73.9	89.1	89.5	73.7	74.0	82.3	82.8	82.3	65.6	60.8	64.9
SR	50.8	51.4	63.5	65.4	55.4	51.4	60.5	60.8	59.6	55.9	52.1	61.6

5.3 Comparison with Baseline Methods

We compare our methods with the baseline trackers to demonstrate the strength of our proposed methods in this part.

Overall Evaluation: To illustrate the effectiveness of our proposed model, we implement the model in DCF and HCF, named DCF_{our} and HCF_{our}, respectively. Table 3 shows the overall evaluation results between DCF, HCF and DCF_{our}, HCF_{our} in OTB-2013. Benefiting from our proposed model, the DCF_{our} outperforms DCF in PR/SR by 1.1%/0.6% and 0.4%/1.9%, respectively. HCF_{our} achieves slight improvement over HCF in PR. But HCF_{our} exceeds 1.9% in SR over HCF. This evaluative performance illustrates that the feature decomposition scheme can improve the CF tracking performance. Table 4 represents tracking performance on several challenging factors between our trackers and the baseline trackers.

Attribute-Based Evaluation: We further analyze the tracking performance under different challenging attributes (*e.g.*, background clutter (BC), occlusion (OCC), fast motion (FM)) annotated in the benchmark OTB-2013. Table 4 shows PR and SR of one-pass evaluation (OPE) for ten main challenging factors. We mainly take DCF_{our} and DCF as an example to analysis the evaluative results. According to Table 4, we have the following observations.

First, our model is effective in handling motion blur (MB). In general, MB results in appearance degradation. And how to learn a good feature is critical to

Table 4. Attributed-based PR/SR on OTB-2013 compared with DCF, HCF, SRDCF, CSR-DCF, SAMF, KCF, DSST trackers.

Trackers	FM	BC	SV	MB	IV	OCC	LR	OPR	DEF	IPR
DCF	55.9/44.1	71.9/52.2	65.4/41.6	58.8/47.0	69.9/48.1	72.6/50.2	71.2/48.8	63.2/54.4	74.0/53.1	70.4/48.7
DCF$_{our}$	59.9/45.4	75.8/53.8	67.5/42.4	65.0/49.2	72.5/49.1	74.4 /51.2	37.8/31.0	72.7/49.5	74.1/53.5	72.8/50.1
HCF	78.4/58.3	88.3/63.4	88.3/60.0	84.6/62.4	84.3/59.9	87.8/64.1	89.7/60.2	86.8/61.5	88.0/65.9	86.7/60.1
HCF$_{our}$	79.0/61.4	85.6/62.8	88.3/62.9	80.5/61.9	85.7/ 61.8	88.1/66.4	87.5/62.3	87.5/63.4	87.3/65.1	85.6/61.2
SRDCF	77.2/58.4	81.4/58.8	77.7/57.5	78.2/59.2	79.0/58.6	83.7/60.9	40.4/33.8	83.3/60.0	84.9/62.8	74.8/54.0
CSR-DCF	68.8/53.0	77.4/55.4	71.2/51.7	74.6/57.4	75.5/56.1	78.7/58.4	40.2/32.3	81.3/57.7	87.5/64.1	78.2/55.7
SAMF	68.2/53.5	72.7/55.4	78.7/55.7	67.2/53.9	69.6/52.0	85.3/62.1	55.4/43.3	79.7/57.5	75.2/58.4	75.4/54.8
KCF	60.2/46.0	75.3/53.5	67.9/42.7	65.0/49.7	72.8/49.4	74.9/51.4	38.1/31.2	72.9/49.6	74.0/53.4	72.5/49.7
DSST	51.7/43.5	69.4/51.7	73.0/54.1	54.7/46.4	73.5/56.3	71.6/53.4	49.7/40.9	73.3/53.5	66.0/51.0	76.5/56.0

address the MB challenge. The DCF$_{our}$ achieves superior performance in PR/SR by 6.2%/2.2% over DCF. In comparison with the baseline, the excellent performance in MB illustrates DCF$_{our}$ can learn the "clean" feature to better represent objects. Second, our method also achieve excellent performance in handling BC and OCC. BC and OCC lead to features to be disturbed by the background. As shown in Table 4, DCF$_{our}$ outperforms DCF in PR/SR in 3.9%/1.6% and 1.8%/1.0% over BC and OCC, respectively. It demonstrates that the learned "clean" feature and noise-aware filter help to suppress the interference of noises. Finally, our method also represents the powerful strength on scale variation (SV) and illumination variation (IV). The critical important point is how to capture the variation of objects caused by the two challenging factors. For SV and IV, DCF$_{our}$ trackers both achieve the superior performance in PR/SR over baseline. Compared the baseline, the DCF$_{our}$ promotes 2.1%/0.8% and 2.6%/1.0% in SV and IV. It demonstrates our model can capture the variation of objects to learn the appearance model and the robust filter.

In addition, our method also achieves excellent performance in other challenging factors (*fast motion (FM), low Resolution (LR)*) and several factors (*in-plane rotation (LPR), out-of-plane rotation (OPR)*). In summary, the learned feature can better represent the object and optimize the filter.

5.4   Comparison with State-of-the-Art Trackers

Overall Evaluation: For comprehensive evaluation, we select several state-of-the-start trackers, including DSST [5], KCF [15], CSR-DCF [24], SRDCF [7], SAMF [23], Struck [13], TLD [16] and SCM [36] to compare with our proposed methods. The Table 3 represents the evaluation results between our proposed model and comparison trackers. As shown in Table 3, DCF$_{our}$ outperforms DSST in PR/SR while DCF achieves lower performance than the DSST in PR/SR. DCF$_{our}$ achieves comparable performance in PR/SR against KCF. This observation strongly illustrates that the proposed feature decomposition can help to improve the performance of CF trackers.

Attribute-Based Evaluation: We also select several trackers, including DSST [5], KCF [15], CSR-DCF [24], SRDCF [7] and SAMF [23] to evaluate the performance on different challenging attributes. Table 4 shows PR and SR of one-pass evaluation (OPE) for ten main challenging factors in OTB-2013 benchmark, including fast motion (FM), background clutter (BC), motion blur (MB), deformation (DEF), illumination variation (IV), low resolution (LR), occlusion (OCC) and scale variation (SV), respectively.

In this experimental analysis, we mainly focus on DCF and DCF_{our} to analyze the performance of this proposed algorithm. According to Table 4, we have the following observations. The first conclusion is that the proposed feature decomposition can benefit the CF trackers to learn a "clean" feature to deal with the MB. In OTB-2013, DCF is lower than DSST in PR but the DCF_{our} is greater than DSST and even outperforms KCF in PR. Although the performance of DCF_{our} is still lower than SAMF, CSR-DCF, SRDCF, HCF, it is explained that DCF_{our} is only based on the DCF to incorporate feature decomposition without any improvement while these extension trackers employ several tricks such as rich features and updating strategy besides their own improvement. The second conclusion is that our method also help to improvement the tracker's performance in BC and OCC. According to Table 4, DCF_{our} even outperforms the SAMF, KCF and DSST in PR for BC. Finally, it is observed that our method also improves the strength in handling SV and IV. In OTB-2013 benchmark, the performance of DCF_{our} outperforms the DSST in PR/SR while the performance of DCF_{our} is still lower than these trackers that have scale processing, including DSST, SAMF, SRDCF, CSR-DCF, HCF. Because the DCF_{our} have no the scale process model and only mixes the feature decomposition scheme to learn "clean" feature and noise-aware filter. In summary, the proposed feature decomposition learns more robust filter and "clean" feature to improve the performance.

6   Conclusion

In this paper, feature decomposition is introduced into correlation filter tracking to learn "clean" feature and noise-aware filter for improve the tracking accuracy and robustness. The "clean" feature and noise-aware filter are jointly optimized in an unified framework to mitigate noises effect in filters and features. The proposed tracking framework utilizes the learned "clean" feature to represent objects and the noise-aware filter to classify the object from background. As a result, it has the advantages of several existing correlation filter trackers such as suppressing the influence of noises. Both qualitative and quantitative evaluations on challenging datasets demonstrate that the effectiveness of proposed tracking algorithm against baseline methods.

Acknowledgment. This work is jointly supported by National Natural Science Foundation of China (61702002, 61472002), China Postdoctoral Science Foundation, Natural Science Foundation of Anhui Province (1808085QF187), Natural Science Foundation of Anhui Higher Education Institution of China (KJ2017A017), and Co-Innovation Center for Information Supply & Assurance Technology, Anhui University.

References

1. Bolme, D.S., Beveridge, J.R., Draper, B.A., Lui, Y.M.: Visual object tracking using adaptive correlation filters. In: Proceedings of IEEE Conference on Computer Vision and Pattern Recognition, pp. 2544–2550. IEEE (2010)
2. Boyd, S., Parikh, N., Chu, E., Peleato, B., Eckstein, J.: Distributed optimization and statistical learning via the alternating direction method of multipliers. Found. Trends® Mach. Learn. 3(1), 1–122 (2011)
3. Candès, E.J., Li, X., Ma, Y., Wright, J.: Robust principal component analysis? J. ACM (JACM) 58(3), 11 (2011)
4. Chen, M., Ganesh, A., Lin, Z., Ma, Y., Wright, J., Wu, L.: Fast convex optimization algorithms for exact recovery of a corrupted low-rank matrix. Coordinated Science Laboratory Report no. UILU-ENG-09-2214 (2009)
5. Danelljan, M., Hager, G., Khan, F., Felsberg, M.: Accurate scale estimation for robust visual tracking. In: Proceedings of British Machine Vision Conference. BMVA Press (2014)
6. Danelljan, M., Bhat, G., Khan, F.S., Felsberg, M.: ECO: efficient convolution operators for tracking. In: Proceedings of IEEE Conference on Computer Vision and Pattern Recognition. IEEE (2017)
7. Danelljan, M., Hager, G., Shahbaz Khan, F., Felsberg, M.: Learning spatially regularized correlation filters for visual tracking. In: Proceedings of IEEE International Conference on Computer Vision, pp. 4310–4318. IEEE (2015)
8. Danelljan, M., Robinson, A., Shahbaz Khan, F., Felsberg, M.: Beyond correlation filters: learning continuous convolution operators for visual tracking. In: Leibe, B., Matas, J., Sebe, N., Welling, M. (eds.) ECCV 2016. LNCS, vol. 9909, pp. 472–488. Springer, Cham (2016). https://doi.org/10.1007/978-3-319-46454-1_29
9. Danelljan, M., Shahbaz Khan, F., Felsberg, M., Van de Weijer, J.: Adaptive color attributes for real-time visual tracking. In: Proceedings of IEEE Conference on Computer Vision and Pattern Recognition, pp. 1090–1097. IEEE (2014)
10. Felzenszwalb, P.F., Girshick, R.B., McAllester, D., Ramanan, D.: Object detection with discriminatively trained part-based models. IEEE Trans. Pattern Anal. Mach. Intell. 32(9), 1627–1645 (2010)
11. Galoogahi, H.K., Fagg, A., Lucey, S.: Learning background-aware correlation filters for visual tracking. In: Proceedings of IEEE Conference on Computer Vision, pp. 1144–1152. IEEE (2017)
12. Gray, R.M.: Toeplitz and circulant matrices: a review. Found. Trends® Commun. Inf. Theory 2(3), 155–239 (2006)
13. Hare, S., Saffari, A., Torr, P.H.: Struck: structured output tracking with kernels. In: Proceedings of IEEE International Conference on Computer Vision, pp. 263–270. IEEE (2011)
14. Henriques, J.F., Caseiro, R., Martins, P., Batista, J.: Exploiting the circulant structure of tracking-by-detection with kernels. In: Fitzgibbon, A., Lazebnik, S., Perona, P., Sato, Y., Schmid, C. (eds.) ECCV 2012. LNCS, vol. 7575, pp. 702–715. Springer, Heidelberg (2012). https://doi.org/10.1007/978-3-642-33765-9_50
15. Henriques, J.F., Caseiro, R., Martins, P., Batista, J.: High-speed tracking with kernelized correlation filters. IEEE Trans. Pattern Anal. Mach. Intell. 37(3), 583–596 (2015)
16. Kalal, Z., Mikolajczyk, K., Matas, J.: Tracking-learning-detection. IEEE Trans. Pattern Anal. Mach. Intell. 34(7), 1409–1422 (2012)

17. Li, C., Cheng, H., Hu, S., Liu, X., Tang, J., Lin, L.: Learning collaborative sparse representation for grayscale-thermal tracking. IEEE Trans. Image Process. **25**(12), 5743–5756 (2016)
18. Li, C., Liang, X., Lu, Y., Zhao, N., Tang, J.: RGB-T object tracking: benchmark and baseline. arXiv preprint arXiv:1805.08982 (2018)
19. Li, C., Lin, L., Zuo, W., Tang, J.: Learning patch-based dynamic graph for visual tracking. In: Proceedings of The AAAI Conference on Artificial Intelligence, pp. 4126–4132. AAAI (2017)
20. Li, C., Lin, L., Zuo, W., Tang, J., Yang, M.H.: Visual tracking via dynamic graph learning. IEEE TPAMI (2018). https://doi.org/10.1109/TPAMI.2018.2864965
21. Li, C., Wu, X., Bao, Z., Tang, J.: ReGLe: spatially regularized graph learning for visual tracking. In: Proceedings of the ACM on Multimedia Conference, pp. 252–260. ACM (2017)
22. Li, C., Zhao, N., Lu, Y., Zhu, C., Tang, J.: Weighted sparse representation regularized graph learning for RGB-T object tracking. In: Proceedings of the ACM on Multimedia Conference, pp. 1856–1864. ACM (2017)
23. Li, Y., Zhu, J.: A scale adaptive kernel correlation filter tracker with feature integration. In: Agapito, L., Bronstein, M.M., Rother, C. (eds.) ECCV 2014. LNCS, vol. 8926, pp. 254–265. Springer, Cham (2015). https://doi.org/10.1007/978-3-319-16181-5_18
24. Lukezic, A., Vojir, T., Cehovin, L., Matas, J., Kristan, M.: Discriminative correlation filter with channel and spatial reliability. In: Proceedings of IEEE Conference on Computer Vision and Pattern Recognition. IEEE (2017)
25. Ma, C., Huang, J.B., Yang, X., Yang, M.H.: Hierarchical convolutional features for visual tracking. In: Proceedings of the IEEE International Conference on Computer Vision, pp. 3074–3082. IEEE (2015)
26. Rifkin, R., Yeo, G., Poggio, T.: Regularized least-squares classification. Nato Sci. Ser. Sub Ser. III Comput. Syst. Sci. **190**, 131–154 (2003)
27. Simonyan, K., Zisserman, A.: Very deep convolutional networks for large-scale image recognition. In: CoRR (2014)
28. Sui, Y., Tang, Y., Zhang, L.: Discriminative low-rank tracking. In: Proceedings of the IEEE International Conference on Computer Vision, pp. 3002–3010. IEEE (2015)
29. Sun, C., Wang, D., Lu, H., Yang, M.H.: Correlation tracking via joint discrimination and reliability learning. In: Proceedings of the IEEE Conference on Computer Vision and Pattern Recognition, pp. 489–497. IEEE (2018)
30. Tang, M., Feng, J.: Multi-kernel correlation filter for visual tracking. In: Proceedings of the IEEE International Conference on Computer Vision, pp. 3038–3046. IEEE (2015)
31. Tang, M., Yu, B., Zhang, F., Wang, J.: High-speed tracking with multi-kernel correlation filters. In: Proceedings of the IEEE Conference on Computer Vision and Pattern Recognition, pp. 4874–4883. IEEE (2018)
32. Valmadre, J., Bertinetto, L., Henriques, J., Vedaldi, A., Torr, P.H.: End-to-end representation learning for correlation filter based tracking. In: Proceedings of IEEE Conference on Computer Vision and Pattern Recognition, pp. 5000–5008. IEEE (2017)
33. Van De Weijer, J., Schmid, C., Verbeek, J., Larlus, D.: Learning color names for real-world applications. IEEE Trans. Image Process. **18**(7), 1512–1523 (2009)
34. Wu, Y., Lim, J., Yang, M.H.: Online object tracking: a benchmark. In: Proceedings of IEEE Conference on Computer Vision and Pattern Recognition, pp. 2411–2418. IEEE (2013)

35. Xu, Y., Yin, W.: A block coordinate descent method for regularized multiconvex optimization with applications to nonnegative tensor factorization and completion. SIAM J. Imaging Sci. **6**(3), 1758–1789 (2013)
36. Zhong, W., Lu, H., Yang, M.H.: Robust object tracking via sparsity-based collaborative model. In: Proceedings of IEEE Conference on Computer Vision and Pattern Recognition, pp. 1838–1845. IEEE (2012)

Research on Urban Street Order Based on Data Mining Technology

Guanlin Chen[1,2], Peipei Tang[1], Canghong Jin[1],
and Zhuoyue Zhu[1,2(✉)]

[1] School of Computer and Computing Science,
Zhejiang University City College, Hangzhou 310015,
People's Republic of China
zhuzy@zucc.edu.cn
[2] College of Computer Science, Zhejiang University, Hangzhou 310027,
People's Republic of China

Abstract. With the promotion of urbanization, more and more people enjoy the happiness brought about by the urban development, but the type of problems and the amount of problems in urban management are increasing. Under the background of new era, the direction and requirement of the city governance has promoted influenced by the "Internet Plus" strategy and big data strategy. In the construction of information and intelligent construction of urban management, a large number of city operation and management data have been emerged and accumulated, which provide favorable basic conditions for urban research. Through literature reading and field research, this paper made an in-depth study of the current situation of urban management in a city of Zhejiang province. In view of the actual demand of a city in the street order, this paper takes the data fusion cleaning of the street sequence data combined with data mining technology. It builds a Street classification and prediction model by using the decision tree C5.0 algorithm, and a high incidence area classification prediction model by using the Apriori algorithm. The information and knowledge of street order are analyzed. On this basis, this paper designs a street order decision support system, and uses web technology and visualization technology to realize the function modules of street order data service, analysis application, scene display and decision support. The research content of this paper is oriented to the actual demand in the city's street sequencing work, which provides a favorable support for the actual business of urban management, and also provides decision support for urban management.

Keywords: Urban management · Data mining · Street order · Visualization
Decision support

1 Introduction

With the development of the new generation of information technology, such as Internet of things, cloud computing and big data, people's production, life and way of thinking have undergone tremendous changes. On the one hand, the new generation of information technology is applied in all walks of life, which brings new opportunities

© Springer Nature Singapore Pte Ltd. 2018
Z. Xu et al. (Eds.): Big Data 2018, CCIS 945, pp. 227–237, 2018.
https://doi.org/10.1007/978-981-13-2922-7_15

for the upgrading and development of the industry. On the other hand, how to make full use of and give full play to the advantages of the new generation of information technology has formed new challenges.

After many years of information construction, the information infrastructure of Zhejiang has been perfected, which provides a good foundation for the construction of "digital city" and "intelligent city". After the construction of "digital city", the relevant data of urban operation can be accumulated. To the "intelligent city" construction stage, with the development of new generation of information technology such as big data and cloud computing, higher requirements for urban governance are put forward. However, in the implementation of this series of measures, there are some problems to be solved. For example, the data and information system need to be integrated, the utilization of information resources need to be improved, and the ability to support the management decision needs to be promoted.

There are many kinds of data resources, including business data, location data, Internet of things monitoring data, video monitoring data, site detection data and field measurement data. Massive data is indeed a valuable resource like oil, but if it is not developed and utilized, it will not be able to give full play to its great value. Big data need to rely on data mining technology to make it effective. Therefore, in the era of big data, data mining technology has been widely applied in the field of urban management. It has become a research hotspot in the field of urban governance to excavate the correlation among various urban data and assist in the exploration and resolution of urban diseases.

Combined with the actual operation of urban management, urban street order is a general term for street order cases in urban management cases. This paper will take the street order in Zhejiang Province as the research object, use data mining technology to analyze the law enforcement data of city management, establish a street order prediction model, design and implement an urban street order decision support system based on data mining technology to provide visual display and decision guidance for the urban street work. The supporting platform helps to realize the scientific enforcement of urban management, innovate the urban governance mode, and promote the construction process of urban management intellectualized.

2 Related Work

Urban management is an important research topic. Urban Street order is one of the organic components of urban management.

Academician Li et al. believes that the intelligent city is the integration of the real world and the digital world based on the digital city, the Internet of things and the cloud computing technology [1]. It can be described vividly by the formula of "Intelligent city = Digital City + IOT + Cloud Computing", and points out that control and intelligence service are the advanced stage of wisdom. Wang et al. have made an in-depth study on the construction of smart cities supported by large data, and think that the use of large data technology and data mining technology to process urban data processing and analysis can produce a large number of useful knowledge [2]. Song et al. put forward the design and application of the mobile application service platform for law enforcement city management, promoting the leap of the digital city tube to the intelligent city management, which is also an important exploration of the innovation of the mobile government administration [3].

After many years of the information construction, urban management related departments have accumulated a large number of urban historical data, and still continue to produce new city operation and management related data. The magnitude of these data is large and the structure is complex. Data itself is a valuable resource. But to transform these data resources into effective information and knowledge, it is necessary to use data mining technology, so that the city data not only has a large scale, but also shows rich connotations.

In order to make full use of the city's large data resources, experts and scholars have launched a rich study, aiming to transform the data resources into information and knowledge, which can help urban development by using a new generation of information technology. Wang et al. point out that the construction of data centric intelligent city will inevitably require great support of data mining technology [4]. Han et al. think that data mining can extract valuable information and knowledge from a large number of noisy and fuzzy data [5]. Wang et al. have use data mining technology to integrate a variety of data in the city, set up a visual query model for multi source city travel data, and use relevant technology to analyze and solve the various problems in the city, such as urban planning and disaster warning [6].

The orderly management of urban streets is one of the effective measures to improve the level of urban management. The case statistics range of the street sequence includes the street order cases in the intelligent city management and the order cases in the supervision system of the law enforcement, such as the non operating business, the disorderly parking of motor vehicles, the unauthorized parking of the motor vehicles, street hanging and so on. In view of urban management law enforcement and street order, experts and scholars have carried out related research. Wang et al. proposed that the integration and development of information and communication technology promoted the innovation and democratization of the knowledge society [7]. Chen et al. put forward a "Smarter" model based on the perspective of intelligent city management, and designed an intelligent management and control system for street order [8]. Chen et al. pointed out that the data analysis system with rich visual function can provide more intuitive data analysis results for researchers on specific research content [9].

3   The Design of Urban Street Order Model

In order to predict the development trend of the comprehensive management of urban event management, this paper will establish a street order model, which involves the data conclusions of a large number of thematic analysis models, such as event distribution trend model, event high area model, personnel efficiency model and so on. According to the content of urban management, this paper analyzes and studies historical data with higher breadth and depth, and explores the inevitable and accidental events between events management development and events type itself, event area, personnel power deployment, personnel working status and so on. To find out the problems and crux in the sequence of the current street order, predict the trend of the management situation, put forward the countermeasures and suggestions on the street order, and provide scientific decision support for the city management.

Through field investigation and other ways, this paper has obtained some data of urban management law enforcement in a city of Zhejiang province. The data used in this study are data from September 21, 2011 to June 25, 2017, with a total of more than 600 million records, with a total of more than 40 attributes. Combined with the actual business situation in the street, data cleaning is used to preprocess the data, and then the street sequential analysis model is established.

In order to improve the quality of data and improve the availability of the model, this paper mainly completed the work of data cleaning, including data format conversion, vacancy value processing, discretization and unrelated attributes cleaning in the data preprocessing stage. In the process of modeling, we filter abnormal data and list the corresponding data quality reports. Among them, take the field of eventsrcid (problem source ID) as an example, its partly data quality analysis is shown in Table 1.

Table 1. The data quality analysis table for field eventsrcid (problem source ID).

Eventsrcid	Percent	Count
1	75.05125828	4561647
34	16.0669176	976554
36	3.556079408	216140
11	2.909522507	176842
2	2.162933392	131464
4	0.08920636	5422
1003	0.001793341	109

After processing the field, this paper reconstructs four new fields, such as year, month, working day, and hour according to the reporting time createtime, specifically adding the year field with formula datetime_year (createtime), adding the month field with the formula datetime_month (createtime), and adding the working day field with the common datetime_weekday (createtime). Use the formula datetime_hour (createtime) to increase the hour field. On this basis, the field weekday is reclassified as a new field, week_day, and the field hour is reclassified as a new field, hour_daynight, using the workday and hourly fields.

In the light of the actual situation of the sequence of the streets in this city, it is found that the situation of the street order is closely related to the road, the team and the process. In order to analyze the order of street scenes, a sequential analysis model is established and expressed in formula (1).

$$SOAM = E_R + E_T + E_P \tag{1}$$

Among them, SOAM represents Street ordinal analysis model. The effectiveness of road work, team work efficiency and process efficiency can be represented by formula (2), formula (3) and formula (4).

$$E_R = \frac{R \times F_R}{P_R} \times E_1 \tag{2}$$

$$E_T = \frac{A \times F_A}{P_A} \times E_2 \tag{3}$$

$$E_P = \left(\frac{C_S}{A} + \frac{C_E}{C_S} \right) \times E_3 \tag{4}$$

Among them, R represents the total number of road events, FR represents road complexity, PR represents road personnel input, $E1$ represents the efficiency coefficient of road work; A represents the total number of regional events, FA indicates regional complexity, PA expresses regional personnel input, $E2$ indicates the efficiency coefficient of the team, CS shows the number of cases in the street case, CE indicates the number of cases in the street ordered case, $E3$ indicates the efficiency coefficient of the process.

By analyzing the data quality of the corresponding cases, we can see that many cases have a very low frequency of subclasses. In order to not affect the results of the analysis, this article only analyzes the case of cases with frequency greater than 1%. In this paper, in the process of establishing a case classification prediction model, considering the data of 2011 and the 2017 data is not a complete year, this paper only analyzes the data from 2012 to 2016, and takes the 2014 data as an example to establish the classification prediction model and the association rule model.

3.1 The Street Order Classification Prediction Model

In this subsection, the decision tree C5.0 algorithm is applied to research the street order model. The set of the city management data of a city during 2012 and 2016 include 6078042 records, each of which contains 40 attributes. Taking the data in 2014 as an example, the prediction model of street type classification is established, and the credibility of the model is improved by the data set in 2015. Among them, the data audit of case data in 2014 is shown in Fig. 1.

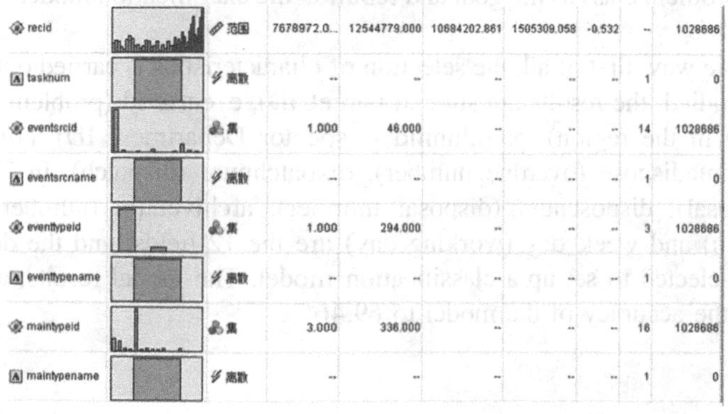

Fig. 1. The schematic diagram of data audits in 2014 case data.

According to the results of the data audit, we remove all 1 values and all the empty values in the modeling process, such as tasknum (task number), eventsrcid (problem source ID), eventtypename (problem type), maintypename (problem class) and so on. In order to ensure the effect of the classification model, this article makes a further selection of the existing attributes, specifically the selection of the 7 fields, such as streetid (street ID), patrolunitid (inspector Department ID), districtid (ID), hour (hour), month (month) and weekday (work day) as characteristic values. Sex is set to be more important than the influence coefficient 0.95. Considering the characteristics of the classification algorithm, this paper uses the decision tree C5.0 to establish a street classification model with the goal of the problem small class. The modeling results are shown in Fig. 2.

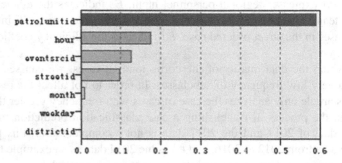

Fig. 2. The street order classification model based on problem subclass

Based on the analysis of the order classification model, the result shows that the accuracy of the classification model is only 48.29%, and the modeling effect is not satisfactory. According to the results of ordinal classification modeling, we can get a classification model based on the problem class with low accuracy. By further analysis of the reasons why the modeling effect is not ideal, it can be summed up that the classification of small classes is too much, and the proportion of most street order problems is very low, which affects the classification effect. Therefore, this paper chooses the problem class as the goal and rebuilds the classification model based on the data in 2014.

In the same way, first of all, the selection of characteristics is carried out. When the problem is studied, the results are streetid (street ID), eventscrid (problem source ID), districtid (ID in the region), patrolunitid (inspector Department ID), hour_daynight (hour), overtimedispose (overdue number), dispatchnum (dispatch), in Timedispose (timely disposal), disposenum (disposal number), archivenum (number of cases), month (month) and week_day (working day) are the 12 fields, and the decision tree C5.0 is still selected to set up a classification model. The model results, as shown in Fig. 3, raise the accuracy of the model to 69.46%.

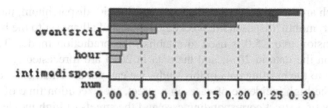

Fig. 3. The street order classification model based on the problem class.

Then, in this paper, we set up a classification model for the problem of street order, in which there are still selected small classes with a proportion of more than 1%. And we use the C5.0 decision tree to generate the classification model. The generated models are shown in Fig. 4.

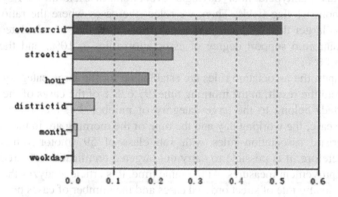

Fig. 4. The classification model for small classes in order to solve the problem of order streets.

3.2 The High Incidence Region Classification Prediction Model

In order to establish a high incidence area classification prediction model, this paper first makes a statistical analysis of the high incidence types in each region, that is, in all cases in the corresponding region, a street order case with frequency of more than 10% is occurring, and the high incidence number is arranged at the frequency from high to low.

According to the frequency of street sequence cases, it can be concluded that 3, 4, 6, 1, 2, 5 (according to the number of cases occurring in a case) are high incidence areas, and the high incidence cases in these areas are 59 (motor vehicle discontinuation), and the number of cases in the region 2, 3, 4, 5, 6 is 59 (motor vehicle disorderly discontinuation). More than 50%. From this, it is possible to control the parking chaos of motor vehicles.

In this paper, we set up a prediction model for the cases of high incidence area 1, 2, 3, 4, 5, 6 and high incidence time 7, 8, 9, 10, 13, 14, and 15. Due to the fine classification of cases, the case is reclassified and the case frequency is lower than that of 6%, and the prediction model is set up. In the selection of feature, 12 fields are selected, which are streetid, overtimedispose, patrolunitid, dispatchnum, intimedispose, districtid, hour, month, disposenum, weekday, rainy, and all are set to the highest level. Then, the decision tree C5.0 is used to establish the prediction model. The model is mainly built on the data in 2014, and the data in 2015 are forecasted.

According to the existing data in this study, we can get the small class of cases in a certain area, the high incidence of cases in a street, the execution time of a street, the execution time of a street corresponding case and the streets of high incidence of small cases. Therefore, the areas and streets can be focused on high incidence of cases, and the poor implementation of the case of the region and the streets to take appropriate measures to improve the efficiency of the execution of the street order cases.

3.3   The Street Order Association Rules Model

In order to establish a street sequential association rule model, this paper first transforms the fields maintypeid, hour_daynight, eventsrcid, districid into 0-1 symbols, and remove the non 0-1 flag fields. Then, we select the cases where the ratio of ID and regional ID is larger than 1%, and then use Apriori algorithm to calculate association rules. The minimum support degree of association rules is 10%, and the minimum confidence is 80%.

In this paper, the association rules are established for the large category of cases in the street. From the result, in the morning time, 97.628% of the cases of the number 34 (video reported) belong to the large category of number 7 (street order), while the source of the case, the working day and the time of the morning are in the video. There are characteristic association rules with sub class of 59 (motor vehicle parking) problem. Therefore, it is possible to carry out targeted rectification of street order and formulate improvement measures. At the same time, this article analyzes the correlation between the closing rate of street ordered cases and the number of cases per capita. First of all, the number of ordered cases in the street from 2013 to 2016 is analyzed. Then, the correlation between the closing rate of Street faces and the number of cases per capita in the past four years is analyzed.

According to the list of the number of personnel, the number of cases and the case of closing cases from 2013 to 2016, it is concluded that the correlation coefficient between the incidence of the cases and the number of cases per capita in each year is 0.5994, 0.5595, 0.6436, 0.5105. According to the correlation coefficient analysis, it can be found that the increase of the number of cases per capita can indeed increase the rate of settlement of the cases in the street. The analysis results can be used in the management decision. Through theoretical training and practice strengthening, we can improve the business quality of the law enforcement personnel in the street order, improve the rate of the per capita settlement, and then improve the rate of the case in the street order. The correlation coefficient is 0.5049, 0.6608, −0.1250, 0.0225, respectively. According to the correlation coefficient analysis, it can be found that increasing the number of personnel can improve the rate of settlement. The results of the analysis are used in the management decision. When the rate of cases in the street sequence is low, the number of additional personnel can be used to improve the rate of the sequence in the street.

4 The Implementation of Street Order Decision Support System

In this paper, we have designed a hierarchical structure of street ordinal data analysis model, as shown in Fig. 5. The hierarchical architecture of street order design is mainly composed of interface layer, integration layer, sharing layer and application layer. Among them, the functions of each layer are described as follows (see Fig. 5).

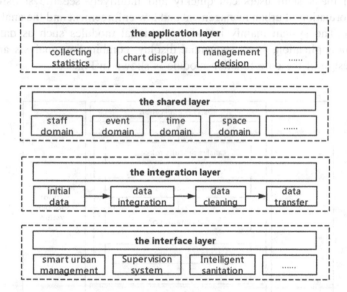

Fig. 5. The hierarchical structure of street ordinal data analysis model.

(1) The interface layer is consistent with the data structure of the source system, and maintains the integrity and consistency of the data and source. It mainly connects the system modules of intelligent city management, inspection and assessment, intelligent sanitation and so on, which can be used to obtain the relevant data needed for the analysis of the street surface.

(2) The integration layer purge the data of the source layer according to certain rules, standardize the transformation, form the standard data format, and mainly complete the process of the original data collection, data integration, data cleaning and data conversion, and through the data fusion cleaning to get the data that conforms to the quality requirement of the street sequence analysis.

(3) On the basis of the source layer, the sharing layer designs the wide table based on the requirements of business and performance, and forms coarse grain data. It is mainly based on the data resources to design the database, and draws the database tables of the personnel domain, event domain, time domain and space domain.

(4) According to the requirements of various applications, the application layer forms the pertinent data through the design of market and wide tables. It mainly combines the actual needs of the street order work, and provides functional modules

such as summary statistics, real-time display, icon display and management decision support.

In this paper, we use SSM (Spring + SpringMVC + MyBatis) framework and MySQL database to design a sequential decision support system based on B/S architecture. Using Java technology to call Weka model package for analysis, and in the background to achieve database calls, Html5 technology to achieve front-end page display.

Through the system users can quickly and intuitively see the statistics of the business information of urban management and law enforcement and the analysis based on big data. The system mainly includes functional modules such as data service, analysis data application, street sequential display, combination analysis and prediction. The design of the core function module is shown in Fig. 6.

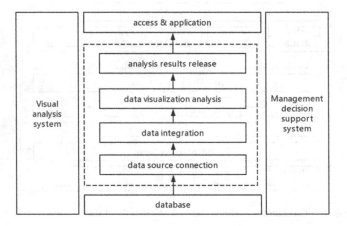

Fig. 6. The design of the core function module.

In this paper, the design of the core function module of the system mainly contains four pieces of content, which are data source connection, data integration, data visualization analysis and analysis results release. Among them, data source connection is the basis of data analysis; data integration is used to realize the multiple table connection of the same data source and data fusion of multiple data sources to meet the different needs of the data. Data visualization analysis is the core module of the data visualization analysis system. The function of the graph, histogram, pie chart, fold line and so on, and has the functions of statistical analysis and spatial and temporal prediction. The analysis result is published on the server, and interactively accessed through the browser or mobile terminal to realize the auxiliary management and decision support function. In terms of performance testing, the concurrency of the system is not high, and the running time of the model is short.

5   Conclusions

This paper builds a street classification and prediction model by using the decision tree C5.0 algorithm, and a high incidence area classification prediction model by using the Apriori algorithm. Based on these models, we design a street order decision support system, and use web technology and visualization technology to realize the function modules of street order data service, analysis application, scene display and decision support. The research content of this paper is closely related to the background of the current "Internet +" and large data, and combines the new generation of information technology with the traditional city management business, responding to the demand for the innovation and upgrading of the government governance model, providing a visual display and supporting platform for the decision guidance for the street order of the city. The implementation of urban management science law enforcement provides a useful reference for the utilization of urban big data and urban governance.

Acknowledgements. This work is supported by the Hangzhou Science & Technology Development Project of China (No. 20162013A08, No. 20171334M12, No. 20170533B22), the Zhejiang Provincial Natural Science Foundation of China (No. LY16F020010) and Zhejiang University City College Scientific Research Foundation (No. J-16003).

References

1. Li, D., Yao, Y., Shao, Z.: Big data in a smart city. Geomat. Inf. Sci. Wuhan Univ. **6**, 631–640 (2014)
2. Wang, Z., He, Y.: Research on the construction of intelligent city based on large data. Internet Things Technol. **8**(1), 46–50 (2018)
3. Song, G., Liu, J., Chen, H., Wei, L., Ding, S.: Design and application of law enforcement management mobile application service platform. E-Government **8**, 56–64 (2015)
4. Wang, J., Li, C., Xiong, Z.: A summary of data centered intelligent city research. J. Comput. Res. Dev. **2**, 239–259 (2014)
5. Han, J., Kamber, M.: Data mining: concept and technology. Machinery Industry Press (2003)
6. Wang, F., Zhang, F., Wu, F.: A visual query model for travel data in multi-source cities. J. Comput.-Aided Des. Comput. Graph. **1**, 25–31 (2016)
7. Wang, L., Song, G.: Cooperation democracy in the 2 perspective of innovation: from consultation to cooperation – taking "wiki" as an example of "I love Beijing". E-Government **4**, 73–81 (2015)
8. Chen, G., Sun, X., Li, S.: Research on the promotion mechanism of smart city management in Zhejiang Province based on the perspective of smart city. Sci. Technol. Econ. **28**(3), 86–90 (2015)
9. Chen, C., Zhang, G., Ma, X., Wang, Y.: Using big data mining and knowledge discovery technology to assist smart city development. Big Data Res. **2**(3), 39–48 (2016)

A Vertex-Centric Graph Simulation Algorithm for Large Graphs

Jingdong Li[1], Jin Li[2](✉), and Xiaoling Wang[1]

[1] Shanghai Key Laboratory of Trustworthy Computing,
East China Normal University, Shanghai, China
lljjdd567@gmail.com, xlwang@sei.ecnu.edu.cn
[2] School of Software, Yunnan University, Kunming 650000, China
lijin@ynu.edu.cn

Abstract. Graph simulation as a well studied model of graph pattern matching problem, has been adopted to reduce the complexity and meet the need of novel applications such as mining potential associations between users in online social networks. In recent years, graph processing frameworks such as Pregel bring in a vertex-centric, Bulk Synchronous Parallel (BSP) programming model for processing massive data graphs and achieve encouraging results. However, developing efficient vertex-centric algorithms for graph simulation model is very challenging, because this problem does not naturally align with a vertex-centric programming model. This paper presents novel distributed algorithms based on the vertex-centric programming model for graph simulation. At the same time, considering the enormous cost of the message passing and the algorithm complexity of the pattern matching in the processing of the massive data graph, the part of message passing in the algorithm is optimized to reduce the communication cost. We experimentally verify the effectiveness and efficiency of these algorithms, using real-life massive data graph.

Keywords: Vertex-centric · Graph simulation · Optimization

1 Introduction

Graph pattern matching is widely used in various modern applications, e.g., protein networks, social networks, knowledge graph [12,14,23]. Graph matching is typically defined in terms of subgraph isomorphism [8]. However, this problem is a NP-Complete problem [24]. The subgraph isomorphism sometimes can't catch sensible matches in real-life applications such as social networks [2,6]. In order to meet the need of modern applications and reduce the response time. Graph simulation [8] has been adopted for graph pattern matching. Compared to subgraph isomorphism, graph simulation is less restrictive and can be determined in quadratic time [11].

As an alternative of subgraph isomorphism, graph simulation and some extensions [7,16] play a critical role for the analysis of social position in social networks

© Springer Nature Singapore Pte Ltd. 2018
Z. Xu et al. (Eds.): Big Data 2018, CCIS 945, pp. 238–254, 2018.
https://doi.org/10.1007/978-981-13-2922-7_16

[2,5]. Nowaday, there is a rapid growth in data volume, so most of the time, we need to evaluate a query on a large dataset. We have to partition and distribute the data to multiple machines, so that the query can be efficiently processed in parallel. MapReduce [4], as a famous associated implementation for processing and generating big data sets with a parallel, distributed algorithm on a cluster, plays an important roles in distributed computation. However, graph simulation needs recursive computations to work out the matching results. MapReduce is typically not fit for this kind of graph algorithms, which needs a series of chained MapReduce invocations [3,18].

Pregel [18] introduces a vertex-centric model, which is typically fit for graph algorithms with recursive computations. However, if we simply delegate the computation tasks to Pregel, it may involve too many rounds of computations [17]. Hence, a natural question raised is how to design an efficient graph pattern matching algorithm, in terms of graph simulation, on large graph process platform?

In order to answer these questions, we have implemented some vertex-centric graph simulation algorithms on the existing large-scale graphing platform GraphX. We have verified the efficiency of these algorithms through experiments on real datasets. The main technical contributions of this article are summarized as follows:

We successfully implement vertex-centric graph simulation algorithms on GraphX, an existing large-scale graph computing platform.

According to the characteristics of these vertex-centric graph simulation algorithms, we propose some optimization strategies, which reduce the communication cost in the message passing process and achieve good results.

The experiments on real-life dataset validate that our graph simulation algorithms with optimization are highly efficient and very suited for vertex-centric model.

The rest of the paper is organized as follows. Section 2 introduces the main concepts used in vertex-centric graph processing and graph pattern matching. Section 3 introduces our improved graph simulation algorithms and some concepts used in query graph analysis. We cover some experiments about the behavior of our algorithms in Sect. 4. We also present some related work in section Sect. 5, followed by conclusion in Sect. 6.

2   Background

In this section, we introduce some concepts that are important to our work, called graph pattern matching model and vertex-centric graph processing.

2.1   Graph Pattern Matching

The goal of a pattern matching algorithm is to find all the matches of a given graph, called query graph, in an existing larger graph, called data graph. To define it more formally, assume that there is a data graph $G(V, E, l)$, where V is the set of vertices, E is the set of edges, and l is a function that maps the vertices to their labels. Given a query graph $Q(V_q, E_q, l_q)$ the task of pattern matching algorithm is to find all subgraphs of G that match the query Q. Subgraph isomorphism is the most famous model for pattern matching. The most distinctive feature of this model is that It preserves all topological features of the query graph. However, this model is an NP-Complete problem [24]. So our work is based on graph simulation model that, as an alterative of subgraph isomorphism, reduces query time to an acceptable level by relaxing some restrictions on matches. Pattern $Q(V_q, E_q, l_q)$ matches data graph $G(V, E, l)$ via graph simulation, denoted by $Q \trianglelefteq_{sim} G$, if there is a binary relation $R \subseteq V_q \times V$ such that (1) if $(u, u') \in R$, then $l_q(u) = l(u')$; (2) $\forall u \in V_q, \exists u' \in V : (u, u') \in R$; (3) $\forall(u, v) \in E_q, \exists(u', v') \in E : (u, u') \in R$ and $(v, v') \in R$ [7].

2.2   Vertex-Centric Graph Processing

In recent years, several platforms, such as Pregel [18] GraphX [9] and Giraph [21], have been designed for processing massive graphs. They are all based on BSP-based vertex-centric programming model. Bulk Synchronous Parallel (BSP) was first proposed by Valiant [25] as a model of computation for parallel processing. In the BSP model, computation proceeds in a series of global supersteps and each superstep consists of three components: (1) concurrent computation: every participating processor may perform local computations, (2) communication: processes exchange data between themselves to facilitate remote data storage capabilities and, (3) barrier synchronization: when a process reaches the barrier, it waits until all other processes have reached the same barrier.

In the vertex-centric programming model proposed in Pregel, these three components can be described like following: (1) each vertex execute the same user-defined function that expresses the logic of a given algorithm. A vertex can modify its state by utilizing received messages sent to it in the previous superstep and send messages to other vertices(to be received in the next superstep), (2) vertices communicate directly with one another by sending messages, each of which consists of a message value and the name of the destination vertex. Moreover, it can provide very high scalability, and by design it is well-suited for distributed implementation. The type of the message value can be specified by the user, (3) when a vertex checks out that it has accomplished tasks, it votes to halt and goes to inactive mode so the Pregel framework will not execute that vertex in subsequent supersteps unless it receives a message. When all vertices become inactive the algorithm terminates [18].

3   The Proposed Algorithm

This paper studies vertex-labeled, directed graphs and assumes a query graph is a connected graph because the result of graph pattern matching for a disconnected query graph is equal to the union of the results for its connected components

[19]. Let $G = (V_d, E_d)$ be a graph, where V_d is a vertex set, and E_d is an edge set. Each vertex $v \in V_d$ has a unique identifier (id for short) $v.id$ and an attribute set $v.A$.

3.1 Query Graph Analysis

Although the query graph analysis is not mentioned in the steps of algorithm, it is still a very important step in our algorithm implement of vertexes and different query graphs needing completely different query strategies and it is necessary to separately introduce several related concepts here:

Definition 1. **Top-vertex and Sub-vertex.** If there is a directed edge from vertex A to vertex B then A is the top-vertex of B and B is the sub-vertex of A. for example, in Fig. 1, c2 is the top-vertex of d3 and c2 is the sub-vertex of b4. These concepts will be mentioned frequently below.

Definition 2. **Relevant vertex and Irrelevant vertex.** If the kind of a vertex (e.g. the kind of a1 is A) in the data graph appears in the query graph, then it is a relevant vertex, otherwise it is an irrelevant vertex, as shown in the Fig. 1, c1 is a relevant vertex and e1 is an irrelevant vertex.

Definition 3. **Start vertex.** Start vertex is a kind of vertex in query graph that does't have a sub-vertex(belonging to the relevant vertex set). In our algorithm, we will initialize start vertex and set their value to true at first. As shown in Fig. 1, these vertices with $v.kind = D$ are start vertex (e.g. d1, d2, etc.). However, sometimes there is no vertex without sub-vertex in query graph. For example, in Fig. 3, you can not find a start vertex in query graph, in such situation, we will choose the intersectant top-vertex B (both in C->B->E->C and B->C->B) as start vertex.

Definition 4. **End vertex.** End vertex is a kind of vertex in query graph that doesn't have a top-vertex(belonging to the relevant vertex set) and end vetex is the end of message passing. If the matching status of all end vertices does not change, the processing of algorithm stops. As shown in Fig. 1, these vertexs with $v.kind = A$ are end vertex (e.g. a1, a2, etc.). However, sometimes there is no vertex without top-vertex in query graph and similarly we will choose the common sub-vertex as start vertex.

Definition 5. **Pattern set.** Pattern set contains the kinds of sub-vertex needed by current vertex to ensure the matching conditions, for example, in Fig. 1, the pattern set of b1 contains D and C, because in query graph, according to the rules of graph simulation, vertex B must have sub-vertex C and D.

Definition 6. **Message set.** For current vertex, its message set contains the kinds of its sub-vertexes in relevant vertex set, for example, in Fig. 1, the message set of b1 contains D and C, because in data graph, vertex b1 has sub-vertex c1 and d1(e1 not in relevant vertex set).

3.2 Acyclic Query Graph Algorithm

This part introduces the steps of the **Acyclic Query Graph Algorithm(AQG)**. It should be noted that the introduction of this part will include the optimization methods proposed by us. The acyclic query graph algorithm proposed in this paper is based on Pregel's BSP model, so the algorithm is executed at each vertex of the data graph. The specific steps include initialization, message broadcast and match trial, as shown in Algorithm 1. We describe the details of each step in the following. In this algorithm, each vertex owns a label named value($v.value \in v.A$) to indicate whether it is matched and $v.value = true$ means matched.

Algorithm 1. Acyclic Query Graph Algorithm

Require: relevant set V, pattern set P_v and message set M_v
Ensure: match state of v
 1: **Initialization:**
 2: **if** v is start vertex **then**
 3:    set $v.value = true$
 4: **else**
 5:    set $v.value = null$
 6: **end if**
 7: Set step as *Message Broadcast*
 8: **Message Broadcast:**
 9: **if** v is active **then**
10:    **if** $v.value == null$ **then**
11:       Wait to receive message from sub-vertex;
12:       Set step as *Match Trail* after receiving all messages;
13:    **else if** this $v.value == true$ **then**
14:       Send message $v.value == true$ to top-vertex;
15:       Set vertex to Inactive and Matched;
16:    **else**
17:       Send message $v.value == false$ to top-vertex;
18:       Set vertex to Inactive;
19:    **end if**
20: **end if**
21: **Match Trial:**
22: **for** sub-vertex.value==true **do**
23:    $m \leftarrow sub - vertex.kind$
24:    $M_v \leftarrow M_v \cup m$
25: **end for**
26: **if** $\forall$ sub-vertex.value!=null & $P_v == M_v$ **then**
27:    Set $v.value = true$;
28: **else if** $\forall$ sub-vertex.value!=null & $P_v != M_v$ **then**
29:    Set $v.value = false$;
30: **else**
31:    Set step as *Message Broadcat*;
32: **end if**

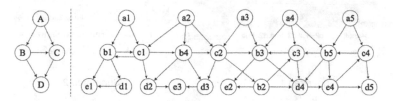

Fig. 1. Data graph with acyclic query graph.

Step1: Initialization. The first step of the algorithm is to initialize the data graph. Each vertex collects and records the value of its sub-vetex (e.g.: c1 records states of its sub-vertexs b1 and b4). More precisely, these vertexs are not start vertexs contained in the relevant point set, because we know that (1) messages sent by neighbor sub-vertexs, that don't belong to relevant vertex set, are useless. (2) according to the definition of the start vertex, its neighbor vertexs must not belong to the set of relevant vertexs. The value of the relevant vetexs except the start vertexs is set to $v.value = null$ which means that the status of match remains unknown.

Algorithm 2. Pro-initialization

Require: pattern set P_v and message set M_v
Ensure: initialization state of v
1: **if** v is start vertex **then**
2:    **if** $P_v == M_v$ **then**
3:        set $v.value = true$
4:    **else**
5:        set $v.value = false$
6:    **end if**
7: **else**
8:    set $v.value = null$
9: **end if**

Step2: Message Broadcast. Whether a vertex v should send messages and which neighbor to send messages to are determined by $v.value$. Specifically: (1) If $v.value = false$, then it doesn't need to receive the message sent by the sub-vertex in the following steps, but it need to send messages to its top-vertex. Such a message will cause some vertexs to get matching results quickly because they will be judged to false for lacking necessary sub-vertex during the first few steps, which will also reduce message passing during the iterations. (2) If $v.value = true$, similar to the case before, it does not need to receive the messages sent by sub-vertexs, it will also send messages to top-vertices. (3) If $v.value = null$, the current vertex matching state is unknown. So it does not need to send its own information to the top-vertices, on the contrary, it will receive message sent by sub-vertexs until its own match state can be determined.

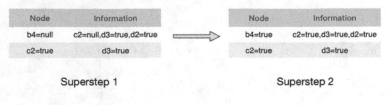

Fig. 2. Example of message passing between supersteps

Step3: Match Trial. After the current vertex has received the message from its sub-vertices, it needs to be judged whether it satisfies the matching condition according to these messages. In other words, whether a vertex is matched is determined by the matching value of sub-vertexs. Taking the b4 vertex in the Fig. 1 as an example. In the first superstep, it receives and saves the d2 matched message ($d2.value = true$) and d3 matched ($d3.value = true$) message. However, the value of b4 will not change, because it has also saved the message that c2 does not match ($c2.value = null$) during the initialization process. Since c2 can't be judged whether the matching condition can be reached at this time, b4 neither can be judged. Therefore, in the first superstep, the matching value of b4 remains null. At the same time, c2 receives the message that d3 matched ($d3.value = true$). According to the pattern of query graph, if a vertex $v(v.kind = C$) has a sub-vertex v' and $v'.kind = D$ then v is matched. So the value of c2 is updated to true at the first superstep. In the second superstep, b4 receives the c2 matched ($c2.value = true$) message, and stored information is set as shown in the Fig. 2. Then b4 has at least a matched sub-vertex c2($c2.kind = C$) and a matched sub-vertex d2($d2.kind = D$), which meets the matching condition, so the b4.value is updated to true at the second superstep(shown in Fig. 2).

Optimization Techniques: In order to reduce the message passing cost during the processing of the algorithm, we propose some optimization techniques by means of pruning some vertexes and edges in advance. The algorithm with optimizations is named **Promoted Acyclic Query Graph Algorithm (proAQG)**. Consider this case:

Case 1: When a vertex's sub-vertexes can't make it meet the matching conditions.

For example, b3 in Fig. 1 only has a valid sub-vertex named d4 and in the query graph, however, the B vertex must have a sub-vertex v with $v.kind = C$ to achieve Matching conditions. So in this situation, you can directly set $b3.value = false$ which means that current vertex can't be matched. Such a process can be summarized as Algorithm 2. In this way, we can filter out some vertexes that can be judged to mismatch without message passing and reduce the communication cost in the process of algorithm.

3.3 Cyclic Query Graph Algorithm

This part introduces the steps of the **Cyclic Query Graph Algorithm (CQG)**. The difference between two subgraph matching algorithms is that in

CQG, there is at least one circle in query graph. Such a circle causes the acyclic query graph algorithm can't be applied to cyclic query graphs. For this reason, we separately design an algorithm that is suitable for cyclic query graphs. The introduction of this section will also include our proposed optimization method. The cyclic query graph algorithm is also based on Pregel's BSP model. The specific steps include vertex initialization, message broadcast and match trail. However, the details of each step are different from what mentioned before. We describe the details of each step below. What is different from the cyclic query graph algorithm.

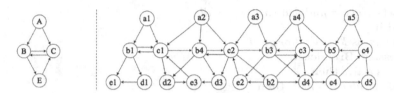

Fig. 3. Data graph with cyclic query graph.

Step1: Initialization. The first step of the algorithm is to initialize the data graph. Each vertex collects and records the value of its sub-vertex in the relevant point set(e.g. in Fig. 3, c1 records its sub-vertexs b1 and b4 without d2). More precisely, these vertices are not start vertices contained in the relevant point set, because we know that messages sent by sub-vertex which is not in relevant vertex set are useless. Different from the initialization strategy of the acyclic query graph algorithm, each vertex maintains two values, one is the matching value and the other is the value used to temporarily indicate the matching state called fake value. Except for the start vertex, the matching value of the relevant vertex value and the fake value are both set to null, and the value of the start vertex in circle graph is initialized to null, the fake value is initialized to true, if there is a start vertex outside the circle graph, both value and fake value are initialized to true. The specific reasons are described in message broadcast and match trial parts.

Step2: Message Broadcast. Whether a vertex v should send messages and which neighbors to send messages are determined by $v.value$ and $v.fake$. The message broadcast step of out-circle vertex is basically the same as acyclic query graph algorithm and for an out-circle vertex v, $v.value$ is updated synchronously with $v.fake$. If a message needs to be sent, vertex will only send $v.value$ to the sub-vertex. The message broadcast rules of the in-circle vertex is more complex and can be summarized to two rules: (1) In-circle vertexs will first pass the fake value in circle. After all vertices update their fake values, they will begin to pass value, for example: If a vertex v in a circle graph, has $v.fake = null$, then it must have $v.true = null$. (2) The message passing between in-circle vertex and out-circle vertex will be processed after the in-circle vertex's value is updated, whatever the message passing direction is.

Algorithm 3. Cyclic Query Graph Algorithm

Require: relevant set V, pattern set P_v and message set M_v
Ensure: match state of v
1: **Initialization:**
2: **if** v is start vertex **then**
3:　　**if** v in a circle **then**
4:　　　　set $v.fake = true$ and $v.value = null$
5:　　**else**
6:　　　　set $v.fake = true$ and $v.value = true$
7:　　**end if**
8: **else**
9:　　set $v.fake = null$ and $v.value = null$
10: **end if**
11: Set step as *Message Broadcast*
12: **Message Broadcast:**
13: **if** v is active **then**
14:　　**if** $v.value == null$ **then**
15:　　　　**if** $v.fake == null$ **then**
16:　　　　　　Wait to receive message from sub-vertex;
17:　　　　　　Set step as *Match Trail* after receiving all messages;
18:　　　　**else if** this $v.fake == true$ **then**
19:　　　　　　Send message $v.fake == true$ to top-vertex;
20:　　　　**else**
21:　　　　　　Send message $v.fake == false$ to top-vertex;
22:　　　　　　Set vertex to Inactive;
23:　　　　**end if**
24:　　**else if** $v.value == false$ **then**
25:　　　　Send message $v.value == false$ to top-vertex;
26:　　　　Set vertex to Inactive and Mismatched;
27:　　**else**
28:　　　　Send message $v.value == true$ to top-vertex;
29:　　　　Set vertex to Inactive and Matched;
30:　　**end if**
31: **end if**
32: **Match Trial:**
33: **for** sub-vertex.value==true **do**
34:　　$m \leftarrow sub - vertex.kind$
35:　　$M_v \leftarrow M_v \cup m$
36: **end for**
37: **if** $\forall$ sub-vertex.value!=null & $P_v == M_v$ **then**
38:　　Set $v.value = true$;
39: **else if** $\forall$ sub-vertex.value!=null & $P_v != M_v$ **then**
40:　　Set $v.value = false$;
41: **else**
42:　　Set step as *Message Broadcat*;
43: **end if**

Step3: Match Trial. After the current vertex has received the message from its sub-vertexes, it needs to be judged whether it satisfies the matching condition according to these messages. In other words, whether a vertex is matched is determined by the matching value of sub-vertices, whether the vertex is in a circle graph and whether the fake value is determined. For a vertex outside circle graph, the judgment rules are the same as those for the acyclic query graph algorithm. For a vertex inside circle graph, the only difference is that this vertex needs to judge the fake value. For after a new run of match trial, the value of the sub-vertex may change.

Algorithm 4. Pro-initialization

Require: pattern set P_v and message set M_v
Ensure: initialization state of v
1: **if** v is start vertex **then**
2:    **if** P_v != M_v **then**
3:        set $v.fake = false$ and $v.value = false$
4:    **else**
5:        **if** v in a circle **then**
6:            set $v.fake = true$ and $v.value = null$
7:        **else**
8:            set $v.fake = true$ and $v.value = true$
9:        **end if**
10:    **end if**
11: **else**
12:    set $v.fake = null$ and $v.value = null$
13: **end if**

Optimization Techniques: We use the same technique as the optimization of the acyclic query graph algorithm and name it **Promoted Cyclic Query Graph Algorithm (proCQG)**, if the current vertex's sub-vertices are not enough to make it meet the matching conditions, for example, the valid sub-vertex of b4 has only c4. However, in the query graph, a vertex v must have matched sub-vertex $v'(v'.kind = C)$ and $u'(u'.kind = E)$, you can directly set $b4.value = false$ and $b4.fake = false$. Such a process can be summarized as Algorithm 4. In this way, you can filter out some vertices that can be judged to mismatch without message passing and reduce the communication cost in the process of algorithm.

4 Experiments and Results

In this section, we show the experiment results of our algorithms. We have used GraphX [9], which can be considered a free distribution of Pregel, to implement our distributed algorithms. GraphX is a open library based on Spark [27] and has an extended Pregel API to provide a type of global communication among vertices and a series of graph operators. Using real-world and synthetic datasets, we conducts five different experiments to evaluate our algorithms: (1) the size of data graph, (2) the pattern of query graph, (3) the degree of data graph, (4) the optimization of algorithm, (5) the distributed environment.

Table 1. Statistics of graph datasets

DataSet	|V|	|E|	Average degree
Amazon0302	262,111	1,234,870	4.7
Amazon0312	400,727	3,200,440	7.9
Amazon0601	403,394	3,387,388	8.4

4.1 Experiment Setup

Algorithms. We implement our proAQG and proCQG algorithms in scala on GraphX [9], a Pregel-like memory-based distributed graph processing framework. In addition, to the best of our knowledge, there are few studies on vertex-centric algorithm for graph simulation models and the existing algorithm [7] can't output entire result set after searching. So the comparative algorithms of our paper are AQG and CQG.

Dataset. The real-world graph datasets used in our experiments are Amazon product co-purchasing network graph[1]. They are based on Customers Who Bought This Item Also Bought feature of the Amazon website. If a product i is frequently co-purchased with product j, the graph contains a directed edge from i to j. The amazon-meta file contains information about 548,552 different products (Books, music CDs, DVDs and VHS video tapes). We extract product id from the dataset and use product group to label these vertices.

There are four large data graphs in this dataset named: amazon0302, amazon0312, amazon0505 and amazon0601. Considering that the data size and average degree of amazon0505 and amazon0312 represents a small difference, the real-life data graphs used in this paper are amazon0302 with 262,111 vertices and 1,234,870 edges, amazon0312 with 400,727 vertices and 3,200,440 edges, and amazon0601 with 403,394 vertices and 3,387,388 edges. Some statistics about these graphs are summarized in Table 1.

In order to more effectively evaluate whether these algorithms can be applied to data graphs with different degrees and sizes, we used graph-tool [1] to synthesize a series of data graphs and the label of each vertex is randomly selected from the label set of Amazon product co-purchasing network graphs.

Pattern Graph. To generate some pattern graphs, we randomly extract a connected subgraph from Amazon dataset and implement a query generator to add edges for more complex query graph. Our query generator has two input

---

[1] http://snap.stanford.edu/data/amazon-meta.html.

Table 2. The result of Experiment 1 running acyclic query graph algorithm

Graph	P-1	P-2	P-3
Amazon0302	5.4	10.4	14.0
Amazon0601	11.8	18.4	25.4

Table 3. The result of Experiment 1 running cyclic query graph algorithm.

Graph	PC-1	PC-2	PC-3
Amazon0302	7.2	9.4	10.0
Amazon0601	14.6	17.2	18.9

parameters: the connected subgraph, the number of edges to be added. Using this generator, we generate three acyclic query graphs with edges increased in order, named: P-1, P-2, P-3 and three cyclic query graphs with edges increased in order, named: PC-1, PC-2, PC-3 as well.

Environment. Most of the experiments were conducted using GraphX on a MacBook Pro with 16 GB DDR3 RAM and 2.2 GHz Intel Core i7 CPUs and distributed environment consists of a cluster of 13 machines. All these nodes are connected to the same 1 GB network switch. Each one has 32 GB DDR3 RAM, two 2.8 GHz Intel Xeon E5-2680 CPUs, each with 2 cores. One of the machines plays the role of master. We use Spark version 2.10 and Hadoop version 1.2.1.

4.2 Experiment Results

In a subgraph matching real-life application scenario, query time is the most important evaluation index. So the experimental results of this paper choose the query time to be the evaluation index.

Experiment 1: The Pattern of Query Graph

We examine the query time of the proposed algorithms using query graphs with different patterns to study their performance and flexibility. As displayed in Tables 2 and 3, query time increases as the size of query becomes more complex, which is not surprising, and the increase of query time keeps pace with the complexity of pattern graph. The behavior remains similar across datasets with different numbers of vertices. This shows that the algorithm proposed in this paper can adapt well to various query graphs.

Experiment 2: The Size of Data Graph

We examine the query time of the proposed algorithms using three data graphs with different sizes to study their performance and scalability. Tables 4, 5 and Figure shows the impact of the size of data graph. As expected, the query time increases with growth in the size of data graph, but the increases of query time and data graph approximately keep a linear relationship. This shows that the algorithm proposed in this paper can adapt well to data graphs with different size (Fig. 4).

Table 4. The result of Experiment 2 running acyclic query graph algorithm

Graph	P-2	P-3
Amazon0302	10.4	14.0
Amazon0312	18.4	25.4
Amazon0601	19.4	27.3

Table 5. The result of Experiment 2 running cyclic query graph algorithm.

Graph	PC-2	PC-3
Amazon0302	9.4	10.0
Amazon0312	18.2	18.0
Amazon0601	17.2	18.9

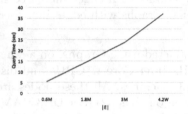

Fig. 4. Query time on different size of data graphs

Fig. 5. Query time on different average degree of data graphs

Experiment 3: The Degree of Data Graph

In this experiment, we use synthesizer to generate a series of data graphs with different average degrees. The input parameters of our graph synthesizer are the number of edges $|E|$, the average degree of every vertex $|D|$ and the label set. It selects an even between 4 and 10 as the average degree $|D|$ and the number of edges $|E|$ is fixed to 3M. The label of each vertex is randomly selected from label set which is the same as real-life datasets. Figure 5 shows that query time decreases as the average degree increases. This is because, for fixed degrees $|E|$, $|V| = |E|/|D|$ the growth of $|D|$ will result in less vertexes and each vertex becomes more decisive in message passing process.

Experiment 4: Optimization

Tables 6, 7 and Figs. 6, 7 show the impact of the optimizations in quering both real-life and synthetic datasets. The optimized algorithms, proAQG and proCQG have great improvements (at least 25%) in decreasing the query time compared to AQG and CQG. This shows that the optimized algorithms proposed in this paper can effectively reduce query time.

Experiment 5: Distributed Environment

We examine the query time of the proposed algorithms with different numbers of workers to study their performance and scalability (Fig. 8). In the exper-

Table 6. The result of Experiment 4 running acyclic query graph algorithm with optimization.

Graph	P-1	P-2	P-3	Optimization
Amazon0302	5.4/9.6	10.4/13.7	14.0/17.0	35.2%
Amazon0312	11.8/17.2	18.4/25.0	25.4/28.4	27.0%
Amazon0601	12.6/19.3	19.4/26.3	27.3/31.2	29.5%

Table 7. The result of Experiment 4 running cyclic query graph algorithm with optimization.

Graph	PC-1	PC-2	PC-3	Optimization
Amazon0302	7.2/10.4	9.4/12.0	10.0/14.8	39.8%
Amazon0312	11.6/15.4	17.6/26.1	18.4/31.4	53.2%
Amazon0601	12.0/16.9	18.8/26.2	20.0/41.2	65.9%

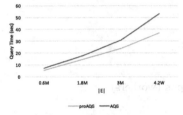

Fig. 6. AQG and proAQG in synthetic graphs (size)

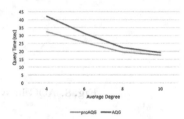

Fig. 7. AQG and proAQG in synthetic graphs (average degree)

iment displayed in Fig. 8, we can conclude that in a relatively small data graph ($|E| = 3M$ $|V| = 0.5M$), proposed algorithms can work well in single worker and there is a tradeoff between distributed environment and message passing cost.

4.3 Summary

From these experiments, we can draw the following conclusions: (1) proAQG and proCQG are efficient and scale well on large and dense data graphs varying query graph pattern, data graph sizes, and considerably outperform AQG and CQG. (2) our optimization techniques are effective, reducing the query time by 25% to 60%.

5 Related Work

In the following, we review some techniques related with graph pattern match and distributed graph mining framework.

5.1   Subgraph Isomorphism

Subgraph isomorphism is a well-studied problem and there exist many different works. Some works aims to find exactly all matched subgraphs for a given query graph. Ullmann [24] firstly proposes a algorithm to solve this problem, then many other methods [10,28] are proposed to optimize this method with different strategies. As the graph pattern match via subgraph isomorphism model is an NP-Complete problem [24], so these methods mentioned above may not work well when the data graph is at large size. However, with the continuous expansion of social applications, new algorithm is needed to support billion-edge graphs. Others pays attention to find a subset of data graphs to lower the subgraph isomorphism checking cost between the query graph and data graph [13,26]. In these methods, a data graph is always small, but the number of data graphs is large.

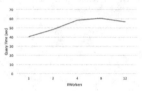

Fig. 8. AQG in synthetic graphs (workers)

5.2   Relaxed Subgraph Match

Many studies focus on the relaxed pattern match. Graph simulation is an alternative way to find the matching relation between a query graph and a data graph in a polynomial time [6,17]. An approximate pattern match method via relaxation on the vertex label mapping is proposed in [22]. Strong simulation [16] and strict simulation [7] are also alternative ways and they pay more attention to keep topological features of data graph.

5.3   Subgraph Match in Distributed Framework

Distributed graph processing framework is effective in handling large graphs. Many works are developed to process large graphs, including Pregel [18], GraphLab [15], GraphX [9] etc. By using such a integrated framework, developers can easily start up distributed clusters to process complex graph mining work by writing some simple scripts. Closer to this work is the graph pattern matching using GPS framework [20]. They implement four kinds of methods of graph pattern matching in distributed environment and get a great results, however they focus more on distributed implements and our work pays more attention to how to reduce the communication cost in message passing.

6 Conclusion

This paper designs and implements graph simulation algorithm based on vertex-centric model using GraphX [9]. According to different patterns of the query graph and the characteristics of message passing during processing of the algorithm, cyclic query graph algorithm and acyclic query graph algorithm are proposed. Experiments on real-life large graph datasets have verified the effectiveness of these algorithms and optimization techniques. The next step is to consider how to integrate incremental model into existing algorithms.

Acknowledgements. The authors acknowledge the financial support from the following foundations: National Key R&D Program of China (No. 2017YFC0803700), National Natural Science Foundation of China (61562091), Natural Science Foundation of Yunnan Province (2016FB110).

References

1. The graph-tool python library. http://figshare.com/articles/graph_tool/1164194
2. Brynielsson, J., Högberg, J., Kaati, L., Mårtenson, C., Svenson, P.: Detecting social positions using simulation. In: 2010 International Conference on Advances in Social Networks Analysis and Mining (ASONAM), pp. 48–55. IEEE (2010)
3. Cohen, J.: Graph twiddling in a mapreduce world. Comput. Sci. Eng. **11**(4), 29–41 (2009)
4. Dean, J., Ghemawat, S.: Mapreduce: simplified data processing on large clusters. Commun. ACM **51**(1), 107–113 (2008)
5. Fan, W., Li, J., Ma, S., Tang, N., Wu, Y.: Adding regular expressions to graph reachability and pattern queries. In: 2011 IEEE 27th International Conference on Data Engineering (ICDE), pp. 39–50. IEEE (2011)
6. Fan, W., Li, J., Ma, S., Tang, N., Wu, Y., Wu, Y.: Graph pattern matching: from intractable to polynomial time. Proc. VLDB Endow. **3**(1–2), 264–275 (2010)
7. Fard, A., Nisar, M.U., Ramaswamy, L., Miller, J.A., Saltz, M.: A distributed vertex-centric approach for pattern matching in massive graphs. In: 2013 IEEE International Conference on Big Data, pp. 403–411. IEEE (2013)
8. Gallagher, B.: Matching structure and semantics: a survey on graph-based pattern matching. AAAI FS **6**, 45–53 (2006)
9. Gonzalez, J.E., Xin, R.S., Dave, A., Crankshaw, D., Franklin, M.J., Stoica, I.: GraphX: graph processing in a distributed dataflow framework. In: OSDI, vol. 14, pp. 599–613 (2014)
10. He, H., Singh, A.K.: Graphs-at-a-time: query language and access methods for graph databases. In: Proceedings of the 2008 ACM SIGMOD International Conference on Management of Data, pp. 405–418. ACM (2008)
11. Henzinger, M.R., Henzinger, T.A., Kopke, P.W.: Computing simulations on finite and infinite graphs. In: 1995 Proceedings of 36th Annual Symposium on Foundations of Computer Science, pp. 453–462. IEEE (1995)
12. Hosoya, H.: Matching and symmetry of graphs. In: Symmetry, pp. 271–290. Elsevier (1986)
13. Khan, A., Li, N., Yan, X., Guan, Z., Chakraborty, S., Tao, S.: Neighborhood based fast graph search in large networks. In: Proceedings of the 2011 ACM SIGMOD International Conference on Management of data, pp. 901–912. ACM (2011)

14. Liu, C., Chen, C., Han, J., Yu, P.S.: GPLAG: detection of software plagiarism by program dependence graph analysis. In: Proceedings of the 12th ACM SIGKDD International Conference on Knowledge Discovery and Data Mining, pp. 872–881. ACM (2006)

15. Low, Y., Gonzalez, J.E., Kyrola, A., Bickson, D., Guestrin, C.E., Hellerstein, J.: GraphLab: a new framework for parallel machine learning. arXiv preprint arXiv:1408.2041 (2014)

16. Ma, S., Cao, Y., Fan, W., Huai, J., Wo, T.: Strong simulation: capturing topology in graph pattern matching. ACM Trans. Database Syst. (TODS) **39**(1), 4 (2014)

17. Ma, S., Cao, Y., Huai, J., Wo, T.: Distributed graph pattern matching. In: Proceedings of the 21st International Conference on World Wide Web, pp. 949–958. ACM (2012)

18. Malewicz, G., et al.: Pregel: a system for large-scale graph processing. In: Proceedings of the 2010 ACM SIGMOD International Conference on Management of data, pp. 135–146. ACM (2010)

19. Martínez, C., Valiente, G.: An algorithm for graph pattern-matching. In: Proceedings of Fourth South American Workshop on String Processing, vol. 8, pp. 180–197 (1997)

20. Salihoglu, S., Widom, J.: GPS: a graph processing system. In: Proceedings of the 25th International Conference on Scientific and Statistical Database Management, p. 22. ACM (2013)

21. Schelter, S.: Large scale graph processing with apache giraph. Invited talk at Game-Duell Berlin, 29 May 2012

22. Tian, Y., Patel, J.M.: Tale: a tool for approximate large graph matching. In: 2008 IEEE 24th International Conference on Data Engineering, ICDE 2008, pp. 963–972. IEEE (2008)

23. Tong, H., Faloutsos, C., Gallagher, B., Eliassi-Rad, T.: Fast best-effort pattern matching in large attributed graphs. In: Proceedings of the 13th ACM SIGKDD International Conference on Knowledge Discovery and Data Mining, pp. 737–746. ACM (2007)

24. Ullmann, J.R.: An algorithm for subgraph isomorphism. J. ACM (JACM) **23**(1), 31–42 (1976)

25. Valiant, L.G.: A bridging model for parallel computation. Commun. ACM **33**(8), 103–111 (1990)

26. Yan, X., Yu, P.S., Han, J.: Graph indexing: a frequent structure-based approach. In: Proceedings of the 2004 ACM SIGMOD International Conference on Management of Data, pp. 335–346. ACM (2004)

27. Zaharia, M., Chowdhury, M., Franklin, M.J., Shenker, S., Stoica, I.: Spark: cluster computing with working sets. In: Proceedings of the 2nd USENIX Conference on Hot Topics in Cloud Computing, HotCloud 2010, p. 10. USENIX Association, Berkeley (2010). http://dl.acm.org/citation.cfm?id=1863103.1863113

28. Zhao, P., Han, J.: On graph query optimization in large networks. Proc. VLDB Endow. **3**(1–2), 340–351 (2010)

Big Data Applications

A Multi-scale Dehazing Network
with Transmission Range Stretching

Jingjing Yao[1(✉)], Xinzhe Han[2], Gang Long[1], and Wen Lu[1]

[1] Xidian University, Xi'an 710071, China
yaojingjing@stu.xidian.edu.cn, longbenlei@163.com,
luwen@mail.xidain.edu.cn
[2] University of Chinese Academy of Sciences, Beijing 100049, China
hanxinzhe17@mails.ucas.ac.cn

Abstract. Image dehazing has become a significant research area in recent years. However, the traditional dehazing algorithms based on statistics priors cannot adaptive to various conditions of natural hazy images. And those algorithms based on Data-driven learning such as some dehazing networks for estimating transmission almost have the problem that the range of the estimated transmission is too narrow for those haze images where hazy density changes largely. So in this paper, we present a novel Dehazing Network to learn the relationship between the hazy image and its corresponding transmission map. It uses jump connection and the layer of Multi-scale features fusion to obtain more feature related to haze density and use both max pooling and average pooling which in turn remove some details of the transmission map and make the gained transmission map more accurate. Moreover, we also propose a linear stretching algorithm based on dark channel prior to extent the transmission range. The experimental result demonstrate that proposed algorithm achieves favorable result against existing dehazing algorithms on both synthetic images and natural images.

Keywords: Dehazing network · Dark channel prior
Adaptively stretching the transmission range

1 Introduction

In recently years, haze has become more and more serious resulting the serious quality decline and obvious distortion of images captured by equipment. Image dehazing algorithms aim to eliminate the haze and enhance the image clarity. In general, image dehazing algorithms can be divided into three main categories: the algorithms based on image enhancement, the algorithms based on the atmospheric scattering model, and the algorithms based on Data-driven learning.

In the early state of image dehazing study, researchers directly applied the image enhancement algorithms such as histogram normalization, wavelet transformation. However, theses algorithms ignored the atmospheric scattering model resulting that the dehazing image is unnatural.

© Springer Nature Singapore Pte Ltd. 2018
Z. Xu et al. (Eds.): Big Data 2018, CCIS 945, pp. 257–267, 2018.
https://doi.org/10.1007/978-981-13-2922-7_17

Another main algorithms of image dehazing were based on the atmospheric scattering model which are detailed description and derivation in [1, 2]. Tan [3] increased the color contrast by maximizing local contrast. The result was visually superior, but over-enhancement could occur in some areas. Based on the assumption that the transmission had nothing to do with the color of the image surface, Fattal et al. [4] divided the image into two parts including foreground and background, and restored them according to the physical model respectively. However, this algorithm tended to underestimate the haze density especially for the images with thick haze. He et al. [5] proposed dark channel prior which was simple and effective in the image dehazing, but in brighter areas especially in the sky area, dark channel priori assumptions would not work. In [6], a fast iterative algorithm was designed to estimate the transmission according to the contextual constraints. Though the contrast of the dehazing results were more obvious and the detail were more prominent, some regions appeared color over-enhancement. Tarel et al. [7] and Gibson et al. [8] replaced the minimization operator with a median operator for dehazing and Lai et al. [9] extracted the optimal transmission map based on hazy model. In conclusion, although these algorithms based on the physical model and statistical priori laws could get a more reasonable and clear dehazing result in most cases, there still have some priori failure situations.

Recently, data-driven learning has also been applied in the field of image dehazing. Tang et al. [10] devoted to get the relationship between the extracted features and the transmission using random forest. Zhu et al. [11] adopted a linear model to learn the depth of the image and then estimate the transmission map. Fan et al. [12] used two layers of Gaussian process regression to regress the extracted features and the neighboring pixels to get more detailed transmission map. The Bayesian dehazing algorithm proposed in [13] converted the block estimation in the traditional method into point estimation, but the computational complexity was higher. Ren et al. [14] proposed a multi-scale convolution neural network, which used two networks of rough estimation and fine-tuning to obtain a fine transmission. Cai et al. [15] proposed DehazeNet. However, those algorithms based on Data-driven learning for estimating transmission almost have the problem that the range of the estimated transmission is too narrow for those haze images whose hazy density changes greatly. In order to overcome these problems, we devised and implemented a novel Dehazing Network and a linear transmission range stretching algorithm based on dark channel prior, which can directly get a more accurate transmission map from hazy images. The experimental result demonstrate that proposed algorithm achieves better results than existing dehazing algorithms on both synthetic images and natural images. Our main contributions are as follows:

(1) We proposed a novel Dehazing Network which is more suitable for image dehazing. The jump connection and the layer of Multi-scale features fusion is more conducive to learning the relationship between the hazy image and its corresponding transmission map; And unlike other dehazing networks which only use the max pooling, we use both max pooling and average pooling. Because the transmission is always constant in a small patches, adding average pooling will remove some details from the feature map which in turn will remove some details of the transmission map and make the resulting transmission map more accurate.

(2) Aiming at the problem that the range of the estimated transmission by Data-driven learning algorithms is too narrow for those haze images where hazy density changes significantly resulting that those algorithms do not work well for near parts without haze and in far parts with dense haze, we design a method of adaptively stretching the transmission range according to the dark channel priori. In this way, the output transmission is further optimized and the haze can be fully removed for the areas with dense haze, and also solve the problem of over-enhanced for haze-free areas.

2 Related Work

2.1 The Atmospheric Scattering Model

Hazy image degradation is caused by the combination of the incident light attenuation and the imaging of ambient light. Therefore, the atmospheric scattering model is actually a combination of the two parts as follows:

$$I(x) = J(x)t(x) + A(1 - t(x)) \tag{1}$$

Where $I(x)$ is the observed value of the hazy image at pixel x; $J(x)$ is the haze-free image; A is the atmospheric light; $t(x) = e^{-\beta(\lambda)d(x)}$ is the transmission which indicate the light attenuation reaching the lens from the target, and $d(x)$ is the depth information.

2.2 Features Related to Haze Density

Hazy image quality decline will show that the overall brightness is higher, both of the visibility of the object and the local contrast are reduced, and the overall gray value of hazy image is higher. There are some features related to hazy density as follows.

(1) Dark channel features

Dark channel features is a statistical priori laws which show that in a haze-free image patch without sky, the minimum brightness of all pixels in three channels always approaches zero.

(2) Local maximum saturation

Image saturation refers to the color purity of the image, the greater the saturation, the more vivid the image. Saturation is mainly determined by the amount of white light in the color. The hazy image degradation process is partly due to the imaging of the white atmospheric light at the receiving device. The local maximum saturation can be defined as follows:

$$S^s(p_i) = 1 - \frac{\left(\min_{p_s \in \Omega_s(p_i)} \left(\min_{c \in \{r,g,b\}} (I^c(p_s))\right)\right)}{\left(\max_{p_s \in \Omega_s(p_i)} \left(\max_{c \in \{r,g,b\}} (I^c(p_s))\right)\right)} \tag{2}$$

(3) The feature of color attenuation

The color attenuation priori [21] is proposed based on the imaging mechanism under hazy conditions. In the hazy images, the higher the haze density, the higher the average luminance and average saturation is lower. Therefore, the difference between the luminance and the saturation can be used to describe the haze density showed as follows:

$$A(x)=I^v(x)-I^s(x) \tag{3}$$

Where I^v and I^s represent the luminance and saturation in the image HSV model, respectively. According to the statistics of a large number of images and the color attenuation prior, we can obtain $d(x) \propto A(x)$ from which the depth and the transmission of the image can be estimated easily.

3 The Proposed Method

In this paper, we present a novel Dehazing Network under the Caffe framework to realize the feature extraction and realize the non-linear regression between the features and the corresponding transmission, then achieve a single image dehazing though the atmospheric scattering model.

3.1 The Design of Dehazing Network

The overall structure of the proposed Dehazing network can be divided into three parts including Feature Extraction, the fusion of Multi-scale features, and Nonlinear Regression as showed in Fig. 1.

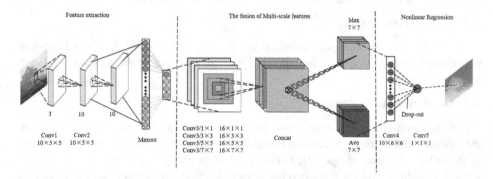

Fig. 1. The structure of proposed dehazing network.

(1) Feature Extraction: After choosing the appropriate size of the filter, the convolution process of the input image is equivalent to automatically extracting feature maps from the hazy image related to its transmission. Similar to the traditional

method that the feature extraction process chooses the maximum value in different color channels, Maxout element is selected as the activation function in our paper which achieves three goals: highlighting important features, reducing feature dimensions and computational complexity, and the most basic nonlinear activation of neurons.

(2) The Fusion of Multi-scale Features: Multi-scale feature fusion has proved to be very effective in traditional Dehazing algorithms. Therefore, the second part convolves the features map output by Maxout with convolution kernels of different sizes including $1 \times 1, 3 \times 3, 5 \times 5, 7 \times 7$, and then parallelizes the multi-scale features to achieve the effect of multi-scale fusion in the traditional method. Moreover, in order to increase the robustness for transmission estimation while suppressing the local noise of the extracted feature maps, which in turn will remove some details of the transmission map and make the resulting transmission map more accurate.

(3) Nonlinear regression: There are two fully connected layers for nonlinear regression In order that different sizes of images can be dehazed by the trained network, we transform the full connection layer into the convolutional layer whose kernel size is the same as the input feature maps.

3.2 The Training of the Dehazing Network

In our paper, we use the Depth Dataset KITTI in [16] and the corresponding depth information for hazy image synthesis according to the atmospheric scattering model (1). This database simulates the natural landscape images whose depth has a wide range, so the training data generated by this database is more reasonable than by those databases consisted of indoor images with small depth range. In order to make the training samples more widely adaptable, three kinds of hazy images with different haze density were synthesized in this paper as training data:

(1) 400 haze-free images are used to generate the hazy images with moderate haze density and the corresponding transmission. Due to the fact that the haze density of mostly hazy images is relatively moderate in the real world, such training data can closely reflect the relationship between the haze density and image information. So these haze images with moderate haze density should be as the main part of the training data.

(2) 50 haze-free images whose transmission values are 1 are second part of our training data. In many cases, the hazy of the hazy image exists only in the deeper depth of field, and there is almost no haze in the near field. In order to improve the accuracy of the regression, add the hazeless sample as a small part of training data.

(3) 50 dense-haze images are used to generate the hazy images with dense haze and the corresponding transmission. In order to improve the accuracy and adaptability of regression, we should increase sample with smaller transmission to expand the range of regression.

We divided the hazy image into small pieces whose size is 16×16, and take the median for each small transmission patch as its label:

$$t^P(x) = \underset{y \in P(x)}{\text{median}}\{t(y)\} \tag{4}$$

Where, $P(x)$ represents a 16×16 small patch.

The purpose of the proposed dehazing network is that the transmission obtained by regression is as close as possible to the true transmission, so we select Euclidean distance as the model loss function.

$$L(\Theta) = \frac{1}{N} \sum_{i=1}^{N} \left\| F(I_i^P; \Theta) - t_i \right\|^2 \tag{5}$$

3.3 Recover the Haze-Free Image

For the estimation of atmospheric light, we using the minimum filter window sliding on each channel, and finally taking the maximum of each channel as the atmospheric light in the corresponding channel value as in [6]. Then, we can get the haze-free image by the following equation.

$$J(x) = \frac{1}{t(x)} I(x) + \left(\frac{1}{t(x)} - 1 \right) A \tag{6}$$

3.4 Method of Adaptively Stretching the Transmission Range

Algorithms based on Data-driven learning for estimating transmission such as [14] and [15] almost have the problem that the range of the estimated transmission is too narrow for those haze images where hazy density changes significantly resulting that those algorithms do not work well for near parts without haze and in those parts with dense haze as Fig. 2.

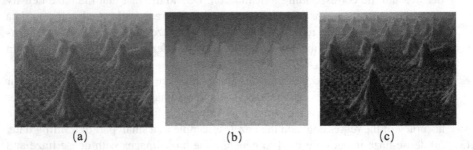

| (a) | (b) | (c) |

Fig. 2. (a) Hazy image. (b) Transmission t(x). (c) The dehazed image.

In order to solve the above problem, we design a method to adaptively stretch the transmission range based on the dark channel priori. Dark channel priori is simple and effective, but does not apply in bright areas such as the sky. Even so, it is more than sufficient to speculate the approximate the haze density range by the range of dark channel. The dynamic range of stretching is small when the distribution of dark channel is relatively concentrated, and is larger when the distribution of dark channel is more dispersed. The transmission stretching method is designed as follows:

$$t(x) = \frac{F_5(x) - \min\{F_5(:)\} + 0.15}{\max\{F_5(:)\} - \min\{F_5(:)\} + 0.15}$$
$$\cdot \min\left\{(\max\{F_5(:)\}) - \min\{F_5(:)\}\frac{\sigma_D}{\sigma_F} + 0.2, 1\right\} \tag{7}$$

Where, F_5 is the estimated transmission by those dehazing algorithms based on data-driven learning, and σ_D, σ_F represent the standard deviation of the dark channel feature and F_5. To avoid over-stretching, we add constant terms to the numerator and denominator respectively. In order to limit the transmission in the range of (0, 1), we take the minimum of the stretch range and 1 as the final stretching range. The dehazing result after transmission stretching is showed as Fig. 3.

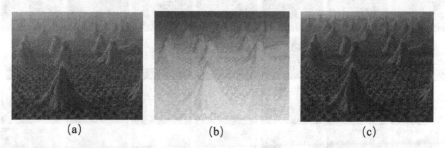

(a)   (b)   (c)

Fig. 3. (a) Hazy image. (b) The transmission after stretching. (c) The dehazed image after transmission stretching.

We can find that the dehazing result has been significantly improved. It can well solve the problem of over-enhanced for haze-free areas and the haze can also be fully removed for the areas with dense haze.

4 Experiments

In order to verify the proposed Dehazing network, we compare it with existing dehazing algorithm [5, 6, 12, 15, 17] from the subjective visual effects and the objective evaluation indicators in synthesis images and in natural images.

4.1 Comparison in Natural Images

The traditional dehazing algorithms based on local statistics prior need to get the local statics features of image patches, and then estimate the transmission map of those patches according to the assumption that the transmission is locally invariable, but patch estimation will inevitably cause damage to the image structure and loss of detail.

Although the proposed uses 16×16 image patches for training, it uses a sliding convolution window of step size 1 in the dehazing implementation stage. So pixel-by-pixel point estimation is achieved, which allows the original image detail information to be retained to the greatest degree. Next in importance, the statistical characteristics will fail under certain circumstances. DehazeNet [15] and GPR [12] which are Data-driven learning algorithms have the problem that the range of the estimated transmission by Data-driven learning algorithms is too narrow for those haze images where hazy density changes significantly. In order to solve the above problems, we design a novel dehazing network and a method to adaptively stretch the transmission range according to the dark channel priori. The comparison in natural image is shown in Fig. 4.

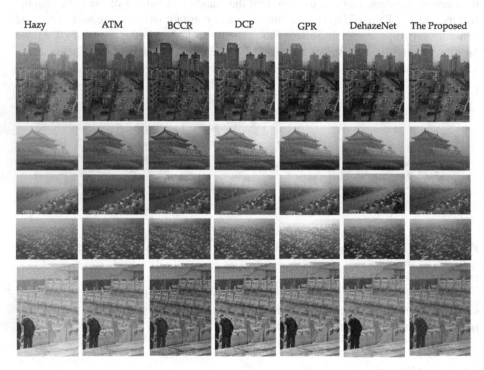

Fig. 4. The comparison with existing algorithms in natural image dehazing. (Color figure online)

DCP [5] performs well for those hazy images without bright sky areas. However, when there are bright regions such as sky in the image, the defects in statistical assumptions of the dark channel priors cannot be avoided resulting that the estimated transmission is significantly lower and significant oversaturation in the sky area such as the second and third row of Fig. 5. ATM [17] is mainly to improve the recovery of atmospheric light by using the statistical law that the brightness of the brightest top 1% pixel in the image is nearly independent of the transmission. Although it is also based on the dark channel algorithm, the distortion in the bright area is much less severe.

However, for some images that do not have a sufficiently large bright area, the relatively dense haze area is also counted as the brightest 1% such as the river area in the third row of Fig. 4, resulting that the estimated transmission of sky area is almost consistent with the transmission of the river, and the sky turns blue causing significant color distortion. BCCR [6] tends to maximize the contrast of the scene, therefore, it performs well in detail recovery and visual clarity. However, it will cause obvious color distortion in those areas that are relatively smooth with little change in depth and transmission. For example, the sky areas of the first and second rows in the Fig. 4 all become colored and have halos. DehazeNet [15] and GPR [12] have a problem that the predicted transmission range is narrow resulting that the predicted transmission is too low and the dahazing result is over-enhanced in near parts without haze and the haze is not fully removed in large depth area with dense haze. Our dehazing results are more visually natural and are more in line with the true color of the original image other than over-enhanced, and the dehazing in large depth area with dense haze are also relatively complete restoring more clear edges and details which demonstrate that the method of proposed Optimization of the transmission map is effective.

4.2 Comparison in Synthesis Images

Test images were synthesized with using (1). Calculate the PSNR, SSIM, and FSIM of dehazing images separately. The objective indicators are shown in Table 1, and the comparison of dehazing results is shown in Fig. 5.

Table 1. The objective indicators comparison results of the proposed algorithm and the existing dehazing methods

Metric	ATM [17]	DCP [5]	BCCR [6]	GPR2 [12]	DehazeNet [15]	The proposed
PSNR	26.7691	34.2208	24.9917	33.0806	32.9955	32.3263
SSIM	0.8243	0.7937	0.7536	0.8049	0.8364	0.8372
FSIM	0.8805	0.8533	0.8381	0.8689	0.8870	0.8836

From Table 1, these three Learning-drive dehazing algorithms [12, 15] and our algorithm have significantly improved the effectiveness than the traditional algorithms [5, 6, 17]. Though DehazeNet has a little better result in PSNR and FSIM, it is obviously not completely dehazed. But the Learning-driven learning for estimating transmission almost have the problem that the range of the estimated transmission is too narrow for those haze images resulting that the haze is not fully removed or over-enhanced in some areas. Compared to GPR2, our dehazing result is more visually closer to the original image which can be seen in Fig. 5. And the biggest problem of GPR is the computational complexity which make it difficult to be applied in practice.

In summary, the proposed network performs better on both synthetic images and natural images.

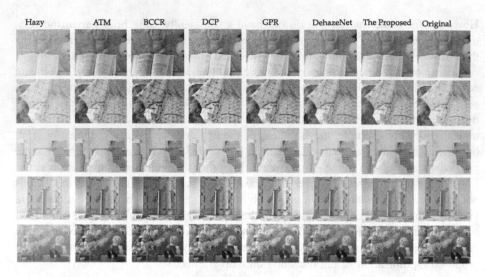

Fig. 5. Comparison of the proposed algorithm and the existing dehazing methods.

5 Conclusions

In this paper, we present a novel Convolutional Dehazing Network to learn the relationship between the hazy image and its corresponding transmission map, which use jump connection and the layer of Multi-scale features fusion to obtain more feature related to haze density and use both max pooling and average pooling which in turn remove some details of the transmission map and make the gained transmission map more accurate. Moreover, we also propose a linear stretching algorithm based on dark channel prior to extent the transmission range. The experimental result demonstrate that proposed algorithm achieves favorable result against existing dehazing algorithms on both synthetic images and natural images.

Acknowledgement. This research was supported partially by the National Natural Science Foundation of China (Nos. 61372130, 61432014, 61501349, 61571343). The authors would like to thank our tutor, Professor Lu Wen, his valuable remarks and suggestions inspired us a lot.

References

1. Nayar, S.K., Narasimhan, S.G.: Vision in bad weather. In: Proceedings of the 7th IEEE International Conference on Computer Vision, Kerkyra, vol. 2. IEEE (1999)
2. Narasimhan, S.G., Nayar, S.K.: Vision and the atmosphere. Int. J. Comput. Vis. **48**, 233–254 (2002)
3. Tan, R.T.: Visibility in bad weather from a single image. In: Proceedings of the 2008 IEEE Conference on Computer Vision and Pattern Recognition, Anchorage, AK. IEEE (2008)
4. Tan, R.: Single image dehazing. In: Proceedings of the 2008 ACM Transactions on Graphics, vol. 27. ACM, New York (2008)

5. He, K., Sun, J., Tang, X.: Single image haze removal using dark channel prior. IEEE Trans. Pattern Anal. Mach. Intell. **33**, 2341–2353 (2011)
6. Meng, G.F., Wang, Y., Duan, J.: Efficient image dehazing with boundary constraint and contextual regularization. In: IEEE International Conference on Computer Vision (2013)
7. Tarel, J.-P., Hautiere, N.: Fast visibility restoration from a single color or gray level image. In: Proceedings of IEEE International Conference on Computer Vision (2009)
8. Gibson, K.B., Nguyen, T.Q.: On the effectiveness of the dark channel prior for single image dehazing by approximating with minimum volume ellipsoids. In: Proceedings of the IEEE International Conference on Acoustics, Speech, and Signal Processing, Prague. IEEE (2011)
9. Lai, Y.-H., Chen, Y.-L., Chiou, C.-J.: Single-image dehazing via optimal transmission map under scene priors. IEEE Trans. Circ. Syst. Video Technol. **25**, 1–14 (2015)
10. Tang, K., Yang, J., Wang, J.: Investigating haze-relevant features in a learning framework for image dehazing. In: Proceedings of IEEE Conference on Computer Vision and Pattern Recognition (2014)
11. Zhu, Q., Mai, J., Shao, L.: A fast single image haze removal algorithm using color attenuation prior. IEEE Trans. Image Process. **24**, 3522–3533 (2015)
12. Fan, X., Fan, Y., Tang, X.: Two-layer Gaussian process regression with example selection for image dehazing. IEEE Trans. Circuits Syst. Video Technol. (2016)
13. Nishino, K., Kratz, L., Lombardi, S.: Bayesian defogging. Int. J. Comput. Vis. **98**, 263–278 (2012)
14. Ren, W., Liu, S., Zhang, H., Pan, J., Cao, X., Yang, M.-H.: Single image dehazing via multi-scale convolutional neural networks. In: Leibe, B., Matas, J., Sebe, N., Welling, M. (eds.) ECCV 2016. LNCS, vol. 9906, pp. 154–169. Springer, Cham (2016). https://doi.org/10.1007/978-3-319-46475-6_10
15. Cai, B., Xu, X., Jia, K.: DehazeNet: an end-to-end system for single image haze removal. IEEE Trans. Image Process. **25**, 5187–5198 (2016)
16. Geiger, A., Lenz, P.: Are we ready for autonomous driving? The KITTI vision benchmark suite. In: IEEE Conference on Computer Vision & Pattern Recognition, vol. 157 (2012)
17. Jia, M., Glatzer, I., Fattal, R.: Automatic recovery of the atmospheric light in hazy images. In: IEEE International Conference on Computational Photography (2014)

LaG-DESIQUE: A Local-and-Global Blind Image Quality Evaluator Without Training on Human Opinion Scores

Ruyi Liu[1,2], Yi Zhang[3], Damon M. Chandler[4], Qiguang Miao[1,2(✉)], and Tiange Liu[5]

[1] School of Computer Science and Technology, Xidian University,
Xian 710071, Shaanxi, China
qgmiao@126.com
[2] Xian Key Laboratory of Big Data and Intelligent Vision,
Xian 710071, Shaanxi, China
[3] School of Electronic and Information Engineering, Xian Jiaotong University,
Xian 710049, Shaanxi, China
[4] Department of Electrical and Electronic Engineering,
Shizuoka University, Hamamatsu, Shizuoka 432-8561, Japan
[5] School of Information Science and Engingeering, Yanshan University,
Qinhuangdao 066004, Hebei, China
http://web.xidian.edu.cn/qgmiao/

Abstract. This paper extends our previous DESIQUE [1] algorithm to a local-and-global way (LaG-DESIQUE) to blindly measure image quality without training on human opinion scores. The local DESIQUE extracts block-based log-derivative features and evaluates image quality through measuring the multivariate Gaussian distance between selected natural and test image patches. The global DESIQUE extracts image-based log-derivative features and image quality is estimated based on a two-stage framework, which was trained on a set of regenerated distorted images with their quality scores estimated by MAD [2] algorithm. The overall quality is the weighted average of local and global DESIQUE scores. Test on several image databases demonstrates that LaG-DESIQUE performs competitively well in predicting image quality.

Keywords: No-reference quality assessment · Derivative statistics
Multivariate Gaussian · Human opinion score

1 Introduction

The ability to assess image quality without need of a reference is a crucial and challenging task for any system that processes images for human viewing. To address this need, numerous algorithms for no reference (NR) image quality assessment (IQA) have been proposed over the past few decades, and these algorithms generally fell into two categories: (1) the distortion-specific approach and (2) the non-distortion-specific (also referred as general-purpose) approach.

© Springer Nature Singapore Pte Ltd. 2018
Z. Xu et al. (Eds.): Big Data 2018, CCIS 945, pp. 268–277, 2018.
https://doi.org/10.1007/978-981-13-2922-7_18

The distortion-specific NR approach operates by assuming that a particular distortion type is known, and thus the image quality is often measured by quantifying specific distortion-related features (e.g., [3–8]). The non-distortion-specific NR approach does not require the prior knowledge of distortions, but assumes that the test image shares some properties with those in the training database, and thus the image quality is often predicted by the trained models/frameworks (e.g., [9–13]). However, these general-purpose NR approaches often require human opinion scores for training, and this process ultimately restrict its further application.

To release algorithms from the dependence on human opinion scores (i.e., MOS/DMOS values), several subsequent work has been conducted. For example, in [14] a latent quality factors (LQFs) based approach was proposed which assesses image quality through computing the difference between pristine and distorted images in terms of the occurrence probability of LQFs. In [15], a natural image quality evaluator was developed to extract local BRISQUE [13] features and image quality is expressed as the distance between the multivariate Gaussian (MVG) fit of the features extracted from the test image and from the corpus of natural images. In [16], Xue et al. proposed an algorithm called QAC, which employs quality-aware clustering and codebook theory to infer image quality.

In this paper, we extend our previous NR IQA algorithm (DESIQUE [1]) to operate both locally and globally without training on MOS/DMOS values. Our algorithm is based on assumption that human judge quality both locally and globally, and thus the image quality should be analyzed by using both the local and global features. For local analysis, block-based DESIQUE features are extracted, and image quality is estimated by computing the local feature distance of sharper regions in natural and test images. For global analysis, image-based DESIQUE features are extracted and image quality is estimated by a two-stage framework [11]. Here, to remove from the dependence on MOS/DMOS, we built another database for training, which consists of 125 natural images from the Berkeley Segmentation database [17], and 4000 distorted images generated by simulating four distortion types at eight different levels. The MAD [2] algorithm was employed to approximately predict the quality scores of these distorted images, which were then used as a replacement of the ground truth quality scores to train the framework. Finally, the image quality is a weighted average of the two aforementioned estimates. We demonstrate the performance of LaG-DESIQUE on various databases.

This paper is organized as follows: Sect. 2 provides details of LaG-DESIQUE algorithm. In Sect. 3, we analyze performance of this algorithm on various image quality databases. General conclusions are presented in Sect. 4.

2   Algorithm

The proposed LaG-DESIQUE algorithm consists of two modules: the local DESIQUE and the global DESIQUE, and its block diagram is shown in Fig. 1.

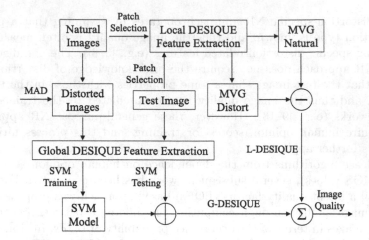

Fig. 1. A block diagram of the LaG-DESIQUE algorithm.

2.1 Local DESIQUE

The local DESIQUE explores the idea in [15] that image quality can be measured in a block-based manner, but operates by using a different set of quality-related features, a different patch-selected approach, and a different model distance measurement. Specifically, we extract log-derivative-based features [1] from the selected image patches based on FISH$_{bb}$ [18] algorithm, and image quality is expressed as the distance between two MVG models that fit features of these selected natural and distorted image patches, respectively.

2.1.1 Local DESIQUE Feature Extraction

The local DESIQUE features consist of two parts: (1) the local spatial-domain-based features and (2) the local frequency-domain-based features. The former part is based on all seven types of log-derivative statistics (denoted by D1 to D7) proposed in [1] at two image scales.

$$D1 : \nabla_x J(i,j) = J(i,j+1) - J(i,j) \tag{1}$$

$$D2 : \nabla_y J(i,j) = J(i+1,j) - J(i,j) \tag{2}$$

$$D3 : \nabla_{xy} J(i,j) = J(i+1,j+1) - J(i,j) \tag{3}$$

$$D4 : \nabla_{yx} J(i,j) = J(i+1,j-1) - J(i,j) \tag{4}$$

$$D5 : \nabla_x \nabla_y = J(i-1,j) + J(i+1,j) - J(i,j-1) - J(i,j+1) \tag{5}$$

$$D6 : \nabla_{cx} \nabla_{cy} J(i,j)_1 = J(i,j) + J(i+1,j+1) - J(i,j+1) - J(i+1,j) \tag{6}$$

$$D7 : \nabla_{cx} \nabla_{cy} J(i,j)_2 = J(i-1,j-1) + J(i+1,j+1) - J(i-1,j+1) - J(i+1,j-1) \tag{7}$$

where $J(i,j) = log[I(i,j)+K]$ is the logarithmic computation of pixel/coefficient values $I(i,j)$, and K is a small constant that prevents $I(i,j)$ from being zero. The latter part is based on five of them (D1 to D4, D6) at the original scale only. Note that the local DESIQUE does not employ point-wise based features, and thus 48 features are extracted from each image patch.

In regards to the fact that human tend to be more sensitive to the sharper regions in an image, a selected number of image patches with sharper content are used for local feature extraction and quality evaluation. Specifically, we measure local sharpness of each image by using FISH$_{bb}$ algorithm [18]. The only difference is that here the local sharpness is computed for images blocks of 48×48 pixel size, resulting into block sizes of 24×24, 12×12, and 6×6 for each of the three discrete wavelet transform coefficient levels to be computed. Finally, those patches with a super-threshold sharpness are selected. Note that this patch selection procedure was applied to both the natural and distorted images with different settings of the threshold T, which is given by

$$T = \delta_{min} + p\left(\delta_{max} - \delta_{min}\right) \tag{8}$$

where p is a fraction number with range $0 < p < 1$; δ_{min} and δ_{max} denote the minimum and maximum value of the patch sharpness within one image. We set p to be 0.75 for patch selection in natural images, and 0.5 for the distorted images.

2.1.2   Local DESIQUE Quality Estimation
To compute local DESIQUE quality, we follow [15] to use the MVG distribution model which is given by

$$f_X\left(x_1, ..., x_k\right) = \frac{exp\left[-\frac{1}{2}\left(x-\mu\right)^T \Sigma^{-1}\left(x-\mu\right)\right]}{(2\pi)^{k/2}|\Sigma|^{1/2}} \tag{9}$$

where $x_1, ..., x_k$ are the computed DESIQUE features for each image block; μ and Σ denote the mean and covariance matrix of the MVG model.

We estimate MVG model parameters on both the natural and test image block features. The 48×48 natural image blocks were selected from 125 natural images in Berkeley image segmentation database [17] with 16 pixels overlap, and the test image blocks were selected in the same way but with 24 pixels overlap and different patch selection standard (i.e., differ in fraction number p). Note that in our work, the sharpness measurement (i.e., Eq. (8)) was applied to both the natural and test image patches (in [15] the natural image patches only), because we found that even in distorted images with loss of sharpness, those relatively sharper regions still weight more on determining image quality. Also, a less number of computing blocks allows less computational complexity. Finally, the quality of test image, denoted by L-DESIQUE, is expressed as the distance between two MVG models that fit features of the selected natural and distorted image patches, respectively:

$$\text{L-DESIQUE} = 0.03d_1 + ln\left(1 + d_2\right) + 0.15d_3 \tag{10}$$

where

$$d_1 = \sqrt{(\mu_1 - \mu_2)^T \Sigma_2^{-1}(\mu_1 - \mu_2)} \tag{11}$$

$$d_2 = \sqrt{(\mu_1 - \mu_2)^T \Sigma_1^{-1}(\mu_1 - \mu_2)} \tag{12}$$

$$d_3 = ln\left(\frac{det\left(\frac{\Sigma_1 + \Sigma_2}{2}\right)}{det\left(\Sigma_1\right) det\left(\Sigma_2\right)}\right) \tag{13}$$

and where μ_1, μ_2 and Σ_1, Σ_2 are the mean vectors and covariance matrices of the distorted image's MVG model and the natural MVG models.

2.2   Global DESIQUE

The global DESIQUE works on entire images and aims to find a mapping that associates feature vector to corresponding image quality. To this end, a two-stage framework was trained on the regenerated database with MAD [2] quality scores to globally assess image quality.

2.2.1   Training Database Generation

We generated the distorted images by using the same 125 natural images. Four most common types of distortions are simulated: JPEG/JPEG2000 compression, Gaussian blur (GBLUR) and Additive Gaussian white noise (AGWN). For each image, we generate its distorted version of each type at eight different levels by controlling the compression ratio (for JPEG and JPEG2000 compression), the Gaussian kernel variance (for GBLUR), and the noise standard deviation (for AGWN). Therefore, a database consisting of 125 reference images and 4000 distorted images was generated.

With access to the reference images, the MAD [2] algorithm was employed to predict the distorted image quality scores. The choice of MAD is based on two reasons: (1) MAD is among the best full reference (FR) IQA algorithms that have been developed, and (2) compared with other FR IQA methods, MAD was found to have better linear relationship with human subjective judgments (before nonlinear fit). Note that a better linear relationship with human subjective judgments allows the regression models to be trained with an equal/balanced consideration on different distortion levels, and thus better prediction performance can be achieved. A subset of the regenerated distorted images, together with their corresponding MAD scores, are shown in Fig. 2.

2.2.2   Global DESIQUE Quality Estimation

With the regenerated training database and the estimated quality scores, a number of 60 DESIQUE features (same as [1]) were extracted for each of the trained distorted images, and a two-stage framework consists of (1) distortion identification followed by (2) distortion-specific quality assessment was learned to assess quality of test image. Let $\mathbf{p}$ denote the n-dimensional vector of probabilities,

Fig. 2. Some examples of the regenerated distorted images, and their corresponding MAD prediction scores.

and $\mathbf{q}$ denote the n-dimensional vector of estimated qualities obtained from n regression models, then the overall estimated quality, denoted by G-DESIQUE, is computed as follows:

$$\text{G-DESIQUE} = \sum_{i=1}^{n} p(i)q(i) \tag{14}$$

where $p(i)$ and $q(i)$ denote elements of $\mathbf{p}$ and $\mathbf{q}$, respectively.

2.3 LaG-DESIQUE Quality Index

With the estimated quality scores of both local and global DESIQUE, a scalar-valued final quality estimation, denoted by LaG-DESIQUE, is obtained by combining them together:

$$\text{LaG-DESIQUE} = \text{L-DESIQUE} + 0.025 \times \text{G-DESIQUE} \tag{15}$$

The weighting factor is chosen to bring two scores to a similar range, and the add operation is chosen because both indices denote quality deviations from the reference ones. Therefore, the smaller value of LaG-DESIQUE, the better quality of the test image.

3   Results and Analysis

To evaluate the performance of LaG-DESIQUE algorithm, four public available image databases were used: (1) the LIVE database [19]; (2) the CSIQ database

Table 1. Performance of LaG-DESIQUE and other FR/NR algorithms on the LIVE database.

	FR/NR	JP2K	JPEG	Gblur	AGWN	ALL
SROCC						
PSNR	FR	0.895	0.880	0.783	0.985	0.875
SSIM	FR	0.961	0.976	0.952	0.969	0.948
LQFs	NR	0.850	0.880	0.870	0.800	0.800
QAC	NR	0.851	0.940	0.909	0.961	0.886
L-DESIQUE	NR	0.913	0.945	0.937	0.964	0.925
G-DESIQUE	NR	0.843	0.955	0.800	0.954	0.893
LaG-DESIQUE	NR	0.928	0.956	0.937	0.970	0.942
CC						
PSNR	FR	0.873	0.865	0.775	0.979	0.858
SSIM	FR	0.893	0.928	0.888	0.958	0.829
LQFs	NR	0.870	0.900	0.880	0.870	0.790
QAC	NR	0.838	0.933	0.906	0.924	0.861
L-DESIQUE	NR	0.914	0.950	0.919	0.969	0.921
G-DESIQUE	NR	0.868	0.968	0.831	0.962	0.896
LaG-DESIQUE	NR	0.932	0.964	0.928	0.975	0.939

[2]; (3) the TID database [20]; and (4) the Toyama database [21]. The predicting performance was measured in terms of (1) the Spearman rank-order correlation coefficient (SROCC) and (2) the Pearson linear correlation coefficient (CC). Note that the global DESIQUE was trained on four distortion types, and thus we tested and compared with other IQA algorithms only on these trained distortions.

3.1 Performance on LIVE

We compared LaG-DESIQUE with two well-known FR IQA algorithms (PSNR [22] and SSIM [23]) and two NR approaches (LQFs [14] and QAC [16], both of which do not require training on human opinion scores) on the LIVE database. The results are shown in Table 1. Also included are the test results of local and global DESIQUE. As shown in the table, LaG-DESIQUE outperforms other two NR IQA methods in terms of SROCC and CC. It also achieves better performance than PSNR, and even challenges SSIM. Compared with L- and G-DESIQUE, we see that the combined local and global analysis improves upon each individual one.

3.2 Performance on Other Databases

We also compared LaG-DESIQUE with general-purpose NR IQA algorithms (e.g., DIIVINE [11], BLIINDS-II [12], and BRISQUE [13]), which usually require

Table 2. Performance of LaG-DESIQUE and other FR/NR algorithms on the CSIQ, TID, and Toyama databases.

		DIIVINE	BLIINDS-II	BRISQUE	QAC	LaG-DESIQUE
SROCC						
CSIQ	JP2K	*0.830*	*0.884*	*0.867*	0.870	0.907
	JPEG	*0.704*	*0.881*	*0.909*	0.913	0.914
	GBLUR	*0.871*	*0.870*	*0.903*	0.848	0.911
	AGWN	*0.797*	*0.886*	*0.925*	0.862	0.873
	ALL	*0.828*	*0.873*	*0.900*	0.863	0.907
TID	JP2K	*0.907*	*0.911*	*0.904*	0.889	0.948
	JPEG	*0.871*	*0.838*	*0.911*	0.898	0.913
	GBLUR	*0.859*	*0.826*	*0.874*	0.850	0.840
	AGWN	*0.834*	*0.715*	*0.823*	0.707	0.688
	ALL	*0.891*	*0.840*	*0.898*	0.870	0.904
Toyama	JPEG	*0.702*	*0.820*	*0.857*	0.672	0.886
	JP2K	*0.612*	*0.627*	*0.867*	0.563	0.886
	ALL	*0.642*	*0.724*	*0.848*	0.519	0.868
CC						
CSIQ	JP2K	*0.893*	*0.912*	*0.896*	0.882	0.928
	JPEG	*0.697*	*0.912*	*0.946*	0.938	0.958
	GBLUR	*0.898*	*0.897*	*0.928*	0.844	0.933
	AGWN	*0.787*	*0.897*	*0.938*	0.874	0.883
	ALL	*0.854*	*0.901*	*0.924*	0.877	0.927
TID	JP2K	*0.879*	*0.919*	*0.906*	0.878	0.952
	JPEG	*0.899*	*0.889*	*0.950*	0.924	0.946
	GBLUR	*0.840*	*0.825*	*0.873*	0.850	0.847
	AGWN	*0.810*	*0.714*	*0.810*	0.720	0.680
	ALL	*0.877*	*0.864*	*0.907*	0.838	0.894
Toyama	JPEG	*0.709*	*0.826*	*0.865*	0.693	0.888
	JP2K	*0.603*	*0.686*	*0.869*	0.616	0.895
	ALL	*0.634*	*0.754*	*0.850*	0.538	0.874

training on the LIVE database. The test results on the CSIQ, TID, and Toyama databases are shown in Table 2. Also include are results of QAC. Note that the italicized entries denote NR algorithms that require training on human opinion scores.

As shown in Table 2, LaG-DESIQUE performs better than QAC, especially when testing on the Toyama database. Even, the performance of LaG-DESIQUE is better than that of DIIVINE and BLIINDS-II, both of which are DMOS-trained NR IQA algorithms. Also, it seems that LaG-DESIQUE and BRISQUE have equivalent performance on these three testing databases. However, Two more factors should be considered here. First, the DIIVINE, BLIINDS-II, and BRISQUE algorithms are all trained on the LIVE database using 779 distorted images, while LaG-DESIQUE is trained on a regenerated database containing 4000 distorted images that cover eight different distortion levels, as well as on the 125 corresponding natural images. Obviously, the more number of images trained, the better predicting performance might be achieved. Second, DIIVINE, BLIINDS-II, BRISQUE are trained on five distortion types while LaG-DESIQUE only four. Consequently, when testing on other databases, a less number of trained distortion types will possibly cause relatively smaller prediction errors. Thus, the results provided in Table 2 are noteworthy. Despite these different training databases and parameters, it is no doubt that the proposed LaG-DESIQUE algorithm has its potential advantage that it could be easily trained on a larger number of distorted images without the assistance of human opinion scores.

4   Conclusion

This paper presented an extension of DESIQUE algorithm to a local-and-global way to blindly estimate image quality without training on human opinion scores. The local DESIQUE operates in a block-based manner that extracts log-derivative features from each image patch, and the local DESIQUE quality is expressed as the distance between two MVG models that fit the block-based features of the natural and test images, respectively. The global DESIQUE extracts log-derivative features based on entire image, and estimates image quality based on a trained two-stage framework. The final image quality is computed as a weighted sum of both the local and global DESIQUE quality scores. We demonstrated the efficiency of LaG-DESIQUE on several image databases.

Acknowledgements. The work was jointly supported by the National Key Research and Development Program of China under Grant No. 2018YFC0807500, the National Natural Science Foundations of China under grant No. 61772396, 61472302, 61772392, the Fundamental Research Funds for the Central Universities under grant No. JB170306, JB170304, and Xian Key Laboratory of Big Data and Intelligent Vision under grant No. 201805053ZD4CG37.

References

1. Zhang, Y., Chandler, D.M.: No-reference image quality assessment based on log-derivative statistics of natural scenes. J. Electron. Imaging **22**(4), 043025 (2013)
2. Larson, E.C., Chandler, D.M.: Most apparent distortion: full-reference image quality assessment and the role of strategy. J. Electron. Imaging **19**(1), 011006 (2010)
3. Brandao, T., Queluz, M.: No-reference image quality assessment based on DCT domain statistics. Signal Process. **88**, 822–833 (2008)
4. Gabarda, S., Cristobal, G.: No-reference image quality assessment through von mises distribution. J. Opt. Soc. Am. (JOSA A) **29**, 2058–2066 (2012)

5. Cohen, E., Yitzhaky, Y.: No-reference assessment of blur and noise impacts on image quality. Signal Image Video Process. **4**(3), 289–302 (2010)

6. Wang, Z., Bovik, A.C., Evans, B.L.: Blind measurement of blocking artifacts in images. In: Proceedings of IEEE International Conference on Image Processing, September 2000, vol. 3, pp. 981–984 (2000)

7. Wang, Z., Sheikh, H.R., Bovik, A.C.: No-reference perceptual quality assessment of JPEG compressed images. In: 2002 Proceedings of IEEE International Conference on Image Processing, vol. 1, pp. 477–480 (2002)

8. Sheikh, H.R., Bovik, A.C., Cormack, L.: No-reference quality assessment using natural scene statistics: JPEG2000. IEEE Trans. Image Process. **14**(11), 1918–1927 (2005)

9. Ye, P., Doermann, D.: No-reference image quality assessment based on visual codebook. In: 18th IEEE International Conference on Image Processing, September 2011, pp. 3089–3092 (2011)

10. Ye, P., Doermann, D.: No-reference image quality assessment using visual codebook. IEEE Trans. Image Process. **21**(7), 3129–3138 (2012)

11. Moorthy, A.K., Bovik, A.C.: Blind image quality assessment: from natural scene statistics to perceptual quality. IEEE Trans. Image Process. **20**(12), 3350–3364 (2011)

12. Saad, M.A., Bovik, A.C.: Blind image quality assessment: a natural scene statistics approach in the DCT domain. IEEE Trans. Image Process. **21**, 3339–3352 (2012)

13. Mittal, A., Moorthy, A.K., Bovik, A.C.: No-reference image quality assessment in the spatial domain. IEEE Trans. Image Process. **21**(12), 4695–4708 (2012)

14. Mittal, A., Muralidhar, G.S., Ghosh, J., Bovik, A.C.: Blind image quality assessment without human training using latent quality factors. IEEE Signal Process. Lett. **19**(2), 75–78 (2012)

15. Mittal, A., Soundararajan, R., Bovik, A.C.: Making a complete blind image quality analyzer. IEEE Signal Process. Lett. **20**(3), 209–212 (2013)

16. Xue, W., Zhang, L., Mou, X.: Learning without human scores for blind image quality assessment. In: 2013 IEEE International Conference on Computer Vision and Pattern Recognition, pp. 995–1002 (2013)

17. Martin, D., Fowlkes, C., Tal, D., Malik, J.: A database of human segmented natural images and its application to evaluating segmentation algorithms and measuring ecological statistics. In: IEEE Conference on Computer Vision, Vancouver, BC, November 2001, vol. 2, pp. 416–423 (2001)

18. Vu, P.V., Chandler, D.M.: A fast wavelet-based algorithm for global and local image sharpness estimation. Signal Process. Lett. IEEE **19**(7), 423–426 (2012)

19. Sheikh, H.R., Wang, Z., Bovik, A.C., Cormack, L.K.: Image and video quality assessment research at LIVE. http://live.ece.utexas.edu/research/quality/

20. Ponomarenko, N., Lukin, V., Zelensky, A., Egiazarian, K., Carli, M., Battisti, F.: TID2008 - a database for evaluation of full-reference visual quality assessment metrics. Adv. Mod. Radioelectron. **10**, 30–45 (2009)

21. Horita, Y., Shibata, K., Kawayoke, Y., Sazzad, Z.M.P.: Subjective quality assessment Toyama database (2008). http://mict.eng.u-toyama.ac.jp/mict/

22. ANSI T1.TR.74-2001: Objective Video Quality Measurement Using a Peak-Signal-to-Noise-Ratio (PSNR) Full Reference Technique (2001)

23. Wang, Z., Bovik, A.C., Sheikh, H.R., Simoncelli, E.P.: Image quality assessment: from error visibility to structural similarity. IEEE Trans. Image Process. **13**(4), 600–612 (2004)

Missing Recover with Recurrent Neural Networks for Video Object Detection

Ranran Shen, Wenzhong Wang, Shaojie Zhang, and Jin Tang[✉]

School of Computer Science and Technology, Anhui University, Hefei, China
srrgo@foxmail.com, jtang99029@foxmail.com
{wenzhong,zhangshaojie}@ahu.edu.cn

Abstract. Despite recent breakthroughs in object detection with static images, extending state-of-the-art object detectors from image to video is challenging. The detection accuracy suffers from degenerated object appearances in videos, e.g., occlusion, video defocus, motion blur, etc. In this paper, we present a new framework called Missing Recover Recurrent Neural Networks (MR-RNN) for improving object detection in videos, which captures temporal information to recover the missing object. First, We detect objects in consecutive frames to obtain the bounding boxes and their confidence scores. The detector is trained for every frame of the video. Then we feed these detections into a Recurrent Neural Network (LSTM [8] or BiLSTM [4]) to capture temporal information. This method is tested on a large-scale vehicle dataset, "DETRAC". Our approach achieves Average Precision (AP) of 68.90 based on SSD detector, an improvement of 2.68 over the SSD detector. Experimental results show that our method successfully detects many objects which are missed by basic detectors.

Keywords: Video object detection · CNN · RNN model

1 Introduction

Object detection is a fundamental problem in computer visions. While there has been a long history of detecting objectios in static images, there has been much lesss research in detecting objects in videos. However, cameras on surveillance systems, behicles, robots, etc., receive videos instead of static images. Thus, for these scenes to recognize the key objects and their interactions, it is critical to equip them with accurate video object detectors.

A straight forward way to detect objects in videos is to detect them in each frame. In recent years, object detection [3] in images has made significant progress. Deep Convolutional Neural Networks (CNNs) [7,12,17,18] are applied to generate a set of feature maps for the input image. A shallow detection-specific network [1–3,13,15] then generates the detection results from the feature maps. These methods achieve excellent results in high quality static images. However, due to deteriorated objects appearances (e.g., occlusion, video defocus, motion

© Springer Nature Singapore Pte Ltd. 2018
Z. Xu et al. (Eds.): Big Data 2018, CCIS 945, pp. 278–288, 2018.
https://doi.org/10.1007/978-981-13-2922-7_19

blur, etc.) in videos, the performance of any detector trained with high quality images would degrade. One way to boost the performance of an detector on these challenging video images is to train it with a lot of degraded images. However, training an accurate object detector requires laborious data collecting and labeling.

As we all know, video has rich temporal and motion information, but the static image does not. Thus, it's important to add these information to improve the accuracy of video object detection. When we detect consecutive frames, we can find the detection results have temporal correlation. And we know that Recurrent neural networks (RNN) model temporal information very well. So we are inspired to process the detection results by RNN.

In this paper, we propose an alternative way to improve the performance of any off-the-shelf detector on challenging videos. The idea is to exploit the temporal correlation in nearby video frames. As shown in Fig. 1, the appearances and positions of objects in consecutive frames are strongly correlated. Given object states in one frame, we can reliably predict their states in the neighboring frames using these inter-frame correlations.

Fig. 1. The R-FCN detection results are shown in blue, vehicles in the dotted line red boxes are missed by the R-FCN detector, and recovered by MR-RNN. (Color figure online)

We describe each video frame using a state vector consists of the detected object positions and confidence scores. Our goal is to learn a prediction model, with which we can infer the state of any frame from a sequence of the states of its nearby frames. We use Recurrent Neural Network to represent this model. Our prediction model can effectively represent the inter-frame correlations. Given an initial state sequence obtained by applying any object detector, we can effectively adjust any state from its neighboring states using this model, therefore improve the overall detection rates.

The contributions of this paper are as follows.

- We present a state prediction model based on Recurrent Neural Network, which exploits the state correlations among consecutive video frames to adjust detection results.
- We combine our prediction model with state-of-the-art object detectors, and effectively boost their performances on challenging videos.

2 Related Work

2.1 Static Image Object Detection

State-of-the-art methods for general object detection [1–3,6,13] are aminly based on deep CNNs [7,12,17,18]. Girshick et al. [3] proposed a multi-stage pipeline called Regions with Convolutional Neural Networks (R-CNN) for training deep CNNs to classify region proposals for object detection. In SPP-Net [6] and Fast R-CNN [2], ROI pooling is introduced to the feature maps shared on the whole image. To speed up the generation of the region proposals, Faster-RCNN [16] proposes the Region Proposal Network (RPN). R-FCN [1] introduce a position-sensitivity ROI pooling operation to replace ROI pooling operation, and the improvement has been made. YOLO [15] derectly predicts bounding boxes and class probabilities with a single network in a single evaluation. And YOLO model allows real-time predictions. Similar to the YOLO model, W. Liu et al. have developed a Single-Shot Detector [13] to predict all at once the bounding boxes and the class probabilities with a end-to-end CNN architecture.

2.2 Video Object Detection

Compared to static image-based object detection, there has been less research in detecting objects in videos. Han et al. [5] proposed a sequencee NMS method to associate static image detections into sequences and apply the sequence-level NMS on the results. Kang et al. [11] generated new tubelet proposals by applying tracking algorithms to static-image bounding box proposals. The class scores along the tubelet were first evaluated by the static image object detector and then re-scored by a 1D CNN model. The same group [9] proposed a tubelet proposal network to efficiently generates hundreds of tubelet proposals simultaneously. T-CNN [10] propagates predicted bounding boxes to neighboring frames according to precomputed optical flows, and then generates tubelets by applying tracking algorithms from high-confidence bounding boxes. Boxes along the tubelets are re-scored based on tubelets classification. FGFA [22] considers temporal information at the feature level instead of the final box level and present flow-guided feature aggregation framework.

2.3 Modeling Sequence Data with RNNs

Recurrent neural networks (RNNS) are connectionist models that capture the dynamics of sequences via cycles in the networks of nodes, and it can use their internal state to process sequencws of inputs. Thus, we can add temporal information to the static image detection through RNN.

In computer vision, RNNs have been used for image captioning, visual attention, action/object recognition, human pose estimation, and semantic segmentation. Recently, Milan et al. [14] proposed an end-to-end RNN module leirining approach for online multi-target tracking. Tripathi et al. [19] adopted RNNs for video object detection. And in their pipeline, the CNN-based detector is first trained, then an RNN is trained to refine the detection outputs of the CNN. Xiao and Lee [21] introduced Saptial-Temporal Memory Networks (STMN) for object detection in videos. They use the STMN as the recurrent computation unit to module long-term temporal appearance and motion dynamics.

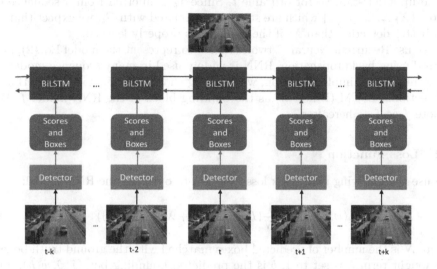

Fig. 2. Our RNN-based state prediction model. Taken consecutive frames as input, we firstly detect objects in each frame, and then feed the detection results (boxes and scores) into a RNN to update the detection results of the reference frames (one reference frame is shown in the first row). Each reference frame t is updated using the detection results from a temporal window centered at t.

3   Approach

3.1   Model

The framework of our model is shown in Fig. 2. Let $\{I_i\}$, $i = 1,...,\infty$, be a sequence of video frames, and $X_i = f_{det}(I_i)$ is the detection result of frame I_i obtained by any object detector f_{det}. The state vector X_i consists of the bounding boxes and confidence scores of up to K objects:

$$X_i = (x_1, y_1, w_1, h_1, c_1^0, ..., c_1^P, ..., x_K, y_K, w_K, h_K, c_K^0, ...c_K^P)_i \tag{1}$$

where (x_j, y_j, w_j, h_j) is the jth bounding box and c_j^p is its confidence score of category p. We conveniently denote the ground truth state vector of frame i as G_i.

Our goal is to learn a prediction model for the state G_t of a reference frame t from its neighoring states:

$$Y_t = f(X_{t-\tau}, ..., X_t, ..., X_{t+\tau}; \theta) \tag{2}$$

where τ controls the temporal window used to estimate the state of frame t, θ is the parameter of this model, and

$$Y_t = (\hat{x}_1, \hat{y}_1, \hat{w}_1, \hat{h}_1, \hat{c}_1^0, ..., \hat{c}_1^P, ..., \hat{x}_K, \hat{y}_K, \hat{w}_K, \hat{h}_K, \hat{c}_K^0, ..., \hat{c}_K^P)_t \tag{3}$$

is the updated state vector of frame t. Since Y_t is inferred from a sequence of states $\{X_{t-\tau}, ..., X_{t+\tau}\}$ which are strongly correlated with G_t, we expect that Y_t is a better detection than X_t if the model f is properly learned.

We use Recurrent Neural Network(RNN) to represent the model Eq. (2), and learn θ using backpropagation. RNN is widely used in many sequence modeling problems. In our implementation, we use long short-term memory (LSTM) and bidirectional LSTM (BiLSTM) as the building block of the RNN model f. We denote f as f_{rnn} hereafter.

3.2   Loss Function

We use the following multi-task loss function to optimize the RNN model:

$$L(\psi, \hat{c}, \hat{b}, \tilde{b}) = \frac{1}{N}(L_{conf}(\psi, \hat{c}) + \lambda L_{loc}(\psi, \hat{b}, \tilde{b})) \tag{4}$$

where N is the number of predicted boxes matched with the ground truth boxes, the weight term λ is set to 1, $\hat{b}$ is the predicted bounding box $(\hat{x}, \hat{y}, \hat{w}, \hat{h})$, and the $\tilde{b}$ is the ground truth box $(\tilde{x}, \tilde{y}, \tilde{w}, \tilde{h})$.

The loss consists of two parts, L_{conf} and L_{loc}. The confidence loss L_{conf} evaluates the class predictions for each bounding box:

$$L_{conf} = - \sum_{i \in Pos}^{N} \psi_{ij}^p \log(\bar{c}_i^p) - \sum_{i \in Neg} \log(\bar{c}_i^p) \tag{5}$$

where $\bar{c}_i^p = \frac{\exp(\hat{c}_i^p)}{\sum_p \exp(\hat{c}_i^p)}$, $\psi_{ij}^p = \{1, 0\}$ indicates whether the i-th prediction box is matched to the j-th ground truth box of category p, and $\hat{c}_i^p$ is the confidence of i-th prediction box for category p.

The localization loss L_{loc} evaluates the predicted bounding boxes, and is defined as (following [15]):

$$L_{loc} = \psi_{ij}^p \sum_{i \in Pos}^{N} \{[(x_i - \hat{x}_j)^2 + (y_i - \hat{y}_j)^2]$$
$$+ [(\sqrt{w_i} - \sqrt{\hat{w}_j})^2 + (\sqrt{h_i} - \sqrt{\hat{h}_j})^2]\} \tag{6}$$

We regress to offset for the center(x, y) and for its width(w) and height(h). To reflect that small deviations in large boxes matter less than in small boxes, we utilize the square root of the bounding box width and height instead of the width and height.

Algorithm 1. Sequential inference algorithm for object detection.

Input: Video frames $\{I_i\}$, aggregation range τ
Output: Detection results $\{Y_i\}$
1: **for** $t = 1$ to $2\tau + 1$ **do**
2:     $X_t = f_{det}(I_t)$
3: **end for**
4: **for** $t = 1$ to ∞ **do**   ▷ for reference frame t
5:     **if** $t <= \tau$ **then**
6:         $Y_t = X_t$ ▷ Use detection results
7:     **else**
8:         $Y_t = f_{rnn}(X_{t-\tau}, ..., X_t, ..., X_{t+\tau})$ ▷ Infer the state of reference frame t
9:         $X_{t+\tau+1} = f_{det}(I_{t+\tau+1})$
10:    **end if**
11: **end for**

Inference. Algorithm 1 summarizes the inference algorithm. Given an input video of consecutive frames $\{I_i\}$ and the specified aggregation range τ, the proposed MR-RNN sequentially processes each frame with a sliding feature buffer on the nearby frames(of length $2\tau + 1$ in general, except for the beginning and the ending τ frames). Initially, the detector network is applied on the beginning $2\tau+1$ frames to initialize the detector result buffer (L1–L3 in Algorithm 1). Then the algorthm loops over all the video frames to perform video object detection, and to update the feature buffer. For the first τ frames, we use the detector result directly (L5–L7). After first τ frames, each frame i as the reference, the detector of the nearby frames in the detector result buffer aggregate into a new feature, and the feature is fed into RNN model. to generate video object detection result (L8). Before taking the $(i + 1)$-th frame as the reference, the detector result is extracted on the $(i + \tau + 1)$-th frame and is added to the feature buffer (L9).

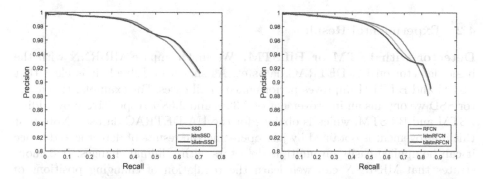

Fig. 3. Precision vs Recall

Training. First we train a detection network on UA-DETRAC dataset. Then we utilize trained detection network to detect on training set and obtain its results. Finally, we utilize the detection results to train our RNN model. In our experiments, we use a 2-layer LSTM or BiLSTM, the hidden size is 2048, and the aggregation range number τ is 8, and the learning rate for training the RNN model is 0.0001.

4   Experiments and Analysis

In this section, we show comparative results and perform sensitivity analysis to show robusstness of MR-RNN. We also conduct specific experiments to compare the improvement of the experiment with LSTM and BiLSTM. And we will discuss the impact of MR-RNN on different detectors.

4.1   Dataset and Experiment Settings

Dataset Selection. To quantitatively evaluate the quality of detection results, we use UA-DETRAC dataset [20], which consists of 100 challenging video sequences captured from real-world traffic scenes (over 14,000 frames with rich annotations, including occlusion, weather, truncation, and object bounding boxes, leading to a total of 1.21 million labeled bounding boxes of objects) for object detection.

Experiment Settings. In this paper, our static image object detectors adopt SSD [13] and R-FCN [1] frameworks. To implement our video object detection system, a computer with core i7-7700k CPU, 32 GB memory and 8 GB memory GTX1080 GPU is employed. The program runs on a 64-bit Ubuntu 16.04.3 LTS operating system with CUDA-8.0, python 2.7.6 installed.

4.2   Experimental Result

Detector with LSTM or BiLSTM. We first compare MR-RNN with the basic detector on UA-DETRAC datas et. As shown in Table 1, it is clear that LSTM and BiLSTM improves performance in all cases. For example, the detector SSD we obtains an improvement of 2.32% and 2.68% respecitively by adding LSTM and BiLSTM, which is obvious for the UA-DETRAC dataset. Note that this improvement is obtained by just operating the results of detector and hence it can be applied easily on multiple detectors with minimal changes. It demonstrates that MR-RNN can well learn the regulation of changing positions of objects over time. For the comparison of LSTM and BiLSTM on SSD, the BiLSTM has a 0.36% improvement with LSTM, it shown that BiLSTM has a stronger learning ability. Figure 3 shows what recall values is MR-RNN performing better than basic detectors.

Table 1. Accuracy of different methods on UA-DETRAC dataset, using RFCN and SSD basic detection networks.

Method	Training data	Testing data	AP(%)
SSD [13]	45 videos	15 videos	66.22
SSD + LSTM	45 videos	15 videos	$68.54_{\uparrow 2.32}$
SSD + BiLSTM	45 videos	15 videos	$68.90_{\uparrow 2.68}$
RFCN [1]	45 videos	15 videos	82.02
RFCN+LSTM	45 videos	15 videos	$83.10_{\uparrow 1.08}$
RFCN+BiLSTM	45 videos	15 videos	$83.12_{\uparrow 1.1}$

Impact of Different Detector. We evaluate MR-RNN on different detectors. In this paper, We use two representative detectors, SSD and RFCN. SSD has a fast detection speed, and RFCN has a good accuracy. From Table 1, we can see that we have an AP of 66.22% and 82.02% respecitively for SSD and RFCN. Note that we do not add bells and whistles like multi-scale training/testing, exploiting context information, model ensemble, etc., in order to facilitate comparison and draw clear conclusions. After LSTM network, we get a 2.32% and 1.08% improvement for SSD and RFCN respectively. We find that detector has lower AP would obtain higher improvement, because lower AP implies more false positive, and MR-RNN will find these false positive from temporal axis.

The Precision-Recall curves of both detectors are shown in Fig. 3. Our model performs better when the recall of basic detectors are higher. This is because there are more helpful information in the neighboring frames when more objects are detected in these frames.

Sensitivity Analysis. MR-RNN has a hyperparameter τ in Eq. 2. We adjust the parameter and measure average precision on the test set of UA-DETRAC dataset for each detector, as shown in Table 2. We set τ to 5, 8 and 11, and the resulting variation in AP for these three cases indicates that MR-RNN has good stability. And we find that $\tau = 8$ achieves the best performance.

Table 2. Accuracy with different value of τ, using SSD as the base detector.

Method	Range τ	AP(%)
SSD + LSTM	5	$68.33_{\uparrow 2.11}$
SSD + LSTM	**8**	$\mathbf{68.54_{\uparrow 2.32}}$
SSD + LSTM	11	$68.20_{\uparrow 1.98}$
SSD + BiLSTM	5	$68.66_{\uparrow 2.44}$
SSD + BiLSTM	**8**	$\mathbf{68.90_{\uparrow 2.68}}$
SSD + BiLSTM	11	$68.55_{\uparrow 2.33}$

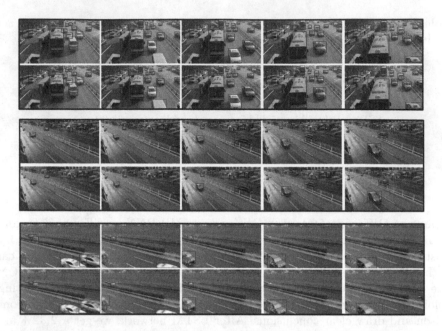

Fig. 4. Detection results of R-FCN and our BiLSTM-based prediction model (first row: R-FCN, second row: MR-RNN). Vehicles in the red boxes are missed by the R-FCN detector, and recovered by MR-RNN. (Color figure online)

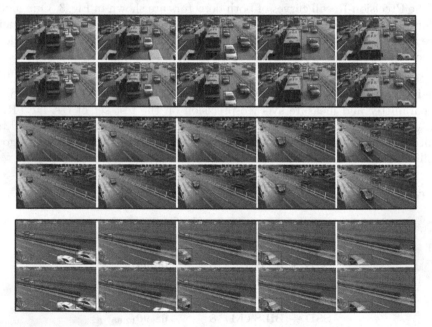

Fig. 5. Detection results of SSD and our BiLSTM-based prediction model (first row: SSD, second row: MR-RNN). Vehicles in the red boxes are missed by the R-FCN detector, and recovered by MR-RNN. (Color figure online)

Qualitative Results. Figure 4 shows the detections obtained by R-FCN and our BiLSTM-based prediction model, and Fig. 5 shows the detections obtained by SSD and our BiLSTM-based prediction model. It can be concluded that MR-RNN performs better than the basic detector. Figure 6 shows two example frames in detail. From the zoomed-in views, we can easily notice that many vehicles missed by R-FCN are recovered by MR-RNN.

Fig. 6. Example detections (blue boxes) obtained by R-FCN and MR-RNN. First row are results of R-FCN, and second row shows the results of MR-RNN. MR-RNN successfully recovered some vehicles missed by R-FCN. (Color figure online)

5 Conclusion

In this paper, we propose an efficient method for object detection in videos. We present an RNN-based state prediction model to exploit the state correlations among consecutive frames, and use this model to improve detection results. Our model can be easily integrated with any object detectors. Our experiments on DETRAC dataset using R-FCN and SSD detector show that our model effectively improves the performances of these detectors.

References

1. Dai, J., Li, Y., He, K., Sun, J.: R-FCN: object detection via region-based fully convolutional networks. In: Advances in Neural Information Processing Systems, pp. 379–387 (2016)
2. Girshick, R.: Fast R-CNN. In: Proceedings of the IEEE International Conference on Computer Vision, pp. 1440–1448 (2015)
3. Girshick, R., Donahue, J., Darrell, T., Malik, J.: Region-based convolutional networks for accurate object detection and segmentation. IEEE Trans. Pattern Anal. Mach. Intell. **38**(1), 142–158 (2016)
4. Graves, A., Schmidhuber, J.: Framewise phoneme classification with bidirectional lstm and other neural network architectures. Neural Netw. **18**(5–6), 602–610 (2005)
5. Han, W., et al.: Seq-NMS for video object detection. arXiv preprint arXiv:1602.08465 (2016)

6. He, K., Zhang, X., Ren, S., Sun, J.: Spatial pyramid pooling in deep convolutional networks for visual recognition. In: Fleet, D., Pajdla, T., Schiele, B., Tuytelaars, T. (eds.) ECCV 2014. LNCS, vol. 8691, pp. 346–361. Springer, Cham (2014). https://doi.org/10.1007/978-3-319-10578-9_23

7. He, K., Zhang, X., Ren, S., Sun, J.: Deep residual learning for image recognition. In: Proceedings of the IEEE Conference on Computer Vision and Pattern Recognition, pp. 770–778 (2016)

8. Hochreiter, S., Schmidhuber, J.: Long short-term memory. Neural Comput. **9**(8), 1735–1780 (1997)

9. Kang, K., et al.: Object detection in videos with tubelet proposal networks. In: Proceedings of CVPR, vol. 2, p. 7 (2017)

10. Kang, K., et al.: T-CNN: tubelets with convolutional neural networks for object detection from videos. In: IEEE Transactions on Circuits and Systems for Video Technology (2017)

11. Kang, K., Ouyang, W., Li, H., Wang, X.: Object detection from video tubelets with convolutional neural networks. In: Proceedings of the IEEE Conference on Computer Vision and Pattern Recognition, pp. 817–825 (2016)

12. Krizhevsky, A., Sutskever, I., Hinton, G.E.: Imagenet classification with deep convolutional neural networks. In: Advances in Neural Information Processing Systems, pp. 1097–1105 (2012)

13. Liu, W., et al.: SSD: single shot multibox detector. In: Leibe, B., Matas, J., Sebe, N., Welling, M. (eds.) ECCV 2016. LNCS, vol. 9905, pp. 21–37. Springer, Cham (2016). https://doi.org/10.1007/978-3-319-46448-0_2

14. Milan, A., Rezatofighi, S.H., Dick, A.R., Reid, I.D., Schindler, K.: Online multi-target tracking using recurrent neural networks. In: AAAI, vol. 2, p. 4 (2017)

15. Redmon, J., Divvala, S., Girshick, R., Farhadi, A.: You only look once: unified, real-time object detection. In: Proceedings of the IEEE Conference on Computer Vision and Pattern Recognition, pp. 779–788 (2016)

16. Ren, S., He, K., Girshick, R., Sun, J.: Faster R-CNN: towards real-time object detection with region proposal networks. In: Advances in Neural Information Processing Systems, pp. 91–99 (2015)

17. Simonyan, K., Zisserman, A.: Very deep convolutional networks for large-scale image recognition. arXiv preprint arXiv:1409.1556 (2014)

18. Szegedy, C., et al.: Going deeper with convolutions. In: Proceedings of the IEEE Conference on Computer Vision and Pattern Recognition, pp. 1–9 (2015)

19. Tripathi, S., Lipton, Z.C., Belongie, S., Nguyen, T.: Context matters: refining object detection in video with recurrent neural networks. arXiv preprint arXiv:1607.04648 (2016)

20. Wen, L., et al.: UA-DETRAC: a new benchmark and protocol for multi-object detection and tracking. arXiv preprint arXiv:1511.04136 (2015)

21. Xiao, F., Lee, Y.J.: Spatial-temporal memory networks for video object detection. arXiv preprint arXiv:1712.06317 (2017)

22. Zhu, X., Wang, Y., Dai, J., Yuan, L., Wei, Y.: Flow-guided feature aggregation for video object detection. In: Proceedings of the IEEE International Conference on Computer Vision, vol. 3 (2017)

Deep Convolutional Neural Network with Feature Fusion for Image Super-Resolution

Furui Bai$^{(\boxtimes)}$, Rui Wang, Xiaopeng Sun, Huxing Sun, and Wen Lu

Xidian University, Xi'an 710071, China
{frbai,wangrui,xpsun}@stu.xidian.edu.cn,
huxsun@163.com, luwen@mail.xidain.edu.cn

Abstract. Recently, deep convolutional neural networks (CNNs) in single image super-resolution (SISR) have received excellent performance. However, most deep-learning-based methods do not make full use of low-level features extracted from the original low-resolution (LR) image, which may reduce the quality of reconstructed image. To address these issues, we propose a method which can connect the low-level features from almost all convolutional layers. Our method use the interpolated low-resolution image as input, employ many skip-connections to combine low-level image features with the final reconstruction process, these feature fusion strategies are based on pixel-level summation operations. After merging the previous convolution features, residual images are used to directly reconstruct high-resolution (HR) images. Experiments demonstrate that the proposed method is superior to the state-of-the-art methods.

Keywords: Single image super-resolution · Residual learning
Feature fusion

1 Introduction

Single image super-resolution is a classic branch of low-level vision, where the aim is to recover a high-resolution (HR) image from a low-resolution (LR) image. Existing super-resolution reconstruction algorithms can be classified into interpolation-based [10, 17], reconstruction-based [18–20], and learning-based methods [21]. In General, learning-based method per-form better than others, especially deep-learning-based methods.

Convolutional neural network (CNN) is an important deep learning model. We do not require artificial design features and the network can autonomously extract features from original data. The model roughly includes convolutional layer, pooling layer, fully connected layer. The convolutional layer is the core of the convolutional neural network and multiple convolution kernels are used to construct multi-channel convolution operations on the input image to obtain a variety of feature maps. These feature maps correspond to different levels of image features. Because the model can learn the features of different levels in the image, it can significantly improve the quality of the reconstructed image, so it has a wide range of applications in the field of super-resolution reconstruction in several years.

Z. Xu et al. (Eds.): Big Data 2018, CCIS 945, pp. 289–298, 2018.
https://doi.org/10.1007/978-981-13-2922-7_20

Dong et al. [3] proposed SRCNN which first learning mapping between LR and HR images, achieving state-of-the-art performance at that time. During experiments he demonstrated that deeper network does not mean better performance. However, this view was quickly overturned by some scholars. Kim et al. [6] used the deep recursive convolutional network (DRCN) to deepen the number of network layers. The recursively-convolved convolution layers use the same weight. This method not only increases the depth of the network layer, but also improves the quality of the reconstructed image. Inspired by He et al. [8], Kim combined the idea of residual learning into super-resolution reconstruction [5]. The network learns high-frequency information between high-resolution images and low-resolution images. The network can converge very quickly during the training phase, so it achieved high reconstruction quality and fast reconstruction speed. However, this method requires a large amount of training data. If training images are incomplete, the image reconstruction quality will be deeply affected. Dong et al. [4] optimizes the network structure and uses the low-resolution images without bicubic interpolation as the input of the net-work directly. After extracting features, it reduces the dimensions and improves the reconstruction speed while ensuring that the reconstruction quality basically unchanged. In the above convolutional neural network models, the output of each layer is the input of the next layer, which is a plain structure. The last layer of the network outputs advanced features of the image, but it will result in the loss of low-level features in images. Scholars have proposed some methods to improve these issues. Lai et al. [7] develop a laplacian pyramid net-work for single image super resolution (LapSRN), which use original LR images as input and progressively reconstruct the high resolution images, achieving excellent performance especially for a larger upscaling factor. Tai et al. [9] propose a deep recursive residual network (DRRN) with more than 50 convolutional layers, which utilizes residual learning and recursive learning respectively, while residual learnings are used both in global and local manners to mitigate the difficulty of training very deep networks, and recursive learning are helpful for controlling the model parameters while increasing the depth. Even these methods can achieve state-of-the-art performance, they do not make full use of low-level features extracted from the original low-resolution (LR) image. In addition, with the deepening of network layers and the increasing of network parameters, such methods require larger datasets to solve the problems of overfitting, gradient explosion and gradient disappearance in training stage. Besides, most of the existing methods can only deal with a single-scale super-resolution reconstruction task, which may need more parameters to train several networks.

To solve above problems, we propose a deep-learning single-frame image super-resolution reconstruction algorithm based on skip-connection. This method takes the bicubic interpolated image as input and passes the convolutional output feature map of each layer. The connection performs a summation operation, and the low-level image features learned by the network are used into the image reconstruction process. At the last layer, residual learning is employed to accelerate the convergence speed of our network, and a high-resolution image is finally reconstructed.

The main contributions of this work are summarized as follows:

(1) Feature Fusion. Owing to the multi-path skip connection of our feature fusion module, sufficient hidden-layers relationships and information could be learned to

represent the LR-to-HR mappings by using our pro-posed FFN method. Experiments demonstrate that, the reconstructed image may have more details with this training strategy. Benefit from the structure of feature associations, our FFN could obtain better performance on several verification datasets.

(2) Multi-Scale. We find a new multi-scale architecture that shares most of the parameters across different scales. In other words, we can deal with 2×, 3×, and 4× super-resolution issues by using single network structure. The proposed multi-scale model can obviously save computing resources, and it can also perform well compared with state-of-the-art methods.

2   Proposed Method

In this section, we first describe the proposed model architecture in two steps of introducing feature fusion module, reconstruction module. Then suggest the training details.

2.1   Network Structure

The network model proposed in this paper is an end-to-end structure. The algorithm is divided into training phase and testing phase. While training, the low-resolution image is firstly interpolated by bicubic interpolation and used as the input of the network. The mean square error (MSE) between the output image and the input image is calculated at the end of our network, and then a stochastic gradient descent (SGD) algorithm is used to update the parameters of the convolutional layer in the network. After many iterations, the optimal weights are obtained. In the test phase, images are input into the network and high-resolution images are directly reconstructed through the trained network model.

For SR image reconstruction, we use a feature fusion network inspired by VDSR. We use a lot of skip-connections to improve our performance instead of one global residual learning in VDSR. As shown in Fig. 1, it consists of two parts: a feature fusion module (FM) and a reconstruction module (RM). The network structure contains a total of 22 convolution layers. We define the convolution of each layer as Conv *(ic, oc, k)*, where the variables *ic, oc, k* represents the number of input channels, the number of output channels and the filter size.

FM: Since the SISR algorithm converting RGB images into YCbCr images and extract the Y channel as input, the number of input channel is 1. The network settings of the second to the 19th convolutional layer are the same, input channel is 64 and output channel is 64, and the size of the convolution kernel is 3×3. Each of convolutional layer is followed by a rectified linear unit (ReLU) activation function, which is omitted here. X_{i-1} denotes input layer, the output of i-th layer X_i can be expressed as:

$$X_i = \max(0, W_i * X_{i-1} + B_i) \tag{1}$$

Where W_i and B_i are weights and bias for i-th layer, * denotes convolution operation.

Image super-resolution not only needs to restore detailed information; low-level feature information is also equally important for reconstructing image quality. Therefore, in order to use the low-level features convolved in the previous layers of the network for image reconstruction, we sum the results of the first 19 layers of convolution as the input of the next layer. The 20-th convolutional layer's input is the sum which can be refer as:

$$X_{sum} = \sum_{j=1}^{19} X_j \tag{2}$$

RM: The 20-th and 21-th layers are Conv (*64, 64, 3*), and the last layer is Conv (*64, 1, 3*). These are used for dimensionality reduction and reconstruction the final HR images. there also has a ReLU activation function behind the other convolutional layers to increase the nonlinearity of the model except the last convolutional layer.

In order to solve the problem of gradient disappearance or gradient explosion in convolutional networks, we suggest the idea of residual learning to add convolved results of the input image and the last layer to obtain the final reconstructed image. When using a deeper network for convolution, the input image is reduced in size after convolution. To make the size of convolved image consistent with the size of input, all convolution strides are set to one and we padding zeros to make sure that the input and the output among our network have the same size [5].

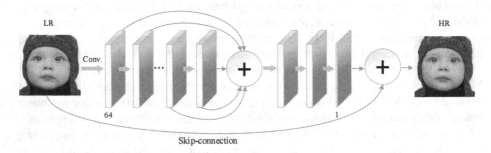

Fig. 1. Architecture of the proposed network.

2.2   Training

Loss and Optimizer: Given a training dataset consists of LR and HR images, N represents the number of samples in the training set. In training phase, we use L2 penalty function to calculate the loss:

$$L(\Theta) = \frac{1}{N} \sum_{i=1}^{N} \|F(x_i; \Theta) - y_i\|_2^2 \tag{3}$$

x_i, y_i denotes i-th low-resolution image and high-resolution image. Θ represent total of parameters. Because of skip-connection we added between input and output. The main output learned by our network can be defined as $r = y - x$, which can also be suggested as residual image. The loss function now becomes:

$$L(\Theta) = \frac{1}{N} \sum_{i=1}^{N} \| F(x_i; \Theta) - (y_i - x_i) \|_2^2 \qquad (4)$$

In addition, W update according to formula (5):

$$L(\Theta) = \frac{1}{N} \sum_{i=1}^{N} \| F(x_i; \Theta) - (y_i - x_i) \|_2^2 \qquad (5)$$

Where i and $\frac{\partial L}{\partial W_i}$ indices iterations and the derivative. α is momentum, and η is the learning rate.

3 Experimental Results

3.1 Benchmarks

In order to make a fair compassion with other methods. For training, we use 291 images which consist of 91 images from Yang et al. and 200 images from the training set of BSD300 datasets [13]. The images in these datasets contain abundant textures and details. For stabilizing the training phase and enhancing the capacity of model, the HR training images are augmented by being flipped horizontally; being downscaled with 0.9, 0.8, 0.7, 0.6, 0.5; and being rotated by 90°, 180° and 270°.

The proposed method is evaluated on four widely used benchmark datasets: Set5 [11], Set14 [12], BSD100 [13], Urban100 [14]. Among these datasets, Set5, Set14 and BSD100 consist of natural scenes and Ur-ban100 contains structured scenes. For assessing the SR results, we apply two objective image quality assessment criterions: PSNR and SSIM. According to our model can only deal with Y-channel image super-resolution. All the criterions are calculated on the Y-channel.

3.2 Implementation Details

Besides, we will use the size of 41 × 41 LR sub-images as network input without overlapping crop, and stride is 41. Since the input of network is a single-channel gray image, the RGB color image in our training set is converted into YCbCr color space, and then the luminance channel Y is used as label. For human vision system is insensitive to the color channel, images of Cb and Cr channels were directly up scaled by 2×, 3×, and 4× using bicubic.

We train all experiments over 80 epochs with batch size 64 by using the SGD optimizer. Learning rate was initially set to 0.1 and then decreased by a factor of 10 every 20 epochs. Momentum and weight decay parameters are set to 0.9 and 0.0001, respectively. We also employ adjustable gradient clipping for maximal speed of convergence. This strategy can control the gradient de-scent caused by each back propagation within a certain range, which is of great significance for the network convergence. The initialization of these convolutional layers is same as the way from He et al. [8] In total, the learning rate was decreased 3 times, and the learning is stopped after 80 epochs. Training our network takes roughly 46 h on Titan X Pascal GPU.

3.3   Network Analysis

To verify the assumption that the summation of the first 19 convolutional layers in our algorithm has an obviously effect on the final image reconstruction, we re-move the summation operations of the first 19 layers in our FFN structure to form a new network and conduct comparison experiments. The comparative experiment adopts training parameters that were same as our method. We upscale the Set5 dataset by a factor of 3 and record the mean PSNR on each Epoch. We can see that in Fig. 2, if the network structure removes the previous convolutional layer summation operation, PSNR is lower than the FFN model, which also verifies that our feature fusion module has a certain effect on the improvement of the performance.

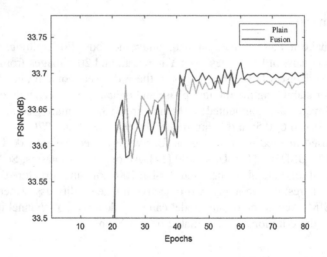

Fig. 2. Comparison of network structures.

We employ MSE loss as the loss function in our net-work structure, while the L1 loss function is another method to minimize the difference about absolute value. In order to verify the effect of different loss functions on the reconstruction results, we changed the loss function to L1 loss and conducted a comparison experiment. The experimental setup and parameters are consistent with above comparison experiment. The test dataset is Set5, and the upscale factor is 3, with different losses. The network structure of the function reconstructs the mean PSNR of the image along with epoch changes as shown in Fig. 3.

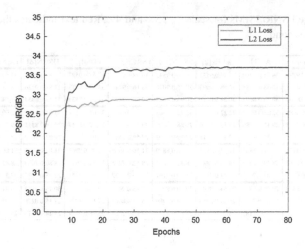

Fig. 3. Comparison of network structures.

We can get conclusion from the figures that the average PSNR of L1 loss function is lower than L2 loss function used in our method. Although in the first few Epoch, the L1 loss function has a good reconstruction effect at the beginning. But when the network finally convergence, training with L2 loss function performance better than training with L1 loss. The reason for this phenomenon may be that the formula for L2 loss is very similar to the PSNR indicator. PSNR is the main indicator to measure the quality of the model.

3.4 Comparison with State-of-the-Arts Methods

We compare the proposed method with several state-of-the-art SR methods, including Bicubic, A+ [1], SelfEx [2], SRCNN [3], FSRCNN [4], VDSR [5], DRCN [6], LapSRN [7]. And compare the performance between different algorithms from objective and subjective quality evaluation. Table 1 shows the average peak signal-to-noise ratio [15] (PSNR) and structural similarity [16] (SSIM) values on four benchmark datasets. Which can evaluate the quality of reconstructed images objectively.

According to Table 1, the PSNR/SSIM evaluated from the proposed FFN is lower than that of LapSRN in 4 × scale factor, while our FFN can achieve slightly better performance in other benchmark datasets. The main reason about this phenomenon is that LapSRN suggests a pyramid structure, which can reconstruct high-resolution image progressively. With this structure, LapSRN can achieve better performance in 4×, 8× scale factor super-resolution reconstruction.

For the subjective quality evaluation of reconstructed images, Figs. 4 and 5 show visual comparisons. the high-resolution images of "253027" in BSD100 and "img099" image in Urban100.

To compare the reconstruction quality more fairly, we select the image area at the same position and size. In Fig. 6, The "zebra" image has many textures and details, several SR methods especially bicubic interpolation may lost high frequency

Table 1. Average PSNR/SSIMs for scale 2×, 3× and 4×. Red text indicates the best and Blue text indicates the second

Dataset	Scale	Bicubic	A+[1]	SelfEx[2]	SRCNN[3]	FSRCNN[4]	VDSR[5]	DRCN[6]	LapSRN[7]	FFN(Ours)
Set5	×2	33.66/0.9299	36.54/0.9299	36.49/0.9537	36.66/0.9542	36.98/0.9562	37.53/0.9587	37.63/0.9588	37.52/0.9591	37.80/0.9603
	×3	30.39/0.8682	32.58/0.9088	32.58/0.9093	32.75/0.9090	33.16/0.9143	33.66/0.9213	33.82/0.9226	33.81/0.9220	33.79/0.9231
	×4	28.42/0.8104	30.28/0.8603	30.31/0.8619	30.48/0.8628	30.70/0.8667	31.35/0.8838	31.53/0.8854	31.54/0.8852	31.42/0.8852
Set14	×2	30.24/0.8688	32.28/0.9056	32.22/0.9034	32.42/0.9063	32.62/0.9094	33.03/0.9124	33.04/0.9118	32.99/0.9124	33.25/0.9147
	×3	27.55/0.7742	29.13/0.8188	29.16/0.8196	29.28/0.8209	29.42/0.8248	29.77/0.8314	29.76/0.8311	29.79/0.8325	29.86/0.8334
	×4	26.00/0.7027	27.32/0.7491	27.40/0.7518	27.49/0.7503	27.59/0.7549	28.01/0.7674	28.02/0.7670	28.09/0.7700	28.07/0.7691
BSD100	×2	29.56/0.8431	31.21/0.8863	31.18/0.8855	31.36/0.8879	31.50/0.8935	31.90/0.8960	31.85/0.8942	31.80/0.8952	32.03/0.8979
	×3	27.21/0.7385	28.29/0.7835	28.29/0.7840	28.41/0.7863	28.52/0.7895	28.82/0.7976	28.80/0.7963	28.82/0.7980	28.85/0.7986
	×4	25.96/0.6675	26.82/0.7087	26.84/0.7106	26.90/0.7101	26.96/0.7132	27.29/0.7251	27.23/0.7233	27.32/0.7275	27.29/0.7254
Urban100	×2	26.88/0.8403	29.20/0.8938	29.54/0.8967	29.50/0.8946	29.85/0.9013	30.76/0.9140	30.75/0.9133	30.41/0.9103	31.12/0.9184
	×3	24.46/0.7349	26.03/0.7973	26.44/0.8088	26.24/0.7989	26.41/0.8061	27.14/0.8279	27.15/0.8276	27.07/0.8275	27.19/0.8297
	×4	23.14/0.6577	24.32/0.7183	24.79/0.7374	24.52/0.7221	24.60/0.7262	25.18/0.7524	25.14/0.7510	25.21/0.7562	25.22/0.7541

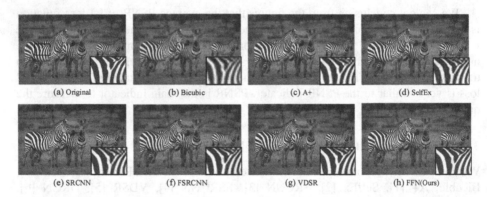

(a) Original (b) Bicubic (c) A+ (d) SelfEx

(e) SRCNN (f) FSRCNN (g) VDSR (h) FFN(Ours)

Fig. 4. The "253027" image from BSD100 with an upscaling factor 3.

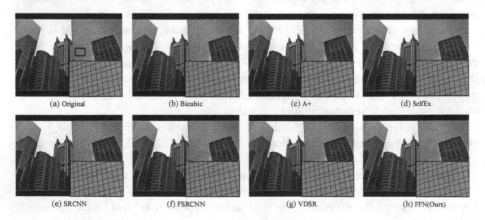

(a) Original (b) Bicubic (c) A+ (d) SelfEx

(e) SRCNN (f) FSRCNN (g) VDSR (h) FFN(Ours)

Fig. 5. The "img099" image from Urban100 with an upscaling factor 3.

information and cause seriously blur. As for Fig. 5, we can obviously see that our method not only reconstruct more clearly details, the image blocks reconstructed at the glass facade are smoother and closer to the original high resolution image.

4   Conclusions

In this paper, we proposed an effective multi-scale super-resolution network via feature fusion. The algorithm joins low-level features learned by the previous layers to the final image reconstruction through skip-connections. Experiments demonstrated that the proposed method is superior to the state-of-the-art methods, it can also effectively recover the missing texture information in the low-resolution images.

Acknowledgement. This research was supported partially by the National Natural Science Foundation of China (Nos. 61372130, 61432014, 61871311). The authors would like to thank our tutor, Professor Lu Wen, his valuable remarks and suggestions inspired us a lot.

References

1. Timofte, R., De Smet, V., Van Gool, L.: A+: adjusted anchored neighborhood regression for fast super-resolution. In: Cremers, D., Reid, I., Saito, H., Yang, M.-H. (eds.) ACCV 2014. LNCS, vol. 9006, pp. 111–126. Springer, Cham (2015). https://doi.org/10.1007/978-3-319-16817-3_8
2. Huang, B.J., Singh, A., Ahuja, N.: Single image super-resolution from transformed self-exemplars. In: Proceedings of IEEE Conference on Computer Vision and Pattern Recognition, pp. 5197–5206 (2015)
3. Dong, C., Loy, C.C., He, K., Tang, X.: Image super-resolution using deep convolutional networks. IEEE Trans. Pattern Anal. Mach. Intell. **38**(2), 295–307 (2016)
4. Dong, C., Loy, C.C., Tang, X.: Accelerating the super-resolution convolutional neural network. In: Leibe, B., Matas, J., Sebe, N., Welling, M. (eds.) ECCV 2016. LNCS, vol. 9906, pp. 391–407. Springer, Cham (2016). https://doi.org/10.1007/978-3-319-46475-6_25
5. Kim, J., Lee, J.K., Lee, K.M.: Accurate image super-resolution using very deep convolutional networks. In: Proceedings of IEEE Conference on Computer Vision and Pattern Recognition, pp. 1646–1654 (2016)
6. Kim, J., Lee, J.K., Lee, K.M.: Deeply-recursive convolutional network for image super-resolution. In: Proceedings of IEEE Conference on Computer Vision and Pattern Recognition, pp. 1637–1645 (2016)
7. Lai, W.-S., Huang, J.-B., Ahuja, N., Yang, M.-H.: Deep laplacian pyramid networks for fast and accurate super-resolution. In: Proceedings of IEEE Conference on Computer Vision and Pattern Recognition, pp. 624–632 (2017)
8. He, K., Zhang, X., Ren, S., Sun, J.: Deep residual learning for image recognition. In: Proceedings of IEEE Conference on Computer Vision and Pattern Recognition, pp. 770–778 (2016)
9. Tai, Y., Yang, J., Liu, X.: Image super-resolution via deep recursive residual network. In: Proceedings of IEEE Conference on Computer Vision and Pattern Recognition, pp. 3147–3155 (2017)

10. Duchon, C.E.: Lanczos filtering in one and two dimensions. J. Appl. Meteorol. **18**(8), 1016–1022 (1979)
11. Bevilacqua, M., Roumy, A., Guillemot, C., Alberi Morel, M.-L.: Low-complexity single-image super-resolution based on nonnegative neighbor embedding. In: Proceedings of British Machine Vision Conference, pp. 1–10 (2012)
12. Zeyde, R., Elad, M., Protter, M.: On single image scale-up using sparse-representations. In: Boissonnat, J.-D., et al. (eds.) Curves and Surfaces 2010. LNCS, vol. 6920, pp. 711–730. Springer, Heidelberg (2012). https://doi.org/10.1007/978-3-642-27413-8_47
13. Martin, D., Fowlkes, C., Tal, D., Malik, J.: A database of human segmented natural images and its application to evaluating segmentation algorithms and measuring ecological statistics. In: Proceedings of IEEE International Conference on Computer Vision, vol. 2, pp. 416–423 (2001)
14. Huang, J.-B., Singh, A., Ahuja, N.: Single image super-resolution from transformed self-exemplars. In: Proceedings of IEEE Conference on Computer Vision and Pattern Recognition, pp. 5197–5206 (2015)
15. Huynh, Q.-T., Ghanbari, M.: Scope of validity of PSNR in image/video quality assessment. Electron. Lett. **44**(13), 800–801 (2008)
16. Wang, Z., Bovik, C.A., Sheikh, R.H.: Image quality assessment: from error visibility to structural similarity. IEEE Trans. Image Process. **13**(4), 600–612 (2004)
17. Keys, R.: Cubic convolution interpolation for digital image processing. IEEE Trans. Acoust. Speech Signal Process. **29**(6), 1153–1160 (1981)
18. Clark, J.J., Palmer, M., Lawrence, P.: A transformation method for the reconstruction of functions from nonuniformly spaced samples. IEEE Trans. Acoust. Speech Signal Process. **33**(5), 1151–1165 (1985)
19. Irani, M., Peleg, S.: Improving resolution by image registration. Graph. Model. Image Process. **53**(3), 231–239 (1991)
20. Schultz, R., Stevenson, R.: Extraction of high-resolution frames from video sequences. IEEE Trans. Image Process. **5**(6), 996–1011 (1996)
21. Ni, K., Nguyen, T.: Image super resolution using support vector regression. IEEE Trans. Image Process. **16**(6), 1596–1610 (2007)

The Application of Big Data in Machine Learning

Text-Associated Max-Margin DeepWalk

Zhonglin Ye[1,3,4], Haixing Zhao[1,2,3,4(✉)], Ke Zhang[2,3,4], Yu Zhu[2,3,4],
and Yuzhi Xiao[2,3,4]

[1] School of Computer Science, Shaanxi Normal University, Xi'an 710119,
China
h.x.zhao@163.com
[2] School of Computer, Qinghai Normal University, Xining 810008, China
[3] Tibetan Information Processing and Machine Translation Key Laboratory,
Xining 810008, China
[4] Key Laboratory of Tibetan Information Processing, Ministry of Education,
Xining 810008, China

Abstract. Most existing network representation algorithms learn the network representations based on network structure, however, they neglect the rich external information associated with nodes (i.e. text contents, communities and label information). Meanwhile, the learnt representations usually lack the discriminative ability for the tasks of node classification and linking prediction. We consequently overcame the above challenges by presenting a novel semi-supervised algorithm, text-associated max-margin DeepWalk algorithm (TMDW). TMDW incorporates text contents and network structures into the network representation learning based on the inductive matrix completion algorithm, and then we use node's category to optimize the learnt network representations based on the mar-margin principle and biased gradient. For integrating the above tasks, we propose a novel and efficient framework of network representation learning, this framework is easy to extend and generate discriminative representations. We then evaluate our model using the multi-class classification tasks. The experimental results demonstrate that TMDW outperforms other baseline methods on three real-world datasets. The visualization task of TMDW shows that our model is more discriminative than the other unsupervised approaches.

Keywords: DeepWalk · Network representation · Network embedding
Maximum margin · Biased gradient

1 Introduction

Data distribution is determined by multiple latent factors. Therefore, the main purpose of representation learning is to deconstruct the latent factors within data variations. Network representation learning aims at generating low-dimensional, compressed and dense vectors for various networks. Network representation learning has been applied to a variety of tasks, such as network classification [1], recommendation system [2, 3], and linking prediction [4].

© Springer Nature Singapore Pte Ltd. 2018
Z. Xu et al. (Eds.): Big Data 2018, CCIS 945, pp. 301–321, 2018.
https://doi.org/10.1007/978-981-13-2922-7_21

The existing representation approaches contain two kinds of implementations, one is based on network structure mining, and the other approaches are based on combination methods of network structure and auxiliary information. The auxiliary information includes text content information, community information of node. DeepWalk [5] first encodes social relationships in a continuous vector space, and it generalizes the unsupervised feature extraction from language modeling to network space. It considers the node sequences in network as the word sequences in language modeling when the random walk algorithm is used to network representation learning based on Skip-Gram algorithm [6], shortly afterwards, the various methods have been presented based on DeepWalk, such as Line [7], GraRep [8], SDNE [9] and node2vec [10]. All of them are based on the network structure features, and they have been effectively verified on several real-world networks. However, the node's text contents can be regarded as a supplementary information in networks, which contribute to node representation learning to some extent.

To deal with the above challenges, some researches utilize text contents, communities and labels to train the network representation models. Such as PTE [11], CECE [12], CNRL [13], and TriDNR [14]. TriDNR algorithm uses two neural networks to learn the representation based on inter-node, node-word, and label-word network relationships. Consequently, TriDNR shows a poor performance in network representation with sufficient node contents. the inter-node relationship would be affected when the edge amount of node-word is significantly greater than that of inter-node, and the proportion factors of three parties are also difficult to weight. CECE approach adopts a neural network method to jointly learn the inter-node and node-sentence network relationships, CENE furtherly decomposes the content into sentence form, and then it applies Wavg, RNN and BiRNN approaches to verify the feasibility and reliability of algorithm.

Matrix factorization approaches are applied to various data mining tasks, such as recommendation system, language representation learning, and dimensionality reduction etc., the traditional matrix factorization approaches factorize the target matrix into the multiplication form of two or three matrices, such as Principal Component Analysis (PCA), Singular Value Decomposition (SVD), Triangular Factorization etc. Inductive Matrix Completion (IMC) [15] is a novel and efficient method applied to data mining. It is the same as the SVD algorithm. IMC also generates the multiplication form of several matrices. But, the main differences are that IMC needs two assistant matrices to factorize the target matrix. In fact, IMC is an algorithm of matrix factorization. The two assistant matrices provide some useful information for factorizing target matrix.

The performance of DeepWalk based on Skip-Gram has been extensively verified on various network representation tasks. Fortunately, it has been found that Skip-Gram with Negative Sampling (SGNS) is to implicitly factorize the word-context matrix M [16], inspired by Skip-Gram, DeepWalk also can be considered as the operation that maximizes the co-occurrence probability in random walk window or factorizes matrix $M \in \mathbb{R}^{|V|*|V|}$. Here $|V|$ denotes the number of nodes. Each entry M_{ij} denotes the average probability that v_i randomly walks to v_j [17]. Based on the matrix factorization approach, some researches have been processed, such as TADW [18] and MMDW [19].

Max-margin approach is usually adopted to some discriminative tasks, such as Support Vector Machines (SVM) [20]. Max-Margin approach is also applied to topic model [21]. It is significant to research how to incorporate text contents and labels into network representation learning and generate discriminative embeddings for classification tasks.

Based on the above analysis, we propose text-associated max-margin DeepWalk algorithm to learn the network representation from text contents, network structures and labels. It is a discriminative network representation learning method. We first delete the stopwords in text titles and build a matrix $N \in \mathbb{R}^{|V| \times |c|}$, $|c|$ denotes the amount of the remaining words, each element represents the presence or absence of words, such as 0 or 1. And then, we use SVD algorithm to gain the text matrix $T \in \mathbb{R}^{|V| \times |k|}$, $|k|$ denotes the representation size of text features. Afterwards, we use identity matrix E and text matrix T to learn the representations based on IMC. Finally, we introduce the idea of the biased gradient proposed by Tu *et al.* [19] and max-margin classifier to optimize the learnt representations. The biased gradient of the representations represents the training direction where the representation vectors should move towards. The procedure can enlarge the margin between two categories, and it is conducted in the gradient form in gradient descent algorithms. Consequently, the trained max-margin classifier can enlarge the distance between support vectors and classification boundaries.

There are three main contributions of this paper: (1) We propose text-associated max-margin DeepWalk algorithm, which incorporates network structures, node's labels and text contents into network representations. (2) We introduce inductive matrix completion, biased gradient and max-margin theory into our algorithm framework and get a better advantage as compared to other baselines. (3) As a semi-supervised learning approach, we conduct node classification experiments on three scientific cooperation networks to verify the performance and effectiveness of TMDW. The experimental results show its superiority with various baseline approaches and training ratios.

However, the integration and application about IMC, biased gradient and max-margin are the main contributions of the works of TMDW.

2 Related Work

Network representation learning aims at exploring how to learn the network representations from network structures and other auxiliary information. The learnt representations can be applied to various tasks, such as node classification, linking prediction, clustering and so on.

Network representation learning consists of neural network-based and non-neural network-based approaches. non-neural network-based approaches are regarded as dimensional reduction technologies under linear algebra perspective, some linear (e.g., PCA) and non-linear dimensional reduction methods have been proposed [22–25]. Hofmann *et al.* [26] first introduces the concept of network representation learning. Inspired by Skip-Gram with negative sampling [6], the DeepWalk algorithm is then proposed to learn the representations from the network structure [5]. Influenced by DeepWalk, node2vec [27] designs a biased random walk algorithm based on Breadth-first sampling (BFS) and Depth-first Sampling (DFS), which sufficiently utilizes the different neighborhoods. Cao et al. [8] and Wang et al. [9] incorporate the global information into network representations when the proposed models and strategies are applied to capture the neighboring nodes. LINE [7] adopts first-order and second-order random walk to learn the network representations by first-order proximity and second-order proximity. CNRL [13], TriDNR [14], CENE [12], M-NMF [28] and HINE [29] incorporate the other information into network representations except inter-node rela-tionships, such as text contexts, network communities and network structures. NEU [30] and WALKLETS [31] aim to explore higher order network representations. GCN [32] is a semi-supervised algorithms of network representation.

Max-margin method has been used for various discriminative tasks, Taskar et al. [33] first proposes the max-margin approach for Markov Network, Zhu et al. [21] then presents maximum margin supervised topic models (MedLDA) [25], and then, max-margin algorithm is widely adopted in some tasks [33, 34]. Inspired by max-margin, MMDW [19] applies the max-margin principle to learn the representations for various networks. Motivated by matrix factorization, TADW [18] incorporates text features into network representation learning based on matrix factorization. However, the net-work representation method also can be based on hypergraph, for example Zhou et al. [35] and Tu et al. [36] propose an algorithm of hypergraph embedding, Li et al. [37] adopts the network embedding approach based on hypergraph to conduct linking prediction research. DepthLGP [38] solves the network representation of dynamic networks.

3 Our Method

3.1 General Framework

The thought of DeepWalk derives from Word2Vec. DeepWalk learns the network representations based on the node sequences generated from random walk strategy, DeepWalk trains the model of the network representation based on Skip-Gram algo-rithm and Hierarchical Softmax [17]. Compared by neural language model, each node generated from random walk is equivalent to the words of the context windows in neural language model, Thus, each sequence of random walk can be regarded as a comprehensive sentence in language model. DeepWalk is a novel approach, but it does not have stronger discriminative ability for classification tasks.

Fortunately, Yang et al. [17] has proved that DeepWalk actually aims at factorizing the matrix $M \in \mathbb{R}^{|V| \times |V|}$, where M_{ij} is the probability value that the node v_i randomly walks to node v_j, the detailed information of matrix factorization of DeepWalk is shown Fig. 1:

As shown in Fig. 1, matrix $M \in \mathbb{R}^{|V| \times |V|}$ is factorized into the multiplication form of the two matrices $W \in \mathbb{R}^{k \times |V|}$ and $H \in \mathbb{R}^{k \times |V|}$, where $k \ll |V|$, note that DeepWalk based on matrix factorization regards the matrix W as the network representations.

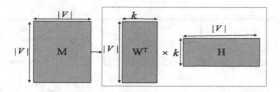

Fig. 1. DeepWalk as matrix factorization

IMC fully utilizes more information of row and column units by incorporating two feature matrices into the objective function. Note that IMC is originally presented to complete gene-disease matrix with gene and disease features, the goal of IMC is quite different from that of our work, but the principle and approach of IMC can be applied to our work. IMC receives two feature matrices as assistant parameters. Therefore, it is usually used to factorize the target matrix M based on the two assistant matrices. In this paper, we recognize the text matrix and identity matrix as the assistant matrices.

Inspired by IMC, we introduce the node's text contents and network structures into network representation learning. As shown in Fig. 2, our approach is to factorize the matrix $M \in \mathbb{R}^{|V| \times |V|}$ into the multiplication form of four matrices $E \in \mathbb{R}^{|V| \times |V|}$, $W \in \mathbb{R}^{|V| \times k}$, $H \in \mathbb{R}^{k \times k}$ and $T \in \mathbb{R}^{k \times |V|}$ based on IMC algorithm, here $T \in \mathbb{R}^{k \times |V|}$ is the external text feature matrix, both the $W \in \mathbb{R}^{k \times |V|}$ and $H \in \mathbb{R}^{k \times c}$ are our target matrices. $E \in \mathbb{R}^{|V| \times |V|}$ is an identity matrix with $|V|$ rows and $|V|$ columns, respectively. Generally, the identity matrix can be replaced as various matrices containing some aspects of network features. For example, we can incorporate the community incidence matrix, the learnt network representations maybe share the community properties. The performance of community detection algorithm affects the performance improvement of network representation learning. However, the TMDW algorithm provides an effective extension framework and possibility.

A simple method for representation learning is to independently train the text features and network structures, and then we concatenate the text representations with the network representations, such as $r = r_{text} \oplus r_{node}$. Here, r is to concatenate two kinds of representations, which results in the loss of relationship between nodes and text titles. Compared the DeepWalk as matrix factorization that uses the W to represent the network representations, our method considers more information about network representation learning. We incorporate the node's text contents, node's labels and inter-node relationships into network representation learning instead of structure-only information. Therefore, we use W and HT to represent the network representations, such as $W \oplus T^T H^T$. The target of TMDW is to incorporate text contents into network representation learning, and generates the discriminative representations. To solve the discriminative classification task, we introduce the maximum margin approach and biased gradient to learn discriminative representations of nodes on real-world scientific collaboration networks. We adopt the learnt representation $W \oplus T^T H^T$ as representations and train a SVM classifier, and then, we use the biased gradient to enlarge the distance between support vectors and classification boundaries, which deeply optimizes the learnt network

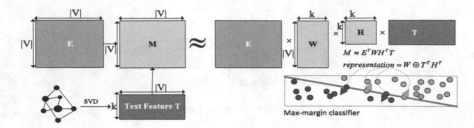

Fig. 2. The algorithm framework of TMDW

representations. TMDW is thus more discriminative and more feasible to network representation tasks, which would be demonstrated in the following parts.

3.2 Formalization

Network representation learning is defined as follows. We suppose that there is a network $G = (V, E)$, where V denotes the node set and E denotes the edge set, i.e. $E \in V \times V$, $|V|$ denotes the size of node set. We want to train the network representation models for various networks. By this kind of model, we can get node representation vector $r_v \in \mathbb{R}^k$ or embedding for a given node $v \in V$, which is also the purpose of network representation learning, here k is the space size of network representation and it is expected much smaller than $|V|$. $l \in \{1, \dots, m\}$ denotes node labels. The learnt representation vectors contain the structural feature and other information of network. Note that $r_v \in \mathbb{R}^k$ is not task-specific and can be applied to diverse tasks, e.g. SVM.

3.3 Inductive Matrix Completion

Compared with vector-space-based approaches, the matrix-space-based methods are lack of scalability, because the space and time complexity cause cubical increasement with the growth of problem size. Therefore, it is much-needed to approximate the target matrix that makes the model more robust, accurate and practical for large-scale machine learning tasks. Inspired by the Support Vector Machine (SVM), Compressed Sensing (CS) and Non-Negative Matrix Factorization (NMF) [39] technologies, some machine learning algorithms based on sparsity and low-rank assumption are proposed.

Roughly speaking, the matrix factorization mainly discusses how to divide the matrix into multiple matrix algebras. The computational model of NMF is the same with the traditional Principal Component Analysis (PCA), Independent Component Analysis (ICA), Singular Value Decomposition (SVD) and so on. It can be formalized to $M = W^T H, W, H \geq 0$. $M \in \mathbb{R}^{d \times n}$ is a matrix, meanwhile, matrix $W \in \mathbb{R}^{k \times d}$ and $H \in \mathbb{R}^{k \times n} (k \ll \{d, n\})$ have a non-negative condition. On the other hand, the advantage of NMF based on sparse and low-rank provides the possibility to conduct efficient and fast matrix computing. Yu *et al.* [3] proposes a matrix approach with a trace norm constraint, its purpose is to minimize the following loss function:

$$\min_{W,H} \sum_{(i,j)\in\Omega} (M_{ij} - (W^T H)_{ij})^2 + \frac{\lambda}{2}(\| W \|_F^2 + \| H \|_F^2), \tag{1}$$

where $\|\cdot\|$ denotes the Frobenius norm, λ is the balance factor. NMF is only based on the low-rank assumption of M. We can adopt the inductive matrix completion approach proposed by Natarajan and Dhillon [15] to incorporate node's text contents. Motivated by the above loss function of NMF. Inductive matrix completion incorporates two feature matrices into loss function as follows:

$$\min_{W,H} \sum_{(i,j)\in\Omega} (M_{ij} - (U^T W^T H V)_{ij})^2 + \frac{\lambda}{2}(\| W \|_F^2 + \| H \|_F^2), \tag{2}$$

where Ω denotes the sample set of matrices $M, U \in \mathbb{R}^{p\times d}$ and $V \in \mathbb{R}^{q\times n}$ are the two feature matrices.

As far as our research are concerned, our goal is to incorporate text features into the task of matrix factorization. Coincidentally, inductive matrix completion is first proposed to complete Gene-Disease matrix using gene and disease feature matrices. Thus, the thought of inductive matrix completion can be used in our research.

3.4 Text-Associated Max-Margin DeepWalk

Importantly, DeepWalk adopts Skip-Gram model to generate network representations based on the various network structures, it captures contexts using random walk strategy. Given a node v_i from random walk sequence S, the objective of DeepWalk aims to maximum the likelihood as follows:

$$L(S) = \frac{1}{|S|} \sum_{i=1}^{|S|} \sum_{-t\leq j\leq t, j\neq 0} \log \Pr(v_{j+i}|v_i), \tag{3}$$

where t denotes the length of random walk, it also can be regarded as the size of context window. $p(v_j|v_i)$ is defined by Softmax function.

$$p(v_j|v_i) = \frac{\exp(c_{v_j}^T r_{v_i})}{\sum_{v\in V} \exp(c_v^T r_{v_i})}, \tag{4}$$

where r_{v_i} and c_{v_j} denote the representations of the current node v_i and its context node v_j. Note that DeepWalk adopts Hierarchical Softmax [40] to speed up the training procedure, instead of Softmax method. Afterwards, Yang et al. [17] proves that DeepWalk is to factorize matrix M, which can be calculated by the following equation:

$$M_{ij} = \log \frac{[e_i(A + A^2 + \cdots + A^t)]_j}{t}, \tag{5}$$

where A denotes the transition matrix, here, we regard it as adjacency matrix, e_i denotes an indicator vector. The time complexity of factorize the matrix M is $o(|V|^3)$. Thus, DeepWalk adopts the random walk strategy to avoid the processing of factorizing matrix M. Fortunately, Yang et al. [18] proposes a simplified calculation approach, finding a threshold between speed and accuracy, the new matrix factorization approach is to factorize the following matrix:

$$M = \frac{A + A^2}{2}. \tag{6}$$

Considered the efficiency and cost, we factorize the matrix M instead of $\log M$, because that the complexity of factorizing matrix M is $o(|V|^2)$ compared with that of $o(|V|^3)$, another reason is that $\log M$ has much more non-zero entries than M [3].

By proving that DeepWalk is equivalent to factorize the matrix $M = W^T H$, thus, DeepWalk is to find $W \in \mathbb{R}^{r \times d}$ and $H \in \mathbb{R}^{r \times n}$ to minimize the likelihood of following equation:

$$\min_{W,H} \left\| M - (W^T H) \right\|_F^2 + \frac{\lambda}{2} (\|W\|_F^2 + \|H\|_F^2), \tag{7}$$

where λ controls the weight of regularization parts. Our method incorporates the text features into network representations. Given a text feature matrix $T \in \mathbb{R}^{c \times |V|}$, $W \in \mathbb{R}^{k \times |V|}$ and $H \in \mathbb{R}^{k \times c}$, we general the above Eq. (7) based on Eqs. (1) and (2) as follows:

$$\min_{W,H} \left\| M - (W^T H T) \right\|_F^2 + \frac{\lambda}{2} (\|W\|_F^2 + \|H\|_F^2). \tag{8}$$

Inspired by the max-margin approach in SVM algorithm and topic model [21], we take the learnt representations $W^T \oplus (HT)^T$ as features to train a SVM classifier which is usually used to various discriminative classification tasks. Suppose the training set is $\tau = \{(x_1, l_1), \ldots, (x_T, l_T)\}$. The SVM algorithm aims to solve the following optimization function:

$$\min \frac{1}{2} \|Y\|_F^2 + C \sum_{i=1}^{N} \varepsilon_i, \tag{9}$$

$$\text{s.t.} \quad y_{l_i}^T x_i - y_j^T x_i \ge e_i^j - \varepsilon, \ \forall i, j$$

here,

$$\varepsilon_i^j = \begin{cases} 1, & \text{if } l_i \ne j, \\ 0, & \text{if } l_i = j, \end{cases} \tag{10}$$

where l_i denotes the element in training set, N is the size of training set, $Y = [y_1, \cdots, y_n]^T$ denotes the weight matrix of SVM, $\varepsilon = \{\varepsilon_1, \cdots, \varepsilon_N\}$ is the slack variable that tolerates errors in the training set. Meanwhile, $y_{l_i}^T r_{v_i} - y_j^T r_{v_i} \geq e_i^j - \varepsilon_i$.

DeepWalk aims at factorizing matrix $M = (A + A^2)/2$, the low-rank matrix factorization can incorporate text features into network representations, max-margin approach contributes to finding the optimal classification boundaries. Thus, TMDW cam optimize the max-margin classifier of SVM and matrix factorization based on DeepWalk, meanwhile, TMDW adopts text features to optimize the network representations for classification tasks. Our objective can be defined as follows:

$$\min_{W,H} \|M - (W^T H T)\|_F^2 + \frac{\lambda}{2}(\|W\|_F^2 + \|H\|_F^2) + \frac{1}{2}\|Y\|_F^2 + C\sum_{i=1}^{N} \varepsilon_i, \tag{11}$$

$$\text{s.t.} \quad y_{l_i}^T r_{v_i} - y_j^T r_{v_i} \geq e_i^j - \varepsilon_i.$$

The training procedure for model in Eq. (11) is based on the literature [19], for example, we adopt an efficient optimization approach proposed by Tu *et al.* [19] for Eq. (11), which divides the objective into two parts and optimizes them separately.

We conduct our optimization procedure as follows.

Firstly, we fix the parameters W, H, and T. The Eq. (11) is equivalent to the form of multi-class SVM classification problem proposed by Crammer and Singer [41]. Consequently, the optimization of the Eq. (11) could be considered as the dual form as follows:

$$\min_Z \frac{1}{2}\|Y\|_2^2 + \sum_{i=1}^{T}\sum_{j=1}^{m} e_i^j z_i^j, \tag{12}$$

$$\text{s.t.} \quad \sum_{j=1}^{i} z_i^j = 0, \forall i \; z_i^j \leq C_{l_i}^j, \forall i, j.$$

where

$$y_j = \sum_{i=1}^{l} z_i^j x_i, \forall j$$

$$C_{y_i}^m = \begin{cases} 0, & \text{if } y_i \neq m \\ C, & \text{if } y_i = m \end{cases}.$$

Secondly, we fix the parameters Y and ε, The Eq. (11) becomes to minimize the square loss of matrix factorization with additional boundary constraints as follows:

$$\min_{W,H} L(W, H; M, T, \lambda)$$
$$= \min_{W,H} \|M - (W^T H T)\|_F^2 + \tfrac{\lambda}{2}(\|W\|_F^2 + \|H\|_F^2). \tag{13}$$

$$\text{s.t.} \quad y_{l_i}^T x_i - y_j^T x_i \ge e_i^j - \varepsilon, \ \forall i, j.$$

Given a node $i \in \tau$, we can get the biased gradient $\frac{\partial L}{\partial x_i} + \eta \sum_{j=1}^{m} a_i^j (y_{l_i} + y_j)^T$, η balances the primal gradient and the bias. The above contents introduce the training procedure for the model in Eq. (11). the following contents introduce our experiments conducted on three real-world datasets.

4 Experimental Results and Evaluations

4.1 Dataset Setup

We conduct our experiments on three real-world citation networks. The overviews of dataset are given in Table 1:

Table 1. Dataset overview

Dataset	Node	Edge	Label	Average degree	Average path length	Graph density	Average clustering coefficient
Citeseer	3312	4732	6	2.857	2.02	0.001	0.080
Cora	2708	5429	7	4.01	4.79	0.001	0.130
DBLP	3119	39516	4	21.07	4.71	0.005	0.221

We conduct our experiments on Citeseer [1], Cora[2] and DBLP-V4 (DBLP[3]) datasets. For the DBLP dataset, we divide conference papers into 4 research sets, such as database, data mining, artificial intelligent and computer vision. After analyzing and mining, we find that it exists a lot of independent nodes without link edges with other nodes in network. Therefore, we delete these nodes from original DBLP dataset. To balance of node amount in each category, we keep the node amount of each category about 800 by deleting these nodes with few connection edges. the average degree of DBLP dataset is 21.07, so the DBLP network is a dense network.

---

[1] http://citeseerx.ist.psu.edu/.

[2] https://people.cs.umass.edu/mccallum/data.html.

[3] http://arnetminer.org/citation.

4.2   Baseline Methods

DeepWalk (DW). DeepWalk [5] is a popular network representation learning algorithm, which learns the representations by using the Skip-Gram model and Hierarchical Softmax proposed by Word2Vec. It learns representations using network structures. We set parameter $K = 5$, walk length $\gamma = 80$, vector size $k = 200$.

MFDW. MFDW is the abbreviated form of DeepWalk as Matrix Factorization, which simulates the DeepWalk by matrix factorization. MFDW factorizes the target matrix $M = (A + A^2)/2$, and then it uses the matrix W to train a SVM classifiers.

LINE. LINE [7] is proposed to learn network representations for large scale networks, which provides 1st LINE and 2nd LINE implementations. We adopt the 2nd LINE to train and learn the network representations. Same as DeepWalk, the vector size is set as 200.

MMDW. Like MFDW, MMDW also factorizes the matrix $M = (A + A^2)/2$ and takes the matrix W to train classifier. MMDW uses the max-margin approach to optimize matrix W, thus the learnt representations possess a discriminative ability.

TEXT. We use text feature matrix $T \in \mathbb{R}^{|V| \times 200}$ as a 200-dimensional representations. The approach of text feature is a content-only baseline.

DW+TEXT. This is a compound method that concatenates features from text features with features from DeepWalk. The text feature matrix $T \in \mathbb{R}^{|V| \times 100}$ is a 100-dimensional representation. The feature from DeepWalk is also a 100-dimensional representation. The final dimension size of representation vectors is 200.

MFDW+TEXT. This method concatenates features from MFDW with text features. The length of the two different features is 100, respectively.

LINE+TEXT. It is same as DW + TEXT, LINE + TEXT concatenates Line's features and text features. The representation length of two components is 100, respectively.

MMDW+TEXT. It concatenates the features from MMDW and text features. The representation length of two components is 100, respectively.

Enhanced Network using DeepWalk (ENDW). we first remove all stopwords in node's text titles, we then recognize the remaining words as separated nodes in the enhanced network. Note that ENDW consists of inter-node relationship networks and node-word relationship networks.

4.3   Classifiers and Experiment Setup

We conduct our experiments on real-world network datasets. We adopt the classification tasks to verify the feasibility of our algorithm. For semi-supervised learning, we choose linear SVM implemented by liblinear [42]. To evaluate our algorithm, we randomly generate a portion of the labeled sample as training set, the rest are testing set. We then evaluate and calculate the accuracy of classification based on the different proportions of training sets. The proportion varies from 10% to 90%. Like DeepWalk, MFDW, DW, we set the representation length to 200. we also set λ as 0.1.

4.4 Experimental Results and Analysis

TMDW algorithm ensembles network structures, node's labels and text contents into network representations using inductive matrix completion, biased gradient and max-margin theory. The detailed algorithm framework of TADW can be found in Sect. 3.4, which introduces how to integrate the text information and label category into the procedure of training model of TADW, the detailed procedure of training can be found at Eq. (11).

The classification experimental results are shown in Table 2 (Citeseer), Table 3 (Cora) and Table 4 (DBLP) as follows, TMDW algorithm consistently and significantly outperforms the baseline algorithms on different datasets at most training proportions, it shows the feasibility of our method. We define a hyper-parameter η to balance the biased gradient and original-gradient. From Tables 2, 3 and 4, we can get the accuracies of each algorithm on three kinds of paper citation networks. Therefore, we have the following interesting observations:

Table 2. Accuracy (%) of vertex classification on Citeseer

Algorithm	10%	20%	30%	40%	50%	60%	70%	80%	90%
DW	48.31	50.36	51.33	52.31	52.85	53.33	52.98	53.47	53.71
MFDW	49.78	54.80	56.66	56.75	57.90	58.32	58.60	58.29	57.07
LINE	39.82	46.83	49.02	50.65	53.77	54.20	53.87	54.67	53.82
MMDW	55.49	60.70	63.66	65.27	66.02	69.14	69.34	69.47	69.72
TEXT	56.24	61.80	62.16	64.74	66.77	68.52	69.89	71.37	70.41
DW+TEXT	59.17	61.93	64.45	66.93	68.21	69.17	70.28	71.77	71.21
MFDW+TEXT	60.82	60.69	63.62	64.22	67.97	68.74	70.16	71.22	70.35
LINE+TEXT	44.18	49.37	56.11	59.29	63.30	63.73	69.44	70.47	67.39
MMDW+TEXT	61.63	62.71	65.82	67.49	69.56	70.01	71.57	72.51	71.45
ENDW	57.26	57.39	62.18	64.47	66.69	69.17	70.24	70.56	68.40
TMDW ($\eta = 10^{-2}$)	62.34	66.95	69.30	69.60	70.97	71.45	72.40	73.05	70.95
TMDW ($\eta = 10^{-3}$)	62.41	67.01	69.30	69.52	71.27	71.45	72.30	72.74	71.56
TMDW ($\eta = 10^{-4}$)	62.64	67.13	69.30	69.65	70.67	71.38	72.61	72.90	70.64

(1) For the Citeseer dataset, TMDW achieves nearly 15% improvement compared with DW, MFDW and LINE with the training ratios varied from 10% to 90%, and TMDW obtains about 5% improvement compared with MMDW. For the Cora dataset, LINE's performance is inferior to the other baseline approaches. TMDW works generally better than MMDW when the training ratio is less than 60%, the rest are nearly same when the training ratios range from 70% to 90%. For the

Table 3. Accuracy (%) of vertex classification on Cora

Algorithm	10%	20%	30%	40%	50%	60%	70%	80%	90%
DW	73.29	75.46	76.19	77.49	77.89	77.83	78.86	79.05	78.62
MFDW	66.38	75.52	78.78	80.54	82.09	81.93	82.62	81.57	83.81
LINE	65.13	70.17	72.20	72.92	73.45	75.67	75.25	76.78	79.34
MMDW	73.61	79.99	80.43	81.92	83.76	84.97	86.39	86.70	87.45
TEXT	69.92	69.56	72.25	73.52	74.10	74.15	74.15	75.25	75.19
DW+TEXT	71.41	75.78	78.82	80.52	82.05	84.01	84.42	86.92	87.38
MFDW+TEXT	73.64	79.02	79.61	82.03	83.06	84.96	84.49	83.79	88.37
LINE+TEXT	69.48	72.33	73.49	74.26	79.61	80.39	79.62	81.53	83.27
MMDW+TEXT	75.13	80.36	81.92	83.01	84.27	85.79	86.54	86.94	87.69
ENDW	72.73	75.39	77.35	80.25	83.11	83.36	85.38	85.92	86.01
TMDW ($\eta = 10^{-2}$)	77.66	81.49	83.36	83.63	84.35	85.87	86.76	86.89	88.56
TMDW ($\eta = 10^{-3}$)	77.66	81.53	83.20	84.56	84.42	85.78	86.89	87.08	88.56
TMDW ($\eta = 10^{-4}$)	77.70	81.44	83.15	83.69	84.49	85.87	87.01	87.08	88.56

Table 4. Accuracy (%) of vertex classification on DBLP

Algorithm	10%	20%	30%	40%	50%	60%	70%	80%	90%
DW	81.84	82.41	83.25	83.74	83.98	84.24	84.55	84.26	83.53
MFDW	75.06	80.82	83.00	83.96	84.70	84.94	85.72	84.62	85.11
LINE	79.13	79.81	80.41	81.22	82.95	83.39	83.04	84.74	83.85
MMDW	79.70	82.05	84.23	84.84	83.45	85.42	84.96	85.78	84.49
TEXT	64.13	70.18	72.35	73.68	73.69	74.61	74.39	74.48	76.08
DW+TEXT	82.22	82.95	83.38	83.84	84.19	84.48	84.49	85.08	84.72
MFDW+TEXT	81.24	81.43	84.24	84.93	84.15	84.74	84.62	84.17	83.66
LINE+TEXT	80.46	80.96	81.51	82.39	83.15	83.61	83.28	84.38	82.87
ENDW	82.35	82.92	84.45	84.62	84.19	84.65	84.64	84.75	83.91
TMDW ($\eta = 10^{-2}$)	83.11	83.97	84.88	85.11	84.60	85.26	84.96	85.30	84.54
TMDW ($\eta = 10^{-3}$)	83.18	84.05	84.89	85.22	84.60	85.27	84.85	85.14	83.54
TMDW ($\eta = 10^{-4}$)	83.11	84.09	84.97	85.01	84.54	85.26	84.86	85.14	83.54

DBLP dataset, the performance of TMDW is slightly better than the other algorithms. The accuracies of the all algorithms fluctuate within in a smaller range, which demonstrates that the difference characteristics of these algorithms can be neglected for this kind of dense network structure.

(2) The performance of MFDW is nearly equivalent to DeepWalk, LINE shows the poorest performance compared with the other baseline methods. We concatenate text features with DeepWalk, MFDW, LINE, and MMDW, respectively. The experimental results show that text features contribute to the performance improvement, learning efficient network representations when the network is a

sparse network. For the dense networks, the traditional network representation algorithms can learn sufficient features from network structures, therefore, the text features have a negligible effect on the performance improvement.

(3) ENDW consists of inter-node relationships and node-word relationships, which considers the key words in titles as special nodes of the enhance network. TMDW algorithm takes the text features as a constraint condition, and then TMDW uses IMC algorithm to factorize the target matrix M. They make full use of the text features, but TMDW achieves the best performance, which shows that TMDW gets more information from text features based on IMC algorithm.

(4) Citeseer, Cora and DBLP are three citation network datasets driven by contents, however, the node's text contents are more necessary for learning the network representations. From Tables (2), (3) and (4), we find that the performance improvement of TMDW is obvious and clear, it has a greater advantage on Citeseer and Cora than DBLP. When the training ratio is less than 50%, TMDW significantly outperform other baseline methods, which demonstrates that TMDW is more robust and especially performs better when the network possesses a smaller amount of nodes or the network is a sparse network.

(5) The difference is not obvious on DBLP dataset, the main reason is that we optimize the DBLP dataset by deleting the separate nodes in network, we make sure that each node has three connection edges as least. There existing four categories in DBLP dataset, we then sample and balance the node's amount for each category, thus there are about 800 nodes in each category. By balancing the node's amount in our work, the performance of the learnt representation is more accurate for the dense networks. Another important observation is that the algorithm selection is very important for a sparse network, and the differences of algorithm are not obvious and clear for dense networks.

TMDW incorporates text features into matrix completion algorithm for generating high quality representations. Meanwhile, TMDW introduces the max-margin and biased gradient to deeply optimize the learnt representations and makes the embeddings more discriminative for various prediction tasks. In our research, the learnt representations are used to learn a SVM classifier, importantly, it can be utilized in many tasks, such as linking prediction, node's classification, node's similarity computing and so on. By comparison with MMDW and ENDW, we find that it can bring about performance improvement when the network representation learning is optimized by text features and biased gradient.

4.5   Parameter Sensitivity

TMDW has four parameters: iteration times, order of magnitude, weight of regularization term (Lambda) and representation length k. We set the training ratio to 0.5, and we then show the influence of iteration times in Fig. 3(a), (b) and (c). We fix the training rations to 0.5, and we then test classification accuracies with diverse order of magnitude in Fig. 4(a), (b) and (c). We also show the accuracy variations using the different representation lengths in Fig. 5(a), (b) and (c) as follows.

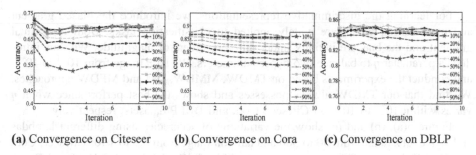

(a) Convergence on Citeseer   **(b)** Convergence on Cora   **(c)** Convergence on DBLP

Fig. 3. Convergence comparison on Citeseer, Cora and DBLP

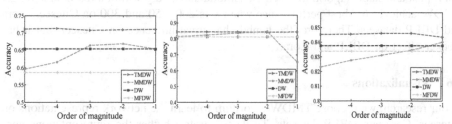

(a) Parameter sensitivity on Citeseer **(b)** Parameter sensitivity on Cora **(c)** Parameter sensitivity on DBLP

Fig. 4. η sensitivity on Citeseer, Cora and DBLP

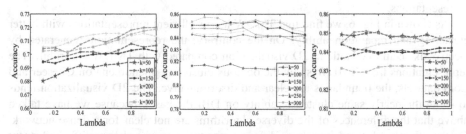

(a) Parameter sensitivity on Citeseer **(b)** Parameter sensitivity on Cora **(c)** Parameter sensitivity on DBLP

Fig. 5. Lambda and sensitivity on Citeseer, Cora and DBLP

Figure 3(a), (b) and (c) show the convergence variations when TMDW is trained with different training ratios. We let training ratio range from 10% to 90% and iteration times vary from 1 to 10 on Citeseer, Cora and DBLP datasets. The convergence speed denotes time complexity of training TMDW algorithm. We find that the performance of TMDW declines sharply after one-time iteration, and then it becomes stable. The accuracies vary within 18.85%, 13.66% and 4.29% for different training ratios and iteration times on Citeseer, Cora and DBLP datasets, respectively. Therefore, the dense network DBLP is more robust and stable compared with Citeseer and Cora-datasets.

For learning the discriminative representations, we introduce the biased gradient and max-margin, which are widely applied to classification tasks. Both the primal gradient and the bias are initially under different orders of magnitude. We thus introduce a parameter η to balance two hyper-parameters. We let η range from 10^{-1} to 10^{-5}, and conduct the experiments based on TMDW, MMDW, DM and MFDW approaches. We find that our TMDW always possesses and shows the best performance when η varies within 10^{-1} to 10^{-5} on Citeseer, Cora and DBLP datasets, respectively.

Figure 5(a), (b) and (c) show the variations of accuracies using different lambdas and k. We let k range from 50 to 300 and lambda range from 0.1 to 1 for Citeseer, Cora and DBLP datasets. The accuracies vary within 3.9%, 4.5% and 2.1% for different k and lambdas on Citeseer, Cora and DBLP datasets, respectively. Therefore, TMDW is a robust and stable algorithm when the lambda and k vary within a reasonable range. TMDW achieves the best performance when k is 200, 250 and 300 on Citeseer, Cora and DBLP datasets. Therefore, the lambda has great influence on Citeseer, but it has weaker effect on DBLP and Cora.

4.6 Visualizations

In our research, we propose TMDW algorithm to learn the network representations on Citeseer, Cora and DBLP datasets. To demonstrate whether the representations generated from TMDW work better than DeepWalk, and show the discriminative ability or not. We randomly choose 4 network categories. Each category contains 150 nodes generated by randomly-selected approach. We introduce t-SNE algorithm to show the 2D representation to verify that TMDW is quite qualified for generating discriminative representations.

As shown in Fig. 6, we find that TMDW learns efficient representations with better clustering and reparation ability. On the contrary, the representations generated by DeepWalk seem to chaotic in 2D visualization compared with TMDW algorithm. The representations from TMDW show the obvious clustering phenomena on Citeseer and Cora datasets, the boundaries are clear and discriminative. The 2D visualization tends to show the nearly same clustering ability on DBLP dataset, because we have talked above that the differences of the diverse algorithms are not clear for a dense network. On the other hand, the papers from different categories maybe share quite similar text titles, which could cause the phenomenon that some points are distributed in other classes for visualization. 2D visualization results demonstrate the effectiveness of our discriminative representation learning algorithm.

4.7 Case Study

Network representation learning is to encode each node into a low dimensional vector. Suppose that we have generated all representations of network nodes, we then set a target node and find the most relevant several nodes using cosine similarity calculating method. To verify the probability and performance of TMDW, we conduct an instance experiment on DBLP dataset, the target title is "**Maximum Margin Planning**". We get the top 5 nearest neighbor nodes by computing the cosine similarity based on learnt representations. We set the representation length is 200, and training ratio as 50%, η as 10^{-2}. The results of case study are in Table 5.

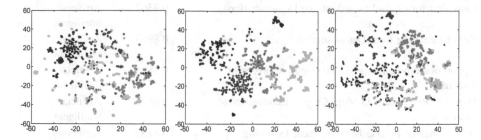

(a) Citeseer visualization using DW (b) Cora visualization using DW (c) DBLP visualization using DW

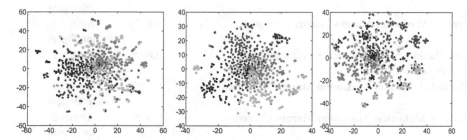

(d) Citeseer visualization using TMDW (e) Cora visualization using TMDW (f) DBLP visualization using TMDW

Fig. 6. 2D Visualization on Citeseer, Cora and DBLP

First, we know that some papers are quoted by the paper "**Maximum Margin Planning**", or some of them quote the paper "**Maximum Margin Planning**". However, all these papers have the same labels "**Artificial intelligent**", which demonstrates that DeepWalk and TMDW can learn the efficient representations from network structure-based features. The top 5 nearest nodes generated from DeepWalk take the network structural similarity into account, therefore, the titles of top 5 nearest nodes are not about the knowledge of "**maximum margin**" or "**planning**". Artificial intelligent category includes these papers extracted from IJCAI, AAAI, NIPS, ICML, ECML, ACML, IJCAI, UAI, ECAI, COLT, ACL, KR and son on. DeepWalk considers and generates the representations from the overview of citation relationships and network structures. TMDW incorporates labels into network representation learning using max-margin and biased gradient, so that the accuracy of classification is desirable. TMDW incorporates the text contents of network nodes into network representation learning, so that the node's similarity slightly tends to text similarity. For example, text titles of top 5 nearest nodes contain the words "**Maximum Margin**". We find that these titles are

Table 5. Top five nearest neighbors generated by DeepWalk and TMDW

Title	Similarity	Label
DeepWalk		
Learning for Control from Multiple Demonstrations	0.9367	Artificial intelligent
Robot **Learning** from Demonstration	0.8953	Artificial intelligent
Apprenticeship Learning via Inverse Reinforcement **Learning**	0.8895	Artificial intelligent
Dynamic Preferences in Multi-Criteria Reinforcement **Learning**	0.8653	Artificial intelligent
Algorithms for Inverse Reinforcement **Learning**	0.8152	Artificial intelligent
TMDW		
Maximum Margin Clustering Made Practical	0.6821	Artificial intelligent
Laplace **Maximum Margin** Markov Networks	0.6778	Artificial intelligent
Fast **Maximum Margin** Matrix Factorization for Collaborative Prediction	0.6620	Artificial intelligent
The Relaxed Online **Maximum Margin** Algorithm	0.6457	Artificial intelligent
Efficient Multi-class **Maximum Margin** Clustering	0.6456	Artificial intelligent

relevant to "**Maximum Margin**", most of the nearest nodes from DeepWalk have no relatedness with topic "**Maximum Margin**", most of them are relevant to the topic "**Reinforcement Learning**". Consequently, TMDW can learn better representations than DeepWalk.

5   Conclusion

In this paper, we propose text-associated max-margin DeepWalk (TMDW), it is a novel and discriminative network representation approach which incorporates network structures, node's labels and text contents into network representation learning based on inductive matrix completion, biased gradient and max-margin classifier. We conduct experiments on three real-world datasets (Citeseer, Cora and DBLP) with the task of node classification. The experimental results show that TMDW is an effective and robust network representation approach compared the other baseline approaches. Meanwhile, the 2D visualization results of the learnt representations generated from TMDW demonstrate stronger discrimination ability. TMDW provides a normalized framework for joint learning with different types of resources via inductive matrix completion instead of concatenating them. For future work, we will extend our approaches to representation learning of large-scale networks, meanwhile, we will explore some new technologies of matrix factorization, such as max-margin matrix factorization and matrix co-factorization.

Acknowledgement. This project is supported by NSFC (No. 61663041, 61763041), the Program for Changjiang Scholars and Innovative Research Team in Universities (No. IRT_15R40), the Research Fund for the Chunhui Program of Ministry of Education of China (No. Z2014022) and the Nature Science Foundation of Qinghai Province (2014-ZJ-721), the Fundamental Research Funds for the Central Universities (2017TS045), and the Tibetan Information Processing and Machine Translation Key Laboratory (2013-Z-Y17).

References

1. Tsoumakas, G., Katakis, I.: Multi-label classification: an overview. Int. J. Data Warehous. Min. **3**(3), 1–13 (2007)
2. Tu, C., Liu, Z., Sun, M.S.: Inferring correspondences from multiple sources for microblog user tags. In: Huang, H., Liu, T., Zhang, H.-P., Tang, J. (eds.) SMP 2014. CCIS, vol. 489, pp. 1–12. Springer, Heidelberg (2014). https://doi.org/10.1007/978-3-662-45558-6_1
3. Yu, H.F., Jian, P., Kar, P., Dhillon, I.S.: Large-scale multi-label learning with missing labels. In: Proceedings of ICML, pp. 593–601 (2014)
4. Liben-nowell, D., Kleinberg, J.: The link-prediction problem for social networks. J. Assoc. Inf. Sci. Technol. **58**(7), 1019–1031 (2007)
5. Perozzi, B., Al-Rfou, R., Skiena, S.: DeepWalk: online learning of social representations. In: ACM SIGKDD International Conference on Knowledge Discovery and Data Mining, pp. 701–710 (2014)
6. Mikolov, T., Sutskever, I., Chen, K., Corrado, G., Dean, J.: Distributed representations of words and phrases and their compositionality. In: Advances in Neural Information Processing Systems, pp. 3111–3119 (2013)
7. Tang, J., Qu, M., Wang, M.Z., Zhang, M., Yan, J., Mei, Q.Z.: Line: large-scale information network embedding. In: Proceedings of WWW, pp. 1067–1077 (2015)
8. Cao, S.S., Lu, W., Xu, Q.K.: GraRep: learning graph representations with global structural information. In: Conference on Information and Knowledge Management, pp. 891–900 (2015)
9. Wang, D.X., Cui, P., Zhu, W.W.: Structural deep network embedding. In: The ACM SIGKDD International Conference, pp. 1225–1234 (2016)
10. Grover, A., Leskovec, J.: Node2vec: scalable feature learning for networks. In: Proceedings of ACM SIGKDD, pp. 855–864 (2016)
11. Tang, J., Qu, M., Mei, Q.Z.: PTE: predictive text embedding through large-scale heterogeneous text networks. In: ACM SIGKDD International Conference on Knowledge Discovery and Data Mining, pp. 1165–1174 (2015)
12. Sun, M.S., Guo, J., Ding, X., Liu, T.: A general framework for content-enhanced network representation learning. arXiv:1610.02906 (2016)
13. Tu, C.C., Wang, H., Zeng, X.K., Liu, Z.Y., Sun, M.S.: Community-enhanced network representation learning for network analysis. arXiv:1611.06645 (2016)
14. Pan, S.R., Wu, J., Zhu, X.Q., Zhang, C.Q., Wang, Y.: Tri-party deep network representation. In: Proceedings of IJCAI 2016, pp. 1895–1901 (2016)
15. Natarajan, N., Dhillon, I.S.: Inductive matrix completion for predicting gene-disease associations. Bioinformatics **30**(12), 60–68 (2014)

16. Levy, O., Goldberg, Y.: Neural word embedding as implicit matrix factorization. In: Advances in Neural Information Processing Systems, pp. 2177–2185 (2014)
17. Yang, C., Liu, Z.Y.: Comprehend deepwalk as matrix factorization. arXiv:1501.00358 (2015)
18. Yang, C., Liu, Z.Y., Zhao, D.L.: Network representation learning with rich text information. In: International Conference on Artificial Intelligence, pp. 2111–2117 (2016)
19. Tu, C.C., Zhang, W.C., Liu, Z.Y., Sun, M.S.: Max-margin DeepWalk: discriminative learning of network representation. In: Proceedings of IJCAI (2016)
20. Hearst, M.A., Dumais, S.T., Osman, E.: Support vector machines. IEEE Intell. Syst. Their Appl. 13(4), 18–28 (1998)
21. Zhu, J., Ahmed, A., Xing, E.P.: MedLDA: maximum margin supervised topic models. In: International Conference on Machine Learning, pp. 2237–2278 (2009)
22. Borg, I., Groenen, P.: Modern multidimensional scaling: theory and applications. Econ. Inst. Res. Pap. 40(3), 277–280 (2005)
23. Tenenbaum, J.B., De Silva, V., Langford, J.C.: A global geometric framework for nonlinear dimensionality reduction. Science 290(5500), 2319–2323 (2000)
24. Roweis, S.T., Saul, L.K.: Nonlinear dimensionality reduction by locally linear embedding. Science 290(5500), 2323–2326 (2000)
25. Blei, D.M., Ng, A.Y., Jordan, M.I.: Latent dirichlet allocation. J. Mach. Learn. Res. 3(5), 993–1022 (2003)
26. Hofmann, T.: Probabilistic latent semantic indexing. In: Proceedings of ACM, pp. 50–57 (2000)
27. Grover, A., Leskovec, J.: node2vec: scalable feature learning for networks. In: ACM SIGKDD International Conference on Knowledge Discovery and Data Mining, pp. 855–864 (2016)
28. Wang, X., Cui, P., Wang, J.: Community preserving network embedding. In: AAAI Conference on Artificial Intelligence (2017)
29. Huang, Z.P., Mamoulis, N.: Heterogeneous information network embedding for meta path based proximity. arXiv:1701.05291 (2017)
30. Yang, C., Sun, M.S., Li, Z.Y., Tu, C.C.: Fast network embedding enhancement via high order proximity approximation. In: Proceedings of IJCAI, pp. 3894–3900 (2017)
31. Perozzi, B., Kulkarni, V., Skiena, S.: Walklets: multiscale graph embeddings for interpretable network classification. arXiv:1605.02115 (2016)
32. Kipf, T.N., Welling, N.: Semi supervised classification with graph convolutional networks. arXiv:1609.02907 (2016)
33. Taskar, B., Guestrin, C., Koller, D.: Max-margin Markov networks. In: Proceedings of NIPS, pp. 25–32 (2003)
34. Pei, W.Z., Ge, T., Chang, B.B.: Max-margin tensor neural network for chinese word segmentation. In: Proceedings of ACL, pp. 293–303 (2014)
35. Zhou, D.Y., Huang, J.Y., Schölkopf, B.: Beyond pairwise classification and clustering using hypergraphs. In: proceedings of NIPS 2006, pp. 1601–1608 (2005)
36. Tu, K., Cui, P., Wang, X., Wang, F., Zhu, W.W.: Structural deep embedding for hypernetworks. In: AAAI Conference on Artificial Intelligence (2018)
37. Li, D., Xu, Z.M., Li, S.: Link prediction in social networks based on hypergraph. In: Proceedings of WWW, pp. 41–42 (2013)
38. Ma, J.X., Cui, P., Zhu, W.W.: DepthLGP: learning embeddings of out-of-sample nodes in dynamic networks. In: AAAI Conference on Artificial Intelligence (2018)
39. Lee, D., Seung, H.S.: Learning the parts of objects by non-negative matrix factorization. Nature 401(6755), 788–791 (1999)

40. Morin, F., Bengio, Y.: Hierarchical probabilistic neural network language model. In: Tenth International Workshop on Artificial Intelligence and Statistics, pp. 246–252 (2005)
41. Crammer, K., Singer, Y.: On the algorithmic implementation of multiclass kernel-based vector machines. J. Mach. Learn. Res. **2**(2), 265–292 (2002)
42. Fan, R., Chang, K., Hsieh, C., Wang, X., Lin, C.: LIBLINEAR: a library for large linear classification. J. Mach. Learn. Res. **9**, 1871–1874 (2008)

A Deconvolution Methods Based on Retinex Prior

Shuzhen Wang[1(✉)], Yang Guo[1], Hong Liu[1], and Yang Fang[2]

[1] School of Computer Science and Technology, Xidian University,
Xi'an 710071, China
shuzhenwang@xidian.edu.cn
[2] School of Electronics and Information, Northwestern Polytechnical University,
Xi'an 710129, China

Abstract. High-quality image deconvolution is required for many image processing applications. Our work concentrates on portraying a new image deconvolution method based on Retinex prior knowledge. We build a new image degradation mode, on the one hand, the data fitting term considers a spatial mask that keeps only pixels that do not depend on the unknown boundaries, on the other hand, the regularization term considers the image prior information and a new cross-channel prior (a prior knowledge of Retinex) information. Then the extended alternating directions method of multipliers (ADMM) is used to solve the image deconvolution convex optimization problem. In experiments, compared with several state-of-the-art deconvolution algorithms, including BM3D, HYP and YUV, our method can not only obtain a color aberration correction effectively to better cope with degraded images and get high-quality restoration results, but also achieve stronger applicability.

Keywords: Image deconvolution · Boundary handling
A prior knowledge of Retinex · ADMM

1 Introduction

Image deconvolution has long been a focus of research and timeless topic for scholars with a broader development perspective. Image deconvolution, a degraded image restoration technology, is used to process degraded image to achieve the high-quality image through the corresponding algorithms. However, there are lots of difficulties during the research process because of its own ill-condition and complexity, up to now studies on the image deconvolution have made a great development, but almost each method has some drawbacks. Therefore, in order to obtain better restoration image, improving the original algorithms and putting forward more effective new methods are especially important.

Considering the general problem: Let $u \in R^N$ be the unknown source image and $b \in R^N$ be the observed degraded image as vectors. Due to the linear nature of optical processes, the image transformation can be expressed as a matrix K operating on the image vector, An express traditional image model as

© Springer Nature Singapore Pte Ltd. 2018
Z. Xu et al. (Eds.): Big Data 2018, CCIS 945, pp. 322–335, 2018.
https://doi.org/10.1007/978-981-13-2922-7_22

$$b = u * K + \mathrm{n} \tag{1}$$

where u is the observed image contaminated by additive noise n, and the noise is usually given Gaussian distributed noise, K is the convolution kernel, or tipically referred to as a point spread function (PSF).

The goal of image deconvolution concentrates on finding the underlying source image u from the degraded image b, to solve both the image deconvolution and the PSF estimation problem, considering the following core minimization problem:

$$prox_\Gamma(b) = \min_u \frac{1}{2}\|b - ku\|_2^2 + \Gamma(u) \tag{2}$$

Where $\frac{1}{2}\|b - ku\|_2^2$ and $\Gamma(u)$ denote the data term regularizers respectively.

The focus of image deconvolution lies in the construction of image degradation model and the selection of regularizers. Baesd on those, we aim to build a new image degradation model considering the boundary handling and cross-channel image priors further.

In our work, We build a new image degradation mode, on the one hand, the data fitting term considers a spatial mask that keeps only pixels that do not depend on the unknown boundary, on the other hand, the regularization term considers the image prior information and a new cross-channel prior (a prior knowledge of Retinex) information, then the extended alternating directions method of multipliers (ADMM) is used to solve the image deconvolution convex optimization problem. Several benefits are concluded as follows:

- favourable color aberration correction efficiency;
- more accuracy, stronger applicability compared with the existing superior techniques.

2 Related Work

During last decades, many algorithms work on image deconvolution have been put forward to pursue more perfect performance. On the whole, most previous methods tackle the problem from a perspective assuming known image noise model and natural image gradients following certain distributions.

The focus of image deconvolution lies in the construction of image degradation model and the selection of regularizers.

Analysis of the image boundary is a challenging problem. Precious deconvolution methods usually assume one of the classical boundary conditions [1], such as periodic, zero and Neumann boundary conditions. However, It can not represent the natural boundary information according with the performance of a real imaging system in the process of image deconvolution. In recent years, some image deconvolution methods avoiding the boundary artifacts have been proposed [2–4], In the approach [3], the algorithm based on variable splitting and quadratic penalization estimates the clear image under non-smooth regularization. In the approach [4], the work explore the

Tikhonov regularization making use of the solution CG to obtain a simplified matrix inversion in FFT domain.

Several algorithms for the selection of regularizers have been proposed [5, 6]. In the approach [5], the CS algorithm with mean shift regularization helps to suppress the convolutive artifacts and reconstruct the solar fine structures, although it has much smaller reconstruction errors than those of other methods, it can not reconstruct fine solar structures. The approach in [6] choose total variation constraint and nonlocal operator and has surprisingly lower reconstruction errors than [5].

The use of deconvolution algorithms for correcting for aberrations date back at least to the original development of the Richardson-Lucy algorithm. In the approach [7], based on edge detection, the algorithm is feasible for removing color fringing in images. However, it leads a bad performance when the blurs are very large. In the approach [8, 9], the algorithms estimate the locally learned gradient statistics for deconvolution. Although the methods have not been attempted in the literature, they are feasible to transfer gradient information from one channel to another. In the approach [10], the method makes full use of the nature that one color channel is focused significantly better than the other channels, which are blurred beyond recognition. So they aim to share information between the deconvolution processed of the different channels, so that frequencies preserved in one channel can be used to help the reconstruction in another channel, and the cross-channel prior is based on the assumption that edges in the image appear in the same place in all channels.

In contrast to all these methods, we proposed a prior knowledge of Retinex. After in-depth study of Retinex theories, we carry out a large amount of experiments using RGB color images as input and find the phenomenon that images are more similar between channels and regard the nature as a prior knowledge named Retinex, which solves the problem of channel information loss. In essence, the prior knowledge of Retinex using SSR algorithm improves the constraints between clear image and blurred images, by this way, bulrry images can obtain better recovery effect with the help of clear image. On this basis, we build a new image degradation model. As for the image degradation model, on the one hand, the data fitting term considers the edge inpainting strategy, on the other hand, the regularization term considers the image prior information and a new cross-channel prior (a prior knowledge of Retinex) information, then the extended alternating directions method of multipliers (ADMM) is used to solve the image deconvolution convex optimization problem.

3 Review of ADMM Algorithm

The ADMM algorithm we study in this paper is summarized in Algorithm 1. Note that the algorithm considers the unconstrained optimization problem

$$\min_{z \in R^n} \sum_{j=1}^{J} g_j\left(H^{(j)}x\right) \tag{3}$$

Where $g_j : \mathbb{R}^{p_j} \to \mathbb{R}$ are proper, closed and convex functions, $H^{(j)} \in \mathbb{R}^{p_j \times n}$ are arbitrary matrices, Let $p = p_1 + \cdots + p_J$, define matrix $g : \mathbb{R}^p \to \bar{\mathbb{R}}$ as

$$g(e) = \sum_{j=1}^{J} g_j(e^{(j)}) \tag{4}$$

Where each $e^{(j)} \in \mathbb{R}^{p_j}$ is a p_j dimensional sub-vector of $e = \left[\left(e^{(1)}\right)^*, \cdots, \left(e^{(J)}\right)^* \right]^*$.

Let $\Upsilon = \text{diag}\left(\underbrace{\mu_1, \cdots, \mu_1}_{p_1 \ elements}, \cdots, \underbrace{\mu_j, \cdots, \mu_j}_{p_j \ elements}, \cdots, \underbrace{\mu_J, \cdots, \mu_J}_{p_J \ elements} \right)$, the core implementa-

tion process of above problem is presented in Algorithm 1.

Algotithm 1

---

1. Initialization:set $k = 0$, choose $\mu_1, \cdots, \ \mu_J > 0$,

e_0, d_0

2. repeat

3. $\zeta \leftarrow e_k + d_k$

4. $x_{k+1} \leftarrow \left(\sum_{j=1}^{J} \mu_j \left(H^{(j)}\right)^* H^{(j)} \right)^{-1} \sum_{j=1}^{J} \mu_j \left(H^{(j)}\right)^* \zeta^{(j)}$

5. for $j = 1$ to $j = J$ do

6. $e_{k+1}^{(j)} \leftarrow \arg\min_{v \in \mathbb{R}^{p_j}} g_j(v) + \dfrac{\mu_j}{2} \left\| v - \left(H^{(j)} x_{k+1} - d_k^{(j)}\right) \right\|_2^2$

7. $d_{k+1}^{(j)} \leftarrow d_k^{(j)} - \left(H^{(j)} x_{k+1} - e_{k+1}^{(j)}\right)$

8. end

9. $k \leftarrow k+1$

10. until stopping criterion is satisfied

---

4   Optimization for Image Deconvolution

4.1   Deconvolution with Unknown Boundaries

This paper extends image deconvolution to the more realistic scenario of unknown boundaries, where the image degradation process is modeled as the composition of a convolution(with arbitrary boundary conditions) with a spatial mask that keeps only pixels that do not depend on the unknown boundaries. Therefore, the paper proposes a model for simultaneous image inpainting and deconvolution. The image observation model is given by:

$$b = MKu + n \tag{5}$$

Where $M \in \{0, 1\}^{m \times n} (m < n)$ is a masking matrix that is to observe only the subset of the image domain in which the elements of Ku do not depend on the boundary pixels.

At this point, our goal is to find the underlying latent image u from the observed image b, so the minimization problem changes to

$$prox_\Gamma(b) = \arg\min_{u \in \mathbb{R}^n} \frac{1}{2} \|b - MKu\|_2^2 + \Gamma(u) \tag{6}$$

Where $\frac{1}{2}\|b - MKWz\|_2^2$ is the data term and $\Gamma(u)$ is the regularization term. And next we describe the choices of regularization operators.

4.2   Choices of Regularization Operators

The estimation and application of image priors are crucial for yielding high-quality image reconstructions. Here, we are particularly interested in image priors and a cross-channel prior named Retinex.

Image Priors

The first regularization term is the $\lambda_c \sum_{a=1}^{2} \|D_a u_c\|_1$, which enforces a heavy-tailed distribution for both gradients and curvature, the convolution matrices $D_{\{1,2\}}$ implement the first derivatives that are used to ensure the smoothness of image and eliminate ringing effect. We employ a l_1 norm in our method rather than a fractional norm to ensure that our problem is convex.

Retinex Cross-Channel Prior

To remove the chromatic aberration and yield improved results to help the other two blurry channels recover better we include the cross-channel prior named Retinex. After in-depth study of Retinex theories, we carry out a large amount of experiments using RGB color images as input and find the phenomenon that images are more similar between channels and regard the nature as a prior knowledge named Retinex, which solves the problem of channel information loss. In essence, the prior knowledge of Retinex using SSR algorithm improves the constraints between clear image and blurred images, by this way, bulrred images can obtain better recovery effect with the help of clear image.

We choose a pair of channels i, k and express our prior knowledge of Retinex as

$$\text{Retinex}(u_i) \approx \text{Retinex}(u_j) \tag{7}$$

Where Retinex is used to compute the different channel image u using the algorithm SSR, a clear form describing the thought is defined as

$$R_i u_i \approx R_j u_j \tag{8}$$

For its analyses we experimented with a large number of RGB color images. Table 1 shows the experimental reports of the evaluation SSIM value between different channel images processed by the algorithm SSR. It is clear that the similarity between image channels is enhanced.

Table 1. Similarity measurement between image channels

Chanel SSIM Images	RGB 3-channels of original image			RGB 3-channels of image processed by SSR		
	(R,G)	(R,B)	(G,B)	(R,G)	(R,B)	(G,B)
lena	0.6072	0.7822	0.8717	0.9429	0.8598	0.9374
tower	0.8629	0.8556	0.7873	0.9160	0.9665	0.9688
clock	0.7873	0.8369	0.9201	0.9688	0.9441	0.9517
wood	0.9369	0.8935	0.9653	0.9793	0.9824	0.9824
bench	0.8529	0.8145	0.9159	0.9426	0.8497	0.9208

4.3 Deconvolution Algorithm

Making full use of the data fidelity operator, image priors and Retinex cross-channel prior, we finally formulate the problem of jointly deconvolving all channels as the optimization problem

$$(u_{1\cdots3})_{opt} = \arg\min_{u_{1\cdots3}} \frac{1}{2} \sum_{c=1}^{3} \|b_c - M_c K_c u_c\|_2^2 +$$

$$\lambda_c \sum_{a=1}^{2} \|D_a u_c\|_1 + \sum_{l \neq c} \beta_{cl} \|R_c u_c - R_l u_l\|_2^2 \tag{9}$$

Considering the problem of selecting the clearest channel and any of channels from a blurred image, the optimization is rewritten as

$$(u_c)_{opt} = \arg\min_{u_{1\cdots 3}} \frac{1}{2}\|b_c - M_c K_c u_c\|_2^2 +$$
$$\lambda_c \sum_{a=1}^{2} \|D_a u_c\|_1 + \sum_{l \neq c} \beta_{cl}\|R_c u_c - R_l u_l\|_2^2 \tag{10}$$

Where $R_l u_l$ is the processing result by Retinex algorithm SSR and can be represented by r_l

$$(u_c)_{opt} = \arg\min_{u_{1\cdots 3}} \frac{1}{2}\|b_c - M_c K_c u_c\|_2^2 +$$
$$\lambda_c \sum_{a=1}^{2} \|D_a u_c\|_1 + \sum_{l \neq c} \beta_{cl}\|R_c u_c - r_l\|_2^2 \tag{11}$$

Problem (11) has the form (4), and

$$J = 4, n = 3$$
$$g_1 : R^m \rightarrow R, g_1(v) = \frac{1}{2}\|b - Mv\|_2^2$$
$$g_2 : R^n \rightarrow R, g_2(v) = \beta\|Rv - r_l\|_2^2$$
$$g_i : R^2 \rightarrow R, g_i(v) = \lambda\|v\|_1, \quad i = 3, \ldots, n+1 \tag{12}$$
$$H^{(1)} \in R^{n \times n}, H^{(1)} = K$$
$$H^{(2)} \in R^{n \times n}, H^{(2)} = I$$
$$H^{(i)} \in R^{2 \times n}, H^{(i)} = D_{i-2}, \quad i = 3, \ldots, n+1$$

Since $H^{(1)} = K, H^{(2)} = I, H^{(i)} = D_{i-2}$ are in the periodic BC case. For matrix

$$\Upsilon = \text{diag}\left(\underbrace{\mu_1, \cdots, \mu_1}_{p_1 \text{ elements}}, \cdots, \underbrace{\mu_j, \cdots, \mu_j}_{p_j \text{ elements}}, \cdots, \underbrace{\mu_J, \cdots, \mu_J}_{p_J \text{ elements}}\right) \text{ we adopt } \mu_2 = \ldots = \mu_{n+1} > 0,$$

and the matrix to be inverted in line four of Algorithm 1 can be written as

$$\sum_{j=1}^{J} \mu_j \left(H^{(j)}\right)^* H^{(j)} \Rightarrow \mu_1 K^* K + \mu_2 I^* I + \mu_3 \sum_{j=1}^{n-1} D_j^* D_j$$
$$= \mu_1 K^* K + \mu_2 I^* I + \mu_3 \left((D^h)^* D^h + (D^v)^* D^v\right) \tag{13}$$

Where $D^h, D^v \in R^{n \times n}$ are the matrices collecting the first and second rows respectively and

$$K = U^* \Delta^K U$$
$$I = U^* \Delta^I U$$
$$D^h = U^* \Delta^h U \quad (14)$$
$$D^v = U^* \Delta^v U$$

So (13) can be written as

$$U^* \left(\mu_1 |\Delta^K|^2 + \mu_2 |\Delta^I|^2 + \mu_3 |\Delta^h|^2 + \mu_3 |\Delta^v|^2 \right)^{-1} U \equiv W \quad (15)$$

Moreover

$$\sum_{j=1}^{J} \mu_j \left(H^{(j)} \right)^* H^{(j)} \Rightarrow \mu_1 K^* \zeta^{(1)} + \mu_2 I^* \zeta^{(2)} + \mu_3 \sum_{j=1}^{n-1} D_j^* \zeta^{(j+2)}$$
$$= \mu_1 K^* \zeta^{(1)} + \mu_2 I^* \zeta^{(2)} + \mu_3 \left((D^h)^* \delta^h + (D^v)^* \delta^v \right) \quad (16)$$

Where $\delta^h, \delta^v \in R^n$ contain the first and second component respectively of all the $\zeta^{(j+2)}, j = 1, \ldots, n$.

The proximity operators are given

$$prox_{g_1/\mu_1}(v) = \arg\min_e \|Me - b\|_2^2 + \mu_1 \|e - v\|_2^2$$
$$= (M^*M + \mu_1 I)^{-1} (M^*b + \mu_1 v)$$
$$prox_{g_2/\mu_2}(v) = \arg\min_e \|Re - r_l\|_2^2 + \mu_2 \|e - v\|_2^2$$
$$= (R^*R + \mu_2 I)^{-1} (R^*r_l + \mu_2 v) \quad (17)$$
$$prox_{g_i/\mu_i}(v) = v - soft\left(v, \frac{\lambda}{\mu_i} \right)$$
$$= v/\|v\|_2 \max\left\{ \|v\|_2 - \frac{\lambda}{\mu_i}, 0 \right\}, i = 3, \cdots, n+1$$

The updates of e

$$e_{k+1}^{(1)} \leftarrow (M^*M + \mu_1 I)^{-1} \left(M^*b + \mu_1 \left(K u_{k+1} - d_k^{(1)} \right) \right)$$
$$e_{k+1}^{(2)} \leftarrow (R^*R + \mu_2 I)^{-1} \left(R^*r_l + \mu_2 \left(I u_{k+1} - d_k^{(2)} \right) \right) \quad (18)$$
$$e_{k+1}^{(j)} \leftarrow v - soft\left(D_{j-2} u_{k+1} - d_k^{(j)}, \frac{\lambda}{\mu_3} \right), j = 3, \cdots, n+1$$

The updates of d

$$d_{k+1}^{(1)} \leftarrow d_k^{(1)} - \left(Ku_{k+1} - e_{k+1}^{(1)}\right)$$
$$d_{k+1}^{(2)} \leftarrow d_k^{(2)} - \left(Iu_{k+1} - e_{k+1}^{(2)}\right) \qquad (19)$$
$$d_{k+1}^{(j)} \leftarrow d_k^{(j)} - \left(D_{j-2}u_{k+1} - e_{k+1}^{(j)}\right), \quad j = 3, \ldots, n+1$$

As discussed above, the core implementation process of our method is presented in Algorithm 2.

Algorithm 2

---

1. Initialization: set $k = 0$, choose $\mu_1, \cdots, \mu_J > 0$,

    e_0, d_0, $j = 1, \ldots, n+1$

2. repeat

3. $\zeta \leftarrow e_k + d_k$

4. $\delta^h = \left[\left(\zeta^{(3)}\right)_1, \ldots, \left(\zeta^{(n+1)}\right)_1\right]^*$

5. $\delta^v = \left[\left(\zeta^{(3)}\right)_2, \ldots, \left(\zeta^{(n+1)}\right)_2\right]^*$

6. $u_{k+1} \leftarrow \mu_1 K^* \zeta^{(1)} + \mu_2 I^* \zeta^{(2)} + \mu_3 \left(\left(D^h\right)^* \delta^h + \left(D^v\right)^* \delta^v\right)$

7. when $j = 1$

8. $e_{k+1}^{(1)} \leftarrow \left(M^*M + \mu_1 I\right)^{-1} \left(M^*b + \mu_1 \left(Ku_{k+1} - d_k^{(1)}\right)\right)$

9. $d_{k+1}^{(1)} \leftarrow d_k^{(1)} - \left(Ku_{k+1} - e_{k+1}^{(1)}\right)$

10. when $j = 2$

11. $e_{k+1}^{(2)} \leftarrow \left(R^*R + \mu_2 I\right)^{-1} \left(R^*r_l + \mu_2 \left(Iu_{k+1} - d_k^{(2)}\right)\right)$

12. $d_{k+1}^{(2)} \leftarrow d_k^{(2)} - \left(Iu_{k+1} - e_{k+1}^{(2)}\right)$

13. for $j = 3$ to $j = n+1$ do

14. $e_{k+1}^{(j)} \leftarrow v - soft\left(D_{j-2}u_{k+1} - d_k^{(j)}, \dfrac{\lambda}{\mu_3}\right)$

15. $d_{k+1}^{(j)} \leftarrow d_k^{(j)} - \left(D_{j-2}u_{k+1} - e_{k+1}^{(j)}\right)$

16. end

17. $k \leftarrow k+1$

18. until stopping criterion is satisfied

---

5 Results and Discussions

To test the correctness and effectiveness of our algorithm, we choose the benchmark images showed in Fig. 1. With three different blurs described in Fig. 2, all of size 19x19, at three different Gaussian noise levels: 0.005, 0.008 and 0.010. Moreover, we apply the performance comparisons against the approach BM3D [11], HYP [12] and YUV [13], we choose the performance evaluation criterion described as following:

$$PSNR = 10\lg\left\{\max[u^2(i,j)]/MSE\right\} \tag{20}$$

(a) River   (b) Tree

(c) Airplane   (d) Pepper

Fig. 1. Benchmark images

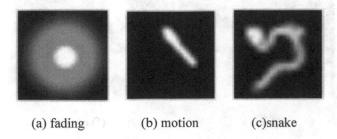

(a) fading   (b) motion   (c)snake

Fig. 2. Blurry images

Where MSE denotes the mean square error and satisfies $\mathrm{MSE} = \sum_{i=1}^{M} \sum_{j=1}^{N} [u(i,j) - u'(i,j)]^2 / MN$, u and u' represent the reference image and the image needing to be assessed respectively.

As illustrated above, the bigger PSNR value, the better the performance is.

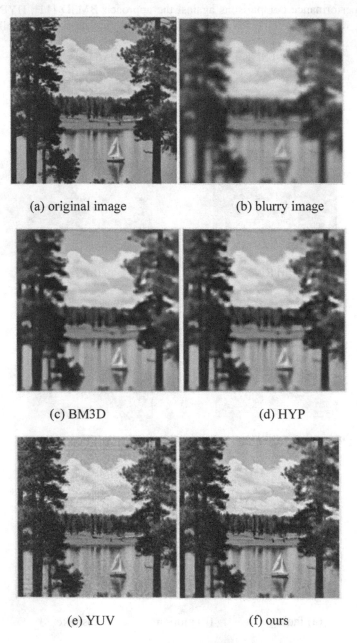

(a) original image   (b) blurry image

(c) BM3D   (d) HYP

(e) YUV   (f) ours

Fig. 3. Restoration effect comparison of Fig. 1(a)

5.1 Comparison of Visual Effect

We choose benchmark images River and Tree that are displayed in Fig. 1 and synthesize noise level 0.08 for each image to test the performance further. As illustrated in Figs. 3 and 4, there are interesting observations we would like to point out, we can get

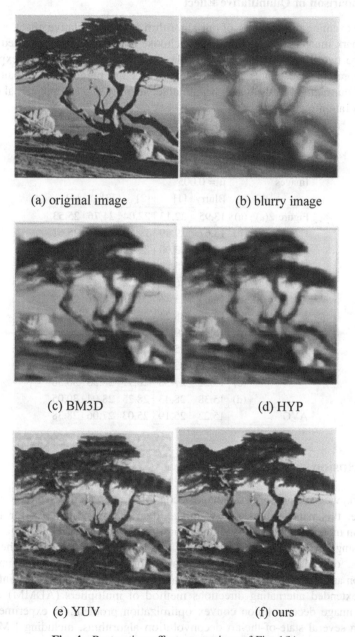

(a) original image   (b) blurry image

(c) BM3D   (d) HYP

(e) YUV   (f) ours

Fig. 4. Restoration effect comparison of Fig. 1(b)

a better mage reconstruction result visually compared with the approaches BM3D, HYP and YUV, The methods BM3D and HYP lead a bad performance, or even blur still exists. Although YUV has a slightly better performance, it appears obvious and large-scale ring effect because of the influence of noise.

5.2   Comparison of Quantitative Effect

In order to examine the behavior of our algorithm from quantitative effect, we capture all benchmark images illustrated in Fig. 1, choose all blur kernels described in Fig. 2 and add the two different Gaussian noise levels 0.005 to continue our experiments. Table 2 represents the results and our method get the biggest PSNR value, which means that our method leads to a higher suitable, stability and computational efficiency image restoration quality.

Table 2. Comparision of PSNR

Images		$\mu = 0.005$				
		Blurry	[1]	[2]	[3]	[4]
Figure 2(a)	(a)	13.95	22.13	22.04	24.76	25.53
	(b)	13.67	19.89	19.80	23.94	25.62
	(c)	18.77	23.00	22.92	28.46	29.66
	(d)	15.54	25.88	25.65	26.65	28.28
Figure 2(b)	(a)	13.78	26.43	26.22	27.47	27.76
	(b)	13.04	23.46	23.51	25.81	25.99
	(c)	18.06	28.21	27.93	30.68	32.14
	(d)	15.36	29.31	29.36	29.40	30.61
Figure 2(c)	(a)	13.69	25.44	25.10	26.17	27.69
	(b)	13.22	22.76	22.62	24.94	26.70
	(c)	18.21	27.33	26.91	28.38	30.58
	(d)	15.38	28.43	28.25	28.04	30.05
AVG		15.22	25.19	25.03	27.06	28.38

6   Conclusion

In our work, we have proposed an image deconvolution method based on Retinex prior knowledge, this paper builds a new image degradation model. As for the image degradation model, on the one hand, the data fitting term considers the edge inpainting strategy using a spatial mask that keeps only pixels that do not depend on the unknown boundaries, on the other hand, the regularization term considers the image prior information and a new cross-channel prior (a prior knowledge of Retinex) information, then the extended alternating directions method of multipliers (ADMM) is used to solve the image deconvolution convex optimization problem. In experiments, compared with several state-of-the-art deconvolution algorithms, including BM3D, HYP and YUV, this method can not only obtain a color aberration correction effectively to better cope with degraded images and get high-quality restoration results, but also achieve stronger applicability.

Acknowledgment. This work was supported by the National Natural Science Foundation of China under Grant 61771369 and 61540028; the Fundamental Research Funds for the Central Universities under Grant No. JB180310and NSIY221418.

References

1. Ng, M.K.: Iterative Methods for Toeplitz Systems, p. 350. Oxford Science Publications, Oxford (2004)
2. Chan, T.F., Yip, A.M., Park, F.E.: Simultaneous total variation image inpainting and blind deconvolution. Int. J. Imaging Syst. Technol. **15**(1), 92–102 (2010)
3. Matakos, A., Ramani, S., Fessler, J.A.: Image restoration using non-circulant shift-invariant system models. In: IEEE International Conference on Image Processing, pp. 3061–3064. IEEE (2013)
4. Sorel, M.: Removing boundary artifacts for real-time iterated shrinkage deconvolution. IEEE Trans. Image Process. **21**(4), 2329 (2012)
5. Wang, S., Wang, Y., Xu, L., et al.: Heliograph imaging based on compressed sensing with mean shift regularization. Sens. Mater. **26**(5), 339–345 (2014)
6. Wang, S.Z., Xu, L.Y., Wang, Y., et al.: Heliograph imaging based on total variation constraint and nonlocal operator. J. Spectrosc. **2014**(4), 1–7 (2014)
7. Kang, S.B.: Automatic removal of chromatic aberration from a single image. In: 2007 IEEE Conference on Computer Vision and Pattern Recognition, CVPR 2007, pp. 1–8. IEEE (2007)
8. Cho, T.S., Joshi, N., Zitnick, C.L., et al.: A content-aware image prior, vol. 23, no. 3, pp. 169–176 (2010)
9. Cho, T.S., Zitnick, C.L., Joshi, N., et al.: Image restoration by matching gradient distributions. IEEE Trans. Pattern Anal. Mach. Intell. **34**(4), 683 (2012)
10. Heide, F., Rouf, M., Hullin, M.B., et al.: High-quality computational imaging through simple lenses. ACM Trans. Graph. **32**(5), 149 (2013)
11. Danielyan, A., Katkovnik, V., Egiazarian, K.: BM3D frames and variational image deblurring. IEEE Trans. Image Process. **21**(4), 1715 (2012)
12. Krishnan, D., Fergus, R.: Fast image deconvolution using hyper-Laplacian priors. In: International Conference on Neural Information Processing Systems, pp. 1033–1041. Curran Associates Inc. (2009)
13. Santhi, V., Thangavelu, A.: DWT – SVD combined full band watermarking technique for color images in YUV color space. Int. J. Comput. Theory Eng. **1**(4), 424 (2009)

Identification of High Priority Bug Reports
via Integration Method

Guofeng Gao, Hui Li[(⊠)], Rong Chen, Xin Ge, and Shikai Guo

Information Science and Technology College, Dalian Maritime University,
Dalian 116026, China
li_hui@dlmu.edu.cn

Abstract. Many software projects use bug tracking systems to collect and allocate the bug reports, but the priority assignment tasks become difficult to be completed because of the increasing bug reports. In order to assist developers to reduce the pressure on assigning the priority for each bug report, we propose an integration method to predict priority levels based on machine learning. Our approach considers the textual description in bug reports as features and feeds these features to three different classifiers. We utilize these classifiers to predict the bug reports with unknown type and obtain three different results. Simultaneously, we set weights to balance the abilities of identifying different categories based on the characteristics of different projects for each classifier. Finally, we utilize the weights to adjust prediction results and produce a unique priority for assigning to each bug reports. We perform experiments on datasets from 4 products in Mozilla and the experimental results show that our approach has a better performance in terms of identifying the priority of bug reports than previous general methods and ensemble methods.

Keywords: Classification prediction · Bug report · Integration method

1 Introduction

With the increasing complexity of software systems, rigorous and sufficient testing becomes difficult to be completed, so the software products often contain various numbers of software bugs. Aiming to address the bugs and improve the quality of the released software systems, the discovered bugs should be recorded and sent to the developers, and hence some bug tracking systems, such as Bugzilla, are obtained to provide the testers an access to submit the bug reports [1]. In the testing process of both open source software projects and commercial software projects (such as Windows operating system), bug reporting can obviously improve the work efficiency and the production quality, but the massive bug reports cannot be handled for the software developers. It is reported that hundreds of bug reports should be collected, triaged and assigned for Mozilla project everyday [2], thus the bug reports should be prioritized to help the developers distinguish the severity of bugs. For instance, 5 priority levels (P1, P2, P3, P4 and P5) are obtained in Bugzilla. Specifically, P1 is the highest priority and the corresponding bugs should be fixed urgently, while P5 is the lowest priority and the corresponding bugs could allow to be fixed after a period of times [3].

© Springer Nature Singapore Pte Ltd. 2018
Z. Xu et al. (Eds.): Big Data 2018, CCIS 945, pp. 336–349, 2018.
https://doi.org/10.1007/978-981-13-2922-7_23

Manually prioritizing bug reports is a time-consuming process [4], in which bug reports information provided by the submitters should be read and compared with the existing bug reports first, and then the reasonable priority can be assigned. In order to reduce the workload of bug report prioritizing, we propose an approach to assist developers to assign priority to bug reports. As we know, textual short and long descriptions are utilized to demonstrate the bugs encountered in the software system. Therefore, the textual features from textual description in bug reports are extracted as training data sets and testing data sets. In addition, due to the various performance that different classification algorithms predict imbalance datasets, we apply three classifiers for predicting which are J48 [5], K-nearest neighbors (KNN) and Random Tree (RT) [6].

The prediction results demonstrate that the obtained classifiers represent different prediction capabilities of discovering different categories. Specifically, a classifier may represent stronger ability in category detection for binary classifications, but represent weaker ability in other category detections. Thus, the method we propose achieve integrating the advantages of the all the classifiers adopted in our method and balancing the ability of each classifier.

Specific for, bigger weight is assigned to the classifier with weaker ability to detect one category, while smaller weight is assigned to the classifier with stronger ability, in order to achieve better classification effect. First, the training data is feed to objective classifier and the unlabeled bug reports among them are predicted hypothetically in weight training phase. The classification probability for each bug report is obtained and the weights are utilized to adjust them respectively. Aiming to achieve higher accuracy, the detection process for appropriate weights is converted to Linear Programming Problem, and the constraint solver, named CPLEX, is used to address that problem [7]. In weight adjustment phase, the probability for each classification is adjusted and the optimized result for each classifier is obtained. Finally, the final result produced by integration method of three classifiers can be got through the principle of the minimum.

The validity of our method is investigated through the experiments on bug reports of 4 products from Mozilla. The experimental results demonstrate that our method achieve better prediction performance than general classification algorithms. Simultaneously, it is proved that our integration method outperforms other classic ensemble methods in most cases and the improvement of F-Measure is 14.89%–44.52%.

The contributions of this work are as following:

(1) We propose an integration method to assist developers to identify the priority of the bug reports. Considering the characteristics of the various experimental data, we balance the ability of detecting different categories for each classifier and take advantages of different classification algorithms. The proposed method treats the weight optimization as Linear Programming Problem, and use CPLEX to get the appropriate weights for higher accuracy of the results.

(2) We evaluate our method through datasets from 4 products of Mozilla. And the results of experiments show our method is feasible and outperforms general classification algorithms and classic ensemble methods.

The rest of this paper is as follows. In Sect. 2, we describe the details of our approach. Section 3 presents the experimental datasets, experimental settings, evaluation metrics and 2 research questions. We illustrate the results of each research question and analysis based on the results of experiments in Sect. 4. Section 5 shows related work. We conclude this paper in Sect. 6 and threats to validity is illustrated in Sect. 7.

2 The Approach

In this section, the proposed method named predicting priority of bug reports by utilizing integration method is described. First, the pre-processing phase before classifier integration is demonstrated. Then, the integration method is divided into three modules. The details of the proposed method are described in the following two subsections.

2.1 Pre-processing Phase

Before utilizing classifiers to predict bug reports, the text features characterizing each bug report should be collected. Firstly, the text descriptions are extracted from the descriptions or summaries of bug reports, because the useful information of the corresponding bug can be provided from the two fields. In the following, word segmentation is used to segment the descriptions into words. For the purpose of noisy data removal, the little meaning words, such as stop words, numbers and punctuation marks, are removed in this step. Finally, Iterated Lovins Stemmer [8] is applied to convert the words with different forms but similar meaning into same root words, since it is noticed that the developers usually utilize semantically related words that consist same root words [3]. After the steps mentioned above, the words contained in each bug report could obviously less than the initial step. The stemmed words appearing in bug reports are treated as textual features, and can be transformed to textual feature vector. Thus the bug report BR can be formed by the following,

$$BR = (t_1, t_2, t_3, \ldots, t_n), \tag{1}$$

where t_i denotes whether the i_{th} word is existing in BR. If the i_{th} word exists, $t_i = 1$; otherwise, $t_i = 0$. Additionally, the number n denotes the number of textual features.

It is known that the prediction ability of a classification algorithm for different datasets is different, while the prediction ability of different classification algorithms for the same dataset is also different. Therefore, three classification algorithms, including J48, KNN and RT, are integrated to improve the prediction effect in this work.

2.2 Integration Phase

Features vectors extracted form training data are fed to the different classifiers and we utilize these classifiers to predict datasets with unknown label. Each classifier gives us probabilities of two categories for each bug report and we will get three groups of prediction probability. Thus, we propose an integration method to take advantages of each classifier and produce an ultimate prediction. The overview of the proposed is illustrated in Fig. 1. We introduce the integration phase through following 3 modules.

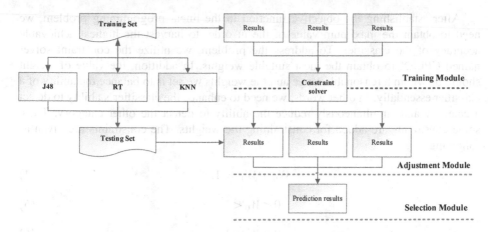

Fig. 1. Framework of the proposed approach.

Training Module. As we know, the ability of different classification algorithms detect minority class is different due to different projects. Thus, we want to infer weights that can adjust the ability of a classifier to detect different categories in order to improve the prediction accuracy of each classifier.

From a probabilistic point of view, the classifier classifies correctly as the probability of real category which a bug belongs to is greater than another category. The objective function we establish is to determine the weights $\{W_0, W_1\}$ of two categories so as to correspond prediction results of bugs to probability distribution above as much as possible. Firstly, we extract the feature of the training data and fed these features to the classifier. Then we convert the category of BR in training data to corresponding number C_i. Specifically, if a bug belongs to minority class, the number of this category $C_i = -1$; otherwise, if a bug belongs to majority class, the number of this category $C_i = 1$. Then, we assume the bugs in training data are unlabeled and utilize the classifier to predict them. Given $\{p_i^0, p_i^1\}$ is defined by the probability of two categories which produced by classifier predicting the i_{th} bug, where p_i^0 denotes the probability of majority class and p_i^1 denotes the probability of minority class. Finally, we establish the sub-objective function as in (2).

$$SubOBJ_i = \frac{(W_0 * p_i^0 - W_1 * p_i^1) * C_i}{\left| W_0 * p_i^0 - W_1 * p_i^1 \right|} \tag{2}$$

Equation (2) shows that the sub-objective function only contains two results. If the prediction result of the i_{th} bug after adjusting is true, the value of $SubOBJ_i$ is 1; otherwise, the value of $SubOBJ_i$ is -1. Thus, we demonstrate the objective function in (3).

$$OBJ = \sum_{i=1}^{n} SubOBJ_i \tag{3}$$

After establishing the objective function in the linear programming problem, we need to obtain the maximum value of the in order to harvest the highest achievable accuracy of the classifier. To address the problem, we utilize the constraint solver named CPLEX to obtain the most suitable weights. In addition, the value of weight should be within a reasonable range and the weights we set is to balance the ability of a classifier essentially. In other words, we need to enhance the classifier's ability to detect a category and simultaneously reduce the ability to detect the other category. Thus, some constraints are added for constraining the weights. The constraints are given as following,

$$W_0 + W_1 = 1, \tag{4}$$

$$0 < W_0 < 1, \tag{5}$$

$$0 < W_1 < 1. \tag{6}$$

Adjustment Module. After training weight, we obtain the most suitable weights W_0 and W_1 for each classifier. Then, the weights are used to adjust the prediction results that generated by corresponding classifiers. We utilize corresponding W_0 to adjust the probability p_i^0 of majority class and utilize corresponding W_1 to adjust the probability p_i^1 of minority class for the i_{th} bug. Table 1 presents the details of adjustment.

Table 1. The rule of utilizing weights to adjust prediction result.

	Majority class	Minority class
Weight	W_0	W_1
Original probability	p_i^0	p_i^1
Adjusted probability	$W_0 * p_i^0$	$W_1 * p_i^1$

After utilizing the corresponding weights to adjust predicted probability generated by three classifiers and we obtain three different prediction results. Then, we need to integrate these optimized results for higher accuracy. We will illustrate the method of integrating in the next subsections.

Selection Module. After the adjustment, we obtain three sets of predicted results and each set contains probabilities of two categories. Thus, there are three different probabilities about majority class and three different probabilities about minority class about a bug. We set p_0^{min} to present minimum value among three different probabilities about majority class and set p_1^{min} to present minimum value of minority class probabilities. Thus, each bug contains p_0^{min} and p_1^{min} about majority class and minority class. At last, our method utilizes the category represented by the maximum value of p_0^{min} and p_1^{min} to determine the final type of this bug.

3   Experiments

In this section, we describe the datasets which our experiments perform on at first. Next, we present our experimental setting to evaluate the performance of our method. And we demonstrate the evaluation measures finally.

3.1   Datasets

To validate the proposed method, we utilize bug reports form the open-source project named Mozilla which adopts Bugzilla as its bug tracking system [9]. Mozilla is an open source project hosting several software products, such as Firefox, Thunderbird. In our experiments, we collect the bug reports submitted in the period of 1998-2007 and we only focus on four products of Mozilla which are Mozilla Application Suite (MAS), Bugzilla, Tech Evangelism (TE) and Firefox.

Many bug report submitters for Mozilla products are from different background and abilities so that a large number of invalid bug reports and bug reports with less meaning exist in all the bug reports. We note that bug reports whose priority is "P3" account for a large proportion of the total bug reports and these bugs reports may confuse the classifiers due to their messy features. Thus, we remove the invalid bug reports and the bug reports belonging to the priority of "P3". After step of removing, the datasets we perform on contain four class of priority which are "P1" "P2" "P4" and "P5". In our experiment, we consider bug reports with priority of "P1" and "P2" as priority bug reports and consider the other bug reports as non-priority bug reports.

Table 2. Details of products.

Product	Total	Priority	Percentage (P)	Non-priority	Percentage (NP)
MAS	4295	3482	81.07%	813	18.93%
Bugzilla	1505	1140	75.75%	365	24.25%
TE	1465	928	63.34%	537	36.66%
Firefox	1335	1088	81.50%	247	18.50%

Table 2 presents the details of datasets of four products after pre-processing, contains the total number of bug reports (Total), the number of priority bug reports (Priority), the percentage of the priority bug reports as a percentage of all the bugs (Percentage (P)), the number of non-priority bug reports (N-Priority) and the percentage of the non-priority bug reports as a percentage of all the bugs (Percentage (NP)).

In this experiment, K-Fold Cross-validation [10] is applied to divide the collection of bug reports into disjoint training and testing sets, then train the classifier by using the training data set. Finally, the classifier is validated by the testing data set and the corresponding accuracy results can be got. The above steps are repeated for K times, and we set K = 5 in this experiment.

3.2 Experimental Setting

Firstly, we focus on some general classification algorithms which are Naïve Bayes (NB), J48, KNN and RT. We explore these algorithms and compare the performances with our method. We adopt the same feature extraction method as our method to ensure the same feature sets are fed to the classifiers built by the classification algorithms mentioned above. Furthermore, we investigate whether the efficiency of integration method proposed in this paper outperforms other classic ensemble method, such as Ada boost, Bagging and Vote. We use Ada boost and Bagging in Weka to integrate the general classification algorithm which performs best respectively. In addition, 3 classification algorithms which are the same as classification algorithms are used to produce the final prediction results by Vote respectively. All experiments are run on a computer with Intel i7-7700HQ 2.80 GHz, 16 GB RAM, and Windows10 operating system.

3.3 Evaluation Metrics

In order to measure the effectiveness of priority prediction methods, we utilize three metrics: precision, recall and F-Measure [11]. The three metrics are widely used for evaluating the classification methods. We use the Table 3 to explain the definitions of precision, recall and F-Measure for classification.

Table 3 shows four results of classification. True positive (TP) denotes a bug is predicted correctly as a non-priority bug. False positive (FP) denotes a bug is predicted mistakenly as a non-priority bug. The definitions of true negative (TN) and false negative (FN) are similar with true positive and false positive. TP, FP, TN, FN in mathematical formulas below express the total number of each result of classification respectively.

Table 3. The definitions of precision, recall and F-Measure for classification.

	Predicted non-priority	Predicted priority
Observed non-priority	TP	FN
Observed priority	FP	TN

Precision signifies the proportion of bugs observed as non-priority bug that are predicted correctly. Precision can be expressed mathematically below:

$$Precision = \frac{TP}{TP + FP}.$$

Recall signifies the proportion of bugs predicted as nonpriority bug to all bugs observed as non-priority bug. Recall can be expressed mathematically below:

$$Recall = \frac{TP}{TP + FN}.$$

F-Measure is the harmonic mean of precision and recall. It is used to balance the discrepancy between precision and recall. F-Measure can be expressed mathematically below:

$$F - Measure = \frac{2 * Precision * Recall}{Precision + Recall}.$$

3.4 Research Questions

RQ1: How accurate is our method as compared with other classification algorithms?

RQ2: Can the integration method we propose outperform the classic ensemble methods named Ada boost, Bagging and Vote?

4 Results and Analysis

In this section, we analyze the two experimental results to demonstrate the answers of two research questions.

4.1 Addressing RQ1

RQ1: How accurate is this method as compared with other classification algorithms? We illustrate the results of different classification algorithms and our method in Tables 4, 5 and 6. Tables 4, 5 and 6 present the precision, recall and F-Measure for these classification methods respectively. We demonstrate each classification algorithms on the horizontal axis, and demonstrate four projects on the vertical axis.

As shown in the Tables 4, 5 and 6, we note that different general classification algorithms have various performance for predicting the projects. RT achieves 0.327, 0.274, 0.484 and 0.203 of F-Measure and achieves 0.356, 0.284, 0.519 and 0.224 of recall, outperforms other general classification algorithms in almost all the projects. However, RT does not remain the best performance in terms of recall and F-Measure at all. In addition, despite the worst performance in term of F-Measure, J48 have the best precision for predicting the products named Bugzilla and MAS. As for KNN and NB, we also note these classification algorithms achieve better performance than other general algorithms in certain aspects. Thus, simply utilizing a classification algorithm to predict many products is unsuitable and may ignore the unique advantages of some classification algorithms.

The result for our method is shown on the last vertical axis. Obviously, our method achieves the best performance in terms of recall and F-Measure for predicting all the projects. Our method achieves 0.427, 0.370, 0.629 and 0.270 of recall for predicting four products, improves the performance of RT 19.82%, 30.35%, 21.19% and 20.99% respectively. Simultaneously, in term of F-Measure, our method achieves 0.341, 0.306, 0.510 and 0.217, improves the performance of RT 4.33%, 11.85%, 5.48% and 6.94%.

The improvement of recall and F-Measure proves that our method can obtain higher accuracy and more stable result of predicting priority.

Table 4. The *precision* for general classification algorithms and our method.

Products	KNN	NB	J48	RT	Our method
Bugzilla	0.279	0.379	**0.497**	0.314	0.296
MAS	0.297	0.371	**0.460**	0.274	0.272
TE	0.444	**0.469**	0.455	0.458	0.434
Firefox	**0.252**	0.176	0.080	0.212	0.214

Table 5. The *recall* for general classification algorithms and our method.

Products	KNN	NB	J48	RT	Our method
Bugzilla	0.195	0.293	0.073	**0.356**	0.427
MAS	0.171	0.120	0.104	**0.284**	0.370
TE	**0.564**	0.411	0.287	0.519	0.629
Firefox	0.074	0.157	0.008	**0.224**	0.270

Table 6. The *F-Measure* for general classification algorithms and our method.

Products	KNN	NB	J48	RT	Our method
Bugzilla	0.219	0.316	0.125	**0.327**	0.341
MAS	0.213	0.173	0.166	**0.274**	0.306
TE	**0.493**	0.436	0.347	0.484	0.510
Firefox	0.098	0.148	0.014	**0.203**	0.217

We also notice that a gap exists between the performances that a classification algorithm predicts different projects. As shown in the Table 6, J48 achieves 0.125, 0.166, 0.347 and 0.014 of F-Measure for predicting various projects. The difference between the best performance and the worst performance is extremely huge, and this phenomenon occurs in all classification methods. We investigate the reason of unstable prediction result and find that the category distribution of different projects is different. The degree of imbalance in the dataset determines the classification effect of the classifiers.

4.2 Addressing RQ2

RQ2: Can the integration method we propose outperform the classic ensemble methods named Ada boost, Bagging and Vote?

We compare our method with three classic ensemble methods named Ada boost, Bagging and Vote. Tables 7, 8 and 9 demonstrate the precision, recall and F-Measure for each integration method.

Table 7. The *precision* for classic ensemble method and our method.

Products	Our method	Ada boost	Bagging	Vote
Bugzilla	0.296	0.332	**0.376**	0.356
MAS	0.272	0.356	**0.486**	0.412
TE	0.434	0.476	**0.546**	0.491
Firefox	0.214	0.263	0.238	**0.319**

Table 8. The *recall* for classic ensemble method and our method.

Products	Our method	Ada boost	Bagging	Vote
Bugzilla	**0.427**	0.184	0.112	0.088
MAS	**0.370**	0.166	0.108	0.140
TE	**0.629**	0.415	0.274	0.372
Firefox	0.270	**0.279**	0.040	0.089

Table 9. The *F-Measure* for classic ensemble method and our method.

Products	Our method	Ada boost	Bagging	Vote
Bugzilla	**0.341**	0.236	0.173	0.141
MAS	**0.306**	0.227	0.177	0.209
TE	**0.510**	0.444	0.365	0.424
Firefox	0.217	**0.271**	0.069	0.139

Despite the worse performance in precision, our method achieves the best performance in almost every projects among all the ensemble method. In terms of recall, our method achieves 0.427, 0.370 and 0.629 of recall for Bugzilla, MAS and TE, improves Ada boost 131.85%, 123.14% and 51.51% respectively. As for the product named Firefox, our method achieves 0.270 of recall and it is only 3.05% lower than the highest value in other methods. In addition, our method achieves 0.341, 0.306 and 0.510 of F-Measure for Bugzilla, MAS and TE, improves the Ada boost 44.52%, 34.95% and 14.89%. Thus, the experimental results indicate the integration method we propose outperforms other classic ensemble method in most cases.

5 Related Work

Menzies and Marcus are the first to predict the severity of bug reports [12]. They analyze the severity labels of various bugs reported in NASA. They propose a technique that analyzes the textual contents of bug reports and outputs fine-grained severity levels – one of the 5 severity labels used in NASA. Their approach extracts word tokens from the description of the bug reports. These word tokens are then preprocessed by removing stop words and performing stemming. Important word tokens are then selected based on their information gain. Top-k tokens are then used as features to characterize each bug report. The set of feature vectors from the training data is then fed into a classification algorithm named RIPPER [13]. RIPPER would learn a set of rules which are then used to classify future bug reports with unknown severity labels.

Lamkanfi *et al.* extend the work by Menzies and Marcus to predict severity levels of reports in open source bug repositories [14]. Their technique predicts if a bug report is severe or not. Bugzilla has six severity labels including blocker, critical, major, normal, minor, and trivial. They drop bug reports belonging to the category normal. The remaining five categories are grouped into two groups-severe and non-severe. Severe group includes blocker, critical and major. Non-severe group includes minor and trivial. Thus they focus on the prediction of coarse-grained severity labels. Extending their prior work, Lamkanfi *et al.* also try out various classification algorithms and investigate their effectiveness in predicting the severity of bug reports [15]. They tried a number of classifiers including Naïve Bayes, Naive Bayes Multinomial, 1-Nearest Neighbor, and SVM. They find that Naive Bayes Multinomial perform the best among the four algorithms on a dataset consisting of 29,204 bug reports.

Recently, Tian *et al.* also predict the severity of bug reports by utilizing a nearest neighbor approach to predict fine grained bug report labels [16]. Different from the work by Menzies and Marcus which analyzes a collection of bug reports in NASA, Tian et al. apply the solution on a larger collection of bug reports consisting of more than 65,000 Bugzilla reports.

Our work is orthogonal to the above studies. Severity labels are reported by users, while priority levels are assigned by developers. Tian *et al.* proposed a new machine learning framework and classification standard. This method utilized a variety of feature information as training sample, and finally selected the appropriate classification algorithm to predict the priority of the bug report [3].Tian et al. also proposed an automated approach that would recommend a priority level based on information available in bug reports [17]. The multiple factors they considered extracted as features which are then used to train a discriminative model via a new classification algorithm that handles ordinal class labels and imbalanced data. Sharma M et al. have evaluated the performance of different machine learning techniques namely Support Vector Machine, Naive Bayes, K-Nearest Neighbors and Neural Network in predicting the priority of the newly coming reports on the basis of different performance measures [18]. They performed cross project validation for 76 cases of five data sets of open office and eclipse projects. Severity labels correspond to the impact of the bug on the software system as perceived by users while priority levels correspond to the importance "a developer places on fixing the bug" in the view of other bug reports that are received [19]. Khomh *et al.* automatically assign priorities to Firefox crash reports in Mozilla Socorro server based on the frequency and entropy of the crashes. A crash report is automatically submitted to the Socorro server when Firefox fails and it contains a stack trace and information about the environment to help developers debug the crash. In our study, we investigate bug reports that are manually submitted by users. Different from a crash report, a bug report contains natural language descriptions of a bug and might not contain any stack trace or environment information. Thus, different from Khomh *et al.*'s approach, we employ a text mining based solution to assign priorities to bug reports.

Antoniol et al. also used text mining techniques on the descriptions of reported bugs to predict whether a report is either a real bug or a request for an enhancement [20]. They used techniques like decision trees, logistic regression and also a Naive Bayesian classifier for this purpose. The performance of this approach on three cases (Mozilla, Eclipse and JBoss) indicated that reports can be predicted to be a bug or an enhancement with between 77% and 82% correct decisions.

There are other lines of work that also analyze bug reports; these include the series of work on duplicate bug report detection [21–24], bug localization [25], bug categorization [26–28], bug fix time prediction [29, 30], and bug fixer recommendation [31, 32]. Our work is also orthogonal to these studies.

6 Threats to Validity

Our method and experiments still contain some threats. We illustrate these threats as following:

(1) *Conclusion Validity*: In our method, we divide the bug reports into two type and ignore the bug reports whose priority is "P3". Even though the number of these bug reports with medium priority is large and these bug reports may confuse the classifiers due to their messy features, they still contain some information that could assist developers to maintain the software system.

(2) *Internal Validity:* In our method, we adopt three classification algorithms to build classifiers. As we known, the performance that classification algorithms predict different projects is various. Thus, the performance of our method is affected by data from different projects.

(3) *Construct Validity:* Three metrics, precision, recall and F-Measure, are used to evaluate the proposed method in the experiments. Although utilizing accuracy to evaluate the prediction performance for imbalanced datasets is lack of practical significance, the metric is still an important measurement to assess the classification ability. Thus ignoring accuracy may not comprehensively evaluate the prediction performance in the experiments.

(4) *External Validity:* In our experiment, four Mozilla products are used as datasets and 5-Fold Cross validation is used in all the experiments. However, the limited datasets in our experiments may not show the performance of our method as a whole.

7 Conclusion

In this paper, we propose an integration approach to assist developers in identifying the priority of bug reports. In our method, we consider textual feature of bug reports and adopt three classifiers which are J48, KNN and RT to build classifiers and predict the testing data respectively. We integrate the results from different classifiers based on the various category distributions of data sets and different classification capabilities of classifiers. At first, our method balances the ability of detecting different categories based on the characteristics of projects and classifiers through setting weights. Then, our method adjusts the probabilities of prediction from different classifiers aiming to improve accuracy of each classifier. Ultimately, we assign a priority to the bug following the principle of the minimum value.

We have utilized several general classification algorithms as baselines to compare with our proposed approach. The experimental results which perform on 4 products from Mozilla prove that our method outperforms the baselines obviously. In addition, we have compared our method with classic ensemble methods and the results indicate our method achieves better performance in most cases.

In the future works, we plan to perform experiments on more products from various open source projects. And we also plan to improve our method in order to more stable and efficient solutions to imbalance datasets.

Acknowledgement. This work is supported by the National Natural Science Foundation of China (No. 61602077, No. 61672122), the Natural Science Foundation of Liaoning Province of China (No. 20170540097), and the Fundamental Research Funds for the Central Universities (No. 3132016348).

References

1. Xia, X., Lo, D., Wang, X., Zhou, B.: Accurate developer recommendation for bug resolution. In: Conference: Reverse Engineering, pp. 72–81. IEEE (2013)
2. Anvik, J., Hiew, L., Murphy, G.C.: Coping with an open bug repository. In: Proceedings of the 2005 OOPSLA workshop on Eclipse technology eXchange, pp. 35–39. ACM, New York (2005)
3. Tian, Y., Lo, D., Sun, C.: DRONE: predicting priority of reported bugs by multi-factor analysis. In: IEEE International Conference on Software Maintenance, pp. 200–209. IEEE (2013)
4. Wang, Q., et al.: Local-based active classification of test report to assist crowdsourced testing. In: IEEE/ACM International Conference on Automated Software Engineering, pp. 190–201. ACM (2016)
5. Neeraj, B., Girja, S., Ritu, D.B., Manisha, M.: Decision tree analysis on j48 algorithm for data mining. J. Adv. Res. Comput. **3**(6), 1114–1119 (2013)
6. Han, J., Kamber, M.: Data Mining: Concepts and Techniques. Morgan Kaufmann, Burlington (2006)
7. IBM ILOG CPLEX Optimizer. https://www.ibm.com/analytics/data-science/prescriptive-analytics/cplex-optimizer/. Accessed 26 Apr 2018
8. Lovins, J.B.: Development of a stemming algorithm. Mech. Transl. Comput. Linguist. **11**, 22–31 (1968)
9. http://bugzilla.mozilla.org. Accessed 26 Mar 2018
10. Hu, J., Zhang, G.: K-fold cross-validation based selected ensemble classification algorithm. Bull. Sci. Technol. **29**, 115–117 (2013)
11. Weng, C.G., Poon, J.: A new evaluation measure for imbalanced datasets. In: Australasian Data Mining Conference, pp. 27–32. Australian Computer Society, Inc. (2008)
12. Menzies, T., Marcus, A.: Automated severity assessment of software defect reports. In: IEEE International Conference on Software Maintenance, pp. 346–355 (2015)
13. Cohen, W.W.: Fast effective rule induction. In: Twelfth International Conference on Machine Learning, pp. 115–123. Morgan Kaufmann Publishers, Inc. (1995)

14. Lamkanfi, A., Demeyer, S., Giger, E., et al.: Predicting the severity of a reported bug. In: Mining Software Repositories, pp. 1–10. IEEE (2010)
15. Lamkanfi, A., Demeyer, S., Soetens, Q.D., et al.: Comparing mining algorithms for predicting the severity of a reported bug. In: European Conference on Software Maintenance and Reengineering, pp. 249–258. IEEE Computer Society (2011)
16. Tian, Y., Lo, D., Sun, C.: Information retrieval based nearest neighbor classification for fine-grained bug severity prediction. In: Reverse Engineering, pp. 215–224. IEEE (2012)
17. Tian, Y., Lo, D., Xia, X., et al.: Automated prediction of bug report priority using multi-factor analysis. Empir. Softw. Eng. 20(5), 1354–1383 (2015)
18. Sharma, M., Bedi, P., Chaturvedi, K.K., et al.: Predicting the priority of a reported bug using machine learning techniques and cross project validation. In: International Conference on Intelligent Systems Design and Applications, pp. 539–545. IEEE (2013)
19. Khomh, F., Chan, B., Zou, Y., et al.: An entropy evaluation approach for triaging field crashes: a case study of Mozilla Firefox. In: Working Conference on Reverse Engineering, pp. 261–270. IEEE Computer Society (2011)
20. Antoniol, G., Ayari, K., Penta, M.D., Khomh, F.: Is it a bug or an enhancement? A text-based approach to classify change requests. In: Proceedings of the Conference of the Center for Advanced Studies on Collaborative Research, CASCON 2008, pp. 304–318. ACM (2008)
21. Runeson, P., Alexandersson, M., Nyholm, O.: Detection of duplicate defect reports using natural language processing. In: International Conference on Software Engineering, pp. 499–510. IEEE (2007)
22. Sun, C., Lo, D., et al.: A discriminative model approach for accurate duplicate bug report retrieval. In: International Conference on Software Engineering, pp. 45–54. IEEE (2010)
23. Sun, C., Lo, D., Khoo, S.C., et al.: Towards more accurate retrieval of duplicate bug reports. In: IEEE/ACM International Conference on Automated Software Engineering, pp. 253–262. IEEE (2011)
24. Tian, Y., Sun, C., Lo, D.: Improved duplicate bug report identification, vol. 94, no. 3, pp. 385–390 (2012)
25. Zhou, J., Zhang, H., Lo, D.: Where should the bugs be fixed? More accurate information retrieval-based bug localization based on bug reports. In: International Conference on Software Engineering, pp. 14–24. IEEE (2012)
26. Gegick, M., Rotella, P., Xie, T.: Identifying security bug reports via text mining: an industrial case study. In: IEEE Working Conference on Mining Software Repositories, pp. 11–20. IEEE (2010)
27. Huang, L.G., Ng, V., Persing, I., et al.: AutoODC: automated generation of orthogonal defect classifications. In: IEEE/ACM International Conference on Automated Software Engineering, pp. 3–46. IEEE (2011)
28. Thung, F., Lo, D., Jiang, L.: Automatic defect categorization. In: Working Conference on Reverse Engineering, pp. 205–214. IEEE (2012)
29. Kim, S., Whitehead, E.J.: How long did it take to fix bugs? In: International Workshop on Mining Software Repositories, MSR 2006, pp. 173–174. DBLP, Shanghai (2006)
30. Weiss, C., Premraj, R., Zimmermann, T., et al.: How long will it take to fix this bug? In: Proceedings of International Workshop on Mining Software Repositories, p. 1 (2007)
31. Jeong, G., Kim, S., Zimmermann, T.: Improving bug triage with bug toss-ing graphs. In: The Joint Meeting of the European Software Engineering Conference and the ACM Sigsoft Symposium on the Foundations of Software Engineering, pp. 111–120. ACM (2009)
32. Tamrawi, A., Nguyen, T.T., Al-Kofahi, J., et al.: Fuzzy set-based automatic bug triaging (NIER track). In: International Conference on Software Engineering, pp. 884–887. IEEE (2011)

An Incremental Approach for Sparse Bayesian Network Structure Learning

Shuanzhu Sun[1], Zhong Han[2], Xiaolong Qi[2,3], Chunlei Zhou[1],
Tiancheng Zhang[2], Bei Song[2], and Yang Gao[2(✉)]

[1] Jiangsu Frontier Electric Technology Co. Ltd., Nanjing 211102, China
15905166613@139.com, 13851845492@163.com
[2] Department of Computer Science and Technology, Nanjing University,
Nanjing 210046, China
763719732@qq.com, qxl_0712@sina.com,
tiancheng_zhang@163.com, songbei07@gmail.com,
gaoy@nju.edu.com
[3] Department of Electronics and Information Engineering,
Yili Normal University, Yining 835000, Xinjiang, China

Abstract. A Bayesian network is a graphical model which analyzes probabilistic relationships among variables of interest. It has become a more and more
popular and effective model for representing and inferring some process with
uncertain information. Especially when it comes to the failure of uncertainty and
correlation of complex equipment, and when the data is big. In this paper, we
present an incremental approach for sparse Bayesian network structure learning.
In order to analysis the correlation of heating load multidimensional feature
factor, we use Bayesian network to establish the relationship between operating
parameters of the heating units. Our approach builds upon previous research in
sparse structure Gaussian Bayesian network, and because our project requires us
to deal with a large amount of data with continuous parameters, we apply an
incremental method on this model. Experimental results show that our approach
is the efficient, effective, and accurate. The approach we propose can both deal
with discrete parameters and continuous parameters, and has great application
prospect in the big data field.

Keywords: Incremental approach · Sparse Bayesian network structure learning
Big data · Correlation analysis

1 Introduction

Bayesian network is a complete model for the variables and their relationship, it can be
used to answer probabilistic queries about them. Furthermore, it has a strong ability to
deal with uncertain problems logically and understandably, and can make inferences
from uncertain large amount of information [1]. Especially with the increasing computing power and the emergence of big data makes Bayesian network increasingly
powerful. It shows great promise in the big data field.

Accordingly, our team's project is to design and develop a correlation model of
heating load multidimensional feature factors and an online diagnose model for the

© Springer Nature Singapore Pte Ltd. 2018
Z. Xu et al. (Eds.): Big Data 2018, CCIS 945, pp. 350–365, 2018.
https://doi.org/10.1007/978-981-13-2922-7_24

monitoring data of large heating units. For the former model, for each specific unit parameter (such as heat supply), we need to describe the interaction between one and the remaining parameters. For the latter model, it should be able to identify outliers in data online. They are correlation analysis problem and detection problem of time sequence outliers, which means we need a model that can answer probabilistic queries about the variables and their relationships, and can make inference from uncertain information to help us retrospect the outliers [2]. Thus, we choose to build a Bayesian network model to solve the analysis problem and we use the autoregressive model and the posterior inspection to carry out the outlier detection. In this paper, we concentrate on the first problem and the solution to it.

A Bayesian network consists of two components: the structure, which is a Directed Acyclic Graph (DAG), for representing the conditional dependencies among variables, and a set of parameters for representing the quantitative information of the dependency [3]. Accordingly, learning a BN from data includes structure learning and parameter learning.

This paper focuses on the structure learning of Bayesian network. We will discuss our method of structure learning of Bayesian network in the part 3, our method of dealing with the data of our project (incremental learning) in part 4 in this paper, and experiments and testing results in part 5 in this paper.

2 Background

In the big data time, the problem of data sparsity still exist. It will be quite difficult to solve such problems with classical statistical methods. Under this circumstance, the Bayesian network provides us a wonderful answer to these complex problems.

A Bayesian network (BN) is composed of a directed acyclic graph (DAG) and a set of parameters. In a DAG $G = (X, E)$, X is a collection of nodes or vertices, and E is a collection of edges. Those nodes of the DAG represent variables in the Bayesian sense: they might be discrete, continuous, known, or unknown [3]. In our project, we deal with the continuous parameters. Edges represent conditional dependencies. Each node is associated with a probability function that takes, as input, a particular set of values for the node's parent variables, and gives (as output) the probability (or probability distribution, if applicable) of the variable represented by the node.

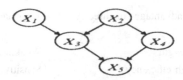

Fig. 1. Example of Bayesian network

As shown in Fig. 1 this is a DAG diagram with P nodes $[X_1, X_2, \ldots, X_p]$ and a directed edge to each other nodes. In DAG, each node represents a free variable, while the opposite side represents the influence of variables. The parent node is "cause" and the child node is effect. By establishing such a relationship, the independence assumption between variables is constructed, that is, a set of parent nodes of a given node, which is independent of all its non-descendant nodes. Therefore, the joint distribution probability of all nodes represented by Bayesian network can be expressed as the product of the conditional probability of each node, namely:

$$P(X_1, X_2, \ldots, X_n) = \prod_{i=1}^{n} P(X_i | X_1, X_2, \ldots, X_{i-1}) = \prod_{i=1}^{n} P(X_i | \pi(X_i))$$

In the graphical structure of Bayesian network, there are three possible types of adjacent triplets allowed in a directed acyclic graph (DAG):

$$X \rightarrow Y \rightarrow Z$$
$$X \leftarrow Y \rightarrow Z$$
$$X \rightarrow Y \leftarrow Z$$

Structural learning is the key point of the whole Bayesian network, which directly influences the subsequent parameter learning and inference. The scientific problem of structure learning of Bayesian network is: how to find a directed acyclic graph structure (DAG) that is corresponding to data distribution in a reasonable time. The challenge is that the structural space is an exponential function of variables. In the face of such a surpass-exponential structure space, it is not feasible to find the global optimal solution in a reasonable time and limited storage space.

The current structure learning method can be divided into two categories: discrete Bayesian network structure learning and continuous Bayesian network structure learning; by the adopted technical methods, it can be divided into: based on the constraints of the Bayesian network structure learning [4–6], based on the scoring-search of the Bayesian network structure learning [7–10], and the mixed method of structure learning; according to the data processing method can be divided into: batch learning and incremental learning [11–13]; by the accuracy of the solution can be divided into: accurate learning and approximate learning.

The comparisons between these approaches can be found in the Tables 1, 2 and 3 below.

Table 1. Advantages and disadvantages of three types of Bayesian network structure learning methods

Methods	Strength	Weakness
Constraint-based	High efficiency	Sensitive to individual detection
Scoring-search-based	Insensitive to individual errors	Large searching-space
Mixed	High efficiency	Inconsistency

Table 2. Advantages and disadvantages of batch learning and incremental learning

Methods	Strength	Weakness
Batch learning	Be able to find global optimal solution	High time cost
Incremental learning	Low time cost	Might end up in local solution

Table 3. Advantages and disadvantages of accurate learning and approximation learning

Methods	Strength	Weakness
Accurate learning	Be able to find global optimal solution	High time cost, high store demand
Approximate learning	Low time cost	Might end up in local solution

The application background is as follows:

(1) We deal with continuous, high-dimensional and large sample data;
(2) The degree of dependence between quantitative variables;
(3) The complexity of reasoning is the exponential of the largest group;

Considering the three reasons, we decided to learn by incremental Sparse Bayesian networks (SBN) to deal with these problems at the same time.

Firstly we introduce SBN [14], it is a kind of continuity for high-dimensional variable method for generating a Bayesian network, different from traditional discrete Bayesian networks, it finally be able to generate a directed acyclic graph, and the corresponding mixture Gaussian distribution. On the specific implementation, the core idea of the algorithm is to maintain a coefficient matrix B. The matrix saves the relation coefficients of each node (attribute characteristics) relative to the parent node, which is continuously updated in the iteration, and finally a relational coefficient matrix is generated, which is further transformed into a directed acyclic graph.

The advantage of using this method to study Bayesian network structure is:

(1) The current mainstream continuous Bayesian network structure learning method mainly includes the conditional Gaussian-Bayesian network and the linear Gaussian-Bayesian network. The conditional covariance matrix is invertible, which limits its application. Linear Gaussian-Bayesian network not only is not subject to this restriction but also combines the sparse technology represented by the current, such as lasso, to reduce the complexity of the model.
(2) It can theoretically guarantee the ring detection in the structure.
(3) Due to the particularity of SBN algorithm, the prior knowledge of domain experts can be easily added, and it is easy to extend to incremental learning.

3   Structure Learning

BN structure learning aims at a NP-hard problem that selecting a probabilistic model that explains a given set of data. The main task of Bayesian network structure learning is to construct a directed graph which conforms to the dependence of the characteristics through the data of the original sample. It is an important basis for subsequent parameter learning and Bayesian inference.

In our project, our team needs to learn large BN structures with high accuracy and efficiency from limited samples. Therefore, a useful strategy is to impose a sparse-constraint of some kind. Many real-world networks are indeed sparse, such as the gene association networks, and our relationship network of heating load factors. When learning the structure of these networks, a sparse constraint helps prevent overfitting and improves computational efficiency.

Meanwhile, considering the common Bayesian Network structure learning is dependent on the characteristics of discretization and continuous raw data, in order to avoid the unreliability caused by human for data discretization as well as the complexity of the reasoning, we adopt a Bayesian network for continuous data structure learning algorithm (Sparse Bayesian Network, SBN) [14]. The method (SBN) itself is aimed at constructing model which has small sample size of continuous data. However, given our sample size is larger, the implementation of SBN based on the further provision of batch training and incremental training, can effectively shorten the training time.

In the SBN algorithm, the following structures are mainly defined. First, the number of variables is p. We use a two dimensional array to represent the DAG we get (a p*p matrix G). $G_{ij} = 1$ represents that there is a directed edge from node i to j. In addition, a p-dimensional matrix is needed to indicate whether there is a path between nodes, namely the relationship closure. $P_{ij} = 1$ represents that there is a path from node i to j, and vice there is none. At last, we need a $(p - 1) * p$ matrix B to record all the coefficients of every node and its parent, since any one node is unlikely to become their own parent node (assuming that there is no the loop), matrix B has one less row compared to the above matrixes. SBN is essentially a model which learn a group of relation coefficients iteratively on each dimension to build a connection between nodes and their parents through data. There is no directed edge between two nodes when the relationship coefficient is zero, the core of algorithm is to optimize a set of formula below:

$$\hat{B} = \min_{i=1}^{p} \left\{ \left(x_i - \beta_i^T x_{/i}\right)\left(x_i - \beta_i^T x_{/i}\right)^T / 2 + \lambda_1 \|\beta_i\|1 \right\}$$
$$\text{s.t } \beta_{ji} \times P_{ij} = 0, i, j = 1, \ldots, p, i \neq j$$

Where β_i is the ith column of matrix B, and $x_{/i}$ represents the sample matrix after remove the variable i in matrix B. The optimization goal of the formula is to minimize the fitting error between real value and the coefficient, then plus the regular penalty term L_1. Moreover, λ_1 is used to control the number of non-zero in matrix B, namely controls the sparsity of network structure. The larger the λ_1, the smaller the number of non-zero in matrix B and the more sparse network structure. In the iterative process, it is necessary to maintain the $\beta_{ji} \times P_{ij}$ constant to zero, so as to guarantee the network structure is not circular.

Unfortunately, the formula above is hard to solve. Therefore, in practical development, we change the optimization to matrix B to aiming at each column. That is, the cumulative process of each variable optimization. The transform formula is:

$$\widehat{B_{ap}} = \min_{B} \sum_{i=1}^{p} f_i(\beta_i)$$

$$= \min_{B} \sum_{i=1}^{p} \left\{ \left(x_i - \beta_i^T x_{/i} \right) \left(x_i - \beta_i^T x_{/i} \right)^T / 2 + \lambda_1 \|\beta_i\| 1 + \lambda_2 \sum_{j \subset X_i} |\beta_{ji} \times P_{ij}| \right\}$$

In the formula, $j \subset X_i$ represents the sample without variable i. We add two penalty terms into this formula, Where λ_1 is still the L_1 regularization penalty term, and it is used to control the sparsity of network structure. Then, λ_2 make $|\beta_{ji} \times P_{ij}|$ close to zero, so as to avoid having a loop in the graph, and by proving that, when λ_2 meets the following condition:

$$\lambda_2 > \frac{(n-1)^2 p}{\lambda_1} - \lambda_1$$

It ensures that no ring is formed during training.

After given the value of λ_1, λ_2, we can calculate it by the BCD algorithm. BCD algorithm is a method of block optimization. For matrix B, BCD algorithm holds all the remaining columns, updating β_i in turn, which is equivalent to optimizing $f_i(\beta_i)$ until the preset convergence conditions are satisfied. Specifically, in our scheme, we adopted the L_2 paradigm change less than 0.001 as the convergence condition. For each optimization of $f_i(\beta_i)$, it's feasible to use a form similar to LASSO optimization:

$$f_i(\beta_i) = \frac{\left(x_i - \beta_i^T x/i \right) \left(x_i - \beta_i^T x/i \right)^T}{2} + \sum_{j \subset X_i} (\lambda_1 + \lambda_2 |\beta_{ji} \times P_{ij}|) |\beta_{ji}|$$

In the above formula, x_i is the sample vector with the characteristic of i. β_i^T is the column that corresponds to feature i in the relational matrix, P_{ij} is the connectivity of feature i and j at this stage. In particular, the judgment of connectivity can be resolved using BFS (depth-first search). For the optimization of $f_i(\beta_i)$, the shooting algorithm can be used to iterate. Finally, we calculate the following formula through constant calculation:

$$\beta_{ji}^{t+1} = \left(\left| \frac{\left(x_i - \beta_{i/j}^t{}^T x/(i,j) \right) x_j^T}{x_i x_i^T} \right| - \frac{\lambda_1 + \lambda_2 |P_{ij}|}{x_i x_i^T} \right) + \text{sign} \left(\frac{x_i - \beta_{i/j}^t{}^T x_j^T}{x_i x_i^T} \right)$$

Then the convergence conditions are going to be approximated. Finally, considering the implementation details of the algorithm, we need to normalize the original samples (keep the mean value 0 and the variance 1).

4 Bayesian Incremental Learning

The aim of incremental learning is for the learning model to adapt to new data without forgetting its existing knowledge, it does not retrain the model. Incremental algorithms are less restrictive than online algorithms, and incremental algorithms process input examples one by one (or batch by batch) and update the decision model after receiving each example. Incremental algorithms may have random access to previous examples or representative/selected examples. In such a case, these algorithms are called incremental algorithms with partial memory.

Typically, in incremental algorithms, for any new presentation of data, the update operation of the model is based on the previous one. Streaming algorithms are online algorithms for processing high-speed continuous flows of data. In streaming, instances are processed sequentially as well and can be examined in only a few passes (typically just one). These algorithms use limited memory and limited processing time per item.

Considering the large sample size (training, using data a year sample size of 500000 or so), although the overall training is able to achieve a more accurate solution, the training process is time-consuming. Based on two characteristics of overall training, we made a few improvements and changes.

First of all, we have to deal with a large number of data samples. The cost of all training to achieve a global solution is to sacrifice the training time, so we take the sliding window as the core. Because is streaming data at the same time, in practical applications, the sample data may not all have, many times the relationship between the variables can cause some changes with time and, this is called online learning based on streaming data. Considering that boiler data satisfies the definition of streaming data, an incremental updating interface is provided based on SBN. Learning through some offline data first, obtain a fairly network structure, and also using the ideas of sliding window, with the passage of time constant sliding window, the data change, once every sliding window, try to calculate the coefficient of a matrix. On the basis of the changes to set a threshold value, and the relationship between the threshold, when the relationship matrix in the numerical change more than threshold, modified, this is to prevent the noise to interfere with the accuracy of the model, and use of the threshold value to determine the relationship between directed edge, when the relation matrix of value across the threshold, the relationship between the network structure of the directed graph updates, because every incoming sample occupies smaller share in the window, it can ensure no dramatic changes, the relationship between matrix when a concept drift, window after drift value will be more and more, makes the relationship matrix gradually closer to the new concept, Thus, an incremental learning method is realized.

In the actual implementation, we take the original sample and store it in a matrix, each row representing a set of boiler data. The number of rows in this matrix is advance given. First, we accept a certain amount of data in the matrix, and then we learn it, and we get the raw matrix B. When the data that we accept exceeds the number of rows in the matrix, we delete the first row of the matrix and add a new set of data to the last row of the matrix, which is the idea of a sliding window. In the process of checking, because our Bayesian network structure is done with linear Gaussian fitting every time, the structure and solution we get corresponding to the next new window are close to the convergence. Furthermore, because the sample size of a single batch is much smaller than the total sample, the relational matrix can converge to a local solution in a short time. With the arrival of the subsequent batch, the relationship matrix is constantly updated. After update, we check the new relationship matrix, and we update the edges according to the value of threshold, to control the sparsity of the structure (Tables 4 and 5).

Table 4. Shows a more detailed description of proposed algorithm.

Input: sample matrix ,SM; number of variable, p; regularization
parameters , $\{\lambda_i\}$i=1,2;initial ,B0 ;stopping criterion , ε .

Initialize:
    Let converge=false;
    Let t=0;
Repeat
    For i=1,2,...,p
    A Breadth-first search on G with Xi being the
    Root node to calculate P_{ij} for
    j=1,...p.
    Using the shooting algorithm[1] to Optimize $f_i(\beta_i)$ and get β^{t+1}_i
    End for
  If $\|B^{t+1}-B^t\| \leq \varepsilon$ and no new sample update SM

    Converge=true;
  Else
    Converge=false;
    Update SM;
  End if
Let t=t+1;
Until converge =true
Output:B^{t+1}

Table 5. Shows sliding window update algorithm.

SM update algorithm:

Input: new data, SM
    Delete first row sample in sample matrix SM;
    Remainder m-1 row move forward first row;
    New a set of data fill last row of SM;

For the SBN algorithm, the time complexity is np^2. n is the iteration number of BCD and p is the number of variables. In most cases, the number of variables is small, no more than 100, and the iteration number can be really big when the data size is huge. For example, for our boiler data from our client, it is usually a hundred thousand orders of magnitude. In this case, the n is too big and the time complexity is high. However, our incremental approach can converge within 10 iterations. In our implementation of the window slide in the project, we set a window in 3000 sets of data, and the time complexity of this method is cp^2. In this case, c3000*10 which is much less than a hundred thousand.

In addition, the experiment found that if the overall distribution of a one-dimensional data is basically unchanged, the first batch of samples can converge the matrix which corresponds to the dimension to an approximate global solution, and subsequent iterations will be greatly reduced.

Whether the data is large or small, the approach of incremental sparse Bayesian network structure learning theoretically takes much less time than other Bayesian network learning method.

5 Experiments and Testing Results

Based on the methods above, we conducted test based on the provided data, because an important idea of BCD + shooting method is to use a specific column, fixed other data to do block optimization. So the order of the original data characteristics will produce certain effect to the learned structure. According to the proof of theory and experiment, the higher the variable sequence, the more the tendency to become the ancestor node. Therefore, it is an important part to find a more reasonable feature order for this method. According to the prior knowledge, choosing a good order of characteristic variables can effectively improve the validity of the network. We conducted experiments based on two variables and different sample sizes.

5.1 Time Efficiency Experiments

We firstly conducted two time efficiency experiments on two data sets which belong to two kinds of variable order. One data set is large and another data set is small. Since we need to deal with data with continuous parameters, we choose batch learning approach to compare with our incremental approach, they both can deal with continuous parameters by using SBN algorithm.

oder1 = [0, 1, 2, 3, 4, 5, 6, 7, 8, 9, 10, 11, 12, 13, 14, 15, 16, 17, 18, 19, 20, 21, 22, 23, 24, 25, 26, 27, 28, 29, 30, 31, 32, 33, 34, 35, 36, 37, 38, 39, 40, 41, 42, 43, 44, 45, 46, 47, 48, 49, 50, 51, 52, 53, 54, 55, 56, 57, 58, 59, 60, 61, 62]

The size of this group of data is 520,000, with 57dimensions (57 variables) (Figs. 2 and 3).

Time efficiency:
Incremental approach : 14h   Batch approach: 29h
Analysis:

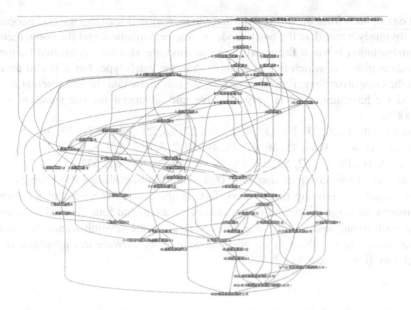

Fig. 2. Structure constructed by incremental learning for order 1 (sample size is 520,000).

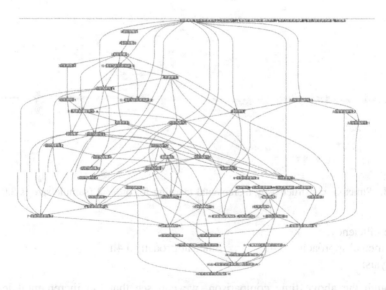

Fig. 3. Structure constructed by batch approach for order 1 (sample size is 520,000)

This group of data has 57 sets of variables, of which 37 groups converge before the 5th iteration, and 14 groups converge in the sixth iteration, and the remaining six groups converge in the eighth iteration. The time complexity is much lower than the batch.

It can be seen from the above time that the incremental learning method adopted by us is obviously better than the batch mode in time performance, and the convergence to the final solution is much faster. At the same time, the structure generated by the two methods is different, which is less accurate than the batch type, but it is still accurate. From the comparison experiment on order1, we can know that the incremental learning method we have improved is very effective and advanced for the situation size of 500,000+.

oder2 = [0, 1, 20, 21, 22, 23, 24, 25, 26, 27, 28, 29, 30, 31, 32, 33, 34, 35, 36, 37, 38, 39, 40, 41, 42, 43, 44, 45, 46, 47, 48, 49, 50, 2, 3, 4, 5, 6, 7, 8, 9, 10, 11, 12, 13, 14, 15, 16, 17, 18, 19, 51, 52, 53, 54, 55, 56, 57, 58, 59, 60, 61, 62]

The size of this group of data is 3, 000, with 30 dimensions (30 variables).

Therefore, the running time is very fast. In the case of incremental situation, 25 dimensions converge in the first five rounds of iteration, and four dimensions converge in the sixth round, and the last one will converge in the seventh round. Because the sample size is small, there is no great advantage for incremental approach in time complexity (Figs. 4 and 5).

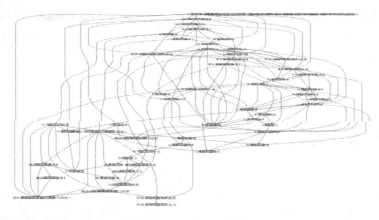

Fig. 4. Structure constructed by incremental learning for order 2 (sample size is 3,000)

Time efficiency:
Incremental approach : 0.4h   Batch approach: 0.4h
Analysis:

Through the above time comparison, we can see that the incremental learning method adopted by us is not significantly different from the batch type in terms of time performance, and the obtained structure is basically the same. This shows that for smaller sample sizes, incremental learning and batch can be equally accurate and fast.

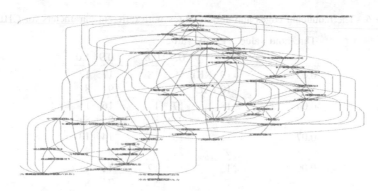

Fig. 5. Structure constructed by batch approach for order 2 (sample size is 3,000)

5.2   Accuracy Experiment of Structure Learning Results

In the first experiment, we managed to prove that the incremental approach for sparse Bayesian network structure learning is more efficient in time than batch learning. In this experiment, we try to conduct a comparison experiment to find the most suitable penalty value and the relation threshold value that can make our method most accurate. At the same time, prove that our method can generate an accurate structure through structure learning.

Before the test, we should explain the two parameters in our system: penalty and threshold. The penalty controls the sparsity and avoid becoming a circle through training (It is explained in detail in part IV. structure learning of this paper), and the relation threshold removed some edge with weak correlation, namely after-pruning (It is explained in detail in part V. Incremental learning of this paper).

It's worth mentioning that the effect of redundant edges on structural correctness is less than absent edges, which means we should pay more attention on decrease the number on absent edges. Moreover, the bigger the threshold is, the more sparse the structure is, and normally the bigger the penalty is, the more sparse the structure is.

Dataset: CHILD

Number of nodes: 20; Number of arcs: 25; Number of parameters: 230; Average Markov blanket size: 3.00; Average degree: 1.25; Maximum in-degree: 2.

The influence of the value of these two parameters can be found in Table 6, we can find a best setting for them when the redundant edges (RE) and absent edges (AE) are least.

As we can observe in chart 4, when the penalty value is 0.8 and the relation threshold is 0.10, the absent edges and redundant edges are the least. It is mathematical that there must be a most suitable setting that the structure is closest to the ground truth, since the two parameters can both regulate the sparsity of the structure.

Apparently, when the threshold or the penalty is too small, the structure can be so tense, in that case, there will be too much redundant edges. On the other hand, when the threshold or the penalty is too big, the structure can be so sparse, which leads to a result that it is more likely that more edges will be missed. Therefore, the balance spot is when the penalty value is 0.8 and the relation threshold is 0.10.

Table 6. The redundant edges and absent edges with different parameters for CHILD

Threshold value	Penalty value				
	0.6	0.7	0.8	0.9	1.0
0.02	19 (RE)	19	18	17	17
	10 (AE)	9	9	9	8
0.04	18	17	17	16	15
	9	9	8	8	8
0.07	16	15	14	14	13
	9	7	6	8	8
0.09	13	13	12	11	11
	7	6	5	7	8
0.10	10	8	8 (right 4setting)	7	6
	5	5		6	7
0.11	10	8	8	9	8
	6	6	5	7	8

Dataset: ALARM

Number of nodes: 37; Number of arcs: 46; Number of parameters: 509; Average Markov blanket size: 3.51; Average degree: 2.49; Maximum in-degree: 4.

Table 7. The redundant edges and absent edges with different parameters for ALARM

Threshold value	Penalty value				
	0.6	0.7	0.8	0.9	1.0
0.02	23 (RE)	23	22	22	20
	13 (AE)	13	13	13	13
0.04	23	22	22	21	19
	11	11	12	12	13
0.07	22	22	21	20	18
	10	11	11	12	13
0.09	20	19	17	16	16
	10	10	11	13	12
0.10	19	17	15 (right 8setting)	14	13
	8	8		9	11
0.11	17	15 (right 8setting)	14	14	13
	8		10	9	10

As we can observe in the Table 7, there are two situations which have the least redundant edges and absent edges. First one is when the penalty value is 0.7 and the relation threshold is 0.10, and the second one is when the penalty value is 0.8 and the relation threshold is 0.09. For a dataset which has 44 edges, a structure with 8 absent edges is considerably accurate in machine learning field.

Compare this setting of the dataset ALARM and the last setting of the dataset CHILD, we conclude that when the penalty value closes to 0.8 and the relation threshold value closes to 0.10, the structure our system learned is the most suitable one.

On the other hand, we need to evaluate the performance between the offline learning and online learning based on sliding windows which we set as 10,000. Due to the size of our data is large, online learning is more efficient than offline method. In order to prevent the influence of individual data on parameter updating, we slide the windows once when every 100 new data coming. What's more, we set penalty based on the size of sliding windows. Due to the lack of correct network based on our data, we assume the offline result as the groundtruth. In order to reduce the influence of data size, we set the size of data as 50,0000. The result is showing in Tables 8, 9 and 10.

Table 8. The comparison of offline learning and online learning Bayesian networks for 30 nodes

	Nodes	Edges	RE	AE
Offline	30	56		
Online_1	30	63	10	3
Online_2	30	60	6	2
Online_3	30	58	3	1

Table 9. The comparison of offline learning and online learning Bayesian networks for 50 nodes

	Nodes	Edges	RE	AE
Offline	50	74		
Online_1	50	83	16	7
Online_2	50	78	8	4
Online_3	50	77	7	4

Table 10. The comparison of offline learning and online learning Bayesian networks for 62 nodes

	Nodes	Edges	RE	AE
Offline	62	89		
Online_1	62	99	19	9
Online_2	62	93	10	6
Online_3	62	93	9	5

Tables 8, 9 and 10 show the comparison of offline learning and online learning Bayesian Networks. First line is the offline groudtruth, second line is the network based on first batch, The third line is the networks based on the half of data, The last line is the networks based on the total data.

From the result of table above, we can find the result of online learning is close to offline learning, but the edges of online learning is more than offline learning, it may because the penalty we set is not reasonable enough. The first batch result has a gap with groundtruth, but with the online updating of the network, the result is slowly approaching the groundtruth. Base on the experiment, we can find the online learning method we proposed is more efficient than offline learning which could maintain accuracy.

6 Conclusion of Experiments

In conclusion, the time efficiency experiments prove that our incremental approach for sparse Bayesian network structure learning more efficient than the existing approach of Bayesian structure learning. The accuracy experiments prove that our method is considerably accurate among machine learning methods.

The results demonstrate the accuracy, efficiency of our approach, and our approach can solve the current intractability that learns structure with continuous parameters. All the advantages above reflect the advancement of our approach.

Last but not the least, those experiments we conduct prove that our incremental approach of Bayesian network structure learning can be wildly applied in big data filed. For example, many situations in big data like the analysis of the data of complex equipment, the origin of galaxies, the pathogenic genes, the operation mechanism of the large heating units, etc. To uncover the laws underlying these problems, we must understand their genetic networks and tease out the intricacies of the events. Due to the invalidation of classical statistics in those problems, our approach appears to be more valuable.

References

1. Borsuk, M.E., Stow, C.A., Reckhow, K.H.: A Bayesian network of eutrophication models for synthesis, prediction, and uncertainty analysis. Ecol. Model. **173**, 219–239 (2004)
2. Agosta, J.M., Gardos, T.R., Druzdzel, M.J.: Query-based diagnostics. In: Jaeger, M., Nielsen, T.D. (eds.) Proceedings of the Fourth European Workshop on Probabilistic Graphical Models, PGM-08, Aalborg, Denmark, pp. 1–8 (2008)
3. Cooper, G., Herskovits, E.: A Bayesian method for the induction of probabilistic networks from data. Mach. Learn. **9**(4), 309–347 (1992)
4. Spirtes, P., Glymour, C., Scheines, R.: Causation, Prediction and Search. Springer, New York (1993). https://doi.org/10.1007/978-1-4612-2748-9
5. Cheng, J., Greiner, R., Kelly, J., et al.: Learning Bayesian networks from data: an information-theory based approach. Artif. Intell. **137**(1–2), 325–331 (1997)
6. Zhang, H., Zhou, S., Zhang, K., et al.: Causal discovery using regression-based conditional independence tests. In: AAAI, pp. 1250–1256 (2017)
7. Alcobé, J.R.: Incremental hill-climbing search applied to Bayesian network structure learning, pp. 1320–1324 (2008)
8. Ko, S., Kim, D.W.: An efficient node ordering method using the conditional frequency for the K2 algorithm. Pattern Recognit. Lett. **40**, 80–87 (2014)

9. Alonso-Barba, J.I., de la Ossa, L., Regnier-Coudert, O., et al.: Ant colony and surrogate tree-structured models for orderings-based Bayesian network learning. In: Proceedings of the 2015 Annual Conference on Genetic and Evolutionary Computation, pp. 543–550. ACM (2015)
10. Heckerman, D., Geiger, D., Chickering, D.M.: Learning Bayesian networks: the combination of knowledge and statistical data. Mach. Learn. **20**(3), 197–243 (1995)
11. Dimkovski, M., An, A.: A Bayesian model for canonical circuits in the neocortex for parallelized and incremental learning of symbol representations. Neurocomputing **149**(4), 1270–1279 (2015)
12. Yue, K., Fang, Q., Wang, X., et al.: A parallel and incremental approach for data-intensive learning of Bayesian networks. IEEE Trans. Cybern. **45**(12), 2890–2904 (2017)
13. Yasin, A., Leray, P.: Incremental Bayesian network structure learning in high dimensional domains. In: International Conference on Modeling, Simulation and Applied Optimization, pp. 1–6. IEEE (2013)
14. Huang, S., et al.: A sparse structure learning algorithm for Gaussian Bayesian network identification from high-dimensional data. IEEE Trans. Pattern Anal. Mach. Intell. **35**(6), 1328–1341 (2013)

Cloud Goods Recognition System Based on PCA and SVM

Jianbiao He, Baojiang Zhang(✉), Ruina Chen, and Changtai Li

Central South University, Changsha 410083, China
1654441578@qq.com

Abstract. In this paper, an intelligent inventory management system for vending machines based on image recognition has been proposed. The outside image of a vending machine goods cabinet is obtained by a camera installed on a lifted mechanism of the machine, whenever a good reloading or a buying action has been done by the routine operator or the costumer respectively. That image is labeled with the vending machine ID and forwarded to a Cloud Goods Recognition Center (CGRC). It is firstly recognized by employing the cloud goods recognition algorithm based on Principal Component Analysis (PCA) and Support Vector Machine (SVM) and then relabeled with the classified item name, items unit price by accessing the item samples database. The relabeled result is finally returned to the vending machine for its inventory updating. The experiments show that excellent classification results with more than 96% of images, correctly.

Keywords: Image recognition · SVM · PCA · Cloud goods recognition

1 Introduction

As a typical representative of new retail, vending machines are getting more and more attention. The application of this new thing has also created a new problem: a prominent problem is the inventory management of vending machines. At present, we have still relied on manual way for inventory management, and the replenishment staff of vending machines must have manually entered the inventory after each replenishment. This method results in significantly longer replenishment time for single machine, low efficiency and it is easy to make mistakes. On the one hand, Consumers are often less satisfied when they buy goods that do not exist. On the other hand, Vending machine owners can not know which machine has sold which goods, and they also can not do the accurate statistics of profits and costs. Not to mention intelligent analytic and value-added services based on big data sales.

With the development of artificial intelligence, the combine between machine learning and intelligent recognition have become possible. This paper presents an automatic inventory management system based on image recognition. The system consists of two parts: the vending machine networking terminal (cluster) which are operated all over the city and the cloud goods recognition system. As is shown in Fig. 1. Replenishment operators only carry out replenishment operations, and the quantity and type of goods are handed over to the intelligent inventory platform. This

© Springer Nature Singapore Pte Ltd. 2018
Z. Xu et al. (Eds.): Big Data 2018, CCIS 945, pp. 366–377, 2018.
https://doi.org/10.1007/978-981-13-2922-7_25

can greatly reduce manual operations, reduce the loss caused by human error, and it is more convenient for the management of vending machines.

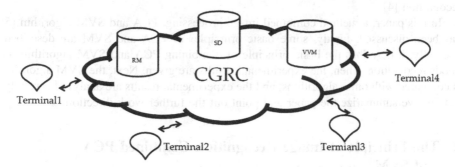

Fig. 1. Automatic inventory management system based on image recognition

The vending machine terminal is equipped with an automatic lift pickup device which is equipped with a miniature camera. Every time the repairman makes up a shipment. The vending machine controls the lift to go through the aisles one by one at a time. The image of goods at the outermost end of each aisle is photographed according to the aisle number and submitted to the cloud through the network for processing and recognition. When an item is sold in a particular aisle, a camera on an elevator receiver takes a picture of the aisle at the same time. Image of the single aisle is sent to the cloud for recognition. The cloud recognition results are sent back in real time to a specific smart vending machine terminal, making the terminal to always perceive whether the outermost end of all its aisles is in stock in real time and the specific of the goods, as well as the sale price of them.

The cloud goods recognition center (CGRC) mainly consists of recognition module (RM), sample database (SD) and virtual vending machine (VVM). Each metadata in the sample database is a good sample image library with a label (product name and price). Virtual vending machine is the copy the terminal which records the details of goods for sale. Each time the recognition module receives an unlabeled (to be recognized) image of the good from the terminal with attributes (attributes that contain the number of the vending machine and the number of the specific aisle of the vending machine), and it is matched with images of labeled goods in the sample database. In the end, the labeled goods to be recognized are returned to a specific vending machine in real time. For an image of 128 × 128, its upload and download can be completed in three seconds and its occupied memory size is 0.046875M. Inventory management system can support 50 vending machines to upload and download pictures at the same time. In the recognition module, we further put forward an algorithm combining PCA and SVM algorithm to recognize various goods.

The principal component analysis (PCA) is the main method of feature extraction for image [1]. It can compress the data set and greatly reduce the computation. Support vector machine (SVM) is suitable for dealing with small sample, non-linear and high dimension [2]. In this article, algorithm combining PCA and SVM innovative are applied to the vending machines [3] in the cloud goods recognition system for image recognition [4].

In this paper, a method combined image processing, PCA and SVM algorithm [5] has been discussed. Firstly, some basic principles of PCA and SVM are described. Secondly, it describes the basic principle of combining PCA and SVM algorithm for goods recognition. Then, the experimental results are given. Next, the SVM algorithm is compared with other algorithms, and the experimental results are analyzed carefully. Finally, we summarize the paper and point out the further work direction.

2 The Principle of Image Recognition Combined PCA and SVM

2.1 PCA

The basic principle of principal component analysis (PCA) is [6]: PCA method which is also named the discrete Karhunen–Loeve transform (KLT) in signal processing, is a statistical procedure that uses an orthogonal transformation to convert a set of sample variables into a set of values of linearly uncorrelated variables called principal components. It has the characteristics such as eliminating correlation and energy concentration, which belongs to the minimum distortion of the transformation under the mean square error measure. And it is a kind of transformation which can mostly remove the correlation of the original data. PCA selects the eigenvectors of the largest K eigenvalues of the covariance matrix to form the k-l transformation matrix.

2.2 SVM

The basic principle of principal component analysis (PCA) is [6]: PCA method which is also named the discrete Karhunen–Loeve transform (KLT) in signal processing, is a statistical procedure that uses an orthogonal transformation to convert a set of sample variables into a set of values of linearly uncorrelated variables called principal components. It has the characteristics such as eliminating correlation and energy concentration, which belongs to the minimum distortion of the transformation under the mean square error measure. And it is a kind of transformation which can mostly remove the correlation of the original data. PCA selects the eigenvectors of the largest K eigenvalues of the covariance matrix to form the k-l transformation matrix.

By nonlinear mapping [8], SVM maps the sample space to a feature space (the Hilbert space) of a high dimensional or infinite dimension, making problem which is nonlinear separable in the original sample space into a linear separable in feature space. The kernel function is used to solve nonlinear classification problems. There are several common kernel functions [9]:

(1) Linear Kernel

$$k(x,y) = x^T y + c \tag{1}$$

(2) Polynomial Kernel

$$k(x,y) = (ax^T y + c)^d \tag{2}$$

(3) Radial Basis Function (RBF)

$$k(x,y) = \exp(-\gamma\|x - y\|^2) \tag{3}$$

2.3 PCA and SVM Are Combined for Identification

Generally, combining the advantages of both PCA and SVM [10]. PCA algorithm is responsible for image feature extraction and multidimensional reduction. Goods image in the training set are used to train the SVM classifier. In the end, forecasting and classifying the goods image in the test set.

3 Goods Recognition Based on PCA and SVM

Step 1: preprocess the image in the training set and in the test set.

In the original image, there are a larger proportion of unrelated backgrounds. The accuracy of goods recognition is seriously affected [11]. Therefore, we first cut the background part of the image, leaving part of the goods information. Then the RGB pixels of images of the training set and the test set are placed in the different data matrices.

Step 2: reducing the dimension for the two data matrices by PCA and carrying out data normalization.

The PCA algorithm for good recognition is described:

(1) A total number of n images are selected as training samples, and each image is arranged as a column vector in the form of X_i data matrix.

$$X = (X_1, X_2, \ldots, X_n) \tag{4}$$

(2) Figure out the mean vector u

$$u = \frac{1}{n}\sum_{i}^{n} X_i \tag{5}$$

data matrix after centralization D

$$D = (X_1 - u, X_2 - u, \ldots, X_n - u) \tag{6}$$

covariance matrix

$$\sum \frac{1}{n} D D^T \tag{7}$$

(3) The eigenvalue of the covariance matrix is obtained by solving covariance matrix, and the maximum K eigenvalues are selected to find the corresponding eigenvector, and K eigenvectors like this are taken as transformation matrix W in terms of columns.

$$W = (e_1, e_2, \ldots, e_k) \tag{8}$$

(4) Calculate the projection of each image (K dimension vector)

$$Y_i = W^T (X_i - u) \tag{9}$$

(5) Calculate the projection of the test image (K dimension vector), Supposed the test image is Z

$$chZ = W^T (Z - u) \tag{10}$$

Step 3: the training set data are trained with SVM algorithm for classification modeling, and the optimal penalty factor C and g of kernel function are determined.

SVM can be divided into hard interval SVM and soft interval SVM. Here, the soft interval c-svm is used to improve the recognition accuracy.

The description about the algorithm of nonlinear classification (Fig. 2):

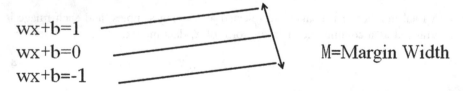

wx+b=1
wx+b=0   M=Margin Width
wx+b=-1

Fig. 2. The definition of maximum interval

As is shown in 2, the interval $M = \frac{2}{\sqrt{w \cdot w}}$. Under the condition of linear segmentation [12], the optimal function is $\min \frac{1}{2} \|w\|^2$. In the case of linear indivisibility [13], The soft interval C-SVM is equivalent to adding a penalty function after the original optimization function (C is the penalty factor), with the constraint condition [14]:

$$\min \frac{1}{2}\|w\|^2 + C\sum_{i=1}^{R}\varepsilon_i \qquad (12)$$

$$s.t., y_i(w^T x_i + b) \geq 1 - \varepsilon_i, \varepsilon_i \geq 0$$
$$y_i = \{-1, 1\}$$

It is transformed into dual problem, then be solved.

Step 4: put the test set data into the training model trained by the training data for prediction.

4 Experimental Results and Analysis

4.1 Experimental Environment

The experiment uses a self-made goods database. All image data of the goods database is collected from vending machines. It consists of 19 kinds of goods. Each kind of good has 10 images, and 190 images in total. Each image has a resolution of $540 \times 960 = 518400$, all of them are RGB tricolor images. Each kind of good in the data set is divided into two groups, the first 7 as the training set and the last 3 as the test set. There are 133 samples in the training set and 57 samples in the test set.

Hardware: DELL, PC, windows 10, intel (i3). Software: Matlab.

4.2 The Experimental Process

Goods Feature Extraction

Image Preprocessing

Numbering photos of each of them by category. The number is the category of each good. The first 7 photos of each product are placed in the training gallery, forming the training sample set. And the last 3 photos are placed in the test gallery, forming the test sample set. 540×960 dimensional goods images are preprocessed into 128×128 format using Matlab image cutting function. As is shown in Fig. 3.

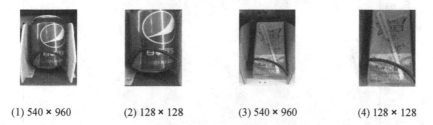

(1) 540 × 960   (2) 128 × 128   (3) 540 × 960   (4) 128 × 128

Fig. 3. The good images for different sizes

Feature Extraction

The fast PCA multidimensional reduction method is used to remove the correlation between pixels, and the main components were extracted from it, and the 128×128 dimensions are reduced to 16 dimensions, so that the product samples are all represented by the 16 dimensional feature vectors. Figure 4. shows that the characteristic images of a kind of good in the training set. From left to right are original image without dimension reduction, feature image with dimension 95, feature image with dimension 60 and feature image with dimension 20.

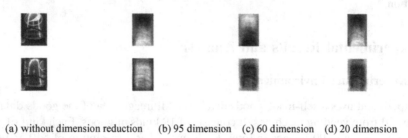

(a) without dimension reduction   (b) 95 dimension   (c) 60 dimension   (d) 20 dimension

Fig. 4. The characteristic image of the good image

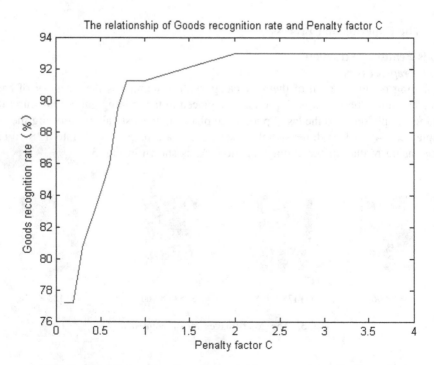

Fig. 5. The relationship of goods recognition rate and penalty factor C

Construct Multiple SVM Classification and Kernel Function Selection

In the multi-purpose SVM training stage, 171 classifiers are generated from 19 types of samples. In the classification stage, the test samples are sorted through these binary classifiers in turn, and finally the number of categories is determined through the voting mechanism.

Using the linear kernel function, the recognition rate of the image is shown in Fig. 5. First, we used the libsvm open source library to find a couple of good values. After that, a lot of experiments were done to compare which value was the best one. When C = 0.1, the correct identification rate of classification in 57 test sets is 77.193%; when C = 1, the correct identification rate is 91.2281%; when C = 3, correct classification accuracy is 92.9825%.

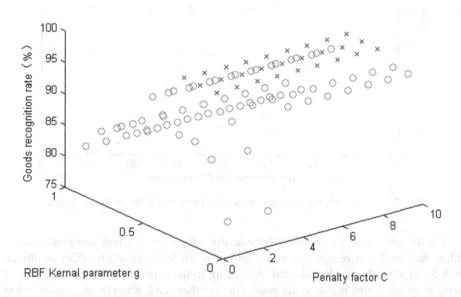

Fig. 6. Goods recognition rate for RBF

Where C is the penalty factor, that is to say, the tolerance to error. The higher the C is, the less tolerance it is for error and function is easy over-fitting. The smaller the C is, the less fitting it is. If C is too large or too small, the generalization ability will be poor.

Using RBF function, the recognition rate of images is shown in the Fig. 6. When C = 4 g = 0.5, the correct identification rate of classification in 57 test sets is 96.4912%; when C = 1 g = 0.5, the correct identification rate is 91.2281%; when C = 4 g = 0.2, the correct classification rate is 94.7368%.

When the punishment factor C in the range of 2–10 and RBF kernel parameter g in the range of 0.3–0.5, the classification of SVM can achieves better results.

Comparison About Recognition Rate of Different Original Image Types
All the steps are consistent. The following is the SVM prediction after dimension reduction between the original RGB diagram and gray scale diagram. The recognition rate obtained is shown Fig. 7.

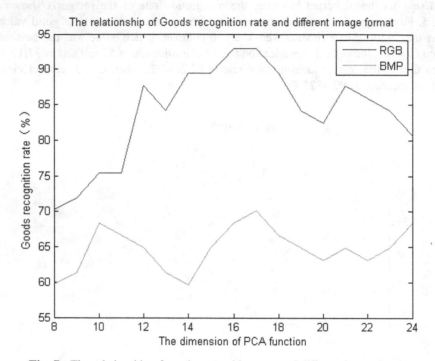

Fig. 7. The relationship of goods recognition rate and different image format

On the one hand, the grayscale is generally adopted in the face recognition algorithm, and the better recognition rate does not reach 95%. Similarly, PCA combined with SVM algorithm is also adopted. According to the experimental data, the effect of using grayscale graph is indeed not good. On the other hand, when the dimension is too low, the image loses too much information, which is not conducive to find the difference of different images. As the dimension increases, the recognition rate becomes better. After crossing the optimal value, with the increase of dimensions, the redundant information is more and more, which is also not conducive to find the difference of images.

Comparison Between Different Classifiers
All the previous steps are consistent. The test samples are identified by Euclidean Distance [15], k-Nearest Neighbor (KNN) [16] and SVM respectively, and the recognition rate is shown in Table 1 below:

Table 1. Classifier name and accuracy for goods recognition

Classifier name	Accuracy
Euclidean distance	26.3157%
KNN	89.4737%
SVM (Linear Kernel)	92.9825%
SVM (RBF Kernel)	96.4912%

When using the European Distance algorithm for recognition, there is little accuracy. KNN algorithm has certain accuracy, but accuracy is low, so there is no possibility of practical application. In SVM, the accuracy of RBF function is higher than linear kernel function, and it has practical application conditions.

Analysis of Experimental Results

(1) Fast PCA algorithm can effectively reduce the dimension of goods image samples and simplify classification calculation rate. In the article, we have reduced the dimension down to 16 dimensions, the accuracy rate 96.4912% is achieved.

(2) The selection of SVM kernel function is very important, and the selection of parameters is closely related to the recognition rate. The selection of optimization parameters is very important. Matlab toolbox can be used to get better penalty factor C and g. We use the libsvm open source library developed by professor chih-jen Lin of Taiwan university to achieve the 96.4912%.

(3) As is shown in Table 1, although there is a large amount of redundant information in images, color is still an important differentiator in image classification. For the same (17) dimension, the accuracy of the grayscale image is 70% when the accuracy of the RGB image is 93%.

(4) The classification effect of SVM, KNN and Euclidean Distance is compared on the premise of the same principal component contribution rate. The results show that when the contribution rate is the same, the recognition rate of SVM is higher than the other two classifiers, that is to say, SVM is a relatively ideal multi-class classifier. The recognition rates shown in Table 1 are respectively 26.3157%, 89.4737%, 96.4912%.

(5) RBF kernel function is better than linear kernel function in small space samples. For the 16 dimensions, the recognition rate of linear kernel function is 92.9825% while the recognition rate of RBF kernel function is 96.4912%. For the 18 dimensions, the recognition rate of linear kernel function is 89.4737% while the recognition rate of RBF kernel function is 93.3582%. For the 19 dimensions, the recognition rate of linear kernel function is 87.7193% while the recognition rate of RBF kernel function is 90.3518%.

(6) For small sample space, SVM is more advantageous than neural network algorithm [17] which has a better effect only when the sample set is large. For the same data, the recognition rate of alexnet is 70.18% and the recognition rate of googlenet is 23.68% (Fig. 8).

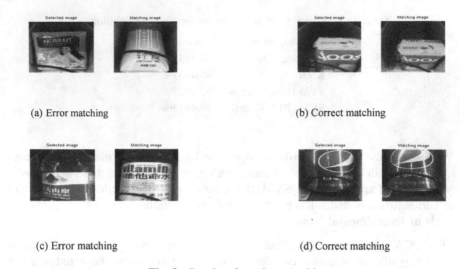

(a) Error matching   (b) Correct matching

(c) Error matching   (d) Correct matching

Fig. 8. Results of goods recognition

5 Conclusion

An intelligent inventory management system for vending machines has been proposed. And joined with the characteristics of PCA algorithm and SVM algorithm, a algorithm of combining PCA and SVM for goods recognition has been proposed. Fast PCA algorithm projects the original image into the feature space directly, without Gray-scale treatment. Not only getting rid of a large number of redundant information of images, retaining the image color information and other useful information at the same time. This algorithm has realized the feature compression and extraction of PCA algorithm, and SVM is used to train multiple classifiers to solve the problem of multiple classification. Different SVM kernel functions are used to train samples for comparison. However, how to effectively implement PCA algorithm and SVM algorithm, and how to effectively select the parameters of SVM kernel function, so as to further improve the goods recognition rate are problems to be further explored.

Acknowledgments. We give thanks to Qidi Liang, Mengshu Jiao for their technical assistances. The work was sponsored by: National Natural Science Foundation of China Grant No. 61272147, and Platform and Talent Planning No. kc1701026.

References

1. Li, C.J.: A research of key technology of dimensionality reduction of high dimensional data. University of Electronic Science and Technology of China, Chengdu, Sichuan, China (2017)
2. Zhang, D.F., Zhang, J.S., Yao, K.M., Cheng, W.M., Wu, W.G.: Infrared ship-target recognition based on SVM classification. Infrared Laser Eng. **45**(01), 179–184 (2016)

3. Korucu, M.K., Kaplan, Ö., Büyük, O., Güllü, M.K.: An investigation of the usability of sound recognition for source separation of packaging wastes in reverse vending machines. Waste Manag. **56**, 46–52 (2016)
4. Liu, J.: Research on automatic identification system of commodities. Beijing University of Posts and Telecommunications, Beijing, China (2016)
5. Xing, X.F., Xu, G.C., Cai, B.L., Qing, C.M., Xu, X.M.: Face verification based on feature transfer via PCA-SVM framework. In: 2017 IEEE International Conference on Internet of Things (iThings) and IEEE Green Computing and Communications (GreenCom) and IEEE Cyber, Physical and Social Computing (CPSCom) and IEEE Smart Data (SmartData), Exeter: 1086-91, UK (2017)
6. Zhang, S.Z., Xu, H.G., Li, S.J.: Applicability of principle components analysis (PCA) to evaluate the dynamic characteristics of cavitation from captured images, Harbin, China (2015)
7. Rahmad, C., Rahmah, I.F., Asmara, R.A.: Indonesian traffic sign detection and recognition using color and texture feature extraction and SVM classifier. In: Information and Communications Technology, Yogyakarta, Indonesia (2018)
8. Deng, N., Tian, Y.: Support Vector Machine. Science Press, Beijing (2009)
9. Liu, H., Su, S.M.: A face recognition algorithm based on EHMM-SVM. Comput. Eng. Sci. **39**(5) (2017)
10. Gonzalez, R.C.: Digitial Image Processing, 3rd edn. Publishing House of Electronics Industry, Beijing (2017)
11. Gao, Q., Yan, D.Q., Chu, Y.H., Xu, L.L.: LLE base on fuzzy clustering and SVM for face recognition. Microcomput. Appl. **34**(6), 56–58 (2015)
12. Guo, Y.N.: Face recognition method on SVM and ELM. Taiyuan University of Technology, Shanxi, China (2012)
13. Hu, M.H.: Face recognition system based on PCA and SVM. Comput. Era 60–63+67 (2017)
14. Hao, T.: Supermarket Goods Recognition Based on Image Processing, Dalian, Liaoning, China (2014)
15. Yang, L., Liu, S.Z.: Endmember extraction based on image euclidean distance and laplacian eigenmaps. Electron. Opt. Control **23**(4), 48–52 (2016)
16. Li, X.F.: Wavelet transform for face recognition based on KNN classifier. In: International Conference on Automation, Mechanical and Electrical Engineering, 10–16, Phuket, Thailand (2015)
17. Akcay, S., Kundegorski, M.E., Willcocks, C.G., Breckon, T.P.: Using deep convolutional neural network architectures for object classification and detection within X-Ray baggage security imagery. IEEE Trans. Inf. Forensics Secur. **13**(9), 2203–2215 (2018)

A Method for Water Pollution Source Localization with Sensor Monitoring Networks

Lan Wu[1]([⊠]), Xiaolei Han[1], and Chenglin Wen[2]

[1] College of Electrical Engineering, Henan University of Technology,
Zhengzhou, China
wulan@haut.edu.cn
[2] Institute of Information and Control, Hangzhou Dianzi University, Hangzhou,
China

Abstract. Research on pollution localization using sensor monitoring networks has important significance for environmental protection. There are some challenges in the detection and localization of water pollution sources due to the particularity and complexity of the water environment. The parameters of the traditional concentration diffusion model cannot be obtained in real time, making it difficult to apply. Complex and uncertain interference, such as time delays, flow velocity, and the fact that propagation relies heavily on propagation distance, makes the positions of pollution sources not constant, and the sensor data cannot reach the information processing center at the same time. Therefore, a nonlinear concentration diffusion model formulated by combining Extreme Learning Machine (ELM) with Partial Least Squares (PLS) has been established to describe the relationship between the concentration diffusion value and the position of the sensor in order to solve the problem that the traditional model parameters cannot be acquired in real time. Furthermore, the Sequential Unscented Kalman Filter (SUKF) combined with the ELM-PLS model is utilized to realize precise localization. These proposed methods could achieve accurate diffusion modeling and localization by addressing any nonlinear diffusion data. This would overcome the existing shortcomings of the standard PLS modeling method and LS localization method. Finally, the effectiveness of these proposed methods are verified by numerical simulation examples.

Keywords: Sensor monitoring network · Extreme Learning Machine (ELM)
Sequential Unscented Kalman Filter · Localization

1 Introduction

A spatial event, such as an earthquake, a tsunami, a toxic plume from a nuclear plant, a hurricane, and a wildfire, is initiated by a significant change in the state of a region in 3-D space over an interval of time. In addition to natural disasters, man-made events can also be modeled as such events, for instance, noxious plume emission, waste water emission, oil spills, etc. Additionally, disease outbreak is another event with its own specific characteristics. Networked sensing is a typical approach for event detection. With technical innovation and the development of sensors, more and different types of sensors are employed. Compared to traditional sensor networks composed of a small

© Springer Nature Singapore Pte Ltd. 2018
Z. Xu et al. (Eds.): Big Data 2018, CCIS 945, pp. 378–391, 2018.
https://doi.org/10.1007/978-981-13-2922-7_26

number of high quality (and expensive) sensors, trends in personal computing devices and consumer electronics have made it possible to build large, dense networks at a low cost. A energy-based method of sound source localization is proposed [1]. This method can accurately determine the appropriate path loss exponent and can improve localization accuracy, it can effectively improve the service life and positioning accuracy of the sensor under low power consumption. Two multi-fault diagnosis methods are proposed for sensor fault detection and identification respectively [2]. This method, combining the support vector machine and error-correcting output codes, regards nonlinear transformation as the input of classifiers to enhance the separability of initial characteristics, which can achieve high identification accuracy. The changes in sensor capability, network composition, and system constraints call for new models and algorithms suited to the opportunities and challenges of the new generation of sensor networks.

For the pollution source localization problem with wireless sensor networks, it should be mentioned that most existing research results have been concerned with gas pollution sources and consist of rough localization and analytical localization algorithms. The rough localization algorithm, represented by the closest point approach (CPA), takes the sensor position with the maximum concentration measurement as the approximate position of the pollution source [3]. The main advantage of the CPA algorithm is simplicity, while its position accuracy is heavily dependent on the node density. It is greatly influenced by the concentration observation error of a single node. The analytical localization algorithm generally, based on a diffusion model, utilizes concentration measurements to estimate the pollution source position. Methods used include Bayesian estimation, maximum likelihood estimation and Least Squares (LS) estimation. Bayesian estimation and maximum likelihood estimation methods are difficult to apply directly because they need to predict the prior probability density of the observed noise. For the LS method, while the observation error is larger, it is more difficult to apply to systems with time delays and packet dropout due to it being limited by the centralized strategy. The nonlinear LS and CPA algorithms for gas source localization have been compared and analyzed [4]. The position performance superiority of the LS algorithm has been verified, and it is noted that the iterative initial value should be selected appropriately in the process of solving for the nonlinear least squares estimate. According to the concentration diffusion model of gas pollution, the nonlinear LS and maximum likelihood estimation methods are used and analyzed in terms of localization results; the former performs better in the case of bigger noise, while the latter performs better in the opposite case [5]. The particle filter algorithm is used to estimate the single point position within one hundred square meters of an outdoor area [6]. With this method, it is difficult to model the density function; the phenomenon of particle degradation can appear with the increase of time and particle number, which could reduce the efficiency even to the point of divergence. The nonlinear LS method is used to estimate gas source position, and the localization results have been analyzed for different cases such as the distribution and number of nodes and background noise [7]. A Kalman filter is used to estimate the gas source position [8], but this method is heavily dependent on the accuracy of the model and the sensor data at the current moment, which requires the uniformity of sampling time. For a class of discrete systems with linear equality constraints consideration is subject to both noises and time-varying constrained conditions [9], a new reduced-order filter is proposed under a mild assumption such that the estimation performance of the proposed filter outperforms those of the traditional filters. A new distributed filter is proposed [10]. In order to reduce the bandwidth consumption and estimator update frequencies, an event-based signal transmission strategy is employed as opposed to the traditional time-based one. Deducing the diffusion model of the pollution source has become a focus of localization research.

With multiple man-made disasters, water pollution has been getting increasingly more attention compared to gas pollution. Localization is of great importance to the environmental protection of water. According to the concentration diffusion model, the localization problem only related to source location in wireless sensor networks is proposed. The main existing methods for the detection and localization of water pollution diffusion sources, such as remote sensing localization, underwater robot localization, and artificial detection, have some limitations. The localization accuracy of remote sensing is limited because it is vulnerable to bad weather. The localization accuracy of underwater robots is higher, but this method cannot work at a large scale and over a long time. Artificial detection has a long collection period, has poor real-time performance, and is unable to realize online monitoring. Using a sensor network as a new monitoring method could realize large scale and all-day real-time working performance independent of geographical location.

A piecewise concentration model has been put forward to analyze offshore plume source diffusion in static water [11]. Combined with different application requirements, the nonlinear LS method and Unscented Kalman Filter (UKF) method were applied to the localization problem of a pollution source that is near an impervious boundary in static water. The simulation results show that the algorithm based on UKF balances computation complexity and estimation accuracy well. According to the features of both the discharged pollutants in offshore lake reservoirs and the two dimensional diffusion model of pollutants, two localization methods to effectively deal with pollutants discharged in offshore areas have been proposed [12]. One is boundary constrained LS, and the other is boundary constrained linear programming with least absolute deviations; the latter performs better in either case.

Summarizing the discussions provided so far, localization appears to be a challenging task with three essential difficulties identified as follows: (1) Compared with a gas environment, the process for pollution diffusion and detection is quite different and difficult, which will lead to the parameters of the traditional diffusion model not being acquirable in real time. (2) The position of static pollution sources is not really static but drifts with disturbances on a small scale. (3) It is difficult for the sensor data to reach the information processing center at the same time because their spread relies heavily on the propagation distance, time delays and water flow. The main motivation of this paper is to provide satisfactory solutions to the three issues mentioned.

In this paper, the modeling and localization for a class of static pollution sources in an area of water based on a sensor monitoring network are investigated. The main contributions are outlined as follows:

This method improves the accuracy of positioning by improving the accuracy of the model and eliminating the uncertainties caused by random disturbances. The concentration diffusion model of water pollution was established by using ELM to solve the problem of accurate modeling in the case of unknown nonlinear diffusion equation. SUKF was used to reduce the dependence on data integrity. The method not only improves the positioning accuracy, but also satisfies the practical application situation. Based on the above ideas of the water pollution localization algorithm, the rest of this paper is organized. Section 2 describes a data driven modeling method of concentration diffusion based on ELM-PLS; Sect. 3 tells the ideas of static pollution sources localization; Sect. 4 simulates above algorithms and analyze; Sect. 5 is conclusion.

2 A Data Driven Modeling Method of Concentration Diffusion Based on ELM-PLS

Consider instantaneous diffusion from a point source in two dimensions. An object with quality M is dropped in water that has no boundary at position (x_0, y_0) at instant t_0. The diffusion coefficient D is isotropic, and the velocity of water is $v = (v_x, v_y)$. Let $C(x, y, t)$ denote the pollutant density at position (x, y) at t time, where t is continuous time. The mathematical model of diffusion is as follows [13]:

$$\begin{cases} \frac{\partial c}{\partial t} + v_x \frac{\partial c}{\partial x} + v_y \frac{\partial c}{\partial y} = D \frac{\partial^2 c}{\partial x^2} + D \frac{\partial^2 c}{\partial y^2} & (-\infty < x, y < +\infty, t > 0) \\ c_{x,y,t|t=t_0} = M\delta(x - x_0)\delta(y - y_0) & (-\infty < x, y < +\infty) \\ \lim_{x \to \pm\infty} c = \lim_{y \to \pm\infty} c = 0 & (t > 0) \end{cases} \quad (1)$$

By solving (1), the following relationship between the position and density at t time is obtained:

$$C(x, y, t) = \frac{M}{4\pi(t - t_0)D} \exp\left\{ -\left[\frac{(x - x_0 - v_x(t - t_0))^2 + (y - y_0 - v_y(t - t_0))^2}{4D(t - t_0)} \right] \right\}$$

$$(2)$$

Equation (2) is difficult to solve because the various values assumed in (1), such as the velocity of water and the diffusion coefficient, are difficult to obtain and because of external disturbance. This will make it difficult to apply this equation. To overcome this problem, attention is paid to the modeling methods based on data because they are more flexible and obtain results closer to the actual values.

For modeling based on linear fitting to data, Partial Least Squares (PLS) [14–16] is a classical algorithm in multivariate statistical analysis. The PLS regression algorithm divides the model into an inner model and an outer model. The outer model converts raw data to the latent variable space and obtains mutually orthogonal score vectors, while the inner model establishes a linear relationship between the scores. Then, quality variables are described that have a strong correlation with the process variables. Finally, the quality variables are monitored and predicted simultaneously. For modeling using nonlinear fitting, the PLS method whose inner model is linear will no longer apply. To overcome the nonlinear fitting problem, ELM is utilized in the process of linear PLS modeling. This not only improves the fitting accuracy but also avoids the problem of limited estimated performance. To obtain a robust model, a nonlinear modeling method based on ELM-PLS is proposed in this paper.

ELM is an effective learning method aimed at single layer feed-forward neural networks [17, 18]. As the related literature notes, choosing the right activation function $g(x)$ can allow the input weights and bias to be obtained through the random probability distribution function independent of the training data. It is noted that the network could converge to any continuous objective function given any non-constant piecewise continuous or integrable activation function [19]. After obtaining the hidden layer output matrix, the output weights can be obtained by solving its generalized inverse. This algorithm has a faster execution speed because its parameters for the model input layer do not require iterative adjustment and wide adaptation [20, 21].

Letting the inputs be the source position, sensor position and sampling time, the output is the corresponding measurement of diffusion density. For three dimensional cases closer to reality, the dimension of the input is 7, and the output dimension is 1. N sensors could obtain N values of sample data at one time.

The modeling steps are as follows:

Step 1: training sample data collection

There are N sensors in the monitoring field, assuming that an instantaneous pollution source is positioned at a central point without loss of generality. Let the measurements of diffusion density be outputs, with the source position, sensor position and sampling time as inputs. Use the input data set $X \epsilon R^{nN \times 7}$ to train the output data set $Y \epsilon R^{nN \times 1}$.

Step 2: the outer model using PLS

X and Y should be decomposed into:

$$X = t_i p_i^T + E_i$$
$$Y = u_i q_i^T + F_i \qquad (3)$$

where t_i and u_i are the i-th principal component score vectors of X and Y, respectively. p_i and q_i are the corresponding loads. E_i and F_i are the corresponding residuals, which are obtained by subtracting the i-th principal component from X and Y. To extract the i-th principle component, the data are i-th preprocessed using linear PLS with (3). Through the outer calculation, we can obtain the scores of two data sets. The initial value of i is 1 (Table 1).

Step 3: the inner model using ELM

Train the ELM model with the scores $u_i = f(t_i)$, where $f(.)$ is a nonlinear mapping function.

(1) Given the activation function $g(x)$ and hidden notes L_0, the input weights and bias are randomly set as (w_i, a_i), i = 1, 2..., L_0;
(2) Map the t_i to the feature space in order to obtain the hidden layer output matrix f;
(3) The weight of output nodes, $\beta = f^+ u_i$, f^+, is the generalized inverse of f.

Table 1. The traditional batch processing PLS algorithm

Given the normalization data(X,Y), initialize $X_0=X$, $Y_0=Y$, $i=0$

1) $i=i+1$, u_i is any row of Y_0

2) Calculate the outer model of PLS until converges $w_i = X_{i-1}^T u_i / u_i^T u_i$, $t_i = X_{i-1} w_i / \| X_{i-1} w_i \|$,

  $q_i = Y_{i-1}^T t_i / \| Y_{i-1}^T t_i \|$, $u_i = Y_{i-1} q_i$

3) The load of X $P_i = X_{i-1}^T t_i / t_i^T t_i$

4) The regression coefficients of inner model $b_i = u_i^T t_i / \| t_i^T t_i \|$

5) Calculate residual $E_i = X_{i-1} - t_i p_i^T$, $F_i = Y_{i-1} - b_i t_i q_i^T$

6) Let $X_i = E_i$, $Y_i = F_i$, return to step 1 until all principal components are extracted

---

(4) Replace score u_i with the prediction value $\hat{u}_i$ of u_i based on ELM. Calculate the residual matrices of two data sets:

$$\begin{cases} E = X - t_i p_i^T \\ F = Y - \hat{u}_i q_i^T \end{cases}, \quad \hat{u}_i = f(t_i) \tag{4}$$

Update the data sets (X, Y), and then, calculate the next principal component.

$$\begin{cases} X = E \\ Y = F \end{cases} \tag{5}$$

Repeat the above process with the new data sets (X, Y) until all the principal components are extracted (Table 2).

Table 2. Modeling by ELM-PLS algorithm

---

1) Given the normalization data (X,Y), initialize $i=1$, u_i is any row of Y_0

2) Iterate the outer model of PLS until converges $w_i = X^T u_i / u_i^T u_i$, $t_i = X w_i / \| X w_i \|$, $q_i = Y^T t_i / \| Y^T t_i \|$,

  $u_i = Y q_i$

3) Calculate the load of X $p_i = X^T t_i / t_i^T t_i$

4) Modelling the inner model by ELM algorithm, Using the score (t_i, u_i) from the outer model to train the network, and obtain the prediction $\hat{u}_i$ of u_i

5) Calculate residual $E_i = X - t_i p_i^T$, $F_i = Y - \hat{u}_i q_i^T$, update the next iteration data $X = E_i$, $Y = F_i$

6) $i=i+1$, return to step 2 until all principal components are extracted, the output prediction $\hat{Y} = U Q^T$

The overall calculation is schematically described in Fig. 1.

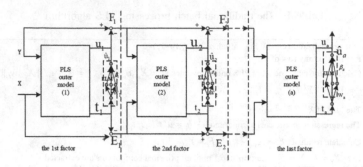

Fig. 1. The flow chart of the new nonlinear PLS modeling

3 The Localization of Static Pollution Sources Based on SUKF

Based on the above model, the diffusion process of a pollution source is nonlinear, as shown by (2). If the position of the pollution source is constant all the time, the LS algorithm could obtain better results. In practice, the position of the static pollution sources can be regarded as a random walk process along with the change of outside disturbance. Its state function is a transcendental equation with strong nonlinearity, which can be transformed to Kalman filter form [22]. Furthermore, it is difficult for the sensor data to reach the information processing center at the same time because of the non-uniformity of the sampling time and time delays in the sensor network, so SUKF [23] is utilized in this paper to realize the localization of the static pollution source by combining it with ELM modeling. In this section, LS and this proposed ELM-SUKF method are introduced and compared.

3.1 LS Localization Method Based on Distance

In the monitoring field, let the sensor positions be (x_i, y_i, z_i), $i = 1..., N$, the pollution source position be (x_0, y_0, z_0), and the distances from each sensor to pollution source be denoted by $d_1, d_2..., d_N$. Thus, from the relationship between them

$$\begin{cases} (x_1 - x_0)^2 + (y_1 - y_0)^2 + (z_1 - z_0)^2 = d_1^2 \\ (x_2 - x_0)^2 + (y_2 - y_0)^2 + (z_2 - z_0)^2 = d_2^2 \\ \vdots \\ (x_N - x_0)^2 + (y_N - y_0)^2 + (z_N - z_0)^2 = d_N^2 \end{cases} \tag{6}$$

we can obtain

$$A[x_0, y_0, z_0]^T = b \tag{7}$$

$$A = 2 \begin{bmatrix} x_N - x_1 & y_N - y_1 & z_N - z_1 \\ x_N - x_2 & y_N - y_2 & z_N - z_2 \\ \vdots & \vdots & \\ x_N - x_{N-1} & y_N - y_{N-1} & z_N - z_{N-1} \end{bmatrix},$$

$$b = \begin{bmatrix} d_1^2 - d_N^2 - (x_1^2 + y_1^2 + z_1^2) + (x_N^2 + y_N^2 + z_N^2) \\ d_2^2 - d_N^2 - (x_2^2 + y_2^2 + z_2^2) + (x_N^2 + y_N^2 + z_N^2) \\ \vdots \\ d_{N-1}^2 - d_N^2 - (x_{N-1}^2 + y_{N-1}^2 + z_{N-1}^2) + (x_N^2 + y_N^2 + z_N^2) \end{bmatrix}$$

(8)

$$[x_0, y_0, z_0]^T = (A^T A)^{-1} A^T b \tag{9}$$

3.2   The Proposed Localization Method Based on ELM-SUKF

Based on the above diffusion model built by ELM-PLS, let $X(k)$ denote the real time pollution location, where w denotes the disturbance at time k. The dynamic model varying with time is described by the following model.

$$\begin{cases} X(k) = f(k, k-1)X(k-1) + w(k, k-1) \\ Z_i(k) = h(k, S_i, X(k)) + v(k) \end{cases} \tag{10}$$

where $f(k, k-1) = I$. $S_i = (x_i, y_i, z_i)^T$ is the position of the i-th sensor. $X(k)$ is the central position of pollution sources at time k, where k is a discrete sample time, $Z_i(k)$ is the density measurement of the i-th sensor at time k, and $h(.)$ is the fitted equation based on ELM-PLS. $w(k, k-1)$ and $v(k)$ are zero mean white Gauss noise that satisfy $w(k, k-1) \sim N(0, Q)$ and $v(k) \sim N(0, R)$.

Assuming the estimated value $\hat{x}_{k-1|k-1}$ at time t_{k-1}, the covariance P_{k-1} and the measurement vectors $Z_i(k)$ ($i = 1...N$) of N sensors at time t_k are known. The abstract process of the algorithm is as follows:

(1) Sigma sampling

Using the symmetric sampling strategy to obtain the Sigma-point set, the corresponding particle weight W^m and the variance weight W^c are as follows.

$$\chi_{k-1} = [\hat{x}_{k-1|k-1}, \hat{x}_{k-1|k-1} - \sqrt{(L+\lambda)P_{k-1}}, \hat{x}_{k-1|k-1} + \sqrt{(L+\lambda)P_{k-1}}] \tag{11}$$

$$W_0^m = \lambda/(L+\lambda) \tag{12}$$

$$W_j^m = 1/[2(L+\lambda)], j = 1, \ldots, 2n \tag{13}$$

$$W_0^c = \lambda/(L+\lambda) + (1 + \alpha^2 + \beta) \tag{14}$$

$$W_j^c = 1/[2(L+\lambda)], j = 1, \ldots, 2n \tag{15}$$

(2) Time update

All particles are nonlinearly mapped through the state equation, and we obtain

$$\chi_{k|k-1} = f(\chi_{k-1}) \tag{16}$$

The state one-step prediction and its corresponding error covariance can be obtained by weighted summation:

$$\hat{x}_{k|k-1} = \sum_{j=0}^{2n} W_j^m \cdot \chi_{j,k|k-1} \tag{17}$$

$$P_{k|k-1} = \sum_{j=0}^{2n} W_j^c (\chi_{j,k|k-1} - \hat{x}_{k|k-1})(\chi_{j,k|k-1} - \hat{x}_{k|k-1})^T + Q \tag{18}$$

Let $\hat{x}_{k|k-1}$ undergo the UT transform again, and obtain the new Sigma point set

$$\chi_{k|k-1}^z = \left[\hat{x}_{k|k-1} \quad \hat{x}_{k|k-1} - \left[\sqrt{(L+\lambda)P_{k-1}} \right] \quad \hat{x}_{k|k-1} + \left[\sqrt{(L+\lambda)P_{k-1}} \right] \right] \tag{19}$$

The Sigma-point sets are nonlinearly mapped through the measurement equation

$$z_k = h(\chi_{k|k-1}^z) \tag{20}$$

The measurement prediction of the system can be computed from the weighted summation

$$\hat{z}_k = \sum_{j=0}^{2n} W_j^m z_{j,k} \tag{21}$$

(3) Measurement update

(a) The cross-correlation matrix and covariance matrix of sensor i are

$$P_{xz} = \sum_{j=0}^{2n} W_j^c \cdot (\chi_{j,k|k-1} - \hat{x}_{k|k-1})(z_{j,k} - \hat{z}_k)^T \tag{22}$$

$$P_z^i = \sum_{j=0}^{2n} W_j^c \cdot (z_{j,k} - \hat{z}_k)(z_{j,k} - \hat{z}_k)^T + R_i \tag{23}$$

The filter gain is

$$K_k^i = P_{xz}(P_z^i)^{-1} \tag{24}$$

Using the measurement z_k^i of sensor i to update the state, the estimated state $\hat{x}_{k|k}^i$ and the corresponding estimated error covariance matrix $P_{k|k}^i$ are

$$\begin{cases} \hat{x}_{k|k}^i = \hat{x}_{k|k-1} + K_k^i(z_k^i - \hat{z}_k) \\ P_{k|k}^i = P_{k|k-1} - K_k^i P_z^i (K_k^i)^T \end{cases} \tag{25}$$

(b) Let $\hat{x}_{k|k-1} = \hat{x}_{k|k}^i$, $P_{k|k-1} = P_{k|k}^i$; the measurement z_k^{i+1} is used to update $\hat{x}_{k|k}^i$ and $P_{k|k}^i$, and the estimated state $\hat{x}_{k|k}^{i+1}$ and the corresponding estimated error covariance matrix $P_{k|k}^{i+1}$ are obtained.

(c) Repeat steps (a) and (b) from $i = 1$ to $i = N$

Finally, the estimated state $\hat{x}_k$ and the corresponding error covariance matrix P_k at time k based on all sensor measurements can be obtained: $\hat{x}_k = \hat{x}_k^N$ and $P_k = P_k^N$.

4 Simulation and Analysis

Example 1: To verify the effectiveness of this new modeling method based on ELM-PLS, the new method and traditional PLS methods are compared. Consider the following numerical example:

$$\begin{cases} x_k = Az_k + e_k \\ y_k = \exp(x_k(1)) + x_k(2) + x_k(3) + v_k \end{cases} \tag{26}$$

$$A = \begin{bmatrix} 1 & 3 \\ 3 & 0 \\ 1 & 2 \end{bmatrix} z_k \sim U(0,1) e_k \sim N(0,q)$$

$$q = \begin{bmatrix} 0.05^2 & 0 & 0 \\ 0 & 0.05^2 & 0 \\ 0 & 0 & 0.05^2 \end{bmatrix} v_k \sim N(0,0.1^2)$$

The data set is (X, Y), $X \in R^{100 \times 3}$, $Y \in R^{100 \times 1}$; the number of ELM hidden nodes is selected as 20. The prediction errors of the two methods are shown in Fig. 2. The prediction RMSE of the two methods is shown in Table 3.

From Fig. 2 and Table 3, it can be seen that the prediction error of this ELM-PLS method is smaller than that of the traditional PLS. When there is a nonlinear relationship between data sets, there will be a nonlinear relationship between principal component scores. The entire model cannot represent nonlinear relationship between data while the linear internal model is still used, as this will make the fitting

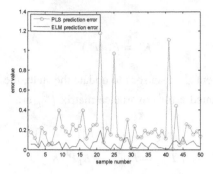

Fig. 2. The prediction errors of the two methods

Table 3. Comparison of prediction RMSE

RMSE	Linear PLS (%)	ELM-PLS (%)
Training data	29.56	6.48
Test data	32.82	6.50

performance worse. Therefore, under the premise of increasing the robustness of PLS, applying ELM to the process of establishing an internal model could better capture the internal nonlinear relationship.

Example 2: Consider two-dimensional cases for simplicity. When $t_0 = 0$, the pollution source with quality $M = 500$ kg is dropped in the center point of an area of water of size 20×20 m^2. The diffusion coefficient $D = 5$ m^2/h and varies isotropically, and the sampling interval $T = 1$ s. The state transition noise variance matrix $Q = [1\ 0;\ 0\ 0.1]$, and the measurement noise variance $R = 0.01$. We initialize $P_0 = [1\ 0;\ 0\ 0.01]$, $\alpha = 1e^{-2}$, $\beta = 2$, and $k_0 = 0$. One hundred Monte Carlo simulations were performed to contrast the localization performances and analyze the localization errors between the SUKF and LS algorithms, shown in Figs. 3 and 4.

In Fig. 3, the blue line composed of points denotes the localization results using the LS method and the red line denotes the localization results using SUKF. It can be seen that the localization errors obtained using SUKF are smaller than those using the LS method as a whole. With increasing number of sensors, the localization errors of the two methods are all gradually reduced, and these errors tend to a stable value. That is, increasing the number of sensors can raise the accuracy of localization. However, more is not better. In practice, the number of sensors should be chosen based on cost and a reasonable demand for accuracy. In Fig. 4, when the number of sensor is 20, the localization result obtained using SUKF is close to the real target, and its localization accuracy is higher compare to LS.

Example 3: There are 20 sensors evenly distributed in an area of water of size 20×20 m^2. When $t_0 = 0$, the pollution source with quality $M = 500$ kg is dropped at the center point of the water area. The diffusion coefficient is $D = 5$ m^2/h and varies isotropically, and the sampling interval $T = 0.2$ s. Recording the sensor position, pollution source position and sampling time as inputs and using the corresponding

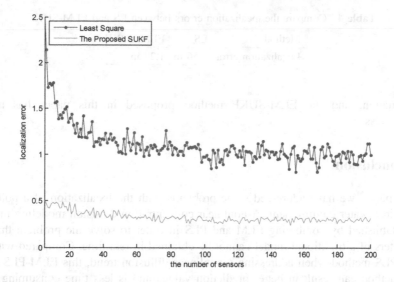

Fig. 3. The localization errors with the increase of sensor number between two methods (Color figure online)

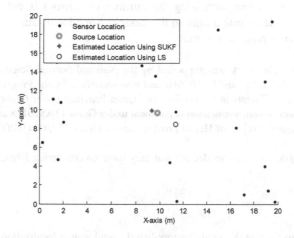

Fig. 4. The localization results between two methods with $N = 20$

measurements as output, ELM is used to train the measurement equation, where the number of hidden nodes is 1000, the activation function $g(x) = tribas(x)$, and the dimensions of the input and output are, respectively, 5 and 1. Two thousand samples of data at 100 sample times are collected to train the measurement equation, and SUKF is used to process the pollution localization. To compare the localization effects between the LS and SUKF methods, 10 Monte Carlo simulations were performed. The results are shown in Table 4.

It can be seen that the localization effect of the LS method is not as good as that of ELM-SUKF because of its poor robustness, being easily affected by the elimination of

Table 4. Compare the localization errors between LS and ELM-SUKF

Method	LS	ELM-SUKF
Localization error	1.56 m	1.29 m

the equation, and the ELM-SUKF method proposed in this paper has higher effectiveness.

5 Conclusion

In this paper, we have addressed some problems with the localization of a pollution source in a water environment. A nonlinear concentration diffusion modeling method was established by combining ELM and PLS in order to solve the problem that the parameters of a traditional model cannot be obtained in real time. Compared with the linear PLS method when addressing a nonlinear diffusion trend, this ELM-PLS modeling method can result in better prediction values and is less time consuming. Furthermore, the localization method, deduced by combining ELM-PLS modeling with the SUKF method, could realize more precise localization compared to the LS method. Meanwhile, it is noted that increasing the number of sensors is not better. Further research topics include investigations on the modeling and localization problems with a continuously moving pollution source.

Acknowledgments. This work was supported by the National Natural Science Foundation of China under Grants 61503123 and U1504616, and was sponsored by the Program for Science & Technology Innovation Talents in Universities of Henan Province under Grant 17HASTIT021, Basic research project of henan education department under Grant 13A510188 and the Basic and frontier science research project of Henan province under Grant 152300410200.

Competing Interests. The authors declare that they have no competing interests.

References

1. Deng, F., Guan, S., Yue, X., et al.: Energy-based sound source localization with low power consumption in wireless sensor networks. IEEE Trans. Ind. Electron. **64**(6), 4894–4902 (2017)
2. Deng, F., Guo, S., Zhou, R., et al.: Sensor multifault diagnosis with improved support vector machines. IEEE Trans. Autom. Sci. Eng. **14**(2), 1053–1063 (2017)
3. Milios, E.E., Nawab, S.H.: Acoustic tracking from closest point of approach time, amplitude, and frequency at spatially distributed sensors. J. Acoust. Soc. Am. **87**(3), 1026–1034 (1990)
4. Michaelides, M.P., Panayiotou, C.G.: Plume source position estimation using sensor networks. In: Proceedings of the 13th Mediterranean Conference on Control and Automation, Cyprus, pp. 731–736. IEEE (2005)
5. Kuang, X., Shao, H.: Study of plume source localization based on WSN. J. Syst. Simul. **19**(7), 1464–1467 (2007)

6. Li, J.G., Meng, Q.H., Wang, Y., et al.: Odor source localization using a mobile robot in outdoor airflow environments with a particle filter algorithm. Auton. Robots **30**(3), 281–292 (2011)
7. Wang, H., Zhou, Y., Yang, X.L., et al.: Plume source localizing in different distributions and noise types based on WSN. In: Proceedings of the 2010 International Conference on Communications and Mobile Computing, Caen, pp. 63–66. ACM (2010)
8. Zoumboulakis, M., Roussos, G.: Estimation of pollutant-emitting point-sources using resource-constrained sensor networks. In: Trigoni, N., Markham, A., Nawaz, S. (eds.) GSN 2009. LNCS, vol. 5659, pp. 21–30. Springer, Heidelberg (2009). https://doi.org/10.1007/978-3-642-02903-5_3
9. Wen, C., Cai, Y., Liu, Y., et al.: A reduced-order approach to filtering for systems with linear equality constraints. Neurocomputing **193**(12), 219–226 (2016)
10. Wen, C., Wang, Z., Geng, T., et al.: Event-based distributed recursive filtering for state-saturated systems with redundant channels. Inf. Fusion **39**, 96–107 (2018)
11. Luo, X., Chai, L., Yang, J.: Offshore pollution source localization in static water using wireless sensor networks. Acta Automatica Sinica **40**(5), 849–861 (2014)
12. Jun, Y., Li, C., Xu, L.: Pollution source localization in lake environment based on wireless sensor networks. In: Proceedings of the 31st Chinese Control Conference, Hefei, China, 25–27 July 2012 (2012)
13. Li, D., Wong, K.D., Hu, H.Y., et al.: Detection classification and tracking of targets in distributed sensor networks. IEEE Sig. Process. Mag. **19**(2), 17–30 (2002)
14. Ding, S.X., Yin, S., Zhang, P., et al.: Study on modifications of PLS approach for process monitoring. Threshold **2**, 12389–12394 (2011)
15. Hu, J., Wen, C.L., Ping, L., et al.: Direct projection to latent variable space for fault detection. J. Franklin Inst. **351**(3), 1226–1250 (2014)
16. Wen, C., Zhou, F., Wen, C., et al.: An extended multi-scale principal component analysis method and application in anomaly detection. Chin. J. Electron. **21**(3), 471–476 (2012)
17. Huang, G.B.: What are extreme learning machines? Filling the gap between Frank Rosenblatt's Dream and John von Neumann's Puzzle. Cogn. Comput. **7**, 263–278 (2015)
18. Huang, G., Huang, G.B., Song, S., et al.: Trends in extreme learning machines: a review. Neural Netw. **61**(1), 32–48 (2015)
19. Huang, G.B., Zhou, H.M., Ding, X.J., et al.: Extreme learning machine for regression and multiclass classification. IEEE Trans. Syst. **42**(2), 513–529 (2012)
20. Alom, M.Z., Sidike, P., Taha, T.M., et al.: State preserving extreme learning machine: a monotonically increasing learning approach. Neural Process Lett. **45**(2), 703–725 (2016)
21. Zhai, J., Shao, Q., Wang, X.: Architecture selection of ELM networks based on sensitivity of hidden nodes. Neural Process Lett. **44**, 471–489 (2016)
22. Wen, C., Ge, Q., Tang, X.: Kalman filtering in a bandwidth constrained sensor network. Chin. J. Electron. **18**(4), 713–718 (2009)
23. Liu, H., Huang, S.: Method for radar target tracking based on sequential unscented Kalman filter. Comput. Eng. Appl. **45**(25), 202–204 (2009)

A Method for Locating Parking Points of Shared Bicycles Based on Clustering Analysis

Guanlin Chen[1,2(✉)], Yukun Shi[1], Huang Xu[1], and Bin Zhang[3]

[1] School of Computer and Computing Science, Zhejiang University City College, Hangzhou 310015, People's Republic of China
chenguanlin@zucc.edu.cn
[2] College of Computer Science, Zhejiang University, Hangzhou 310027, People's Republic of China
[3] E-Government Office of Hangzhou Municipal People's Government, Hangzhou 310020, People's Republic of China

Abstract. In order to solve a series of problems caused by the disorderly cycling of bicycle-sharing, the modeling problem, forecasting the needs of the shared bicycle and selecting the location for the bicycle-sharing parking point, is optimally solved by the improved algorithm. It provides an important theoretical basis for the management planning of bicycle-sharing. First, a bicycle-sharing demand forecasting model based on Canopy-Kmeans is proposed according to a large number of bicycle-sharing data collected. Then, the multi-objective location models for bicycle-sharing parking point is established with the minimum construction cost and the shortest total travel distance as the objective function according to demand points and requirements for planning parked points. It can effectively ensure the effectiveness and rationality of the location. Finally, the improved NSGA-II algorithm is used to solve the location model proposed in this paper. The results show that the model proposed in this paper can provide a more scientific basis for decision-makers to choose parking points for bicycle-sharing.

Keywords: Bicycle-sharing · Clustering analysis · Demand forecasting
Location model · Multi-objective optimization

1 Introduction

At present, with the development of society and the improvement of living standards, people's travel awareness has also changed. Under the background of the rapid development of public bicycles, the product of the fusion of modern technology and public bicycles has emerged—shared bicycles. Because people can borrow vehicles at any time, shared bicycles quickly occupy the core position of the market, which brings many benefits to society and users. However, problems such as the free parking of shared bicycles, severe damage, and failure to clean up in time have affected people's lives seriously. To solve these problems, more feasible programs are basically using electronic fences to plan parking spots. The electronic fences technology is to circle a

© Springer Nature Singapore Pte Ltd. 2018
Z. Xu et al. (Eds.): Big Data 2018, CCIS 945, pp. 392–407, 2018.
https://doi.org/10.1007/978-981-13-2922-7_27

virtual parking area. The user must park the car in this area to complete the returning of the car, otherwise it will continue to charge. It can be seen that the management of shared bicycles still need to return to the fixed-point parking mode of public bicycle systems. It is particularly important to properly plan the parking spots for shared bicycles. If the site selection is unreasonable, it will have the same problems as traditional public bicycles. Based on the characteristics of shared bicycles and the study of public bicycle system demand planning, the paper will study the demand point prediction of shared bicycles and site selection planning of parking spots. Through accurate planning of demand forecasting and parking location methods, it provides a theoretical basis for government decision makers to make more rational decisions, thereby promoting the healthy development of shared bicycles and promoting green and healthy traveling modes.

2 Related Work

From the initial fixed-point borrowing mode to sharing bicycles, public bicycles have experienced a long process of development. In the early stages of its development, because the traffic pressure was relatively small, the concept of green and low-carbon travel has not yet been popularized, which has led to serious obstacles to its development. With the ever-changing concept of travel, people gradually focus on this issue, and the public bicycle rental system has also been applied, but it is still in the development stage. DeMaio et al. have focused on whether bicycles can be used as public transportation and whether they can be accepted by the U.S. nationals, and initially determine the direction of development of public bicycles [1]. García-Palomares et al. pointed out that the relationship between bicycle stations and potential factors is one of the key factors for planning sites [2]. A GIS-based solving method was established to calculate the spatial distribution of potential demand, and to determine the site usage patterns and site capacity. Finally, it discusses the relationship between the increase in the number of vehicles and station coverage and user satisfaction. Chemla et al. adopted a static rebalancing strategy in the scheduling of shared bicycles to ensure that the demand for shared bicycles in the region was balanced [3]. Deng et al. analyzed the bicycle sharing data that had been generated and put forward suggestions for the layout of shared bicycle parking facilities [4]. Zhao used the shared bicycle network as the research object [5]. According to the characteristics of the network and the analysis of the clustering subgroups, a model for optimal scheduling of shared bicycles was established. It establishes a different demand scheduling model for different objective functions to find the optimal scheduling strategy. Through the research of existing methods, it has been found that research on shared bicycles at home and abroad mainly stays at the level of theoretical analysis, but there are not many specific theoretical studies on the operation management, scheduling and planning of shared bicycles. Therefore, the study of public bicycle systems and shared bicycles is not systematic enough and lacks scientific theories and methods. Based on the historical data of shared bicycles, this paper forecasts the demand points and establishes a site selection model between the demand points and the planned parking spots to solve the model and provides a basis for the selection of parking locations for shared bicycles.

3 Demand Forecasting for Bicycle-Sharing

Because of GPS positioning system mounted on bicycle-sharing, massive user data which can truly reflect the use of bicycle-sharing is generated. In this paper, a demand model is built to get the demand for bicycle-sharing according to analyze these data collected. The source of the data set is the traveling data of bicycle-sharing users in Chengdu in January 2017 from a bicycle-sharing company with a relatively high domestic market share in China. As shown in Table 1, the data includes time, bicycle number, bicycle type, GPS location information, etc. As shown in Fig. 1 the obtained data set is shown in the way of map, and it can clearly see the distribution of location information. A number of areas of demand are formed by the way of clustering analysis. Then the cluster center point in demand area is used as the demand point, and the number of shared bicycles in the demand area is used as the demand point.

Table 1. Data set partial data

Time	Type	Number	Longitude	Latitude
2017-01-19T23:43:37	1	280018590	104.126928	30.783269
2017-01-19T23:43:37	2	286046586	104.126985	30.782266
2017-01-19T23:43:37	2	286047583	104.126212	30.782425
2017-01-19T23:43:37	1	280023534	104.127407	30.781737
2017-01-19T23:43:37	1	280013690	104.12612	30.782319
2017-01-19T23:43:37	1	280028989	104.127	30.78427
2017-01-19T23:43:37	1	280022652	104.127044	30.781257
2017-01-19T23:43:37	2	286056292	104.125227	30.782448
2017-01-19T23:43:37	2	286006531	104.124544	30.782998
2017-01-19T23:43:37	2	286006110	104.12834	30.785123
2017-01-19T23:43:37	2	286007251	104.124162	30.783977
2017-01-19T23:43:37	1	280048499	104.13086	30.7818
2017-01-19T23:43:37	1	280047966	104.131123	30.783975
2017-01-19T23:43:37	1	280048516	104.131636	30.783611
2017-01-19T23:43:37	1	280042657	104.129583	30.779117
2017-01-19T23:43:37	1	280025242	104.129548	30.779026
2017-01-19T23:43:37	1	280026185	104.132329	30.78322
2017-01-19T23:43:37	1	280010678	104.131435	30.785551
2017-01-19T23:43:37	2	286052598	104.123329	30.785809

From the obtained data distribution, it can be seen that the parking location of the bicycle-sharing is mainly based on the citizens' travel habits. So, some data points which are disorderly and disorderly are filtered out to get the demand point and demand number which is closer to the reality. And a demand point forecasting model based on Canopy-Kmeans algorithm [5] is established.

The K-means algorithm needs to preset K value and the initialization cluster center is randomly generated, which results in the different results of each running K-means

Fig. 1. Partial shared bicycle distribution

algorithm, so choosing the cluster center reasonably has a great influence on the clustering effect [6]. In view of the shortage of clustering in K-means algorithm and considering the rationality of clustering results and the clustering effect, the K-means algorithm is improved in conjunction with Canopy algorithm [7]. First, Canopy algorithm is used to initialize the data, and then the number of clusters and initial cluster centers needed by K-means is got. Finally, the data are clustered with K-means.

The framework of the demand point prediction model presented in this paper is shown in Fig. 2

(1) The two thresholds of Canopy are set according to the actual situation of the demand point, that is, T1 is the maximum distance between the demand points, and T2 is the maximum range of each demand point.

(2) The Canopy algorithm is executed to get the number of demand points and the location of the demand points.

(3) The generated demand points are screened and then the new data set is got by removing the outliers with less demand.

(4) The number of remaining demand points is used as the K value, and the location of the demand point is used as the initial cluster center, and then the steps are iterated through the K-means algorithm. Finally, the clustering results are obtained.

The combination of Canopy and K-means algorithm can overcome the uncertainty caused by artificial selection of K and avoid the local optimization and the instability of the algorithm caused by the random selection of the initial cluster center. In addition, it reduces the influence of outliers on clustering results, so the clustering performance of K-means algorithm is greatly improved.

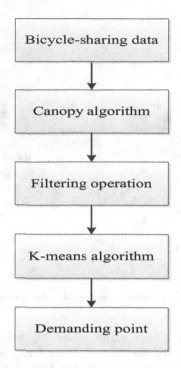

Fig. 2. Demand point prediction model framework

According to the above demand forecasting model, some data sets are processed in this paper. Figure 3 is the distribution state of the original data and Fig. 4 is an effect map after clustering. In Figs. 3 and 4, the points of the same color represent a cluster and the clustering results of the demand point. It can be clearly seen that, a better clustering of the original data is carried out and some isolated points are removed

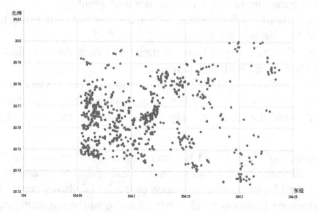

Fig. 3. Data point distribution before clustering

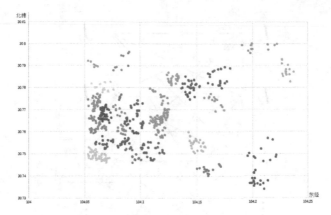

Fig. 4. Data distribution after clustering

through the Canopy-Kmeans algorithm. It is good enough to achieve the prediction results of bicycle-sharing demand points.

4   Shared Bicycle Parking Sites Selection Model

From the above, it can be seen that the problem of the location of the shared bicycle parking sites belongs to the NP-hard problem. In the process of solving this problem, it is impossible to obtain a specific site selection result. Therefore, when establishing the model of shared bicycles parking sites, this paper makes some assumptions for the model to improve the feasibility of the model [8]. Through the analysis of this paper, it is found that the location of shared bicycle parking sites has the following character-istics: The choice of shared bicycle parking sites is limited by cost. In addition to cost limitation, how to improve the traveler's convenience is also the key factor to deter-mine the location of the parking sites. This article assumes that when a traveler is using a shared bicycle, he will definitely choose the nearest parking sites. In order to fully meet the needs of different travelers, the distance from the demand point to the parking sites of the shared bicycle is the shortest [9].

In order to optimize the model, the location of all shared bicycle in the demand area is considered as the location of the demand points [10]. And the center of the electronic fence is considered as the location of the shared bicycle parking sites. In order to make the distance between all bicycles in each demand area and the shared bicycle parking sites as close as possible. The model uses the total distance from the bicycle to the planned parking sites in all areas of demand as the optimization goal. Minimize the distance between all demand points and parking sites.

The problem of the location planning of the shared bicycle parking sites is to allocate the optimal quantity between the demand points and the planned parking sites. And to obtain the number of allocated bicycles allocated to each planned parking sites by the demand points of shared bicycles. Design diagram, as shown in Fig. 5.

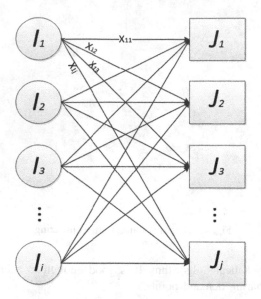

Fig. 5. Target assignment model diagram

The model is based on the optimization goal of minimizing the total construction cost of the shared bicycle parking sites and the shortest total distance of the users [11]. The specific mathematical model is expressed as:

$$\min f(x) = \sum_{i=1}^{I} \sum_{J=1}^{J} x_{ij} d_{ij} \tag{4.1}$$

$$\min f(y) = \sum_{j=1}^{J} (y_j M + a_j Y) \tag{4.2}$$

$$s.t \sum_{j=1}^{J} x_{ij} = n_i \, \forall i \tag{4.3}$$

$$\sum_{i=1}^{I} x_{ij} = c_j \, \forall j \tag{4.4}$$

$$y_j = \begin{cases} 1 & c_j > 0 \\ 0 & c_j = 0 \end{cases} \tag{4.5}$$

$$a_j = \begin{cases} c_j = 0 & c_j - c > 0 \\ 0 & c_j - c < 0 \end{cases} \tag{4.6}$$

In the formula

I: Indicates the set of demand points $\{1, 2, 3...i\}$;

J: Indicates the set of planned parking sites $\{1, 2, 3...j\}$;

n_i: Indicates demand for bicycles at demand point i;

d_{ij}: Denotes the distance from the demand point i to the candidate planning parking sites j;

x_{ij}: Indicates the number of bicycles assigned to the candidate planned parking sites j by the demand point i;

c_j: The number of the total number of bicycles assigned by the parking sites after allocation;

M: Indicates the capital construction cost for each candidate planned parking sites;

c: Indicates the number of basic bicycles planned by each candidate for planning parking sites. Each additional number of basic bicycles will increase construction and management costs Y;

y_j: Indicates whether to build this candidate planned parking sites;

a_j: Indicates the number of planned parking sites beyond the number of basic bicycles;

Among them, the objective function (4.1) minimizes the total distance from the demand point of the bicycle to the candidate planned parking sites [12]. The objective function (4.2) minimizes the total cost of parking sites. Formula (4.3) indicates that the shared bicycles at the demand point are allocated to the parking sites. Formula (4.4) is used to calculate the number of bicycles after the allocation of parking sites. Formula (4.5) shows that if the number of parking sites after the allocation is 0, the planned parking sites will not be constructed. Formula (4.6) represents the number of basic bicycles that exceed the planned parking sites.

From the analysis, it can be seen that if the total distance is the shortest, then all bicycles can be assigned to the nearest parking sites, but this will lead to a large cost. In the same way, in order to reduce the cost, it should be as little as possible to establish a shared parking sites on the premise that the number of basic parking sites is not more than possible. Therefore, the model is a multi-objective optimization model. Although the two objective functions of the model can not be minimized at the same time, there will be a Pareto optimal solution and a Pareto optimal solution set.

5   A Solving Algorithm Based on Improved NSGA-II Model

This paper improves the multi-objective model established in the previous chapter on the basis of the NSGA-II algorithm to solve the model [13]. The real matrix coding method is mainly used for the model, and the crossover and mutation operator are redesigned according to the coding method. The adaptive method of the genetic algorithm is introduced to solve the problem that the crossover probability and the mutation probability cannot be adaptive. Because the NSGA-II algorithm does not pay enough attention to the infeasible solution, this paper proposes an improved NSGA-II algorithm (IA-NSGA-II) which expands the search for global solution by adaptive crossover operation of feasible and infeasible solutions.

5.1 Coding Methods

The NSGA-II algorithm uses real coding and binary coding methods. One-dimensional real coding and binary coding cannot reflect the various combinations of individuals in the model. In this paper, according to the model of parking spots of shared bikes, we propose a real matrix coding method to encode every sites, as shown in Eq. (5.1):

$$
\begin{aligned}
P_k &= [V_1, V_2, \ldots, V_i, \ldots, V_M]^T \\
&= [R_1, R_2, \ldots, R_j, \ldots, R_N] \\
&= \begin{bmatrix}
X_{1,1} & X_{1,2} & \cdots & X_{1,j} & \cdots & X_{1,N} \\
X_{2,1} & X_{2,2} & \cdots & X_{2,j} & \cdots & X_{2,N} \\
\vdots & \vdots & \cdots & \vdots & \cdots & \vdots \\
X_{i,1} & X_{i,2} & \cdots & X_{i,j} & \cdots & X_{i,N} \\
\vdots & \vdots & \cdots & \vdots & \cdots & \vdots \\
X_{M,1} & X_{M,2} & \cdots & X_{M,j} & \cdots & X_{M,N}
\end{bmatrix}
\end{aligned}
\tag{5.1}
$$

In the formula, P_k represents the k-th individual in the population. $X_{i,j}$ is the i-th row and the j-th column element corresponding to the encoding matrix, which means the number of bicycles that the i-th demand point assigned to the j-th parking point. V_i represents he allocation assignments of the i-th demand point to the parking spots. R_j indicates the allocation situation of the j-th parking spot from various demand points.

5.2 Crossover Operator and Mutation Operator

Since the population uses the real matrix coding, the crossover operator and mutation operator of NSGA-II algorithm are redesigned. The NSGA-II algorithm uses a fixed crossover operator and mutation operator. Because the crossover probability P_c and P_m are fixed value, it does not satisfy the dynamic demand of these parameters in the change process of population. A new crossover operator and mutation operator are proposed based on these problems.

The traditional crossover operator generally adopts the alone point crossing and the two-point crossover, but the gene flow between individuals in the population is not enough. In this paper, crossover operation is performed on a column in the matrix. Two population individuals needed to cross are as follows:

$$
P_1 = [R_1^{P_1}, R_2^{P_1}, \ldots, R_N^{P_1}]
\tag{5.2}
$$

$$
P_2 = [R_1^{P_2}, R_2^{P_2}, \ldots, R_N^{P_2}]
\tag{5.3}
$$

The individual generated through crossover operation is C_1, C_2, and its expression is as follows:

$$C_1 = [R_1^{P_1}, R_2^{P_1}, \ldots, (1-\lambda)R_j^{P_1} + \lambda R_j^{P_1}, \ldots, R_N^{P_1}] \tag{5.4}$$

$$C_2 = [R_1^{P_2}, R_2^{P_2}, \ldots, \lambda R_j^{P_1} + (1-\lambda)R_j^{P_2}, \ldots, R_N^{P_2}] \tag{5.5}$$

In the formula, i is a randomly generated crossing point, and it is between 1 and N,

$$\lambda = \frac{P_1.rank}{P_1.rank + P_2.rank} \tag{5.6}$$

$P_1.rank$ represents the individual P_1's non-dominated ranking level and $P_2.rank$ represents the individual P_2's non-dominated ranking level. This paper associates the crossover operator's parameter with each individual's Pareto non-dominated ranking level in the population. In the early stage of the algorithm, since the proportion of individuals with small Pareto non-dominated ranking values is larger in the offspring, the value of λ will be relatively great. But as the algorithm runs, individuals tend to be in the same Pareto production frontier, and the value of λ gradually approaches 0.5. It is good to inherit the better genes in the parent class and increase the diversity of individuals by using this crossover operator.

Based on the traditional mutation operator, and the combination between set model and encoding method, this paper carries out a mutation operation on a column with P individuals, P is as follows:

$$P = [R_1^P, R_2^P, \ldots, R_N^P] \tag{5.7}$$

After the variation, Q is generated as follows:

$$Q = [R_1^P, R_2^P, \ldots, R_i, \ldots, R_N^P] \tag{5.8}$$

R_i is a column of data which is randomly generated and used to replace the original column i of data.

The specific process of the crossover operator and the mutation operator can be known through the above description. Among the parameters of the genetic algorithm, the key to the performance of the genetic algorithm is mainly the selection of the crossover probability P_c and the mutation probability P_m. The larger the crossover probability P_c, the faster the new individual may be generated. If P_m is too large, it will increase the possibility that the genetic model will be destroyed. If P_c is too small, the search process will be slow. For different optimization problems, P_c and P_m are needed to determine by repeated experiments and it is difficult to find the best values for each problem. Since the NSGA-II algorithm uses a fixed crossover probability and mutation probability, this paper introduces an adaptive genetic algorithm introduced by Srinivas et al. [14].

In this strategy, when the individual' fitness is less than the average fitness of the population, it can be determined that the individual has poor performance, and it should be given a greater crossover rate and mutation rate to promote the emergence of individuals with new models. When the individual' fitness is greater than or equal to the average fitness, it can be determined that the individual has a superior model gene, and it should be given a smaller crossover rate and mutation rate to ensure that the superior model gene in the population is not destroyed. Correspondingly, the model is given as follows. Formula (5.9) is an adjustment function of cross-rate, and Formula (5.10) is an adjustment function of mutation rate.

$$p_c = \begin{cases} k_1 \frac{f_{max}-f'}{f_{max}-f_{avg}} & f' \geq f_{avg} \\ k_2 & f' < f_{avg} \end{cases} \tag{5.9}$$

$$p_m = \begin{cases} k_3 \frac{f_{max}-f}{f_{max}-f_{avg}} & f \geq f_{avg} \\ k_4 & f < f_{avg} \end{cases} \tag{5.10}$$

In the formula, P_c represents the crossover rate of an individual to be crossed. P_m represents the mutation rate of an individual to be varied. f_{max} represents the maximum individual fitness of the population. f_{avg} is the average individual fitness of the population, the maximum fitness of the two individuals to be crossed, and f is the individual fitness of one to be varied. The parameter of the adjustment function of the crossover rate is k_1 and k_2. The parameter of the adjustment function of the mutation rate is k_3 and k_4. In general, $k_1 = k_2$, $k_3 = k_4$.

5.3 Improvement of Constraint Optimization

In practical applications, optimal solutions for some constrained multi-objective optimization problems are likely to be found near the constraint boundaries [15]. The objective functions of infeasible solutions near these constraint boundaries are often better than those of feasible solutions in the feasible region. But using these highly infeasible solutions could improve the search speed to the feasible region. The constraints of model will lead to infeasible solutions in the mutation process of population. In order to fully consider the impact of infeasible to the population at the same time, this paper proposes that every few generations select preferred feasible solutions and infeasible solutions to be genetically manipulated.

Two generation recombine the infeasible solutions and feasible solutions through an adaptive strategy. Because the mutation process is towards to the feasible domain and the optimal solution, the number of feasible solutions will be more and more in the process. If there are too many genetic operations on feasible and infeasible solutions in the later stages of mutation, it is possible that these operations will lead to the search performance of the algorithm in the feasible region. Therefore, this paper adopts a method can gradually reduce the number of infeasible solutions and feasible solutions during the mutation process. In the process of performing genetic operations, we set up a generation that can adaptively adjust the intersection of feasible solutions and infeasible solutions. It means when the evolutionary generation of the population is k, reorganization crossover will be performed.

$$\begin{cases} K=round(k) \\ k=\{k \leq T|k=e^t, t=\in[0,1,2...]\} \end{cases}$$ (5.11)

In the formula (5.11), T is the total evolutionary generations of the population. It can be seen from the formula that as the evolutionary generation of the population increases, the operations for feasible solutions and infeasible solutions are gradually reduced.

The algorithm solving steps for this model are as follows

Step 1: Read the original data, the demand point set, the set of facility candidate points, the demand for bicycles for each demand point, the distance of each demand point to each candidate planned parking spot, and the basic construction cost of each candidate planned parking spot;

Step 2: Using the matrix coding method to encode individual individuals, initialize individual populations within the range of the available values of the variables, and generate populations containing N individuals.

Step 3: Calculate the two objective function values for each individual in the population. According to the individual's fitness value, the individuals are quickly non-dominated sorting.

Step 4: Calculate the crowding degree of individuals in the population according to the congestion degree calculation method.

Step 5: According to the self-adaptive crossover operator and the mutation operator improved in this paper, the crossover probability and the mutation probability of each individual are determined, and then the population is selected, crossed, and mutated to generate a new offspring population.

Step 6: Using the Elite Strategy, merge the parent and offspring populations to form a large population with 2N populations.

Step 7: Quickly non-dominated sorting and crowding degree calculations are performed on the merged populations to find the better N individuals to form a new generation of parental populations.

Step 8: Repeat step 5.

Step 9: Based on the adaptive adjustment of the feasible solution and the infeasible solution, determine whether to perform the recombination crossover.

Step 10: Repeat step 6 to get a new generation of offspring populations.

Step 11: Determine whether the evolution algebra exceeds the maximum number of iterations or satisfies the termination condition. If yes, the program ends, otherwise, t = t + 1, go to step 7 to continue execution.

The algorithm execution flow is shown in Fig. 6

5.4   Experimental Analysis

This section uses the software environment Window 10 operating system, using Java language to implement the algorithm on IntelliJ IDEA programming software. The hardware uses a dual-core 2.5 GHz processor and 4 GB of memory on a PC.

This section adopts the improved IA-NSGA-II algorithm and the basic NSGA-II algorithm to solve the above-mentioned shared parking spot location model,

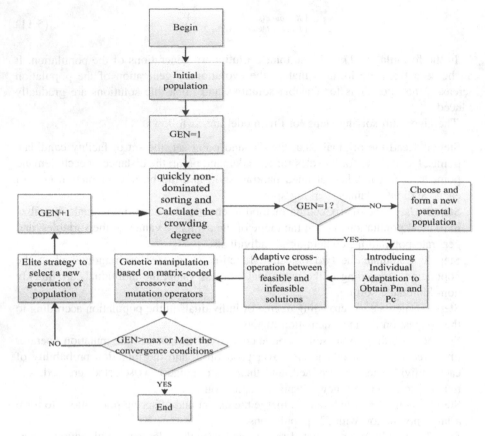

Fig. 6. Algorithm flowchart

10 demand points are used to assign the location of 20 parking spots to be planned. The initial population size is N = 100, the initial crossover and mutation probability, and the maximum number of iterations of the algorithm is max = 100;

The IA-NSGA-II algorithm is used to optimize the parking model of the shared parking spot, and some non-dominated solutions are shown in Fig. 7.

The distribution results of the three non-dominated solutions are selected in Fig. 7. It can be seen from the figure that the experiment allocates 20 planned parking points. After the evolution of the individual individuals, the final distribution results tend to be stable, and the distribution results of some parked points are less floating up and down, and the individual parking points are slightly floating, which can well reflect the distribution results of the model.

Table 2 shows the results of the NSGA-II algorithm and IA-NSGA-II algorithm in solving the model. The table compares the construction cost, travel distance, optimization time, and global optimal solution algebra. From the perspective of the two objective function values of construction cost and travel distance, the performance of IA-NSGA-II algorithm is better than that of NSGA-II algorithm, which can make the

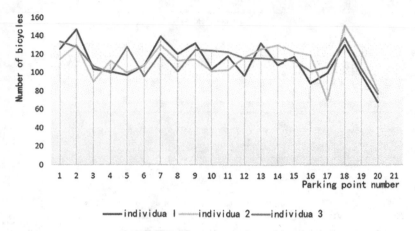

—— individua l ——— individua 2 —— individua 3

Fig. 7. A comparison of the results of partially nondominated solutions

two objective function values more apt to minimize the closest to the true Pareto frontier. From the perspective of convergence, the IA-NSGA-II algorithm has a faster convergence rate, and the optimal solution set is found around the 56th generation, while the NSGA-II algorithm can only obtain the results after about 64 generations. At the same time, the IA-NSGA-II algorithm is slightly better than the NSGA-II algorithm in view of the running speed of the algorithm.

Table 2. Comparison of results between NSGA-II and IA-NSGA-II

	Construction cost/yuan	Travel distance/km	Optimization time/s	Running algebra
IA-NSAG-II	57628	193.271	30.153	56
NSGA-II	63245	235.639	39.649	64

It shows the Pareto frontier distribution of the NSGA-II algorithm and IA-NSGA-II for the final result of the shared bike parking point model in Fig. 8. It can be seen from the figure that the two algorithms can not minimize the construction cost and travel distance of the two objective functions of the shared bicycle parking point model at the same time. If the construction cost is low, the travel distance will increase; the same reason, if you want to travel a short distance, the cost of construction will increase. In the actual site selection, the two objectives should be analyzed and adjusted according to the actual situation. At the same time, it can be seen from Fig. 8 that the construction cost of IA-NSGA-II algorithm is lower when the travel distance is the same. With the same construction cost, the travel distance obtained by the IA-NSGA-II algorithm is shorter. Therefore, the IA-NSGA-II algorithm improved in this paper can achieve a better Pareto frontier and have a better optimization effect.

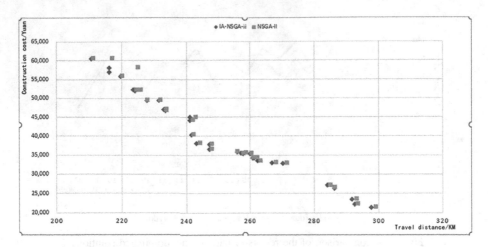

Fig. 8. Comparison of pareto frontier between NSGA-II algorithm and IA-NSGA-II algorithm

6 Conclusions

As an innovative product of urban public transportation, shared bicycle has effectively solved people's short distance travel. At the same time, the problems of shared bicycle management confusion and unreasonable planning have also seriously affected urban traffic. Therefore, based on the characteristics of shared bicycle and some research results of public bicycle, this paper proposes a shared bicycle demand model and location model based on user traveling data. An improved multi-objective optimization algorithm NSGA-II is used to solve the model.

In view of the characteristics of shared bicycle, combined with the characteristics of cluster analysis, a demand forecasting model based on user traveling data is adopted to solve the shortage of k - means of clustering algorithm, and the model is solved by combining the clustering algorithm. Combined with demand points and programmable parking points, a multi-objective shared bicycle parking point location model is proposed.

The improved NSGA-II algorithm is used to solve the model, and the algorithm is optimized based on the constraints and coding methods, the individual adaptive method and so on. The comparison of the model solving data shows that the improved algorithm has a good effect on the model solving.

Acknowledgements. This work is supported by the Hangzhou Science & Technology Development Project of China (No. 20162013A08) and the Zhejiang Provincial Natural Science Foundation of China (No. LY16F020010).

References

1. DeMaio, P.: Will smart bikes succeed as public transportation in the United States. J. Public Transp. **7**(2), 23–27 (2004)
2. Garcia-Palomares, J.C., Gutierrez, J., Latorre, M.: Optimizing the location of stations in bike-sharing programs: a gis approach. Appl. Geogr. **35**(1), 235–246 (2012)
3. Chemla, D., Meunier, F., Calvo, R.W.: Bike sharing systems: Solving the static rebalancing problem. Discret. Optim. **10**(2), 120–146 (2013)
4. Deng, L.F., Xie, Y.H., Huang, D.X.: Research on shared cycling facility planning based on space-time data of cycling. Planner **33**(10), 82–88 (2017)
5. Zhao, M.: Research on analysis and optimal scheduling of shared single cycle network. Shandong University of Science and Technology (2017)
6. Mccallum, A., Nigam, K., Ungar, L.H.: Efficient clustering of high-dimensional data sets with application to reference matching. In: ACM SIGKDD, pp. 169–178 (2000)
7. Li, Y., Zheng, Y., Zhang, H.: Traffic prediction in a bike-sharing system. In: SIGSPATIAL International Conference on Advances in Geographic Information Systems, p. 33 (2015)
8. Santoni, M., Santoni, M., Meenen, M.: Incentivizing users for balancing bike sharing systems. In: Twenty-Ninth AAAI Conference on Artificial Intelligence, pp. 723–729 (2015)
9. Kaspi, M., Raviv, T., Tzur, M.: Bike-sharing systems: User dissatisfaction in the presence of unusable bicycles. IISE Trans. **49**(2), 144–158 (2017)
10. Caulfield, B., O'Mahony, M., Brazil, W.: Examining usage patterns of a bike-sharing scheme in a medium sized city. Transp. Res. Part A Policy Pract. **100**, 152–161 (2017)
11. Xu, P.F., Wang, L., Guan, Z.Y.: Evaluating brush movements for Chinese calligraphy: a computer vision based approach. In: International Joint Conferences on Artificial Intelligence (IJCAI), pp. 1050–1056 (2018)
12. Gong, M.G., Jiao, L.C., Yang, D.D.: Research on evolutionary multi-objective optimization algorithms. J. Softw. **20**(2), 271–289 (2009)
13. Deb, K., Pratap, A., Agarwal, S.: A fast and elitist multiobjective genetic algorithm: NSGA-II. IEEE Trans. Evol. Comput. **6**(2), 182–197 (2002)
14. Srinivas, M., Patnaik, L.M.: Adaptive probabilities of crossover and mutation in genetic algorithms. IEEE Trans. Syst. Man Cybern. **24**(4), 656–667 (2002)
15. Tomoiagă, B., Chindriş, M., Sumper, A.: Pareto optimal reconfiguration of power distribution systems using a genetic algorithm based on NSGA-II. Energies **6**(3), 1439–1455 (2013)

Social Networks and Recommendation Systems

Asynchronous Bi-clustering for Image Recommendation

Yankun Jia, Yidong Li$^{(\boxtimes)}$, and Jun Wu

Computer and Information Technology, Beijing Jiaotong University, Beijing, China
{jyankun,ydli,wuj}@bjtu.edu.cn

Abstract. Collaborative filtering (CF) plays a key role in various recommendation systems, but its effectiveness will be limited by the highly sparse user-image click-through data when CF deploys for image recommendation applications. Some existing methods apply clustering techniques to mitigate the sparseness issue. However, there is still a big room to elevate the recommendation performance, because little is known in taking both click-through data and image visual information into account. In this paper, we propose an Asynchronously Bi-Clustering (ABC) CF approach to improve the CF-based image recommendation. Our ABC approach consists of two coupled clustering solutions. Concretely, it first implements image clustering based on image click-through and visual feature, and then conducts user clustering in a low-dimensional subspace spanned by the image clusters. The final recommendation is accomplished based on both user clusters and image clusters by a similarity fusion strategy. An empirical study shows that our ABC approach is beneficial to the CF-based image recommendation, and the proposed scheme is significantly more effective than some existing methods.

Keywords: Collaborative filtering · Multi-view clustering
Image recommendation

1 Introduction

The emergence of Web 2.0 technology along with the prevalence of mobile devices leads to an explosion of images being upload and shared online. Without doubt, the dominating images require modern recommender system, as an essential supplementary of image retrieval system, to sift through massive image for users in a highly dynamic environment [8,16].

CF is one of the most widely adopted and successful recommendation approach. A general CF technique takes advantages of user of crowd wisdom to establish recommendations. It's suggested that the user feedback signals, such as the user click-through data, can be effective in bridging the semantic gap caused by image retrieval algorithms [13]. The assumption behind CF based

Supported by NSFC of China.

© Springer Nature Singapore Pte Ltd. 2018
Z. Xu et al. (Eds.): Big Data 2018, CCIS 945, pp. 411–426, 2018.
https://doi.org/10.1007/978-981-13-2922-7_28

methods is that if a group of users have the same preference of some items, then they are possible to have same interests about other items. The primary areas of CF are neighborhood based methods and latent factor models [9]. The key idea of neighborhood methods is to calculate the similarity between users or items to build its neighbors, then make predictions based on those neighbors' preference. Differently, latent factor models apply matrix factorization techniques to learn low-rank representations of users and items from the user-item matrix. Latent factor models often have highly expressive ability to describe different aspects of data. However, neighborhood methods have advantages in providing high flexibility to integrate other models, as well as being intuitively explainable for real recommendation [2].

CF algorithms are required to have the ability to deal with high sparse data, to scale with the increasing numbers of users and items, to establish recommendations. However, due to the data sparsity challenge [14], especially the cold start problem, the performance of CF algorithms are really challenged because of lack of information. One of promising solutions to mitigating the sparseness and scalability issues is the clustering. Existing cluster based CF methods are including regular clustering approaches and bi-clustering approaches. One typical regular cluster based CF approach [18] is firstly grouping users into k clusters, then looking for neighbors only in clusters that are closest to the target user. The bi-clustering approaches [4] are to simultaneously obtain user clusters and item clusters via bi-clustering algorithms, then generate the prediction based on those user-item subgroups. Actually, it is more natural to utilize bi-clustering approaches to design CF algorithms, because one users interests are only concentrative on part of categories [1], but not dispersive over all items.

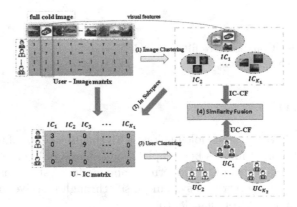

Fig. 1. Motivation illustration of this paper, the high sparsity problem of user-image matrix via bi-clustering melting

However, despite this success, traditional bi-clustering solution is limited by two major shortcomings when deployed for image recommendation. It is often the case that the image view of the user-image matrix suffers from the missing of some data, since a few images might have never been clicked by any user. Meanwhile, due to data sparsity, the user view of the user-image matrix often suffers from the curse of dimensionality, which effectively affect the performance of clustering.

In this paper, we propose an Asynchronously Bi-Clustering (ABC) approach to address the above problems (Fig. 1). Firstly, for the column view of user-image matrix, we group images into multiple clusters utilizing the idea of multi-view clustering algorithm, which is to alleviate the data sparsity problem by integration of the visual features extracted from deep convolutional neural network. Secondly, for the row view of user-image matrix, inspired by the bag-of-visual-word model, we utilize the image clustering results to reduce the dimension of user-image clicking matrix, which is to alleviate the data sparsity problem using the idea of subspace clustering. Finally, we use the similarity fusion algorithm to integrate the image clustering results and the user clustering results to obtain satisfactory recommendations. An empirical study demonstrates the effectiveness and superiority of our proposed models compared with state-of-the-art methods.

The remaining of this paper is organized as follows. In Sect. 2, we introduce some related work and background techniques about our work. Section 3 describes our proposed model for personalized image recommendation in details. The experimental settings and results are presented in Sect. 4. Finally, Sect. 5 presents the conclusions.

2 Related Works

2.1 Collaborative Filtering

Generally, collaborative filtering approaches can be roughly categorized into neighborhood based methods and latent factor models.

The typical neighborhood based method calculates the similarity between users or items to find the similar neighbors of corresponding entities, then make predictions based on the historical ratings of those neighbors. The neighborhood based CF methods obtain good results under relatively high density conditions, but their performance decrease under sparsity conditions. Wang [15] proposes a similarity fusion methods which considers both the relations between users and between items to obtain better recommendation.

The latent factor model is one of most successful CF models, which apply matrix factorization techniques to learn low-rank representations of users and items from the information in the user-item matrix. In recent years, many latent factor models have adopted, such as the Singular Value Decomposition (SVD) [11], Non-negative Matrix Factorization (NMF), Probabilistic Matrix Factorization (PMF) [10]. Those latent factor models reduce the dimensions of representations of users and items, which is helpful to scalability of algorithms.

2.2   Clustering-Based CF

The most related CF model to this paper is the clustering based CF model. A typical traditional clustering based approach is partitioning the whole data samples into several clusters, where each cluster has similar features or close relationships. Then utilizing clusters to find user neighbors or item neighbors to establish recommendations. Xue et al. [18] proposes a user cluster based CF approach, it groups the users into clusters to reach two objectives: approaching the unknown ratings with cluster mean to increase the density of data, and looking for neighborhoods only in the closest clusters to the target user to increase scalability. However, those traditional cluster based approaches are usually difficult to obtain high-quality clusters when encountering the highly sparse data. Yang et al. [19] propose a heterogeneous transfer learning model for image clustering to alleviate data sparsity problem.

Recently, many bi-clustering based approaches are adopted. Those typical works are [4] simultaneously obtaining user clusters and item clusters via a bi-clustering algorithm. Bu et al. [1] propose a Multiclass Co-clustering model (MCOC) for collaborative filtering, which achieve fairly good performance. However, those approaches only utilizing user-item matrix, which still suffer from data sparsity problem.

3   The Proposed ABC Approach

3.1   Preliminaries

Let $U = \{U_1, \ldots, U_m, \ldots, U_M\}$ denotes a set with M users, and $I = \{I_1, \ldots, I_n, \ldots, I_N\}$ denote a set with N images. The user-image clicking matrix associating with U and I can be represented by a binary matrix $X^{(1)} \in \{0,1\}^{N \times M}$, whose element $x_{n,m}^{(1)}$ indicates the clickthroughs behavior of user U_m on image I_n, and $x_{n,m}^{(1)} = 1$ indicates that the n-th image is clicked by m-th user, '0' otherwise. In some sense, each row $X_{n*}^{(1)}$ can be regarded as the clickthrough feature vector of image I_n performed by users.

The visual feature set corresponding to the image set I can be represented by a real value matrix $X^{(2)} \in \mathbb{R}^{N \times D}$, where N is the number of images and D is dimensionality of image's features, each row $X_{n*}^{(2)}$ denotes the visual feature of image I_n.

In most of cluster based CF approaches, the measurement of similarity between users or images has a major influence on the clustering results. In user-image clicking matrix $X^{(1)}$, its elements are all binary values, so we use the Jaccard coefficient as our similarity measure in calculation. The Jaccard coefficient between $X_i^{(1)}$ and $X_j^{(1)}$ is represented as:

$$Jaccard(X_i^{(1)}, X_j^{(1)}) = \frac{|\mathbf{X}_i^{(1)} \cap \mathbf{X}_j^{(1)}|}{|\mathbf{X}^{(1)} \cup \mathbf{X}_j^{(1)}|} \tag{1}$$

Algorithm 1. Asynchronously Bi-Clustering Collaborative Filtering

---

Input:

    Two View data matrixes: $\mathcal{D} = \{\mathbf{X}^{(1)} \in \{0,1\}^{N \times M}, \mathbf{X}^{(2)} \in \mathbb{R}^{N \times D}\}$;

    Target User: U_m

    Parameters: γ, K_1, K_2

Output:

    Rank-Score Vector: $\mathbf{P_m}$

1: Construct the image clusters $C^{(i)} = \{C_1^i, C_2^i, \ldots, C_{K_1}^i\}$ by applying the Robust Multi-view K-Means algorithm on both of the user-image clicking matrix $\mathbf{X}^{(1)}$ and visual feature matrix $\mathbf{X}^{(2)}$;

2: Construct the newly reduced user-image matrix $U^{(r)} = \{U_1^{(r)}, U_2^{(r)}, \ldots, U_N^{(r)}\} \in \mathbb{R}^{N \times K_1}$ based on image clustering results using Eq 3, as illustrated in Fig 2;

3: Calculate the user clusters $C^{(u)} = \{C_1^u, C_2^u, \ldots, C_{K_2}^u\}$ by applying the classical K-means algorithm in the reduced matrix $U^{(c)}$;

4: Choose the most similar user cluster $C_{k_1}^{(m)}$ using Eq 4, and select the top-K most similar users as user neighbors to calculate the probability $P^{(m,i|UC)}$ based on UC data using Eq 5;

5: Similarly, the nearest image cluster $C_{k_2}^{(i)}$ for the i-th image are chosen, then we compute the probability $P^{(m,i|IC)}$ of the i-th image that the user U_m is interested based on IC data using Eq 6;

6: Calculate the final probability $P(m,i|UC,IC)$ by fusion of $P(m,i|UC)$ and $P(m,i|IC)$ using Eq 7 for each unseen image of U_m, and recommend the top-K images with highest probability scores.

In traditional bi-clustering approaches, they are only able to utilize the user-image clicking matrix, however, due to the highly sparsity problem, the performance of those algorithms are challenged. To integration of the visual features of images, we proposed an Asynchronously Bi-Clustering (ABC) CF approach to improve the image recommendations. Firstly, we apply the multi-view clustering algorithm to obtain more accurate image clusters by integrating user-image clicking matrix and image feature matrix together. Secondly, inspired by the BoW model, we utilize the image clustering results to reduce the dimension of user-image clicking matrix, which is to alleviate the data sparsity problem using the idea of subspace clustering. Finally, we use the similarity fusion algorithm to integrate the image clustering results and the user clustering results to obtain satisfactory recommendations. The framework of our Asynchronously Bi-Clustering (ABC) CF approach is shown in Algorithm 1.

3.2 Multi-view Image Clustering

In most of traditional clustering based approaches and bi-clustering based approaches, it is often the case that the image view of the user-image matrix suffers from the missing of some data in real image recommendation systems, since a few images might have never been clicked by any user. And most of traditional clustering approaches are difficult to obtain high-quality clusters because of lack of information. In recent years, a large number of recommendation scenarios have emerged in which various additional information sources are available in addition to the U-I matrix. In such cases, CF can be enhanced to improve recommendation performance further.

In image recommendation, people are usually aware of the visual patterns they prefer, and users would likely favor the images with similar patterns to those images they already liked. It's suggested that the visual features information could be utilized to alleviate the data sparsity problem, and be enhanced to improve recommendation performance further.

To integrate the heterogenous data features to improve the performance of data categorizations, Cai et al. [3] proposed a new robust Multi-view K-Means Clustering (RMKMC) algorithm. This RMKMC method is based on classical K-means clustering algorithm, which can be easily parallelized and processed for big data clustering.

In our scenario, user-image clicking matrix can be regarded as the click-through feature of images performed by users, which captures users personal preferences. So the click-through features and visual features can be regarded as two representations of images. In this paper, we use RMKMC algorithm to find out accurate image clusters, by integrating those two-view features of images.

Previous work has showed that relaxed K-means clustering is equivalent to G-orthogonal matrix factorization. To solve the large-scale multi-view clustering problem, the RMKMC method extends single-view K-means algorithm. As $X^{(1)} \in \mathbb{R}^{M \times N}$ indicates the user-image clicking matrix, and $X^{(2)} \in \mathbb{R}^{M \times D}$ indicates the visual feature matrix. Let $F^{(1)} \in \mathbb{R}^{K_1 \times N}$ and $F^{(2)} \in \mathbb{R}^{K_1 \times D}$ be the centroid matrix for the click-through matrix and the visual feature matrix, respectively, and $G \in \{0,1\}^{M \times K_1}$ be the unique clustering indicator matrix, where K_1 is the number of image clusters. In RMKMC method, The click-through features and visual features are simultaneously utilized to obtain accurate image clusters, so those two-view matrixes shares the consensus clustering indicator matrix G, and G should satisfy the 1-of-K_1 coding scheme, which means the $G_{i,k} = 1$ only when the image I_i is assigned to k_1-th cluster, and $G_{i,k} = 0$ otherwise.

Meanwhile, in order to reduce the effect of outlier data points respect to a fixed initialization, the RMKMC method uses the structured sparsity-inducing norm, $\mathcal{L}_{2,1}$-norm, in the clustering objective function. The multi-view image clustering results can be obtained by solving:

$$\min_{F^{(v)}, G, \alpha^{(v)}} \sum_{v=1}^{V} (\alpha^{(v)})^{\gamma} ||X^{(v)} - GF^{(v)^T}||_{2,1} \tag{2}$$

$$s.t. G_{ik} \in \{0,1\}, \sum_{k=1}^{K_1} G_{ik} = 1, \sum_{v=1}^{V} \alpha^{(v)} = 1$$

where $\alpha^{(v)}$ is the weight factor of the v-th view and γ is the parameter to control the weights distribution of different views. We can learn the weights for two different types of features, such that the important feature will obtain larger weight during multi-view clustering.

To tackle the optimization difficulty of loss function introduced by the non-smooth norm, they derive an efficient algorithm, which is solved by alternatively update $F^{(v)}$, G as well as $\alpha^{(v)}$ iteratively until the process is converged. For more details of optimization about the RMKMC algorithm can be seen in [3].

In regular clustering algorithm or bi-clustering algorithm, they only use the user-image clicking matrix, however, when encountering cold start problem, the clustering performance are challenged. In this paper, we use the multi-view clustering algorithm for image set, this solution can integrate both click-through features and visual features, which alleviates the data sparsity problem of user-image clicking matrix. In RMKMC method, we can learn the cluster indicator matrix G, and by transforming G, and we can group all image into K_1 clusters as $\{C_1^i, C_2^i, \ldots, C_{K_1}^i\}$.

3.3 User Clustering in Sub-space

For user view of user-image clicking matrix, with the large amount of images, we usually cannot obtain accurate user clusters due to the curse of dimensionality problem. One of promising solution is to using the subspace clustering algorithm to alleviate data sparsity and high dimensionality. In this part, we use the previous image clustering results to enhance user clustering results inspired by the bag-of-words model (BoW model).

The BoW model is applied to image classification, which treats image features as words. A definition of the BoW model can be the histogram representation based on independent features. The BoW model is used for object categorization by constructing a large vocabulary of many visual words and representing each image as a histogram of the frequency words that are in the image.

In this paper, our intuition is very similar, we represent each user as a bag of words - patches that described by image clusters. The click-through feature of each user can be represented by the histogram of the image clusters. By applying this idea, the user-image clicking matrix $U \in \mathbb{R}^{M \times N}$ can be transformed into $U^{(r)} = \{U_1^{(r)}, U_2^{(r)}, \ldots, U_M^{(r)}\} \in \mathbb{R}^{M \times K_1}$, where K_1 is the number of image clusters. For each element $U_{m,k}^{(r)} \in U^{(r)}$,

$$U_{m,k}^{(r)} = \sum_{m \in C_k^i} x_{n,m}^{(1)} \tag{3}$$

where C_k^i is the k-th image clusters obtained from previous step.

A more intuitive explanation of this bag-of-words solution can be seen in Fig. 2. Note that the number of image clusters is much smaller than the number of images, so we can effectively reduce the dimension of user-image matrix by this solution. Compared with the original matrix U, the newly obtained matrix $U^{(r)}$ becomes more dense. Each element $u_{n,k}^{(r)} = c_k$ in $U^{(r)}$ represents the number of images have been clicked by m-th user in k-th image cluster, which can provide more discriminative clicking patterns of user about images. Based on this reduced clicking matrix $U^{(c)}$, we use the classical K-means algorithm to calculate user clusters.

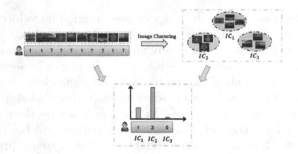

Fig. 2. An illustration of the reduction of user-image clicking matrix. This is an example that converts the i-th user clicking vector U_i into new vector $U_i^{(r)}$ by dimension reduction based on previous image clusters results. As the i-th user have clicked 3 images in cluster 1, so $U_{i,1}^{(r)} = 3$. Similarly, $U_{i,2}^{(r)} = 1$, $U_{i,3}^{(r)} = 0$.

Through this dimension reduction solution, user clusters can be more accurate and easily obtained with lower computational cost. The user clustering results can be represented as $\{C_1^u, C_2^u, \ldots, C_{K_2}^u\}$, where K_2 is the number of user clusters.

3.4   Prediction with Similarity Fusion

Most of existing clustering based CF method based on user clusters or image clusters, which only use part of information in the user-image matrix. However, when encountering highly sparse data, the performance of recommendation are challenged. It would be desirable to use as much information as possible given that the amount of information in the matrix is highly sparse. In this subsection, we propose to accomplish the final recommendation based on both user clusters and image clusters by a similarity fusion strategy, which complements each other under the data sparsity problem.

Firstly, we introduce how to obtain image recommendation based on user clusters (UC) or image clusters (IC) independently as discussed in [18]. Taking user clusters as an example, the process of user clusters based prediction can be divided into three steps: nearest user clusters selection, user neighbors selection and prediction.

Given the target user u, we calculate the similarity between the target user u with all user clusters, and choose the most nearest user cluster C_k^u as candidates set respect to u.

$$sim(u, C_k^u) = Jaccard(u, c_k^u) \qquad (4)$$

where c_k^u is the centroid of user cluster C_k^u.

Then we select $K^{(u)}$ most similar users $K^{(u)}$ as neighbors of u from the candidates set C_k^u based on the same similarity function. To make the prediction scores $Pred(u, i|UC)$ of unseen images for the target user u, we use a weighted aggregate of preferences of $K^{(u)}$ neighbors. The prediction are computed as the weighted average from the neighbors's preferences:

$$P(x_{u,i}^{(1)} = 1|UC) = \frac{\sum_{k=1}^{K^{(u)}} sim(u, C_k^u)sim(u, u_k)(x_{u,i}^{(1)})}{\sum_{k=1}^{K^{(u)}} sim(u, C_k^u)sim(u, u_k)} \tag{5}$$

where u_k is one of the top $K^{(u)}$ similar users respect to u.

Similarly, we computed the image clusters based prediction scores $P(x_{u,i}^{(1)} = 1|IC)$ of each unseen images for target user u:

$$P(x_{u,i}^{(1)} = 1|IC) = \frac{\sum_{k=1}^{K^{(i)}} sim(i, C_k^i)sim(i, i_k)(x_{u,i}^{(1)})}{\sum_{k=1}^{K^{(i)}} sim(i, C_k^i)sim(i, i_k)} \tag{6}$$

Relying on user clusters (UC) or image clusters (IC) data only is undesirable, especially when the predictions from those two source are quite not often available. Similar to the SF algorithm [15], we accomplish the final recommendation based on both user clusters and image clusters by a similarity fusion strategy. We introduce a binary variable O_1, that corresponds to the importance of UC and IC. $O_1 = 1$ indicates that the final prediction depends completely upon UC data, while $O_1 = 0$ depends completely upon IC data. The final prediction is computed as:

$$\begin{aligned} P(x_{u,i}^{(1)} = 1|UC, IC) \\ = \sum_{O_1} P(x_{u,i}^{(1)} = 1|UC, IC, O_1)P(O_1|UC, IC) \\ = \lambda \times P(x_{u,i}^{(1)} = 1|UC) + (1 - \lambda) \times P(x_{u,i}^{(1)} = 1|IC) \end{aligned} \tag{7}$$

where λ is the weight factor, which controls the relative importance of predictions from UC data and IC data. And a bigger λ emphasizes user correlations, while smaller λ emphasizes image correlations.

4   Experiments

In this section, we conduct a set of experiments to certify the performance of our proposed framework in terms of image recommendation.

4.1   Dataset

In our experiments setting, we employ the '10K Images dataset'[1] which is publicly available on the web to make our experiments reproducible. In this '10K Images dataset', user-image clicking matrix $X^{(1)} \in \{0, 1\}^{10000 \times 1000}$ consists of 10,000 images and 1,000 users. All images are from 100 semantic categories, with

---

[1] http://www.datatang.com/data/44353. The dataset was firstly used in [17].

Table 1. Characteristics of 10K Images dataset

Statistics	10K Images dataset
Numbers of users	1000
Numbers of images	10000
Numbers of clicks per user	26.373
Numbers of clicks per image	2.637
Clicks sparsity	0.263%

about 100 images per category. Note that (Table 1), there only include 0.26% clicking ratings in the user-image image matrix, which suggests that the highly sparsity of the user-image clicking matrix. Meanwhile, there are about 26.68% images that are not clicked by any user in the user-image clicking matrix, which is called cold start problem in image view. Because of this, we can't use traditional k-means algorithm for images clustering.

In '10K Images dataset', images are available information for recommendations. The feature extraction of images has developed from low-level hand-crafted features (such as HOG, SIFT) to current high-level features (such as DCNN features [7]), which is extracted from shallow architecture to deep convolutional neural network in recently years. In this paper, we extracted the image's visual features from the well-known VGG16 convolutional neural network [12]. This VGG16 network is the pretrained model using ImageNet dataset, which is implemented by Caffe DL framework. We extracted the last full connected layer features as the representations of images, and each image can be represented as a 4096-dimensions feature vector, the visual features matrix can be represented as $X^{(2)} \in \mathbb{R}^{10000 \times 4096}$.

Many measures have commonly adopted to evaluate both recommendation and retrieval performance, such as precision and recall. In top-K image recommendation scenario, Precision are used to evaluate the quality of the top K images recommend list, Recall is the proportion of relevant items in the top K images recommend list. Considering that some users may click a large number of images in the test data while some other users just have a few, so F_1-score may be more reasonable measure for Top-N recommendation performance. The F_1-score is defined:

$$F_1 = \frac{2PR}{P+R} \tag{8}$$

In order to explain the clustering performance of our scheme, we use one standard clustering metrics, that is Purity, which is more useful for the analysis of image recommendation.

Because of the availability of the categories information in our scenario, we don't need to split into train/test dataset, which can use the categories information to evaluate the recommend performance. For one example, if one user belongs to category 1, then any images in category 1 are recommend to this user can be seen as accurate.

4.2   Compared Methods

In order to show the effectiveness of our proposed approach, we further compare our recommendation results with the following methods:

- U-CF: For user based algorithm, due to the user-image clicking matrix is one binary matrix, we use the Jaccard Similarity as metrics to measure the user-user similarities and use the user based model in [2].
- I-CF: For item based algorithm, similar to user based algorithm, we use the Jaccard Similarity as metrics to measure the similarities between images and use the item based model in [2].
- SF [15]: The Similarity Fusion algorithm is an alternative method based on U-CF and I-CF, which considers both between users and between-items relations. In sparse matrix contexts, this algorithm obtains better predictions.
- UC-CF [18]: User clustering based algorithm like U-CF, with exception that first groups users into clusters, then looking for neighbors in the closest clusters. We use the K-means method as the base clustering algorithm, and use the Jaccard Similarity as metric between users.
- IC-CF [6]: Item clustering based algorithm partition the set of items based on user-image clicking data. Similarly, We also use the K-means method as the base clustering algorithm, and calculate the Jaccard Similarity between images.
- MCoC [1]: Multiclass Co-clustering (MCoC) model extends traditional CF models by co-clustering both users and items into multiple subgroups, and use them to improve CF based recommendation. For simplicity, we make the prediction based on those subgroups with U-CF and I-CF. The MCoC model only able to use the information of user-image interaction data.
- Tri-CF [16]: The Tri-CF is an advanced method based on MF technology, which comprehensively considers the characteristics of the click-through data from the view of low rank, user consistency and image correlation.

4.3   On the Clustering of Images with Two-View

It is often the case that the image view of the user-image matrix suffers from the missing of some data, since a few images might have never been clicked by any user. Traditional clustering algorithm cannot obtain good image clusters with those cold images, we use the multi-view clustering algorithm to alleviate the sparsity of user-image clicking matrix. In this part, we will conduct several experiments to compare our two-view clustering solution with I-CF and IC-CF.

In multi-view image clustering method, there are two importance parameters need to tuned, the exponent of weights factors γ and the number of image clusters K_1. In this part, we use the 5-fold cross validation to tune these two parameters, and the best settings are $\gamma = 20$ and $K_1 = 120$.

In Table 2, we compared the image clustering results using single view K-means algorithm or multi-view K-means algorithm. We can observe that the two view clustering solution can obtain more accurate image clustering results, which alleviate the high sparsity existing in image view of user-image clicking matrix.

Table 2. Comparison of different image clustering results

	K-means	Multi-view K-means
Purity	78.95%	87.99%

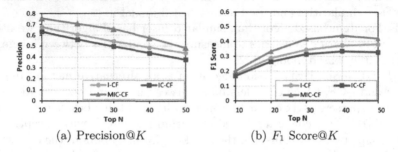

(a) Precision@K (b) F_1 Score@K

Fig. 3. The Precision@K and F_1-Score@K curves of our Multi-view Images Clustering (MIC-CF) solution compared with I-CF approach and IC-CF approach.

In Fig. 3, we compared the Precision@K and F_1-Score@K curves of those three compared methods, some interesting observations are revealed. First, our multi-view image clustering method (MIC-CF) outperform both item-based CF and item clustering based CF significantly, for one example about the Precision@K, our approach outperforms 6.6% than item based CF, and it outperforms 9.6% than item clustering based CF in top-10 image recommendation, which verifies the superiority of our multi-view clustering solution by integration of user-image matrix and visual features.

Second, due to the missing of data in image view of the user-image clicking matrix, in which is about 26% images whose have not clicked by any user, the clustering result of image view is highly degrades, which causes the performance of item clustering based CF is weaker than item based CF. However, in our image clustering solution, we utilize the multi-view clustering algorithm to integrate the visual features information to alleviate the data sparsity problem in image view of the user-image clicking matrix, which significantly improved the performance of image recommendations as shown in Fig. 3.

4.4   On the Clustering of Users in Sub-space

Due to the large amount of images in image recommendation, the user view of user-image clicking matrix often suffers from the curse of dimensionality when simply applies classical K-means algorithm. We use a novel solution to reduce the dimension of image view of user-image clicking matrix utilizing previous image clustering results. In this part, we will conduct several experiments to compare our reduced user clustering solution with some similar CF approaches, including user based CF and user clustering based CF. For the proposed user bi-clustering

Table 3. Comparison of different user clustering results

	K-means	K-means in Sub-space
Purity	67.36%	90.59%

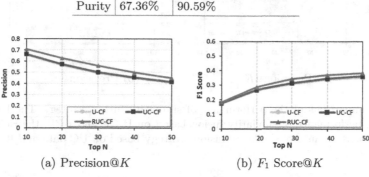

(a) Precision@K (b) F_1 Score@K

Fig. 4. The performance of our user clustering solution (RUC-CF) in subspace compared with two existing approaches, U-CF approach and UC-CF approach.

method, we also use the 5-fold cross validation to tune the parameter of the number of user clusters K_2, and the best settings of K_2 is 110.

In Table 3, we compared the user clustering results using classical K-means algorithm with or without matrix reduction. We can observe that our solution can obtain more distinguishable representations of users, which alleviate the high dimensions existing in user view of user-image clicking matrix.

Figure 4 shows the Precision@K and F_1-Score@K curves of all three comparing methods. As observed, our user clustering solution in subspace outperforms both user based CF and user clustering based CF significantly, for one example our approach outperforms 4.0% than user based CF, and 3.5% than user clustering based CF in terms of Precision@10, which verifies the effectiveness of our solution. One observation need to be noted, user based CF and user clustering based CF almost have same performance in terms of precision and f_1-score, this maybe because of the sparsity of user-image clicking matrix and different characteristics of our dateset compared with other works [18]. Experimental results proved that our user clustering solution within subspace can produce more accurate image recommendations based on the previous image clustering results.

4.5   On the Fusion Between Image and User Clustering Results

To combine user clusters data and image clusters data, we use one similarity fusion strategy to improve the performance of image recommendations. In this part, we conduct experiments to compare its performance of fusion of UC and IC data with simpler SF algorithm.

Figure 5 shows the performance of similarity fusion strategy, from left figure we observed that this SF strategy based on U-CF and I-CF just has similar performance with I-CF in terms of precision. However, when we applies this simple

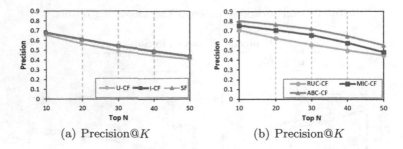

(a) Precision@K (b) Precision@K

Fig. 5. Comparisons of the performance of similarity fusion strategy. The left figure shows the performance of similarity fusion based on U-CF and I-CF, the right figure shows the performance of similarity fusion strategy based on UC data and IC data.

fusion strategy to UC data and IC data, it can improve the performance of recommendation significantly, which outperform 4.0% about Precision@K compared with those UC based or IC based methods that only using part of information of clustering results as shown in right figure. Those observations verity the usefulness of our solution by similarity fusion of UC data and IC data.

4.6 On the Comparison of Different Image Recommendation Approaches

In this part, we will compare our ABC-CF approach with those CF methods summarily to verify the effectiveness of our model. U-CF, I-CF and SF algorithm are traditional neighborhood based CF methods, UC-CF and MCoC algorithm are cluster based CF methods, Tri-CF algorithm is MF based method. For our ABC-CF approach, we tune those hyper-parameters through 5-fold cross-validation, and the best settings are $\gamma = 20$, $K_1 = 120$ and $K_2 = 110$.

The Table 4 shows that the comparisons among those different image recommendation approaches in terms of Precision@K. Firstly, compared with traditional neighborhood based methods, cluster based methods are usually have the

Table 4. Comparisons of different image recommendation approaches

Method	top10	top20	top30	top40	top50
U-CF	0.656	0.563	0.492	0.447	0.406
I-CF	0.677	0.608	0.542	0.488	0.436
SF [15]	0.677	0.609	0.543	0.488	0.435
UC-CF [18]	0.662	0.571	0.498	0.453	0.415
MCoC [1]	0.683	0.627	0.569	0.521	0.467
ABC-CF	**0.803**	**0.765**	**0.720**	**0.644**	**0.553**
Tri-CF [5]	0.750	0.706	0.659	0.609	0.541

better recommend performance. Compared with other cluster based methods, our approach utilizes images' visual features to alleviate the data sparsity of user-image clicking data, which can enhance image recommendation effectively. MF based models often have highly expressive ability to describe different aspects of data, however, our approach can also have the best performance compared with Tri-CF method. It is impressive that the performance of our ABC-CF approach is always the best among all compared methods, which verifies the usefulness and effectiveness of our approach.

5   Conclusions

In this paper, we propose an Asynchronously Bi-Clustering (ABC) approach to improve the CF-based image recommendation. Our ABC approach consists of two coupled clustering solutions. Concretely, we use the convolutional neural network to extract the visual features of images, so as to alleviate the data sparisity problem by utilizing the idea of multi-view clustering, then we reduce the dimension of row view of user-image matrix by utilizing the column view clustering results, and construct user clusters in the reduced subspace. The final recommendation is accomplished based on both user clusters and image clusters by a similarity fusion strategy. Experimental results has validated significantly the effectiveness of our ABC approach compared with several existing collaborative filtering approaches. It is a promising scheme for improving image recommendation. Furthermore, given the fact that one image maybe belong to multiple categories, we are planning to exploit the fuzzy multi-view bi-clustering algorithm into our approach in future.

Acknowledgments. This work is supported by NSFC61671048 and NSFC61672088 and NSFC61790575.

References

1. Bu, J., Shen, X., Xu, B., Chen, C., He, X., Cai, D.: Improving collaborative recommendation via user-item subgroups. IEEE Trans. Knowl. Data Eng. **28**(9), 2363–2375 (2016)
2. Cacheda, F., Formoso, V.: Comparison of collaborative filtering algorithms: limitations of current techniques and proposals for scalable, high-performance recommender systems. ACM Trans. Web **5**(1), 1–33 (2011)
3. Cai, X., Nie, F., Huang, H.: Multi-view K-means clustering on big data. In: International Joint Conference on Artificial Intelligence, pp. 2598–2604 (2013)
4. George, T., Merugu, S.: A scalable collaborative filtering framework based on co-clustering. In: IEEE International Conference on Data Mining, p. 4 (2005)
5. He, Y., Wu, J., Wang, H.: Clickthrough refinement for improved graph ranking. In: International Joint Conference on Neural Networks, pp. 3288–3295 (2017)
6. Herlocker, J.: Clustering items for collaborative filtering. In: Proceedings of the ACM SIGIR Workshop on Recommender Systems (2001)

7. Lecun, Y., Bengio, Y., Hinton, G.: Deep learning. Nature **521**(7553), 436–444 (2015)
8. Liu, L.: A sparse image recommendation model using content and user preference information. In: IEEE/WIC/ACM International Conference on Web Intelligence, pp. 232–239 (2016)
9. Luo, X., Ouyang, Y., Xiong, Z.: Improving K-nearest-neighborhood based collaborative filtering via similarity support. Int. J. Digit. Content Technol. Its Appl. **5**(7), 248–256 (2011)
10. Salakhutdinov, R., Mnih, A.: Probabilistic matrix factorization. In: International Conference on Neural Information Processing Systems, pp. 1257–1264 (2007)
11. Sarwar, B., Karypis, G., Konstan, J., Riedl, J.: Application of dimensionality reduction in recommender systems. In: ACM WebKDD Workshop (2000)
12. Simonyan, K., Zisserman, A.: Very deep convolutional networks for large-scale image recognition. Computer Science (2014)
13. Slaney, M.: Web-scale multimedia analysis: does content matter? IEEE Multimed. **18**(2), 12–15 (2011)
14. Su, X., Khoshgoftaar, T.M.: A survey of collaborative filtering techniques. Hindawi Publishing Corporation (2009)
15. Wang, J., Vries, A.P.D., Reinders, M.J.T.: Unifying user-based and item-based collaborative filtering approaches by similarity fusion, pp. 501–508 (2006)
16. Wu, J., He, Y., Guo, X., Zhang, Y., Zhao, N.: Heterogeneous manifold ranking for image retrieval. IEEE Access **PP**(99), 1 (2017)
17. Wu, J., Shen, H., Li, Y.D., Xiao, Z.B., Lu, M.Y., Wang, C.L.: Learning a hybrid similarity measure for image retrieval. Pattern Recognit. **46**(11), 2927–2939 (2013)
18. Xue, G.R., et al.: Scalable collaborative filtering using cluster-based smoothing. In: SIGIR: International ACM SIGIR Conference on Research & Development in Information Retrieval, pp. 114–121 (2005)
19. Yang, Q., Chen, Y., Xue, G.R., Dai, W., Yu, Y.: Heterogeneous transfer learning for image clustering via the social web. In: Joint Conference of the Meeting of the ACL and the International Joint Conference on Natural Language Processing of the AFNLP, vol. 1, pp. 1–9 (2009)

Integrating LDA into the Weighted Average Method for Semantic Friend Recommendation

Jibing Gong[1,2,3], Shuai Chen[1,2(✉)], Xiaoxia Gao[1,2], Yanqing Song[1,2], and Shuli Wang[4]

[1] School of Information Science and Engineering, Yanshan University,
Qinhuangdao 066004, China
chenshuaiysu@outlook.com
[2] The Key Laboratory for Computer Virtual Technology and System
Integration of Hebei Province, Yanshan University, Qinhuangdao 066004, China
[3] Key Laboratory for Software Engineering of Hebei Province, Yanshan University,
Qinhuangdao 066004, China
[4] The Department of Applied Mathematics, Yanshan University,
Qinhuangdao 066004, China

Abstract. Friend recommendation is a fundamental service for both social networks and practical applications. The majority of existing friend-recommendation methods utilize user profiles, social relationships, or static post content data, but rarely consider the semantic intentions and dynamic behaviors of users. In this paper, we propose FRec++, a friend recommendation method based on semantic and dynamic information. In FRec++, the first plus stands for new friend recommendation using a semantic model based on LDA while the next plus stands for the use of dynamic user attributes (e.g., behaviors or positions). More specifically, we first use the LDA method to generate semantic topics for user interests, and then compute the topic similarities between the target user and candidate friends. Next, we calculate the similarities of dynamic behaviors (i.e., forwarding, making comments, liking, and replying) and investigate the static attributes of users to measure the relevance of preferences. Finally, the weighted average method is used to integrate the above factors. We conducted experiments on the Weibo dataset and the results show that FRec++ outperforms several existing methods.

Keywords: Semantic friend recommendation
Hybrid recommendation architecture · Social behavior analysis
LDA-based framework

Supported by the Hebei Natural Science Foundation of China (Grant No. F2015203280) and the Graduate Education Teaching Reform Project of Yanshan University (Grant No. JG201616).

Z. Xu et al. (Eds.): Big Data 2018, CCIS 945, pp. 427–441, 2018.
https://doi.org/10.1007/978-981-13-2922-7_29

1   Introduction

Nowadays, with the rapid development of social networks, friend recommendation has attracted the attention of an increasingly large number of researchers and can be widely applied in other fields (e.g., product recommendation or online shopping). There are several traditional recommendation methods: collaborative filtering, tag-based recommendation, and content-based recommendation, but these methods have the following limitations: (1) a lack of deep semantic analytics, (2) consideration of only some user attributes, and (3) a unified framework for modelling all types of user features including profile information, post data, real-time location, dynamic behaviors, and social ties. Here, profile information refers to static attributes (such as topics, geographical location, age, gender, birthdate, and birthplace) and dynamic behaviors refer to forwarding and liking posts as well as making comments or replying to them.

Generally, there are two intuitive concepts regarding friend recommendation: (1) the inclusion of more features in the proposed method will improve performance, and (2) semantic analytics can help the recommendation method achieve better recommendation results. Thus, there is a trend of combining both static and dynamic features in the semantic understanding of social text data to construct a unified model for semantic friend recommendation that address the above semantic and attribute limitations. However, some challenges still exist:

- Rapidly expanding social applications have diversified user features and now yield increasingly large amounts social data. Thus, how to utilize these features and the massive amounts of data needed to build a unified model are substantial challenges.
- Normally, social text data (post data, comments, and comment replies) conceal social users interests, hobbies, and viewpoints. Hence, how to semantically analyze this data for recommendation is another challenge.
- The dynamic nature of user behaviors and variety of static user attributes makes it difficult to combine and measure these features for friend recommendation.

To address these challenges, we propose FRec++, a friend recommendation method based on semantic and dynamic information. We use two main approaches: (1) we systematically investigate how to perform semantic friend recommendation in social networks, and (2) we formally measure which factors/features have the greatest influence on friend recommendation results. FRec++ utilizes the LDA framework to obtain semantic information from users social text data. Combining semantic information and dynamic user behaviors, FRec++ calculates the similarity between users using the weighted average method. In addition, we consider geographical location and common friends. Generally, two users in the same city and those with a high number of common friends are more likely to become friends. In addition, we introduce dynamic behaviors, which not only indicate interactions between users but also play an important role in maintaining and establishing the links between users. Inspired by these two intuitions, we formally describe the factors and features that influence recommendation results and then analyze which are the most important for friend recommendation. Finally, after calculating the similarity between users via these features, we obtain the top-N candidate friends and recommend them to the target user.

In this paper, we systematically investigate how to combine the semantic understanding of users social text data and the dynamic behaviors of users for friend recommendation in social networks. Specifically, the main contributions of our work are as follows.

- We design a hybrid friend recommendation architecture based on the LDA algorithm and weighted average method.
- We perform an in-depth analysis of multiple user features (semantic information, dynamic behavior, and static attributes) and model these features simultaneously to improve recommendation performance.
- In contrast to previous methods, we not only consider the structure of the social network (i.e., dynamic behaviors), but also static user attributes (semantic information, geographical location, and common friends).

The rest of the paper is organized as follows: Sect. 2 describes related work. Section 3 presents our proposed recommendation method, and Sect. 4 describes the experiments and results. We conclude the paper in Sect. 5.

We summarized the notations used in this paper in Table 1.

Table 1. Notations

Symbol	Description
$Tsim(u,v)$	Topic similarity between u and v
$Gsim(u,v)$	Geography similarity between u and v
$Cf(u,v)$	Common friend similarity between u and v
$D(u,v)$	Dynamic behaviors similarity between u and v
$Fd(u,v)$	Correlation of forwarding between u and v
$Cm(u,v)$	Correlation of comments between u and v
$Re(u,v)$	Correlation of reply behaviors between u and v
$Le(u,v)$	Correlation of liking behaviors between u and v
$SD(u,v)$	Similarity degree between u and v
E_1	Sum of $Fd(u,v)$, $Cm(u,v)$, $Re(u,v)$ and $Le(u,v)$
T	Sigmoid function value of $Tsim(u,v)$
G	Sigmoid function value of $Gsim(u,v)$
C	Sigmoid function value of $Cf(u,v)$

2   Related Work

Recently, increasingly more researchers have developed individual friend recommendation methods that combine several different algorithms based on social data such as relations, interests, and location. Location-based recommendation (RWCFR) [1] is a typical state-of-the-art friend recommendation that is similar to FRec++. However, RWCFR' s architecture is based on a random walk (RW), instead of a hybrid architecture based on LDA and weighted averages. Both methods are location-based, but RWCFR performs better in mobile social networks [2], while FRec++ is suited to online social networks. In addition, the FRec++ model considers dynamic behaviors and static attributes while RWCFR only employs check-in data.

Inspired by the developments in text mining, some researchers have used LDA in friend recommendation. For instance, [3] generates a user' s subjects and interests using LDA and further recommends friends by finding other users' with similar subjects. Others [4] model the daily lives of users as life documents, their life styles as topics, and their activities as words to achieve satisfactory recommendation results. These studies prove that there are many gaps in friend recommendation using LDA.

Another important component in friend recommendation is computing similarities among users. There exist numerous methods for this task [5]; examples are common neighbors (CN) [6], Google PageRank ranking [7], and the Pearson correlation coefficient in collaborative filtering [8,9]. The authors of [6] consider the nodes within two hops of a target node to model dynamic social networks based on CN. However, methods that only consider a single factor are not effective. In terms of breadth, it has become a trend in friend recommendation to consider several attributes such as geographical location, tags, and interests instead of one attribute [10,11]. Furthermore, it is equally important to reveal more user-related information about each attribute, i.e., in-depth analysis is necessary.

There are other methods for semantic friend recommendation. Traditional methods like collaborative filtering [12,13] and content-based recommendation [14–16] were widely used in the early development of social networks. While the former method is unsuitable for high-dimensional data and suffers from the cold-start problem, the latter considers few attributes. Individual friend recommendation includes tag-based, emotion-based, and location-based methods. Tags can improve the accuracy of individual recommendation by scope list [17]. Emotion-based recommendation models extract a users emotional words to obtain his/her emotional tendencies and compute his/her similarity with other users [18]. Based on the approaches the above studies, we propose FRec++ to improve the accuracy of recommendation.

3   Proposed Method: FRec++

3.1   FRec++ Framework

In this section, we first describe static user attributes and then give the details of dynamic behaviors. We next illustrate how to combine them in FRec++. The architecture of FRec++ is given in Fig. 1.

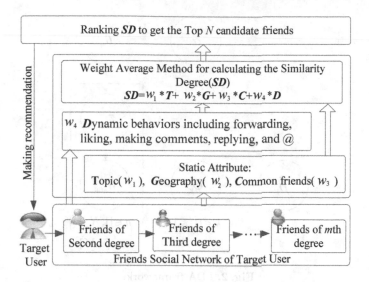

Fig. 1. FRec++ framework.

In this study, we employ the static attributes of topics, geographical location, and common friends. These features are explained in detail below.

Topics: Common topics implicitly reflect similar interests among users and are found using LDA. The higher the topic similarity between two users, the more likely they are to become friends. For example, if one of the hobbies of a user is traveling, he or she might prefer to become friends with another travel enthusiast. Additionally, two users in the same industry might wish to talk about their professional field. People are also inclined to talk with their peers.

Geographical Location: Geographical information reflects a users spatial characteristics. Social networking has broken down some geographical restrictions, but related studies show that two people who live in the same city have a higher probability of becoming friends than others. Additionally, people are willing to share amusing things that happen around them, and users in the same city may find these messages interesting or also have personally experienced them.

Common Friends: Structural balance theory explains the social network well. In this theory, interests, ages, and other features are transitive among friends, so the friends of a users friends are also potential friends for him or her. Normally, if they have more common friends, two users are more likely to establish friendship.

3.2 Semantic Topics

Social text data implies a users interests, hobbies, and views. In this study, we use LDA to discover semantic information using the algorithm shown in Fig. 2.

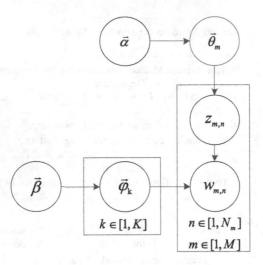

Fig. 2. LDA framework.

In Fig. 2, m denotes a document with N_m words, M is the total number of documents, n denotes a word in the document m, k stands for a certain topic, and K is the number of topics. The detailed steps of the LDA algorithm are given as follows.

Step 1. $\vec{\alpha} \to \vec{\theta_m} \to z_{m,n}$: According to the Dirichlet prior parameter $\vec{\alpha}$, a multinomial topic distribution $\vec{\theta_m}$ corresponding to document m is generated. Then, topic $z_{m,n}$ is selected for word n $\vec{\theta_m}$.

Step 2. $\vec{\beta} \to \vec{\varphi_k} \to w_{m,n}|k = z_{m,n}$: Word multinomial distribution $\vec{\varphi_k}$ is obtained from Dirichlet prior parameter $\vec{\beta}$ and topic $z_{m,n}(k)$. Further, word n is selected from multinomial distribution $\vec{\varphi_k}$.

Here, $\vec{w} = (\vec{w_1}, ..., \vec{w_M})$ denotes the corpus, and the words in document m form the vector $\vec{w_m}$. In addition, $\vec{z} = (\vec{z_1}, ..., \vec{z_M})$ represents the topic matrix corresponding to $\vec{w}$, where each word only corresponds one topic. The process of generating each topic is independent, so we can calculate the probability of the topic in the corpus as follows.

$$p(\boldsymbol{z} \mid \boldsymbol{\alpha}) = \prod_{m=1}^{M} p(\boldsymbol{z_m} \mid \boldsymbol{\alpha}). \tag{1}$$

Words with the same topic k are clustered as $\vec{w_{(k)}}$, $\vec{w}' = (\vec{w_{(1)}}, ..., \vec{w_{(K)}})$ denotes all words that have been classified. In addition, $\vec{z}' = (\vec{z_{(1)}}, ..., \vec{z_{(K)}})$ is the topic matrix corresponding to $\vec{w}'$, where each component in $\vec{z_{(k)}}$ is K. The process of generating words is independent and it can be expressed by Eq. (2).

$$p(\boldsymbol{w} \mid \boldsymbol{z}, \boldsymbol{\beta}) = \prod_{k=1}^{K} p(\boldsymbol{w_{(k)}} \mid \boldsymbol{z_k}, \boldsymbol{\beta}). \tag{2}$$

Using Eqs. (1) and (2), we can deduce the joint distribution probability.

$$p(w, z \mid \alpha, \beta) = p(w \mid z, \beta) p(z \mid \alpha). \tag{3}$$

Gibbs sampling is used to calculate the parameters that we can then use to predict the topics of users' posts.

3.3   Dynamic Behaviors

Interactions between users help form a link structure among users in social networks, so we define dynamic behaviors as interactions. Forwarding or liking posts, mentioning other users, and making or replying to comments are taken into account. We present the details below.

Forwarding can be classified into two categories. One category is when user u forwards user v's posts when user u is not one of friend of user u and the other category is when user u and user v forward the same posts. The former represents a direct behavior between users, the latter represents an indirect behavior among users. Direct behavior is more important than indirect behavior with respect to friend recommendations. The number of these behaviors indicates the degree of intimacy between users. Commenting, replying to comments, liking posts, and mentioning users are behaviors that are defined in the same way as forwarding behavior. Usually, two strangers with more dynamic behaviors in common are inclined to become friends.

Dynamic behaviors in social networks have a time-based characteristic. Hence, time is considered when calculating the similarity of dynamic behaviors. Specifically, compared with a dynamic behavior that has just happened, a dynamic behavior that happened three months ago is not important for recommendation in the social network. Function $Fd(u, v)$, the similarity of forwarding behaviors between users, is expressed as follows.

$$Fd(u, v) = \frac{\sum_{Ti} \phi_t (\alpha \times Fb(u, v) + \beta \times Fa(v, v))}{\sum_{Ti} \phi_t \times Fc(u) + \sum_{Ti} \phi_t \times Fc(v)}, \tag{4}$$

where ϕ_t is a time decay function and Ti is the time series. Moreover, $Fb(u, v)$ is the number of direct forwarding behaviors, $Fa(u, v)$ is the number of indirect forwarding behaviors, $Fc(u)$ is the number of forwarding behaviors of user u, and $Fc(v)$ is the number of forwarding behaviors of user v. Parameters α and β are the weights of the direct and indirect forwarding behaviors. Related studies show that direct forwarding behaviors contribute more than indirect forwarding behaviors to friend recommendations. Hence, α is greater than β and the sum of the weights is one. Parameter α is heuristically set to 0.7. The numerator sums the related forwarding behaviors of two users, and the denominator expresses the sum of all their forwarding behaviors.

Similarly, $Cm(u, v)$, the relevance of comments between users, can be calculated by the following.

$$Cm(u, v) = \frac{\sum_{Ti} \phi_t (\alpha \times Cb(u, v) + \beta \times Ca(u, v))}{\sum_{Ti} \phi_t \times Cc(u) + \sum_{Ti} \phi_t \times Cc(v)}, \tag{5}$$

where ϕ_t is a time decay function, where the value of previous comment behaviors decreases over time. In addition, $Cb(u,v)$ is the number of direct comment behaviors between users, $Ca(u,v)$ is the number of indirect comment behaviors between users, $Cc(v)$ is the number of comment behaviors of user v, and $Cc(u)$ is the number of comment behaviors of user u. Over a period Ti, we sum the number of common comment behaviors between users to determine the numerator, and the denominator is the sum of all their comment behaviors.

For the similarities of reply behaviors $Re(u,v)$ and liking behaviors $Le(u,v)$, we adopt the following equations.

$$Re(u,v) = \frac{\sum_{Ti} \phi_t \times Rc(u,v)}{\sum_{Ti} \phi_t \times Rc(u) + \sum_{Ti} \phi_t \times Rc(v)}, \tag{6}$$

$$Le(u,v) = \frac{\sum_{Ti} \phi_t \times Lc(u,v)}{\sum_{Ti} \phi_t \times Lc(u) + \sum_{Ti} \phi_t \times Lc(v)}, \tag{7}$$

where $Re(u,v)$ denotes the number of reply behaviors between users: $Rc(u)$ denotes the number of user u's reply behaviors and $Rc(v)$ is the number of user v's reply behaviors. In Eq. 7, the numerator is the number of common liking behaviors, and the denominator is the sum of their total liking behaviors.

3.4   FRec++ Implementation

In this section, we obtain the similarities between users according to static attributes and dynamic behaviors. Then, the weighted average method is employed to calculate the overall similarity degree. Finally, we obtain the top-N candidate friends to recommend. Using the LDA algorithm, we can infer a user's interests, age, social status, and other characteristics. More similar topics indicate that two users may have more of the same interests and they could become friends with a higher probability. Topic similarity $Tsim(u,v)$ between u and v is defined as follows.

$$Tsim(u,v) = \frac{\sum_T Tu \bigcap Tv}{\sum_T Tu \bigcup Tv}, \tag{8}$$

where T is the set of topics and Tu is the set of posts user u interacted with. In this paper, when a user interacts with a post, it means that the post is the user's original post, or it has been associated with the user through dynamic behaviors. The numerator is the number of common topics in which u and v interact, and the denominator is all posts with which u and v interact.

Simultaneously, we calculate the geographical location similarity $Gsim(u,v)$. The equation is expressed as follows.

$$Gsim(u,v) = \frac{1}{dist(u,v)}. \tag{9}$$

To determine $Cf(u, v)$, the similarity of common friends, we simply determine the number of common friends of two users and calculate their similarity as follows.

$$Cf(u, v) = \frac{Fu \cap Fv}{Fu \cup Fv}, \tag{10}$$

where Fu and Fv are the friend sets of users u and v, respectively. The numerator determines the number of common friends of two users, and the denominator is the size of the intersection of their friend sets.

After calculating the similarity of the static attributes of two users, we can determine their unified dynamic behavior similarity $D(u, v)$.

$$D(u, v) = \frac{1}{1 + e^{-E_1}}, \tag{11}$$

where $E_1 = Fd(u, v) + Cm(u, v) + Le(u, v) + Re(u, v)$, and the sigmoid function is to used scale the feature.

Finally, by leveraging the weighted average method, we combine static attributes with dynamic behaviors and get the final similarity score $SD(u, v)$.

$$SD(u, v) = w_1 T + w_2 G + w_3 C + w_4 D, \tag{12}$$

where w_1, w_2, w_3, and w_4 are the weights of features, and these weights indicate the importance of the attributes in recommendations. Further, these weights satisfy $w_1 + w_2 + w_3 + w_4 = 1$. T is expressed by $\frac{1}{1 + e^{-Tsim(u_i, u_j)}}$. G is $\frac{1}{1 + e^{-Gsim(u, v)}}$, and C is calculated by $\frac{1}{1 + e^{-Cf(u, v)}}$. The sigmoid function is again utilized to scale the features. Then, the similarity scores between users are obtained. Rating the final similarity score SD of each user, we can determine the top-N candidate friends and recommend them to the target user.

4   Experiments and Evaluations

4.1   Experimental Setup

The experimental dataset consists of the information of 1,800,000 Weibo users, including their post data, likes, forwards, and comments. The statistics of these user behaviors are illustrated in Fig. 3.

We adopted several metrics (P@k, MAP, recall, and F1-measure) to evaluate the performance of FRec++ from different perspectives. The equations of these metrics are given as follows:

$$P@k = \frac{the\ number\ in\ the\ first\ k\ results}{k}, \tag{13}$$

$$MAP = \frac{\sum_{k\ is\ relavent} p@k}{times\ of\ recommendation\ cases}, \tag{14}$$

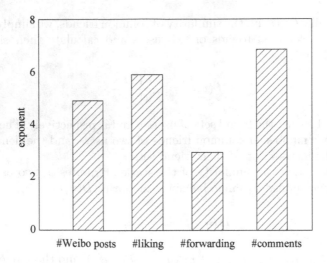

Fig. 3. Statistics of user behaviors in the dataset.

$$Recall = \frac{the\ number\ in\ the\ first\ k\ results}{the\ number\ of\ newly\ added\ friends}, \tag{15}$$

$$F1 = 2P \times R / (P + R), \tag{16}$$

where P denotes precision, and R is recall. The F1-measure combines the precision with the recall.

We compared FRec++ with several other existing friend recommendation methods. We selected the RW, LDA-based similarity (LDAS), RWCFR [1], common neighbors (CN), and Friend++ [2] as comparison algorithms. A brief description of these methods is given below.

RWCFR [1]: In addition to check-in data, we consider position information and user profile data to obtain a ranked recommendation list.

CN: This method finds and measures the common friends of the target user and his or her second-degree friends.

LDAS: Specifically, this LDA-based method only considers the topic attribute and ignores other information to model the recommendations. Then according to the topic similarity, it recommends friends to target users.

RW: This method utilizes an individual intimacy feature to compute the similarity between target users and the candidate users through a RW and then obtains the top-N recommended friends.

Friend++ [2]: Two features (network and node features) are considered in the Friend++ method. It employs an RW algorithm to recommend friends.

4.2 Factor Analysis

In FRec++, we divide features into four factors: topics (T), geographical location (G), common friends (C), dynamic behaviors (D). To analyze the importance of each factor, we designed four comparative experiments, each removing one of these factors. The method without topics is denoted by FRec++-T, and the methods without geographical location, common friends, and dynamic behaviors are denoted by FRec++-G, FRec++-C, and FRec++-D respectively. Here, we used the same weight for all factors. Figure 4 shows the performance of these different methods in terms of F1-measure.

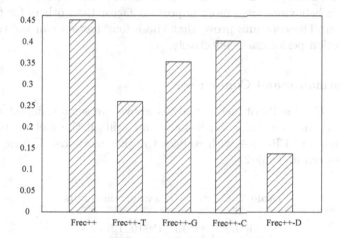

Fig. 4. Values of F1-measure in the dataset.

Here, FRec++ outperforms the other four methods. The second-best method is FRec++-C, followed by FRec++-G, then FRec++-T. FRec++-D performs the worst. The lack of dynamic behaviors causes a clear drop in performance. The F1-measures of FRec++-T and FRec++-G decrease to 0.26 and 0.35, respectively. Simultaneously, the F1-measure of FRec++-C also decline slightly. These results indicate that the performance of dynamic behaviors is the most influential factor. Thus, we set weights of C, G, T, and D to 0.1, 0.2, 0.3, and 0.4, respectively, for the remainder of the experiments.

The experimental results reveal that FRec++ clearly outperforms the modified methods. This indicates that the proposed form of FRec++ does improve the precision and the accuracy of friend recommendation. We see that not only static attributes but also dynamic behaviors help improve the performance of friend recommendations. By combining more features with the weighted average method, FRec++ improves the recommendation performance (by 19%, 10%, 4%

and 28%, respectively, for FRec++-T, FRec++-G, FRec++-C and FRec++-D in terms of F1-measure).

In summary, each factor is useful for improving the friend recommendations in FRec++. Additionally, the dynamic behaviors of interactions have the largest impact on recommendation results followed by common interests and geographical location in turn, while the common friends factor has the weakest impact. Some reasonable explanations for this are as follows: Active interactions may be more important than common interests. That is, if a network is not active, those users who have the same interests or topics would find it very difficult to be friends. In contrast, compared with close geographical location and common friends, those users who have common interests find it easier to be friends. In real society, two people may have many common friends and are close to each other, but their interests may not overlap and they are unwilling to communicate with each other, let alone be friends. In other words, these results indicate that dynamic behaviors are a more important factor than others for friend recommendation. These results prove that these four factors can improve friend recommendation performance effectively.

4.3   Performance and Comparison

According to the results of factor analysis above, we set a different weight for each factor for the remaining experiment and utilized the metrics to evaluate the performance of FRec++ with respect to other methods. The experimental results are shown in Table 2.

Table 2. Comparative evaluation results

Method	P@5	P@10	P@15
RW	0.761	0.464	0.277
CN	0.786	0.512	0.214
LDAS	0.627	0.426	0.2608
Friend++	0.793	0.523	0.273
RWCFR	0.451	0.337	0.257
FRec++	0.806	0.546	0.284

Table 2 lists the P@k results of the five baseline methods and the proposed method on the dataset. The results clearly show that our proposed FRec++ outperforms all five baseline algorithms in terms of P@k. Furthermore, we compare the recall, MAP, and F1-measure of FRec++ with those of the five baseline methods in Fig. 5.

Figure 5 clearly shows that FRec++ outperforms the five comparable methods with respect to all three metrics: recall, MAP, and F1-measure. In terms of recall, FRec++ achieved improvements over RW, CN, RWCFR, LDAS, and

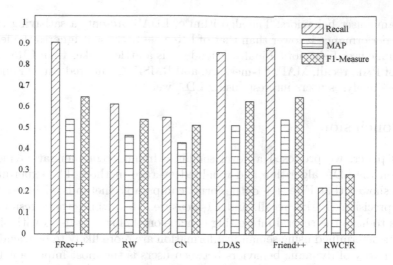

Fig. 5. Experimental results.

Friend++ of 31%, 34%, 67%, 4%, and 1% on the dataset. FRec++ achieves an average 10.94% improvement over the other algorithms with respect to MAP. Furthermore, in terms of F1-measure, the compared methods underperform FRec++ by an average of 15.02%. These results imply that the overall performance of FRec++ is better than that of the other methods. Finally, the root-mean-square error (RMSE) is used to measure the recommendation quality of FRec++. The results are shown in Table 3.

Table 3. RMSE of the experimental results

Method	RMSE
RW	0.31256
LDAS	0.31524 ± (0.00264)
CN	0.35513 ± (0.00248)
Friend++	0.28651 ± (0.00251)
RWCFR	0.28521 ± (0.00235)
FRec++	0.26415 ± (0.00315)

Table 3 show that the RMSE of FRec++ is lower than that of the baseline methods. Further, it also indicates that FRec++ accurately and reliably outperforms other friend recommendation methods.

RWCFR is a popular friend recommendation method based on the RW in mobile social networks, and its poor performance indicates that it is not suited to our online social dataset. Moreover, only considering check-in data is a limitation of RWCFR. RW and CN show similar results, as both lack in-depth consideration of dynamic user behaviors. The algorithm of LDAS presents a satisfying result, but its performance is lower than that of FRec++ because it ignores the feature of dynamic user behaviors. Finally, Friend++ is a little weaker than FRec++ in terms of P@k, recall, MAP, F1-measure, and RMSE. Compared with Friend++, FRec++ analyses users interests using LDA well.

5   Conclusion

In this paper, we proposed a novel semantic friend recommendation method based on the LDA algorithm and weighted average method. The experimental results showed that FRec++ outperforms comparable methods with respect to MAP, precision (P@k), recall, and F1-measure. Meanwhile, these results also help us to have the several interesting conclusions below: (1) users with similar interests determined using semantic information are more likely to be friends, (2) the similarity of dynamic behaviors between users is the most important factor for friend recommendation, (3) geographical location and common friends more weakly impact the results, and (4) dynamic behaviors may be the key factor in helping to find the most appropriate potential friends. Although FRec++ has achieved better performance, two points are worth further study: (1) the extraction of user interests with LDA can be further improved, and (2) more reasonable weights for each factor should be determined.

References

1. Bagci, H., Karagoz, P.: Context-aware friend recommendation for location based social networks using random walk. In: Proceedings of the 25th International Conference Companion on World Wide Web (WWW 2016), pp. 531–536. ACM, New York (2016)
2. Gong, J., et al.: Integrating a weighted-average method into the random walk framework to generate individual friend recommendations. Sci. China Inf. Sci. **60**(11), 110104 (2017)
3. Xu, K., Zheng, X., Cai, Y., et al.: Improving user recommendation by extracting social topics and interest topics of users in unidirectional social networks. Knowl.-Based Syst. **140**, 120–133 (2018)
4. Wang, Z., Liao, J., Cao, Q., Qi, H., Wang, Z.: Friendbook: a semantic-based friend recommendation system for social networks. IEEE Trans. Mob. Comput. **14**(3), 538–551 (2015)
5. Feng, S., Zhang, L., Wang, D., Zhang, Y.: A unified microblog user similarity model for online friend recommendation. In: Zong, C., Nie, J.Y., Zhao, D., Feng, Y. (eds.) Natural Language Processing and Chinese Computing. CCIS, vol. 496, pp. 286–298. Springer, Heidelberg (2014). https://doi.org/10.1007/978-3-662-45924-9_26
6. Yu, W., Lin, G.: Social circle-based algorithm for friend recommendation in online social networks. Chin. J. Comput. **37**(4), 801–808 (2014)
7. Liu, Q., Li, Z.G., Lui, J.C., et al.: PowerWalk: scalable personalized pagerank via random walks with vertex-centric decomposition. In: Proceedings of the 25th ACM International on Conference on Information and Knowledge Management, pp. 195–204. ACM, New York (2016)

8. Yun, Y.D., Hooshyar, D., Jo, J., Lim, H.: Developing a hybrid collaborative filtering recommendation system with opinion mining on purchase review. J. Inf. Sci. **44**(3), 331–344 (2018)
9. Cai, Y., Leung, H., Li, Q., et al.: Typicality-based collaborative filtering recommendation. IEEE Trans. Knowl. Data Eng. **26**(3), 766–779 (2014)
10. Zhu, J., Lu, L., et al.: From interest to location: neighbor-based friend recommendation in social media. J. Comput. Sci. Technol. **30**(6), 1188–1200 (2015)
11. Wu, B.-X., Xiao, J., Chen, J.-M.: Friend recommendation by user similarity graph based on interest in social tagging systems. In: Huang, D.-S., Han, K. (eds.) ICIC 2015. LNCS (LNAI), vol. 9227, pp. 375–386. Springer, Cham (2015). https://doi.org/10.1007/978-3-319-22053-6_41
12. Liu, J., Tang, M., Zheng, Z., Liu, X.F., Lyu, S.: Location-aware and personalized collaborative filtering for web service recommendation. IEEE Trans. Serv. Comput. **9**(5), 686–699 (2016)
13. Ma, H., Jia, M., Zhang, D., et al.: A content-based recommendation algorithm for learning resources. Inf. Sci. **13**, 325–337 (2016)
14. Ullah, F., Lee, S.: Social content recommendation based on spatial-temporal aware diffusion modeling in social networks. Symmetry **8**(9), 89 (2016)
15. Erheng, Z., Nathan, L., Yue, S., et al.: Building discriminative user profiles for large-scale content recommendation. In: Proceedings of the the 21th ACM SIGKDD International Conference on Knowledge Discovery and Data Mining, pp. 2277–2286. ACM, New York (2015)
16. Shu, J., Shen, X., Liu, H.: A content-based recommendation algorithm for learning resources. Multimed. Syst. **24**(2), 163–173 (2018)
17. Ma, H., Jia, M., Zhang, D., Lin, X.: Combining tag correlation and user social relation for microblog recommendation. Sci. China Inf. Sci. **385**, 325–337 (2017)
18. Habibi, M., Popescu-Belis, A.: Keyword extraction and clustering for document recommendation in conversations. IEEE/ACM Trans. **23**(4), 746–759 (2015)

SRE-Net Model for Automatic Social Relation Extraction from Video

Lili Zhou[1,2(✉)], Bin Wu[1,2], and Jinna Lv[1,2]

[1] Beijing University of Posts and Telecommunications, Beijing 100876, China
Lily_super12580@163.com, {wubin,lvjinna}@bupt.edu.cn
[2] Beijing Key Laboratory of Intelligent Telecommunications Software
and Multimedia, Beijing, China

Abstract. Videos spread over the Internet contain a huge knowledge of human society. Diversified knowledge is demonstrated as the storyline of the video unfolds. Therefore, realization of automatically constructing social relation network from massive video data facilitates the deep semantics of mining big data, which includes face recognition and social relation recognition. For face recognition, previous studies are focus on high-level features of face and multiple body cues. However, these methods are mostly based on supervised learning and clustering need to specify clusters k, which cannot recognize characters when new video data is input and individual and its numbers are unknown. For social relation recognition, previous studies are concentrated on images and videos. However, these methods are only concentrated on social relations in same frame and incapable of extracting social relation of characters that are not present in the same frame. In this paper, a model named SRE-Net is proposed for building social relation network to address these challenges. First, MoCNR algorithm is introduced by clustering similar-appearing faces from different keyframes of video. As far as we know, it is the first algorithm to identify character nodes using unsupervised double-clustering methods. Second, we propose a scene based social relation recognition method to solve challenges that cannot recognize social relations of characters in different frames. Finally, comprehensive evaluations demonstrate that our model is effective for social relation network construction.

Keywords: Deep learning · Face recognition · Scene segmentation
Social relation

1 Introduction

Nowadays, with the explosive development of Internet, video, as major way of getting information, occupies the Internet at an exponential growth rate. The analysis of social relation network [5] can be of great benefit to understand the interaction between people for us. How to construct social relation network efficiently from the massive video data has become an urgent need. The aim of social network construction is to train a model which can figure out the characters in video and predict social relation between the characters, named face recognition and social relation recognition.

© Springer Nature Singapore Pte Ltd. 2018
Z. Xu et al. (Eds.): Big Data 2018, CCIS 945, pp. 442–460, 2018.
https://doi.org/10.1007/978-981-13-2922-7_30

In the past years, some previous researches have exactly achieved face recognition using high-level features of face and multiple body cues [10, 11, 17–19, 21, 30]. However, few studies implement an unsupervised automatic face recognition method. Moreover, there are still great challenges: First, these existing face recognition methods are mostly based on supervised learning which requires us to do a lot of manual tagging to get a more robust character recognition. Furthermore, this type of model can only identify people who appear in the training set. Therefore, this kind of model cannot be applied to new video data which contains characters that are not present in the training set. Second, there are also one-clustering based video person recognition methods [21] which requires the number of clusters k to be specified in advance. Therefore, this method also has certain limitations in automatically recognizing character of video. In order to solve challenges that existed in supervised method and one-clustering method, this paper proposes a method of character nodes recognition (MoCNR, method of character node recognition) which adopts unsupervised double-clustering algorithm for character nodes recognition.

The mining of social relations of people helps to better understand the relationship between people, some previous works have recognized social relations from texts, images and videos [1, 2, 13, 14, 22, 23, 28]. However, few researches recognize social relations based on scene which can solve the challenges that character relations in the different keyframes cannot be mining accurately. Thus, a scene segmentation based social relation recognition method is proposed in order to solve these challenges that existed in previous works.

In view of these challenges in the face recognition and social relation recognition, SRE-Net model for automatically constructing social relation network from video is proposed in this paper, and the excavation of social relations from the shallow to the deep. SRE-Net model concludes identification of character nodes and the recognition of character relations. Firstly, this paper introduces MoCNR method in order to solve challenges that characters and their numbers are unknown in the video: (1) extract keyframes using FFmpeg (Fast Forward Mpeg); (2) mtcnn [3] is introduced for face detection and face alignment; (3) deep network is adopted to obtain DeepID features of faces; (4) character nodes and its numbers are obtained using Single-Pass and K-Means clustering algorithm, meanwhile, character feature is represented by clustering center. As we all known, it is the first algorithm to identify character nodes using unsupervised double-clustering methods. Secondly, we propose a scene [4] based relation recognition method in order to recognize the social relations between characters in the different frames of video: (1) introduce a scene segmentation algorithm to get the scene boundaries of the video by semantic similarity between keyframes; (2) deep network model is applied to extract the deep semantic features of the scene; (3) a classifier is proposed for identifying character relations by scene features which takes advantages of ensemble learning. Finally, social relation network of video is constructed. This network contains nodes and edges, in which, nodes represent characters appearing in the video and edges represent characters' relationship.

In summary, the contributions of this paper can be concluded as follows:

1. We propose MoCNR algorithm for unsupervised character recognition. This algorithm draws on unsupervised double-clustering method to identify the

characters in the video, which solves the disadvantages of supervised learning and one-clustering method in previous works. To the best of our knowledge, this is the first algorithm for unsupervised identifying character nodes from video.

2. A scene segmentation based social relation recognition method is proposed. Based on scene segmentation, the relation between the characters in the scene can be fully excavated, and this method incorporates ensemble learning. Thus, social relation network of video can be constructed completely and accurately.

3. A SRE-Net model is proposed, which implements an automatic social relation network construction of video. SRE-Net model blends identifying character nodes from video and recognizing social relations between characters. Firstly, the model identifies the character nodes in the video by unsupervised double-clustering algorithm. Secondly, sence segmentation is implemented. Finally, the model extracts social relations between characters based on scene, so that we can dig out social relations of characters without the same keyframe but within the same scene.

The rest part of this paper is organized as follows. Section 2 introduces the relevant work of several social network construction. Section 3 describes the proposed SRE-Net model in detail. Experiments analysis in Sect. 4, and Sect. 5 draws a conclusion.

2 Related Work

2.1 Deep Learning

Deep learning was proposed by Hinton in 2006 and is an emerging subfield of machine learning [29]. Deep learning extracts high-level abstract features [6] of data utilizing deep computational models composed by multiple layers. This high-level abstract feature can express the inherent structural features of data that cannot be acquired by traditional machine learning methods, obtaining better experimental results. Thereby, it has been widely exploited in the field of traditional artificial intelligence, such as transfer learning [7], natural language processing [8], computer vision [9] etc.

The analysis and mining of video data is an important content in the field of multimedia data mining. It shows significant advantages in the field of image processing, and it is increasingly drawing attention to experts and scholars at home and abroad, such as face recognition [10, 11], trajectory tracking [12], autopilot, social relation extraction [13, 14], image description [15], etc. Therefore, the SRE-Net model also utilizes deep learning to extract high-level features of the characters as well as the features of the video scenes, thereby excavating the deep semantic features in the video.

2.2 Face Recognition

In recent years, face recognition [16, 17] has attracted much attention and its research has rapidly expanded since it has many potential applications in computer vision communication.

The identification of character nodes in the social network is essentially face recognition or face verification. First of all, existing face recognition methods are mostly based on supervised learning methods and require training with a large number of tagged datasets. such as, Yi et al. [18] train a classifier to recognize about 10, 000 face identities in the training set which comes from DeepID feature learned through deep ConvNets. Yaniv et al. [19] train a deep neural network for multi-class classification, in which feature of fully connected layer fc7 is extracted to serve as the face representation. Seong et al. [11] solved the problem of person recognition by PIPER method which provides an in-depth analysis of multiple cues. Chang et al. [20] adopt the two-layer C-RNN model for face recognition in order to recognize the characters in the video. However, this supervised learning model requires us to do a lot of manual tagging to get a more robust character recognition model. Meanwhile, this type of model can only identify people who appear in the training set. Thus, this supervised method can only deal with videos which all characters are present in the training set.

Second, there are also one-clustering based video person recognition methods. For example, Jeremiah et al. [21] propose an active clustering method to face recognition. However, this one-clustering based person recognition method requires the number of clusters k to be specified in advance. Therefore, this method has challenges in automatically constructing a social relation network of characters. In view of the above challenges, this paper proposes the MoCNR method to identify the characters appearing in the video. MoCNR employs unsupervised double-clustering algorithm to recognizing of the character nodes, which not only avoids the problems of the supervised learning model, but also does not need to specify the number of clusters k in advance. Therefore, this method can process any video data specified by users, is an efficient and universal character node recognition method.

2.3 Social Relation Extraction

The mining of social relations of people helps to better understand the relationship between people. Qianru et al. [13] build the model based on domain-based theory that predicts social domain and relation from images. Jeremiah et al. [21] construct social network by active clustering. Zhanpeng et al. [22] introduce a model for social relation prediction from face images. However, this method is just suitable for mining the relations between people appearing in the same image. It is not difficult to find that there is such a phenomenon in video data: two people with social relations often appear in different keyframes, but appear in adjacent keyframes. In this case, the above relation extraction model [13, 21, 22] cannot be applied to extract social relations from video. Therefore, a model is required that can extract the relations between the characters in the video.

In previous work [28] published in 2017, a parallel method is proposed for social network construction of the role relation in unstructured data based on multi-view in order to accurately construct social network from unstructured data and speed up the efficiency of data processing and improve storage efficiency. The method constructs a social network based on video keyframes and video plot respectively, and then fuses the two social network, thereby effectively constructing a character social network from unstructured data. Based on that work [28], a multi-stream fusion model [2] is proposed in order to extract the social relation between the characters in the video that takes into account the temporal feature, spatial feature, and audio feature of the video, and a first social relationship video dataset SRIV is created. Junnan et al. [23] propose a

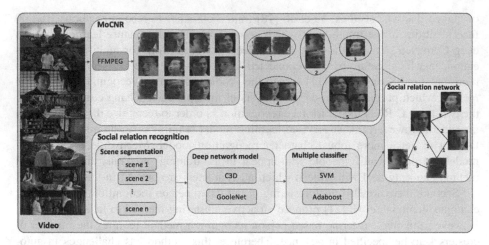

Fig. 1. An overview of SRE-Net for social relation extraction

dual-glance model for social relation recognition of video which deploys attention mechanism [24]. However, these models need to manually capture some video clips which contain a wealth of information, and cannot automatically extract the character relationships of the entire video data.

In view of those challenges in the social relations recognition model, we propose a scene segmentation based video character relation extraction method. The method first divides the video into scenes using scene segmentation method. Then the deep network model is applied to extract the high-level semantic features of the video scenes. Finally, ensemble learning is introduced to predict the social relations of the characters.

3   SRE-Net Model for Social Relation Extraction

In this section, we describe our model SRE-Net for social relation extraction.

There are many characters in video, and the relation between them is complex. Also, the relation between different characters tends to change with the development of the story. Therefore, we need a model to construct more accurate network automatically. Figure 1 shows our SRE-Net model for social relation recognition.

Firstly, we propose a MoCNR method which recognizes character nodes by two phase clustering. In the first phase clustering, single-pass clustering algorithm is applied to find optimal parameter k. Meanwhile, K-Means is introduced for character node recognition in the second phase clustering, whose parameter k comes from the previous clustering algorithm.

Secondly, social relation recognition method based on scene is proposed. In this paper, scene segmentation is introduced to divide the video into segmented video clips. Then the deep network model is applied to extract the high-level semantic features of the video. Finally, multiple classifiers are trained to extract social relations of characters through ensemble learning.

Algorithm 1 MoCNR

Input: v, t
    video data, threshold of first clustering
Output: dataSet
    face feature and its label of every characters
//extract keyframes
Kframe <- v convert to keyframes with FFMPEG
//face detection, face feature extraction
for all frame $\in$ Kframe **do**:
    faceImg <- face detection and alignment using mtcnn.
    faceAttr <- face feature extraction using DeepID
end for
//first clustering: find k
k = 0
// cluster assignments: clu_list
clu_list <- create and initialize clu_list[k] with faceAttr(0)
for all f $\in$ faceAttr(1:) **do**:
    index, min_dis <- nearest cluster and distance from f
    if min_dis < t:
        add f into clu_list[index]
    else: //create a new cluster assignment
        k <- k+1
        cluster_list <- create and initialize clu_list[k] with f
    end if
end for
//second clustering: clusterAssment
centroids, clusterAssment = kmeans(faceAttr, k+1)
labels <- a column vector with dimension (k x 1), its value is 0 to k from top to bottom
dataSet <- centroids and labels merge by column
return dataSet

Finally, we combine MoCNR and social relation recognition method based on scene segmentation for construction of social relation network.

3.1 MoCNR for Character Nodes Recognition

The method of character node identification in previous works are always based on supervised learning. Obviously, the supervised learning model cannot identify people who are not present in the training set. In order to solve this challenge, we propose the MoCNR method, which is different from the previous supervised learning method. It clusters people's facial features to identify the number of people and characters that appear in the video. This method successfully avoids a lot of manual annotation problems. Moreover, this unsupervised method greatly reduce the difficulty of applying to other issues.

Algorithm 1 describes processing of MoCNR method from keyframe extraction, face detection, face alignment, face feature extraction to clustering. In this section, first, video is converted to keyframes using FFmpeg. Second, we put to use mtcnn to face detection and face alignment. Third, high-level face representation DeepID is extracted using ConvNets. Forth, we conduct the first phase clustering to get the best cluster number. Fifth, K-Means is carried out for character nodes recognition, in which, every nodes is represented by its cluster center, meanwhile, every nodes was labeled in order to identify nodes that have social relations extracted by social relation recognition section.

MoCNR uses two kinds of metrics to compute the semantic similarities between characters. To achieve this, we first compute feature representation for the characters of the video. Feature representation are computed by taking all the faces from all the video frames and applying double-clustering on the DeepID features extracted from those faces. Now a character C_i can be represented as a N dimension vector, using this convention, a character is represented by

$$C = f_1, f_2, \ldots, f_j, \ldots, f_N, \tag{1}$$

where f_j gives the value of jth dimension for the character feature, and N is the size of the feature dimension. With this representation of the character, we use Cosine distance and Euclidean distance to find the distance between the characters. Cosine distance between i^{th} character and j^{th} character is given by

$$D_{ij}^c = \frac{\langle C_i, C_j \rangle}{||C_i|| \cdot ||C_j||} = \frac{\sum_{n=1}^{N} f_{in} \cdot f_{jn}}{\sqrt{\sum_{n=1}^{N} f_{in}^2} \cdot \sqrt{\sum_{n=1}^{N} f_{jn}^2}}, \tag{2}$$

where N is the size of the feature dimension, $\langle C_i, C_j \rangle$ is the inner product between C_i and C_j, $||C_i||$ is the norm of C_i, and f_{in} is the value of nth dimension of C_i.

Euclidean distance between i^{th} character and j^{th} character is given by

$$D_{ij}^e = \sqrt{\sum_{n=1}^{N} \left(f_{in} - f_{jn} \right)^2}, \tag{3}$$

The semantic similarity between the characters will be inversely related to these two kinds of distance. Thus, the similarity between i^{th} character and j^{th} character is given by respectively.

$$Sim_{ij}^c = \frac{1}{D_{ij}^c}, \tag{4}$$

$$Sim_{ij}^e = \frac{1}{D_{ij}^e}, \tag{5}$$

where Sim_{ij}^c is the similarity between i^{th} character and j^{th} character by considering Cosine metric, and Sim_{ij}^e is the similarity between i^{th} character and j^{th} character by considering Euclidean metric.

Fig. 2. The phenomenon existing in video frames

3.2 Method for Social Relation Extraction

In this paper, the relations of characters are divided into 8 categories according to Obj-Relation classes [2]: leader- member, peer, service, parents-offspring, lover, sibling, friend and enemy.

Scene Segmentation. Scene is more considerable for semantic analysis of the video since it catches hold of one complete unit of action. As shown in Fig. 2, it reveals a phenomenon that is ubiquitous in video: two interacting people appear in different key frames, or only one person's positive face is seen even in the same key frame. Therefore, the existing model for extracting human relationships from images does not apply to our video data. In this paper, we propose a new scene segmentation based social relation recognition algorithm which gets scene boundaries by computing semantic similarity between keyframes in video. The method of scene segmentation can detect scene changes in videos, and automatically split the video into separate clips, named scenes.

High-Level Feature. In order to obtain better social relation recognition results, we need to extract the high-level semantic features that can reflect relationships between two people in the video. In this paper, we extract two kinds of feature of video which includes audio feature and three-dimensional convolution features. We apply the GoogleNet model [27] to extract the audio features of the video. And in order to better capture the time sequence information of video, we employ the C3D [26] model to extract the three-dimensional convolution features of video. We rely on SRIV [2] dataset to finetune the C3D model and GoogleNet model.

Ensemble Learning. Ensemble learning employs multiple weak classifiers to predict label of the data, and the classification accuracy is improved by combining the results of multiple classifiers. Boosting is a machine learning algorithm that can be applied to reduce bias in supervised learning. Here, we use several ensemble classifiers to achieve social relation prediction: AdaBoost (Adaptive Boosting), XGBoost, GBDT and LightGBM.

AdaBoost. AdaBoost is a typical boosting algorithm. At the beginning of training, the AdaBoost algorithm assigns equal weight to each training sample, and then puts the algorithm to train the training set for t iterations. After each training process, the training samples which are predicted errors are given a greater weight. That is, the algorithm pays more attention to the samples which get the wrong label for learning after each learning, so as to obtain multiple prediction functions. At the same time, the weight of each weak classifiers has a positive correlation with its accuracy rate to the final prediction result.

Suppose we have trained a total of N weak classifiers, the final prediction output of AdaBoost is:

$$f(\mathbf{x}) = \sum_{n=1}^{N} \alpha_n f_n(x), \tag{6}$$

where $f_n(\mathbf{x})$ is the nth weak classifier, α_n is the weight corresponding to this classifier $f_n(\mathbf{x})$. And the value of α_n is related to its classification accuracy. Its formula is as follows:

$$\alpha_n = \frac{1}{2} \ln \frac{1 - e_n}{e_n}, \tag{7}$$

where e_n is the error rate of classifier $f_n(\mathbf{x})$:

$$e_n = P\left(f_n\left(x^{(i)}\right) \neq y^{(i)}\right) = \sum_{i=1}^{m} w_{ni} \mathbf{I}\left(f_n\left(x^{(i)}\right) \neq y^{(i)}\right), \tag{8}$$

where m is the number of samples in the training set, i denotes the ith sample, w_{ni} Represents the weight of the ith sample in the nth classifier $f_n(x)$. $\mathbf{I}()$ is indicator function notation:

$$\mathbf{I}(true) = 1, \tag{9}$$

$$\mathbf{I}(false) = 0, \tag{10}$$

As we can see from (7), the greater the error of the classifier, the lower its weight in the final decision.

XGBoost. The model of XGBoost is given by

$$f(\mathbf{x}) = \sum_{n=1}^{N} f_n(x), \tag{11}$$

where N is the number of weak classifiers, and $f_n(\cdot)$ is the nth weak classifier.

The loss function is defined as

$$L(\theta) = \sum_{i=1}^{m} l(y_i, f(x_i)) + \sum_{n=1}^{N} \Omega(f_n) \tag{12}$$

where m is the number of samples in the training set. From (12), it is not difficult to find that the loss function of XGBoost consists of two parts: degree of fitting to data and tree complexity. The first formula $l(y_i, f(x_i))$ represents the prediction error of the model on the data, the more the number of samples with the correct prediction result, the smaller of this loss value. The second formula $\sum_{n=1}^{N} \Omega(f_n)$ is the tree complexity, the simpler the tree structure, the smaller this loss value.

GBDT. GBDT is also a boosting algorithm, and its weak classifier can only use CART (Classification And Regression Tree) regression tree. The core idea of GBDT is: every iteration is to reduce the residual of the last tree. Its loss function of nth iteration is given by

$$L(y, f_n(x)) = L(y, f_{n-1}(x) + g_n(x)), \tag{13}$$

where $g_n(x)$ is CART regression tree get from n^{th} iteration. We can see that this round of iterations finds the decision tree and makes the loss as small as possible. Thus, the model of GBDT can be represented as

$$f_N(x) = \sum_{n=1}^{N} f_n(x), \tag{14}$$

where N is the number of CART weak classifiers, and also number of iterations.

LightGBM. LightGBM is a gradient boosting algorithm. It has the following advantages: faster training speed, better accuracy, GPU supported, etc. LightGBM adds support for category features so that category features are not converted to one-hot feature before training. And LightGBM uses leaf-wise tree growth strategy, this method can converge much faster comparing with depth-wise growth strategy.

3.3   Construction of Social Relation Network

There are many characters in film and TV series, and the relationship between them is complex. Also, the relationship between different characters tends to change with the development of the story. Therefore, we need a method to construct more accurate network. In this paper, we propose a SRE-Net model to automatically construct the social relation network in video.

Definition 1. Character node set C. Nodes of social relation network consists of character of video. We use an unsupervised double-clustering method to identify and label the characters in the video, thus forming the set of character nodes, $C = \{c_1, c_2, \ldots, c_n\}$, n is the number of nodes in the social relation network, and is also the number of characters in the video.

Definition 2. Social relation set R. Social relations are divided into three classes: working relation, kinship relation, and other relation. Working relation includes leader-member, peer and service. Kinship relation includes parents-offspring, lover, sibling. Other relation includes friend and enemy. Thus, social relation sets R is given by

$$R = \{r_1, r_2, \ldots, r_8\}, \tag{15}$$

where $r_1 = 1$, $r_2 = 2$, and $r_8 = 8$. Each number represents a social relation type that indicates the relations between the nodes connected to the same edge: leader-member, peer, service, parents-offspring, lover, sibling, friend and enemy.

Definition 3. Social relation network $G = \langle C, E, R \rangle$. Where C is the character sets which are obtained by face recognition, face alignment and clustering algorithms. E is the edges between characters, and R is the social relation corresponding the edges: leader-member, peer, service, parents-offspring, lover, sibling, friend and enemy.

SRE-Net model first conducts scene segmentation. For scene segmentation, its input is video data V, its output is a series of scenes, $V = \{s_1, s_2, \ldots, s_k\}$. Then, SRE-Net model analyzes the relationship of characters in each scene and statistics to get the character relationship of the entire video. Finally, SRE-Net recognizes characters and number of characters in the video by our MoCNR method. Therefore, social relation network G is successfully constructed from video.

4 Experiments

4.1 Implementation Detail

MoCNR. We evaluate our algorithm on LFW dataset which is a database of face photographs for face recognition. The data set consists of more than 13,000 face images with different orientations, expressions and lighting environments. There are more than 5,000 people, of which 1680 have 2 or more face images. Each face image has its unique name ID and serial number to distinguish it.

We propose double-clustering method to identify character nodes of video for MoCNR method. For the first round clustering, MoCNR method uses SinglePass algorithm to estimate the number of clusters that should be clustered, the number of clusters also indicates the number of characters appearing in the video. In our experiment, two distance matrix method are applied for clustering: Cosine distance and Euclidean distance, the parameter t is set to 0.71 and 100.34, respectively. For the second round clustering, K-Means algorithm is applied for clustering. In our experiment, the parameter k is set to the result of the first round clustering.

Social Relation Extraction. We evaluate our social relation extraction model on SRIV dataset. SRIV dataset contains 3,124 videos collected from TV dramas and movies, each video ranges from 5 s to 30 s, and it contains two types of relation labels: subjective relation labels and objective relation labels which is our previous work published in MMM. In our experiment, we just use objective relation classes, SRIV dataset is divided into two parts: training set and testing set. The training set has 2,343 videos and the testing set has 781 videos.

Our experiments analyze two aspects. First, we employ SVM multiple classifier to predict social relations by fusion different features: audio feature, C3D feature (temporal-spatial feature). Second, different SVM classifiers are trained using different features respectively, then linear blending is applied for model fusion. Third, we utilize ensemble learning to predict social relations by audio feature, temporal-spatial feature and fusion features. For SVM classifier, parameters are set as follows: the *kernel* is set to "rbf", and *decision_function_shape* is set to "ovo". For AdaBoost, parameters are set as follows: the *criterion* is set to "gini", *max_depth* is set to 64, *max_features* is set to "auto", *n_estimators* is set to 50, *learning_rate* is set to 0.01, and *algorithm* is set to "SAMME.R". For XGBoost, parameters are set as follows: *learning_rate* is set to 0.01, *n_estimators* is set to 100, *n_jobs* is set to 4. For GBDT, parameters are set as follows: the *loss* is set to "devirance", *learning_rate* is set to 0.01, *n_estimators* is set to 50, *subsample* is set to 0.5, *min_samples_split* is set to 2, *min_samples_leaf* is set to 1, and *max_depth* is set to 32. For LightGBM, parameters are set as follows: *boosting_type* is set to "gbdt", *num_leaves* is set to 31, *n_estimators* is set to 50, *objective* is set to "multiclass", and *learning_rate* is set to 0.01.

Audio features are extracted by GoogleNet. First, we make some adjustments to the structure of GoogleNet model, the output layer unit number changed from 1,000 to 8 because we need to identify 8 kinds of social relations between characters. Second, we finetune GoogleNet model pre-trained from [25] using SRIV dataset which is the first video dataset for social relation recognition in order to extract audio feature of video. In our experiment, parameters are set as follows: the batch size is set to 32, the learning rate is set to 0.001 and decreases to its 0.0002 every one epoch, and the max iteration is 2000.

Temporal-spatial features are extracted by C3D model. We use SRIV [2] dataset to finetune the C3D network. In our experiment, parameters are set as follows: the batch size is set to 64, the learning rate is set to 0.001, and decreases to its 0.001 every one epoch, and the max iteration is 1000.

SRE-Net Model. As far as we all know, this is the first work to construct social relation network form video, which meanwhile extracts concrete social relations between characters of video. The analysis data of our experiment is popular TV drama, such as, "Nirvana In Fire" and "House of Cards". The "House of Cards" has a total of 13 episodes, about 650 min of video and 40 dominating figures. The "Nirvana In Fire" has a total of 50 episodes, about 2000 min of video and 50 dominating figures. With the progress of the video, there are new characters and new relations between characters, and there are also the disappearance of the existing characters and the division and alienation of the existing relationship between characters. In this paper, we construct the role relationship network by analyzing the relations between the main characters in the video, and evaluate our experimental results using macro-averaging and micro-averaging.

4.2   Evaluation Protocol

Macro-averaging firstly counts statistical values for each class, and then calculates arithmetic average for all classes. This evaluation method does not take into account the sample size of each, it gives the same weight for each class.

Micro-averaging establishes a global confusion matrix which does not pay attention to the class of data, and then calculates the corresponding indicators. Micro-averaging gives the same weight for each sample.

In our experiment, we consider both evaluation method for more comprehensive evaluation of our model.

Accuracy is given by

$$accu_{mi} = \frac{n_{corrected\_prediction}}{n_{total}}, \tag{16}$$

in (16), $n_{corrected\_prediction}$ is the number of corrected prediction, n_{total} is the number of total samples.

Precision is given by

$$precison_{mi} = \frac{TP_1 + TP_2 + \ldots + TP_m}{TP_1 + FP_1 + TP_2 + FP_2 \ldots + TP_m + TP_m}, \tag{17}$$

$$precision_{ma} = \frac{precision_1 + precision_2 + \ldots + precision_m}{m}, \tag{18}$$

in (17), TP_i is the number of true positive in i-th class, FP_i is the number of false positive in i-th class. In (18), $precision_i$ is the precision of i-th class.

Recall is given by

$$recall_{mi} = \frac{TP_1 + TP_2 + \ldots + TP_m}{TP_1 + FN_1 + TP_2 + FN_2 \ldots + TP_m + FN_m}, \tag{19}$$

$$recall_{ma} = \frac{recall_1 + recall_2 + \ldots + recall_m}{m}, \tag{20}$$

in (19), FN_i is the number of false negative in i-th class. In (20), $recall_i$ is the recall of i-th class.

F1 is given by

$$F1_{mi} = \frac{2 \times recall_{mi} \times precision_{mi}}{recall_{mi} + precision_{mi}}, \tag{21}$$

$$F1_{ma} = \frac{2 \times recall_{ma} \times precision_{ma}}{recall_{ma} + precision_{ma}}, \tag{22}$$

4.3 Algorithm Analysis

MoCNR. Table 1 shows the comparison results of MoCNR with evaluation metrics on LFW dataset. The accuracy of our MoCNR method using Cosine metric is 87.35%, and the accuracy of MoCNR method using Euclidean metric is 85.47%. We can find that this unsupervised double-clustering method greatly reduces the difficulty of scaling to other issues, which also achieves results that are similar to supervised methods.

Table 1. Accuracy performance of character recognition on LFW dataset

Method\Metric	Accuracy (%)
DeepFace [19]	97.35
FaceNet [30]	98.97
MoCNR (Cosine metric)	87.35
MoCNR (Euclidean metric)	85.47

Table 2. Performance of social relation extraction on SRIV dataset

Method\Metric	$accu_{mi}$	$presion_{ma}$	$F1_{ma}$
Multi-stream [2]	0.6136	-	0.6386
Fusion+SVM	0.7029	0.6552	0.6725
Linear blending	0.7029	0.6528	0.6655
Audio+AdaBoost	0.7209	0.6724	0.6855
C3D+AdaBoost	0.7093	0.6645	0.6774
Fusion+AdaBoost	0.7375	0.6899	0.7086
Audio+XGBoost	0.7119	0.6637	0.6806
C3D+XGBoost	0.7094	0.6611	0.6761
Fusion+XGBoost	0.7376	0.6846	0.6980
Audio+GBDT	0.7120	0.6610	0.6753
C3D+GBDT	0.7081	0.6617	0.6792
Fusion+GBDT	0.7260	0.6687	0.6817
Audio+LightGBM	0.7170	0.6659	0.6775
C3D+LightGBM	0.7157	0.6650	0.6813
Fusion+LightGBM	**0.7452**	**0.7029**	**0.7203**

Social Relation Extraction. Table 2 shows the comparison results of social relation extraction with evaluation metrics on SRIV dataset. We can find that our social relation extraction model can significantly improve the performance by using ensemble learning, such as LightGBM, XGBoost, etc. from Table 2. From the results, it demonstrates that action information and audio cues are definitely crucial for social relation extraction. For example, $accu_{mi}$ achieves 0.7452 (+13.16%) on SRIV dataset, $F1_{ma}$ achieves 0.7203 (+8.17%) on SRIV dataset, and $presion_{ma}$ achieves 0.7029 on SRIV dataset.

Figures 3, 4 and 5 show the relation extraction performance of the different relation categories with our ensemble learning method. We can find that the fusion of different features can improve performance of social relation extraction. And in these ensemble algorithms, LightGBM achieves the best performance. However, some classes have not better performance after ensemble learning, because this type of class is not very noticeable comparing with other relations categories.

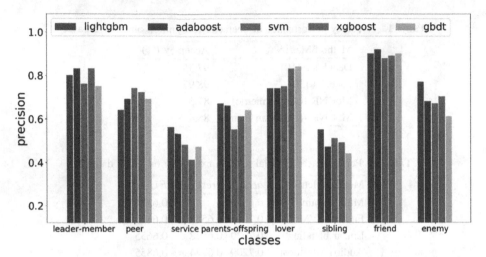

Fig. 3. Precision performance of different methods on SRIV dataset

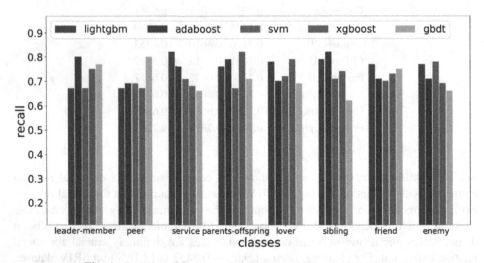

Fig. 4. Recall performance of different methods on SRIV dataset

Table 3. Performance of social relation network construction on "HOUSE OF CARDS" and "NIRVANA IN FIRE"

Video\metric	$accu_{mi}$	$presion_{ma}$	$F1_{ma}$
"House of Cards"	0.6543	0.6107	0.6218
"Nirvana In Fire"	0.6722	0.6236	0.6368

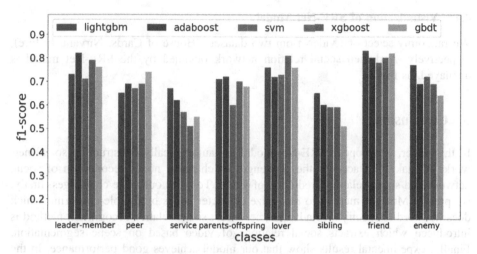

Fig. 5. F1 performance of different methods on SRIV dataset

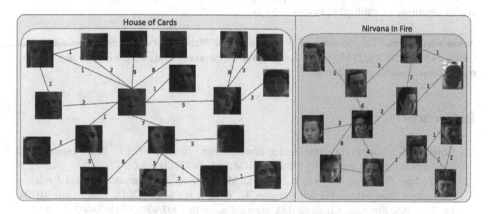

Fig. 6. Social relation network of House of Cards and Nirvana In Fire

SRE-Net Model. Table 3 shows the experiment results of social relation network construction with evaluation metrics on randomly selected several episodes. Table 3 shows the evaluation results of "House of Cards" and "Nirvana In Fire". From the results of Table 3, we can find that our SRE-Net model using ensemble learning can achieve good performance of social network construction. Temporal-spatial feature which generates from C3D network model has a relatively low evaluation results. And the evaluation metrics can get more accurate values by fusing audio feature and temporal-spatial cues. It demonstrates that SRE-Net model can achieve better performance on two video datasets. The evaluation metrics achieves 0.6543, 0.6107 and 0.6218 respectively on dataset "House of Cards". The evaluation metrics achieves 0.6722, 0.6236 and 0.6368 respectively on dataset "Nirvana In Fire".

4.4   Visualization of SRE-Net Model

We randomly select one video from two datasets (House of Cards, Nirvana In Fire), respectively. And then social relation network obtained by the SRE-Net model is displayed, as Fig. 6.

5   Conclusion

In this paper, we propose SRE-Net model for automatically constructing social network. We take into account the challenges of character nodes recognition of social network and social relation prediction of video. To work out these challenges, firstly, we present MoCNR method to recognize character nodes by double-clustering which does not need to specify k in advance; secondly, social relation recognition method is introduced which extracts social relations of video based on scene segmentation. Finally, experimental results show that our model achieves good performance. In the future work, we are going to shift this model to distributed deep learning architecture, such as, TensorflowOnSpark, for speed performance. Furthermore, we will research social relations of multiple person.

Acknowledgment. This research is supported by the National Social Science Foundation of China under Grant 16ZDA055. We are grateful to the anonymous reviewers for their careful reading and valuable suggestions.

References

1. Gopalan, R.: Image clustering under domain shift. In: 2017 IEEE Third International Conference on Multimedia Big Data (BigMM), pp. 74–77. IEEE (2017)
2. Lv, J., Liu, W., Zhou, L., Wu, B., Ma, H.: Multi-stream fusion model for social relation recognition from videos. In: Schoeffmann, K., et al. (eds.) MMM 2018. LNCS, vol. 10704, pp. 355–368. Springer, Cham (2018). https://doi.org/10.1007/978-3-319-73603-7_29
3. Zhang, K., Zhang, Z., Li, Z., Qiao, Y.: Joint face detection and alignment using multitask cascaded convolutional networks. IEEE Signal Process. Lett. **23**(10), 1499–1503 (2016)
4. Kumar, N., Rai, P., Pulla, C., Jawahar, C.V.: Video scene segmentation with a semantic similarity. In: IICAI (2014)
5. Amato, F., Moscato, V., Picariello, A., Sperlí, G.: Recommendation in social media networks. In: 2017 IEEE Third International Conference on Multimedia Big Data (BigMM), pp. 213–216. IEEE (2017)

6. Schmidhuber, J.: Deep learning in neural networks: an overview. Neural Netw. **61**, 85–117 (2014)
7. Wang, M., Deng, W.: Deep visual domain adaptation: a survey (2018)
8. Young, T., Hazarika, D., Poria, S., Cambria, E.: Recent trends in deep learning based natural language processing (2017)
9. Schmidhuber, J.: Deep Learning in Neural Networks. Elsevier Science Ltd. (2015)
10. Li, S., Ma, H.: A siamese inception architecture network for person re-identification. Mach. Vis. Appl. **28**(7), 725–736 (2017)
11. Zhang, N., Paluri, M., Taigman, Y., Fergus, R.: Beyond frontal faces: improving person recognition using multiple cues. In: Computer Vision and Pattern Recognition, pp. 4804–4813. IEEE (2015)
12. Wojke, N., Bewley, A., Paulus, D.: Simple online and realtime tracking with a deep association metric. In: IEEE International Conference on Image Processing, pp. 3645–3649. IEEE (2017)
13. Sun, Q., Schiele, B., Fritz, M.: A domain based approach to social relation recognition. In: IEEE Conference on Computer Vision and Pattern Recognition, pp. 435–444. IEEE Computer Society (2017)
14. Zhang, Z., Luo, P., Chen, C.L., Tang, X.: From facial expression recognition to interpersonal relation prediction. Int. J. Comput. Vis. **126**(5), 1–20 (2018)
15. Rohrbach, A., Rohrbach, M., Tang, S., Oh, S.J., Schiele, B.: Generating descriptions with grounded and co-referenced people (2017)
16. Minoi, J.L., Jupit, A.J.R., Gillies, D.F., Arnab, S.: Facial expressions reconstruction of 3D faces based on real human data. In: IEEE International Conference on Computational Intelligence and Cybernetics, pp. 185–189. IEEE (2012)
17. Oh, S.J., Benenson, R., Fritz, M., Schiele, B.: Person recognition in personal photo collections. In: IEEE International Conference on Computer Vision, pp. 3862–3870. IEEE Computer Society (2015)
18. Sun, Y., Wang, X., Tang, X.: Deep learning face representation from predicting 10,000 classes. In: IEEE Conference on Computer Vision and Pattern Recognition, pp. 1891–1898. IEEE Computer Society (2014)
19. Taigman, Y., Yang, M., Ranzato, M., Wolf, L.: DeepFace: closing the gap to human-Level performance in face verification. In: IEEE Conference on Computer Vision and Pattern Recognition, pp. 1701–1708. IEEE Computer Society (2014)
20. Nan, C.J., Kim, K.M., Zhang, B.T.: Social network analysis of TV drama characters via deep concept hierarchies, pp. 831–836 (2015)
21. Barr, J.R., Cament, L.A., Bowyer, K.W., Flynn, P.J.: Active clustering with ensembles for social structure extraction. In: Applications of Computer Vision, pp. 969–976. IEEE (2014)
22. Zhang, Z., Luo, P., Loy, C.C., Tang, X.: Learning social relation traits from face images. In: IEEE International Conference on Computer Vision, pp. 3631–3639. IEEE (2015)
23. Li, J., Wong, Y., Zhao, Q., Kankanhalli, M.S.: Dual-glance model for deciphering social relationships. In: IEEE International Conference on Computer Vision, pp. 2669–2678. IEEE (2017)
24. Vaswani, A., Shazeer, N., Parmar, N., Uszkoreitet, J., Jones, L., Gomez, A.N., et al.: Attention is all you need (2017)
25. Deng, J., Dong, W., Socher, R., Li, L., Li, K., Li, F.: ImageNet: a large-scale hierarchical image database. In: IEEE Conference on Computer Vision and Pattern Recognition, CVPR 2009, pp. 248–255. IEEE (2009)
26. Tran, D., Bourdev, L., Fergus, R., Torresani, L., Paluri, M.: Learning spatiotemporal features with 3D convolutional networks, pp. 4489–4497 (2014)

27. Szegedy, C., Liu, W., Jia, Y., Sermanetet, P., Reed, S., Angueloval, D., et al.: Going deeper with convolutions, pp. 1–9 (2014)
28. Zhou, L., Lv, J., Wu, B.: Social network construction of the role relation in unstructured data based on multi-view. In: IEEE Second International Conference on Data Science in Cyberspace, pp. 382–388. IEEE Computer Society (2017)
29. Lecun, Y., Bengio, Y., Hinton, G.: Deep learning. Nature **521**(7553), 436 (2015)
30. Schroff, F., Kalenichenko, D., Philbin, J.: Facenet: a unified embedding for face recognition and clustering. In: Proceedings of the IEEE Conference on Computer Vision and Pattern Recognition, pp. 815–823 (2015)

Visual Analysis of Scientific Life of Scholars Based on Digital Humanities

Wei Liu, Xinying Han, Xiaoju Dong$^{(\boxtimes)}$, Zhiwen Qiang, and Xuwei Chen

Shanghai Jiao Tong University, Shanghai, China
{liuwei-cs,xjdong}@sjtu.edu.cn, myj_1234567@126.com,
qlightman@163.com, xwchen@lib.sjtu.edu.cn

Abstract. Digital humanities is an inter-discipline of digital technologies and humanities. The main direction of current digital humanities researches is to use the latest digital technologies to present the existing historical relics in a more intuitive way, yet there are relatively few studies on how users can more effectively and efficiently get information from multiple data levels and dimensions. In this work, we present a visual interactive system designed to study the academic career of well-known scholars. We provide the detailed design and apply it to the dataset of Tsung-Dao Lee's academic papers. In the case study, our proposed system can effectively enhance user's understanding of the underlying relationships in data and help to explore the scholar's scientific life and academic activities. A key contribution of this paper is to provide a novel way of applying visualization technology to digital humanities research.

Keywords: Visualization · Digital humanities · Scientific life
Temporal and spatial distribution

1 Introduction

Digital humanities, also known as Humanities Computing, ofis a typical cross-discipline combining the fields digital computing and the disciplines of the humanities. It involves the synthetical use of digital data in the humanities field and the reflection on their applications [5,13]. The main intention of the computational discipline is not to accelerate the progress of the humanities, but to provide new research methods for the long-term problems in the field of humanities research. The terminological transformation from humanities computing to digital humanities should owe to John Unsworth who attempted to prevent the field being reckoned as mere digitization [7]. Because the real meaning of digital humanities is the new independent research object formed by the combination of digital and humanities, and the methods or modes of thinking it brings.

Scholars in the field of digital humanities are committed to integrating technology into academic research, such as text analysis, GIS, interactive games and multimedia applications in such disciplines as history, philosophy, literature, religion and sociology.

© Springer Nature Singapore Pte Ltd. 2018
Z. Xu et al. (Eds.): Big Data 2018, CCIS 945, pp. 461–473, 2018.
https://doi.org/10.1007/978-981-13-2922-7_31

The development of digital humanities research in developed countries is early. Relatively speaking, the relevant research on digital humanities in China is late. Even so, the research results obtained in China are still quite abundant. Numerous universities have established the research activities of digital humanities and conducted practical research. At the same time, a large number of digital databases have been completed and are free to the public. Digital technology also plays a positive role in Archaeology and cultural relic restoration. For example, using digital technology, the frescoes of Mogao Grottoes [14] in Dunhuang are perfectly preserved.

Compared to traditional humanities projects, digital humanities projects are more likely to involve a lab or a team, and the main relevant research institutions contain archives, museums galleries and libraries. Among the above research institutions, library has much more data and is closer to the public. Hence library has greater influence in the digital humanities field. Under the background of digital humanities, the model of library services has shifted from knowledge service to data analysis service. The main research direction of digital humanities work in library is literary text analysis. WordSeer [9] is a software system based on user-centered design and evaluation and with automated processing of text by interactive techniques and it supports exploratory, interpretive and note-taking forms. Janicke et al. [10] summarize the previous research on visualization supporting close and distant reading of text and compare the methods in both ways to support multifaceted data analysis.

Although the digitized humanistic way has effectively enhanced the understanding and exploration of humanities data, users still lack in-depth analysis on the data. Therefore, it is necessary to integrate visualization into digital humanities. Fortunately, there is a tendency of integrating the visualization tools into digital humanities projects in recent years. Previous researches on scholarly papers mainly focused on analyzing the content of the paper itself, while visualizations for scientific life of a scholar as cultural heritage are seldom been found in visualization communities [4].

In this paper, we propose an interactive visual analysis system which based on the publication papers list of a scholar, and our system aims to explore the unique scientific life of the scholar through the analysis of the relationship between the various data items of the paper list in both space and time. Besides, we add a variety of interactive techniques to make users spontaneously explore the connection between attributes.

The main contributions of this work are as follows:

- Unlike the traditional scheme for analyzing scholar in digital humanities, our system focus on individual scholar and each data unit characterizes value of different attributes of the scholar.
- By introducing interactive technology, we develop a visual interaction tool to help users analyze the potential relationships between the diverse attributes and explore scholar's scientific activity patterns.

2   Related Work

In recent years, visualization has been increasingly applied to literary data. To explore the relationship between literature and human geography space, interactions of dragging or scaling timeline and scaling geospatial have been provided. Some researchers [1] have drawn the poet's travel map and through the path map to analyze the poet's life of poetry. It has performed well to judge poet's personality by visualization. Traditionally, the understanding of author's psychological features mainly origin from subjective interpretation and qualitative analysis. Boyd [8] utilize text analysis methods, using the words people write or speak to extract information of characteristics and individual difference to infer character psychology. Drucker [6] reckons researchers should from a humanistic critical perspective to describe visualization and interface when introduce graphical approaches into digital humanities. Ke et al. [11]presents the visual graph using the InfoVis contest dataset which contains author, abstract, keyword, source, reference, number of pages and year of publication. It displays the citation network and co-author relation graph which is helpful to the understanding of relational data. However, it just makes use of one attribute and lacks connecting with other attributes. Bao [2] proposed a method to visualize 350 library databases and analyze different bibliographic fields to offer a better library instruction. It uses basic view such as bar and column charts, doughnut charts, pie charts, matrices, tables and treemaps to present fields information. By using clicking or highlight form, it realizes the interaction between different attributes.

3   Design Guidelines

This section describes the system's tasks and the detailed visual analytic design, furthermore, it contains the corresponding visualization presentation methods to convey the relationship between attributes of scholar's paper data and illustrate the interaction between views. This system supports filtering in both space and time that enable the user to explore the specific spatiotemporal data.

3.1   Task Analysis

The paper data would come with title, year and location of publisher or place of publication, category, keywords, cited count, co-author and link of paper. Through the understanding of paper data and visual interaction techniques in some previous works, the following tasks should be accomplished:

Temporal and Spatial Distribution. What's the scholar's education journey? What's the spatial distribution pattern of the paper? What are the characteristics of the number and influence of papers in a period of time? These temporal and spatial are info in favor of exploring activities patterns of scholar.

Research Direction Transformation Pattern. Through analyzing the number of papers in each category, we infer the involvement factor of scholar in different periods and understand the scholar's main research focus.

Selection of Specific Visualization Techniques. We need to confirm which visualization and interaction technique is appropriate to corresponding view by considering the possible relationship between dataset.

3.2   View Introduction

Overview. As shown in (see Fig. 1), we can combine any multiple subviews to observe the connections between attributes. The system's interactive interface consists of the following components:

(1) As the main view of our system, bar view (a) adopts spatial multiplexing techniques to switch reference and classified views. It also contains many filter widges.
(2) Right bottom view (b) looks like a sun, which sums up total contribution of specified category in certain time periods.
(3) It is easy to achieve papers' spatial distribution info by connecting map view (c) and main view.
(4) Relation view (d) helps user to understand the affinities between scholar and other collaborators.
(5) If users are familiar to the academic field, they will obtain more implicit information from wordle module (e).

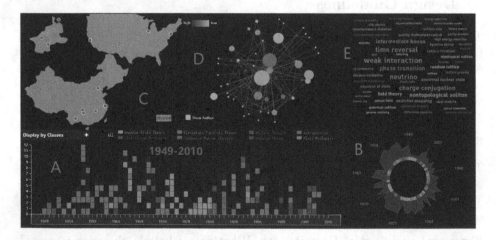

Fig. 1. System overview:(a) The main view presents all the lists of papers with filter widget.(b) The sun view shows the total and classified contribution in certain time period. (c) The map view displays the spatial distribution of publication papers. (d) The force-directed graph exhibits the relationship between cooperators. (e) The wordle module used to analyze the frequency of subject words.

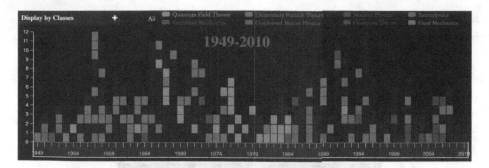

Fig. 2. The distribution of papers in 8 categories.

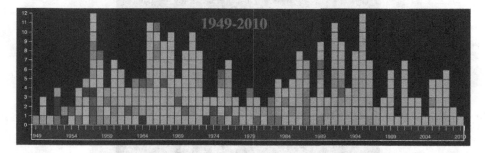

Fig. 3. Switch citation mode to present the citation information of each paper.

Bar View. As the main view of our whole system, bar view(see Fig. 2) can effectively provide users with over-all information of the scholar's papers. It is mainly composed of a set of grids which are arranged based on the timeline. In the coordinate system, x-axis represents the year when each paper is published, while y-axis shows the number of published papers each year. In this way, users can intuitively determine the variation tendency of the scholar's annual output quantity. Then we divide the bar into a set of grids, letting each grid stand for one paper. When the mouse moves over a grid, the city where the scholar wrote the corresponding paper will be highlighted in the map view. On clicking this grid, only the partner of the corresponding paper remains in the relation view and other authors are hidden. Similarly, only relevant keywords are remained in wordle and others are hidden. That is to say, with single click users can find out who wrote this paper and what it is mainly about. When the grid is double-clicked, users will be directed to a page with further information of this paper on a third party website.

The papers are displayed by classes in default. In this mode, the grids are colored according to the category of the papers. Legends with corresponding colors are placed right above. Users can click the legend to hide the papers of other categories. At the same time, the irrelevant authors and keywords are also hidden, and the cities where the papers of this category were published are highlighted. Users can switch to citation mode via the selector above. In citation mode(see Fig. 3), grids are colored from deep to light according to how many time the corresponding papers are cited. It offers users an efficient way to find the most famous papers of the scholar. The legend buttons are also available for users to explore the citation information of a certain category.

There is also a timeline brush on x-axis. Users can filter the data of a certain period of time by dragging the brush. The chosen time period will be displayed in the middle, and the data that are screened out will be either dimmed out or hidden in each view.

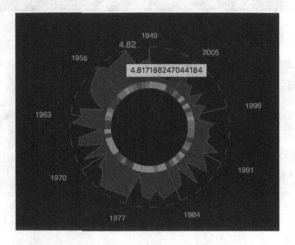

Fig. 4. Inner ring displays which field contributed most in each year, outer ring is a circled line chart which sums up the citation value in all categories.

Sun View. Sun view (see Fig. 4) functions as a scale of the scholar's annual contribution. Due to the limitation of view size and the exist of outliers, we should neither change the initial data nor ignore the weight proportion. It is necessary to determine the scaling functions according to the corresponding values. As for a scholar, most of his/her papers are quoted below 100 times and few references more than thousands. The logarithmic function satisfies the depiction of this property. To make it more intuitive, we define a scholar's contribution of the year:

$$value_c = \log_{100}(\sum_{i=1}^{n} ct) + 3 \tag{1}$$

where ct stands for the time each paper is cited.

In this way, we can decrease the huge citation times to a rather small number, which is much easier to be shown. Now we can display the contribution trend of the scholar, and if he/she has some massive contributions in some certain years, users can also see clearly in this view.

The outer ring is a circled line chart showing cited $value_c$. Axises are placed every few years to indicate the time. When the mouse moves over a certain point on the line, $value_c$ of this year is displayed. Meanwhile, an equivalence circle (like an isoline) is generated to help users identify whether $value_c$ of other years is higher or lower than that of this certain year. In default, there is only one ring containing data of all papers. When the legend in the bar view is clicked, another ring that contains merely the data of the corresponding category is generated. The second ring, whose color is the same as the legend, represents the scholar's annual contribution in a certain field. Displayed together with the first ring, it also allows users to compare the scholar's contribution in one field with that in all fields. The inner ring is like a pie chart. Each arch is colored based on the field in which the scholar contributed most in the corresponding year. Through this inner ring users can see the changing trend of the scholar's research direction.

Map View. As the mapping of geographic feature, map helps user spot spatial patterns of publication papers directly and see where do we need to pay attention. The position cities are abstracted as nodes in corresponding latitude and longitude. To reduce the region occupied by useless information, we replace the global map with the countries where the paper had been published in the dataset. Moreover, for the sake of displaying path of scholar's academic career, we add some necessary city locations and corresponding introductory data.

As shown in (see Fig. 5), the radius and color of circle are used to encode whether the nodes are selected or not, where green and greater radius circle indicates the node is selected, red and smaller radius circle represents that it is not selected. When the mouse is over the node, a text message box on the right will be displayed to show the overview information of scholar in this city. Users can click on the circle to filter the city they are interested in. There is a button in the bottom right partial of the map view as a switch to control the animation that demonstrates the path of academic career and work locations of the scholar. As for line we use the Bezier curve. Dynamic display forms will convey migration information more vividly. To avoid the visual errors caused by the overlap of circles, we append scaling and translation interactions.

Cooperator Relation View. We adopt a force directed graph (see Fig. 6) to show the relationship between the papers' collaborators. For each paper, there have different numbers of collaborators, connection links depicted between any two copartners. We utilize the citation to represent the influence of the paper.

The radius of circle represents the number of cooperative papers. To observe the quoted count and balance the relationship view box size and the value, we draw the scaling formula:

$$r = 2 * \log_e(e + d^3) \tag{2}$$

where d indicates the count of papers.

At the same time, the color of circle indicates the average influence of each paper. The color indicator is in the top left of the view, which adopts the way

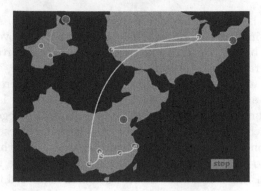

Fig. 5. A dynamic visualization of scholar' learning paths. (Color Figure Online)

of binary interpolation gradients, the deeper color of the circle illustrates the greater influence between scholar and collaborator. To avoid the inconspicuous effect caused by outliers, scaling function should be formulated according to specific data. In contrast to the impact of each paper, the distance of each pair of circles illustrates the whole influence scores of referred person. In consideration of visual confusion caused by overlaps, we set a box which enabling user to choose whether to show or hide the author name. When users drag nodes, the layout will change and eventually balance.

Wordle View. Wordle (see Fig. 7) is a container for all the keywords of the scholar's papers. Word size is larger when it appears more frequently, and its color is the same as the category which has most papers containing this keyword. When users click on a certain word, the grids representing the irrelevant papers are hidden in bar view, and only authors of the remaining papers are displayed

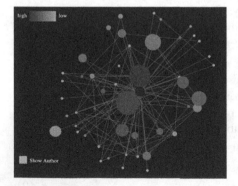

Fig. 6. Force directed graph displays the academic intimacy between co-authors. (Color Figure Online)

Fig. 7. Word Cloud view is used to present statistical information of keywords by connecting with field categories. (Color Figure Online)

in relation view. At the same time, the cities where the remaining papers were published are highlighted in map view. Wordle can effectively help users know what the scholar's papers are mainly about.

3.3 Interactive Technology

One of the innovations of our system is integrating visual interaction technology into traditional digital humanities. It must be interesting to enable library user to explore scientific life of scholar spontaneously by interacting with system. Generally speaking, interactive technology plays a important role in visualization method, it also occupies a large proportion in the design process in our system. The interactive technologies involved in this paper are summarized as follow:

- **Translation and scaling techniques:** applied to map module and wordle module to support user zoom in for view map at different scales.
- **Interactive dynamic filtering technique:** applied to main view, to select consecutive years by dragging the filter bar. By clicking specific category of academic field to filter the relative data.
- **Global and detail techniques:** almost applied to all views, mouse over to achieve current focus element and click to associate multiple views.

3.4 System Implementation

Our system is based on the client-server model and presented by webpage. The client side is implemented with web techniques. The server side is build by Apache HTTP Server and controlled by jQuery. As for visualization module, we use D3 [12] to implement interactive operation.

4 Case Study

In this section, we applied the dataset of Tsung-Dao Lee's papers to our system for assessing its effectiveness and evaluating the practicability of interaction. This dataset includes 321 papers published by Tsung-Dao Lee from 1949 to 2010 and it is supported by Tsung Dao Lee Archives Online [3]. The proposed analysis tasks are completed.

4.1 Overall

The majority of the papers are classified into 8 categories, while some still remain unclassified. Through the bar view (see Fig. 2), we can find that Elementary Particle Theory and Quantum Field Theory, which is represented by blue and yellow respectively, are the two main directions of Tsung-Dao Lee's research. More detailed, Quantum Field Theory was a main research direction of him between 1966 and 1984. He was active in academic field mainly between 1957 and 1996, with a maximum of 12 papers published in 1957 and 1995. From (see

Fig. 5), we can also see the track he moves. T.D. Lee's route of study is consistent with the background of the war age at that time in China. His main research location is in the United States while his later efforts shifted to Beijing's research institutions.

Then we switch to the citation mode (see Fig. 3). We can identify the papers with large citation times are mostly published before 1977. When hovering over these papers, we can find on the map that more than 90% of these papers were published in New York when he was studying or working at Columbia University. Having read the information on map view, we infer that his experience in Columbia University has a rather great impact on him.

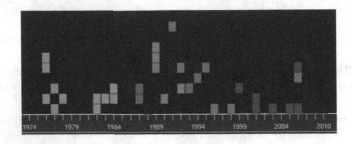

Fig. 8. The bar view after clicking R. Friedberg circle in relation view.

Through relation view (see Fig. 6), we determine the authors with the closest cooperation: C.N. Yang (cooperation:1949–1963) and R. Friedberg (cooperation:1976–2010), Y. Pang (cooperation:1986–1996), G.C. Wick (cooperation:1961–1979) and so on. Among them the most successful papers (with the largest citation times) are written in cooperation with C.N. Yang, R. Friedberg has the longest cooperation time (34 years). What's more, we found that most of the collaborators are basically working with him for a relatively short period of time. You will discover in (see Fig. 8) that R.Friedberg who worked with T.D. Lee for the longest time has different fields of cooperation in specific periods.

In wordle view (see Fig. 7) we find out the keywords mentioned most frequently are *weak interaction, time reversal, Neutrino* and so on. From the color we can know that the most significant keywords are all about Elementary Particle Theory, indicating his main research direction. In sun view, by judging the colored area, we find that Tsung-Dao Lee's contribution degree is higher in early times. After 1992 there was an obvious decline in his contribution.

4.2   Certain Category

From overview we infer that Elementary Particle Theory is Tsung-Dao Lee's most significant research area. So we click on the legend to further explore this category.

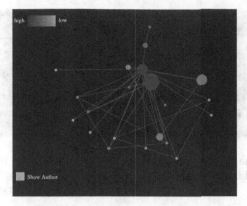

Fig. 9. T.D. Lee and C.N. Yang worked closely together in this period no matter evaluated by single paper's influence or overall cooperation.

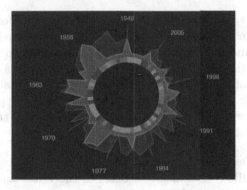

Fig. 10. Elementary Particle Theory (blue curved line) weight a large proportion in so many years. (Color figure online)

In this area, C.N. Yang's relation (see Fig. 9) with him is much closer than others. We infer that C.N.Yang has the greatest influence on Tsung-Dao Lee and his academic career.

It is also obvious in sun view (see Fig. 10) that his contributions in Elementary Particle Theory predominate his contributions in all areas of Physics.

4.3 Interaction with Filter Widge

We are interested in his latest works, so we adjust the brush below to filter the time (see Fig. 11). In this period of time, Tsung-Dao Lee still has published lots of papers, while there are only few cooperators (see Fig. 12). He no longer works with C.N. Yang (actually, we adjust the brush again and find out they had stopped cooperating early in 1960s).

After decades of studying and researching abroad, he finally came back to Beijing. Though the number of published papers declined, his research still covers a wide area. At that time he is already over 70. We can feel his passion in research work deeply.

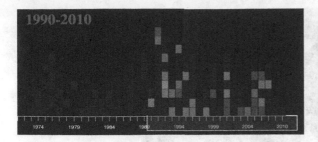
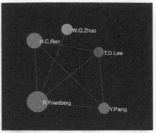

Fig. 11. Dragging the brush bar located in 1990-2010 to analyze scholar's latest works.

Fig. 12. From 1990 to 2010, there are only 4 co-authors with T.D. Lee.

5 Discussion

Compared with the traditional method to analyze the text information or explore multiple authors, our system aims to set scholars as the research center to explore their scientific life and activities patterns, append interaction techniques to help users easily understand how the scholar changed his/her focus of work both in temporal and spatial, observe the transformation in contact with co-authors to analyze the work patterns. The above involved research perspectives rarely appear in previous work for researching scholars. However, there are still some deficiencies remained to be improved.

- In the relation module, we adopt the force directed graph to display the cooperation. While it is difficult to observe the relationship between other authors. A feasible method is gathering partners who are in the same field by clustering. It may be more conductive to analysis research direction of scholars.
- When switching the bar view to citation mode, the citation counts are represented by a set of gradient color which is incomprehensible for users to compare any two papers in close colors. A ranking of citation view should be appended to our system because users may want to analyze the paper which has highest citations.

6 Conclusion and Future Work

In this paper, we find inspiration from the library and present a visual analytic system to explore and analyze scientific life of a scholar. This system provides various interactive techniques for user to link multiple views together for the purpose of demonstrating the relationship between different attributes of paper and further analyzing the unique academic behavior of scholar. Finally, case study shows that this system can effectively acquire potential information from distinct data attributes. Innovation point lies in putting forward a new scheme to analyze scholars' scientific life based on their publication paper through the combination of visualization and digital humanities.

In the future work, we will attempt to append more recent elements that relative to the digital humanities and improve the current interactive techniques in our system.

Acknowledgements. This work was supported by National Key Research and Development Program of China (Grant No. 2017YFB0701900) and National Nature Science Foundation of China (Grant No. 61100053)

References

1. CSDN Article. https://blog.csdn.net/yuanziok/article/details/75220809. Accessed 13 June 2018
2. Bao, X.-M.: Visualization of Library Databases. Digit. Humanit. **5** (2017). https://scholarship.shu.edu/dh/5
3. T.D. Lee Archives. http://tdlee.sjtu.edu.cn/. Accessed 8 May 2018
4. Bradley, A.J., Mehta, H., Hancock, M., Collins, C.: Visualization, digital humanities, and the problem of instrumentalism. Master's thesis, The University of Waterloo, Canada (2016)
5. UCLA Center for Digital Humanities. http://dh101.humanities.ucla.edu/?page_id=13. Accessed May 2018
6. Drucker, J.: A New Companion to Digital Humanities, pp. 238–250. Wiley Online Library, Hoboken (2015). https://doi.org/10.1002/9781118680605.ch17
7. Fitzpatrick, K.: The humanities, done digitally. The Chronicle of Higher Education, May 2011
8. Hai-Jew, S.: Data analytics in digital humanities, pp. 161–189. Springer, Cham (2017). https://doi.org/10.1007/978-3-319-54499-1-7
9. Hearst, M.: WordSeer: a text analysis environment for literature study. University of California, Berkeley (2016)
10. Janicke, S., Franzini, G., Cheema, M.F., Scheuermann, G.: On close and distant reading in digital humanities: a survey and future challenges. In: The Eurographics Association, pp. 83–103, May 2015. https://doi.org/10.2312/eurovisstar.20151113
11. Ke, W., Börner, K., Viswanath, L.: Major information visualization authors, papers and topics in the ACM library. IEEE (2004). https://doi.org/10.1109/INFVIS.2004.45
12. Bostock, M., Ogievetsky, V., Heer, J.: D3 data-driven documents. IEEE Trans. Vis. Comput. Graph **17**(12), 2301–2309 (2011)
13. Terras, M.: Quantifying digital humanities. UCL Centre for Digital Humanities, December 2011
14. Zhang, J., Kang, K., Liu, D., Yuan, Y., Yanli, E.: Vis4heritage: visual analytics approach on grotto wall painting degradations. IEEE Trans. Vis. Comput. Graph. **19**(12), 1982–1991 (2013). https://doi.org/10.1109/TVCG.2013.219

Parallel Computing and Storage of Big Data

RUPredHadoop: Resources Utilization Predictor for Hadoop with Large-Scale Clusters

Shangming Ning[1], Fei Teng[1(✉)], Yunshu Li[1], Zhe Cui[1], Lei Yu[2], and Shengdong Du[1]

[1] School of Information Science and Technology, Southwest Jiaotong University, Chengdu, China
ningsm624@gmail.com, {fteng,sddu}@swjtu.edu.cn, liyunshur@163.com, cuizherr@163.com
[2] Sino-French Engineering School, Beihang University, Beijing, China
yulei@buaa.edu.cn

Abstract. Apache Hadoop is a widely used distributed system in large-scale production environment. With the increasing size of data volume and cluster scale, its performance is limited by inappropriate resources utilization. This paper introduces a resources utilization predictor (RUPredHadoop) to predict utilization of cpu, memory, read/write rate of disk and network, especially for large-scale Hadoop clusters. In terms of the similarity of data and workflow in Hadoop, the pattern of resource utilization for a single task is proposed, and then formulized by a single task model. Besides that, the distribution of fine-grained runtime is studied, so that a parallel-batch-tasks-based model could regenerate the whole Mapreduce job by migrating the single task model from the minimum cluster to a large-scale production cluster. With RUPredHadoop, we can locate the resource bottleneck for Hadoop clusters, meanwhile we can agilely configure clusters for applications with massive data. The performance of RUPredHadoop is validated by a test cluster with 35 nodes and a production cluster with 80 nodes. Results show that the normalization error is below 10% for benchmark applications with maximum 100 TB data.

Keywords: Predictor · Performances · Data center · Hadoop

1 Introduction

With the increasing data volumes and pursuit of superior computing performance, companies invest to build Hadoop data center to respond to that

This paper is partially supported by the National key research and development program of China (No. 2017YFB1400300), the National Natural Science Foundation of China (No. 61573292), State Key Laboratory of Rail Transit Engineering Informatization (FSDI) (No. SKLK16-04) .

demand [1]. However, owing to the increasing cost caused by cluster's growing size, people are gradually taking into account the performance optimizing and resources utilization issue of Hadoop. For the long-term execution of Hadoop-MapReduce applications, operations staffs have to retune the cluster or programs once mistakes appear during the job running. However, traditional rule of thumb is a high-cost and risky method for performance optimization of large-scale Hadoop cluster. As a result, people pay close attention to forecast resources utilization of Hadoop-MapReduce in large-cluster clusters [2], and with this as a reference for Hadoop performance tuning.

Although there are already many researches about optimizing MapReduce performance, few of them are concerned with the resources utilization prediction. The existing works are most focused on forecasting runtime or monitoring performance of MapReduce applications. Runtime prediction are classified into two basic types, one is the time simulation model based on MapReduce execution flow, such as SimMR [3], YARNSim [4] and SimMapReduce [5], etc., the other is complex mathematical based model like Starfish [6] and machine learning-based model [7]. For the former, these models only consider applications' runtime, and some of them lack a detailed description of HDFS and network [4]. Mathematical based models have troubles with modeling cost and transportability [8] for different clusters.

In addition to simulators mentioned above, there are some performance monitoring tools for Hadoop cluster. Ganglia [9] and Nagios [10] are usually deployed for offering solutions of performance monitoring under the high-performance distributed computing environment. Ambari [11] and Dr. Elephant [12] are designed to manage and monitor lifecycle and resources consumption of Hadoop. However, these tools cannot provide forecasting support to help us to predict the resources utilization and pre-optimize applications.

In general, for the large-scale data center, the main way to get reliable reference points for Hadoop optimization is to predict Hadoop performances in the target cluster of the given application. But above all, comprehensive resources utilization prediction is more important rather than only runtime or performance monitoring.

Based on above analysis, this paper makes the first effort to propose the resources utilization prediction model for Hadoop2 (RUPreHadoop). Unlikely recent works, RUPreHadoop is able to predict a whole set of resources indicators (cpu/memory utilization, disk read/write rate, and network read/write rate), particularly under the large-scale cluster environments. The main contributions of our work are indicated as follows.

1. A novel resources utilization pattern. We carry on the baseline test to propose a resources utilization prediction model for a single task, named datum-task-based model. The model establishing is based on fine-grained workflow, hence, it could depict the resources utilization characteristics in detail.
2. Runtime distribution of tasks. We propose and test one hypothesis about runtime distribution of tasks to build the parallel-batch-tasks-based model. The

model extends datum-task-based model to a multi-tasks scenario by regenerating and scheduling tasks along the timeline.
3. Real production testing environments. We evaluate RUPredHadoop in several large-scale production clusters, and input data volumes get into the multi-terabyte level.

The chapters of the paper are organized as follows: Sect. 2 introduces and analyzes the pros and cons of the existing research results. In Sect. 3, we mainly elaborate on model construction methods. The validation of the model prediction accuracy will provide a more complete case and error analysis in Sect. 4. Finally, we summarize the work of the paper and propose future plans.

2   Related Work

In the past few years, Hadoop-MapReduce has been becoming one of the most popular distributed computing frameworks. The focus of Hadoop performance optimization has also became the key issue in the industrial community. Recent researches are most similar to us are Hadoop simulators. These works are divided into two types: one is workflow based simulating, the other is mathematical based models, which rely on a large number of experiments' logs to complete applications runtime prediction. However, as the name hints, the forecast object of these results is mostly just the time.

Flow based simulator. Wang et al. [13,14] come up with MRPerf based on fine-grained simulation, the weakness of that is cannot support resource competition and time delay situations. Actually, these conditions are widespread in any distributed system. Mumak [15] and SimMR [3] are limited by the plentiful completed job logs. As the Apache's official MapReduce simulator, Mumak estimates applications runtime from real test logs by different scheduling algorithms, and a log processing tool – Rumen [17] offers logs collection function. SimMR also relies on log data, and it lacks network simulation. Complexity of the network in shuffle phase is one of the key for simulation precision. However, SimMR, SimMapReduce and YARNSim cannot consider transmission delay over the network in shuffle stage. SimMapReduce [5] is convenient to be used as a toolkit by other packages. It takes into account data locality and dependence between maps and reduces. For the second generation resource scheduling mode, simulating for YARN is harder than first generation Hadoop. Liu et al. [4] proposed YARNSim to simulate resource management and scheduling at the task level, and it allows flexible configuration of the parameters for each stage in MapReduce jobs. Another simulator is MRSim, which simulates network and workload for MapReduce job by using SimJava [18] and GridSim [19], but it lacks of more detailed description for HDFS, because it only simulates data distribution of local rack.

For the mathematical models, these works are usually based on a mass of previous tests' logs, because some historical data are extracted from testing logs should be needed to build the models, such as training machine learning models.

Starfish is the specific implementation of cost-based model [6], which relies on the profiler model to dynamically capture the performance changes of the cluster during the measurement process [20]. Although the profiler component of Starfish is able to estimate resources consumption for the various phases [21], it only considers the average resources consumption for each phase. Starfish likes a kind of white-box system [7], and we should know more underlying system structures to deploy it. Hence, it cannot be easily built and matched to any cluster environments. Different from the cost-based model, machine learning-based model predicts runtime in the way of training and studying samples by different machine learning algorithms [7]. But this method takes runtime as the only indicator, and does not take into account the resources variation of the cluster during job execution. In addition, due to the training data is derived from a mass of actual measurements, it is a time-consuming model.

The existing researches have common defects. Firstly, coarse-grain phase division of MapReduce directly leads to these models cannot get a good predictive ability for each detailed sub-stage. Secondly, they only pay attention to runtime, and neglect resources utilization. Third, these models cannot apply to large-scale clusters, so, they are far unable to meet the demand of the variable cluster size under the background of big data. By contrast, RUPredHadoop is verified to better solve these problems. From a process modeling viewpoint, we mainly consider combining the flow based modeling and mathematical method to implement our model. Based on MapReduce workflow, datum-task-based model considers more detailed sub-phases division. As such, RUPredHadoop could build shuffle, merge and reduce-sort more accurately. For mathematical method, parallel-batch-tasks-based model makes datum-task-based model available to concurrent multi-tasks situation. Combining with the above two submodels, RUPredHadoop is able to predict six resources utilization for each detailed MapReduce phase, and the forecast results could provide overall reference values of the given application in all directions for us.

3   Model Establishment

In this section, the overall modeling process will be introduced firstly, and then the modeling method of the datum-task-based model (DT) and the parallel-batch-tasks-based model (PBT) will be introduced in detail.

3.1   Model Overview

RUPredHadoop predicts six resources utilization (RU) of MapReduce applications (MRApp) under large-scale clusters. Modeling procedure in Fig. 1 can be divided into three stages. Firstly, we come up with a novel pattern of RU to complete the datum-task-based modeling by a baseline test. DT is designed to depict the change rules between data volume and RU in each stage for the single task. Secondly, the parallel-batch-tasks-based model confirms that the single task runtime follows a certain distribution pattern. Based on that observation, tasks could be packaged with resources utilization and runtime. These tasks could be regenerated and scheduled based on the MapReduce concurrent characteristics by PBT. Eventually, visual results could show the six resources consumption of the MRApp under the target cluster.

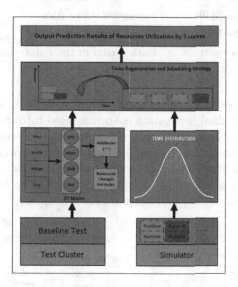

Fig. 1. Overall modeling process

3.2   Datum-Task-Based Modeling

A pattern of resources utilization is proposed to depict RU changing law of a single task in each MapReduce phase, and the datum-task-based (DT) model is the specific implementation of that. We consider that RU of an entire MapReduce job is affected by every single task, and we build datum-task-based model with the baseline test. According to the baseline test, we obtain some data stream variation information from Hadoop logs and overall resources consumption by monitoring tool.

Baseline test (BT) is a pretest method and it is executed in a small cluster (three nodes). The small-scale cluster is part of the target large cluster called the baseline test cluster (BTCluster). The major testing flow of BT is to run a fixed benchmark template on the BTCluster, which will be provided in section IV.

DT model simulates MapReduce resources utilization in each phase more comprehensive for a single task than existing researches. Therefore, it is necessary to have a deep look for RU characteristics of subdivision phases in MRApps.

A Mapreduce job can be split into a series of phases, including map, shuffle and reduce [22]. Especially for shuffle-spill to disk, shuffle-merge and reduce-sort are the most important phases, because of complicated resources utilization in these stages. The meaning of each stage as shown in Table 1

Table 1. Meaning of each phase

Phase	Process
Map	Deal with local data
Shuffle to memory	Pull data to memory
Shuffle-Spill	Memory data to disk
Shuffle-Merge	Merge small files on disk
Reduce-sort	Data sorted in the reduce phase

(1) The map task goes through read, collection, spill, sort, and merge phase in turn [23]. The original input data is transformed into the form of key-value pairs throughout the process, and the number of intermediate files in map phase is finally determined by the number of reduce. From the monitored RU data by Ganglia, during sort & merge stage, the disk reads/writes performance fluctuates at stages. What's more, according to MapReduce working mechanism, network read and write rate are both zero.

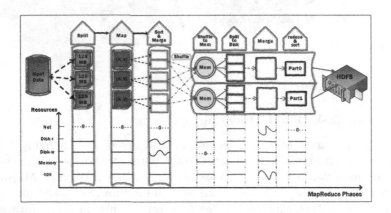

Fig. 2. Data flow and features of resources consumption

(2) The reduce task includes two basic phases: shuffle and reduce-sort. Actually, shuffle is the process of extracting data from the map side by reducers over the network [23], data are pulled into the reduce node memory firstly, and then, spill operation writes on-memory data to disk [24]. Hence, shuffle is divided into two parts that are written into memory and written to disk. The changes of RU is relatively stable in the process of spill, because the stage of written data to memory does not occur in the phase of writing data to disk, so in addition to writing data to disk, the other RU indicators are relatively stable. After shuffle, some files on disk are merged. As a result, cpu utilization and network are regular fluctuations at merge stage. For network performance, there is no network in reduce and sort phase according

to MapReduce work flow [23]. The pipeline of MapReduce and features of resource consumption shown in the Fig. 2. The figure not only reflects the data flow in each phase, but also shows the variation tendency of RU roughly. Due to the similar trends in overall stages what the network read and write have, the figure only shows the net to describe them.

The resources utilization characteristics of various stages are also validated during Hadoop operation logs analyzing. More precisely, according to the data volume changes of each stage in BT logs and resource usage given by monitoring tools, we propose a modeling methodology for variation regularity of resources utilization in different stages. To utilise the DT model, we define some important attributes as shown in Table 2. These attributes are acquired in the way of testing BT to match any different servers. The construction of the modeling methodology is surrounded by these relevant attributes mentioned above. With the detailed division of the MapReduce phases and a large number of measured verifications, we design and implement three algorithms as the core of the DT model.

Table 2. The parameters meaning of datum model

Attributes	Meaning
M/R_Occu{..}	Resources occupation of single map/reduce task
map/red_cpu (%)	The cpu usage of a single map/reduce task
map/red_mem (%)	The memory usage of a single map/reduce task
Sys_Occu{..}	Resources occupation of irrelevant instances
system_cpu (%)	System and other unrelated processes occupied by cpu
system_mem (%)	System and other unrelated processes occupied by memory
Mer_Para{..}	Parameters set about Merge phase
merge_file_size	The size of the merged files during Merge phase
merge_time	The duration of Merge phase one time
spill_size	Data volume of Spill once
Float_Para{...}	Parameters set about volatilities
merge_cpu_change	Floating value of cpu during Merge phase
spill_down_ratio	The rate of decline in disk read during spill phase
net_down	The drop value in net read during Merge phase

Algorithm 1 shows the RU modeling for map phase. Owing to no data transmission over the network, the net read and write rate are both null. In addition, overall cpu and memory utilization and cpu and memory occupancy of the single map task have a linear correlation, and the correlation coefficients is the number of map tasks (M). For the disk read/write rate, it is determined by input data size (D) and the single map task running time (T_m). Noted that

---

Algorithm 1. DT - Map

Require: Paras:$\{M, D, T_m, k, system\_cpu, map\_mem, map\_cpu, Mem\_node,$
$system\_mem\}$
Ensure: Prediction results–$Map\_Result\{\}$:
  $\{pr\_cpu, pr\_mem, pr\_diskR, pr\_diskW, pr\_netR, pr\_netW\}$
 1: **if** $phase = map$ **then**
 2:   $pr\_cpu = map\_cpu * M$
 3:   $pr\_mem = (map\_mem * M)/(Mem\_node + system\_mem)$
 4:   $pr\_diskR = D/T_m$
 5:   $pr\_diskW = (D * K)/T_m$
 6:   $pr\_netR = 0$
 7:   $pr\_netW = 0$
 8: **else**
 9:   $continue$
10: **end if**
11: **return** $Map\_Result\{\}$

---

the map output data written back to disk are compressed by Hadoop parameter ($mapred.map.output.compression.codec$), which is represented by K in the model. It's important to be sure that, some time variables (like map time T_m) in the algorithms should be obtained by parallel-batch-tasks-based model.

Shuffle, merge and reduce-sort are the most complicated stages what we study emphatically. Algorithm 2 describes the specific change of RU in shuffle.

Algorithm 2. DT - Shuffle&Merge

Require: Paras:$\{red\_mem, R, system\_mem, spill\_size, red\_cpu, system\_cpu,$
$merge\_file\_size, merge\_time, T_s, spill\_down\_ratio, net\_down, merge\_cpu\_change,$
$Mem\_node, w\}$
Ensure: Prediction results–$Shuffle\_Merge\_Result\{\}$:
  $\{pr\_cpu, pr\_mem, pr\_diskR, pr\_diskW, pr\_netR, pr\_netW\}$
 1: **if** $phase\ is\ Shuffle$ **then**
 2:   $pr\_mem = (red\_mem * R/Mem\_node) + system\_mem$
 3:   **if** $phase\ is\ Merge$ **then**
 4:     $pr\_cpu = red\_cpu * R + system\_cpu$
 5:     $pr\_diskR = (merge\_file\_size/merge\_time) + (D_r * spill\_down\_ratio)/T_s$
 6:     $pr\_diskW = (merge\_file\_size/merge\_time) + (spill\_size * m) *$
$spill\_down\_ratio/w$
 7:     $pr\_netR/W = D_r/T_s - (net\_down * R)$
 8:   **else**
 9:     $pr\_cpu = (red\_cpu + merge\_cpu\_change) * R + system\_cpu$
10:     $pr\_diskR = D_r/T_s$
11:     $pr\_diskW = (spill\_size * m)/w$
12:     $pr\_netR/W = D_r/T_s$
13:   **end if**
14: **end if**
15: **return** $Shuffle\_Merge\_Result\{\}$

Shuffle is complex because of data transmission by network. Specifically, merge operation is immediately responded since the data size ($merge\_file\_size$) reaches a certain threshold, which will result in network read rate decrease. The descender value (net_down) is fixed in the given cluster, and it is too puny to ignore. Based on previous analysis of RU changes laws, disk read/write rate in the shuffle process is affected regularly by the spill operation.

Data from maps to memory and write data to disk occur concurrently, D_r is the data size of a single reduce, T_s is shuffle time, which can be calculated by parallel-batch-tasks-based model. Hence the disk read rate can be obtained by D_r/T_s before merge. Since the disk read and write are performed at the same time, so, once the number of files exceeds a certain threshold on disk, disk write rate is decreased by merge, and the drop ratio ($spill\_down\_ratio$) is fixed in the fixed cluster. $mapreduce.task.io.sort.factor$ determines threshold of the merge operation startup, and the duration of the merge process ($merge\_time$) could be obtained by logs analysis with baseline test. In addition, since the spill process triggered by shuffle is relatively stable, the disk write rate of the shuffle stage can be gotten by calculating the number of spill occurrences within w seconds, and the spill frequency m is able to computed by $(w * D_r)/(spill\_size * T_s)$.

In reduce-sort phase, the disk read/write and computational resources exist only in this stage, as shown in Algorithm 3, and the remaining indicators calculation are similar to the map phase.

Algorithm 3. DT - Reduce-sort

htbp

Require: Paras:{$system\_cpu$, red_mem, red_cpu, R, D, T_{rs}, Mem_node, $system\_mem$}
Ensure: Prediction results–$Reduce - sort\{\}$:
    { pr_cpu, pr_mem, pr_diskR, pr_diskW, pr_netR, pr_netW}
1: **if** $phase\ is\ Reduce - sort$ **then**
2:    $pr\_cpu = red\_cpu * R + system\_cpu$
3:    $pr\_mem = (red\_mem * R)/Mem\_node + system\_mem$
4:    $pr\_diskR = D/T_{rs}$
5:    $pr\_diskW = D/T_{rs}$
6:    $pr\_netR = 0$
7:    $pr\_netW = 0$
8: **else**
9:    $continue$
10: **end if**
11: **return** $Reduce - sort\{\}$

3.3   Parallel-Batch-Tasks-Based Modeling

Datum-task-based model is designed for single task scene, and we hope that DT could be extended to batching and parallelization mode of MapReduce.

This section expounds the theoretical guidance for the parallel-batch-tasks-based (PBT) model which mainly supported by tasks runtime distribution.

The vast and authentic time samples should be required as the support for the study of runtime distribution. Hence, with the help of YARNSim [4], we obtain the time samples needed from diverse simulation logs of MapReduce jobs. Meanwhile, some time variables and configuration information for the DT model are also extracted from logs under the target job simulation as shown in Table 3. Some of them, *Mean* and *Std* are employed to build runtime distribution.

Table 3. Task runtime parameters of PBT

Parameters	Meaning
$T_m(s)$	Runtime of single task in Map
$T_s(s)$	Runtime of single task in Shuffle
$T_{rs}(s)$	Runtime of single task in Reduce-sort
$Mean\_T(s)$	The mean of $T_m/T_s/T_{rs}$ separately
$Std_{M/R/S}$	The Std of $T_m/T_s/T_{rs}$ respectively

MapReduce tasks runtime distribution is excavated by classic Pearson Chi-Squared Test (χ^2 Test), which is a kind of distribution hypothesis test [26] method. We firstly assume that the single task runtime obey Gaussian distribution ($H0$) after observing runtime scatter diagrams, and confirm this hypothesis based on analyzed result of the χ^2 Test. According to this finding, the time parameters involved in the DT model can be delivered by the normal distribution. In other words, the time arguments will be endowed with concurrent randomness of MapReduce, and each task is activated by PBT model.

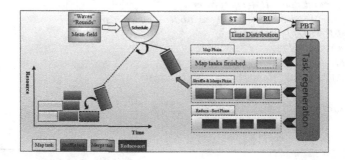

Fig. 3. Tasks regeneration and scheduling mechanism

With above contents, the single task could be regenerated with resources utilization attributes and runtime information for any phases. However, we would like to catch the holistic resources utilization in the context of parallel MapReduce scenario. Therefore, a task scheduling strategy is proposed to be integrated in PBT model, which draws on the idea of Mean-field model [25] and is implemented based on concurrent characteristics of MapReduce.

PBT scheduling strategy takes into account two kinds of tasks status – "parallel" and "waves". According to Mapreduce work flow, both the map tasks and the reduce tasks have the characteristics of "parallel", but there is still "waves" status. For any phases, corresponding tasks could be generated randomly and put into the waiting queue by the strategy firstly. The number of parallel tasks determined by Hadoop parameters (*mapreduce.reduce.memory.mb*, *mapreduce.reduce.cpu.vcores*). If the idle resources are able to meet the required resource consumption for the next task, then the new task will be scheduled in a timeline from the waiting queue. In the case of ensuring the maximum parallelism, tasks will be added in the horizontal axis for the new "wave". The horizontal axis represents the execution time of the job, and the vertical cumulative values can reflect the specific resources consumption of overall job operated in the given cluster until all tasks are generated and dispatched in the timer shaft. The production and dispatching process of the scheduling strategy is shown in Fig. 3.

4 Experiment Evaluation

The following experiments evaluate the effect of RUPredHadoop to validate whether it can output an accurate forecasting result under the large production cluster and multi-terabyte level data size pressure.

4.1 Experiment Environment

We deployed Hadoop-2.7.2 on two clusters with 35 (Environment 1) and 80 (Environment 2) servers respectively, which have the same hardware configuration. Particularly, the Environment 2 is allocated from a real production environment. Each node is connected through 2×10GE S6700-24-EI Switch and equipped with four 3.00 GHZ Decuple-core Intel Xeon(R) E5-2690 v2 CPUs, 378 GB memory and 34 TB disk array. The software environment is based on SuSE11 SP3 as the OS, and Ganglia is deployed for cluster performance monitoring, the underlying Hadoop cluster has a namenode, a secondary namenode, and the rest are datanode under a replication factor of 3. What's more, Table 4 shows the key configuration parameters of Hadoop.

4.2 Experimental Procedure

The experiment flow is divided into three steps:

1. Benchmarks configurations. We produce appropriate data volume for the two different size clusters. We state that the number of maps is uniquely determined by Eq. (1), and HDFS block size is 128 MB. According to the concept

Table 4. The key configuration parameters of Hadoop

Parameters	Value
yarn.nodemanager.resource.cpu-vcores	80
yarn.nodemanager.resource.memory-mb	300 GB
yarn.app.mapreduce.am.resource.mb	15 GB
yarn.app.mapreduce.am.resource.cpu-vcores	20
mapreduce.map.memory.mb	4 GB
mapreduce.map.cpu.vcores	1
mapreduce.reduce.memory.mb	10 GB
mapreduce.reduce.cpu.vcores	16
mapreduce.map.java.opts	2 GB
mapreduce.reduce.java.opts	3 GB

of the "waves" for the batch tasks, we set the different number of reduce to test the two cases in the target clusters. In order to avoid the influence of nonparametric factors as much as possible, we tested use cases for each group with five times and obtain the mean for each result.

$$map\_num = \frac{D}{block\_size}. \tag{1}$$

2. Determining model parameters. Completing the baseline test with three nodes to determine the actual parameters values (Table 2) in the target clusters environment for DT model. And then, Hadoop simulator emulates the target application, we obtain the simulation results as the time parameters values for PBT model.

3. Metric indicator. We adopt absolute normalization error (ANE) to measure forecast error, and it is defined by Eq. (2). N represents number of points what we want to collect, which is decided by simulation runtime T and monitoring resolution X. The normalized references (M) are the maximum values of the target resources indicators, which are determined by hardware configuration of clusters. Table 5 shows the references values of our experimental environment. In addition, Y_{pi} and Y_{ti} are represent predictions and actual values.

$$ANE = \frac{\sum_0^N |Y_{pi} - Y_{ti}|}{N \times M} \quad \left(N = \lceil \frac{T}{X} \rceil\right) \tag{2}$$

4.3   Baseline Test

The tests template of baseline test as shown in Table 6, each case should be tested ten times to get the mean values of parameters. For the observed params, *net_down*, *system_mem/cpu*, *merge_cpu_change*, *spill_down_ratio*,

Table 5. Normalized extremum

Resources indicators	Normalized extremum
Disk read rate	234 MB/S
Disk write rate	378 MB/S
CPU	100%
Memory	100%
Net read/write rate	80 MB/S

$merge\_time$, and $map(red)\_mem/cpu$ are taken by log analysis and RU monitored data, especially $map(red)\_mem/cpu$ can be calculated by Eq. (3). Some params are reflected by Hadoop configuration, $start\_merge\_file$ is assigned by $mapreduce.task.io.sort.factor$; $spill\_size$ is decided by $mapreduce.reduce.shuffle.input.buffer.percent$ ($Per1$), $mapreduce.reduce.shuffle.merge.percent$ ($Per2$) and $mapreduce.reduce.java.opts$ ($javaOpt$), which might be got by Eq. (4). Therefore, $merge\_file\_size$ may be gained in the way of Eq. (5).

$$map(red)\_mem(cpu) = \frac{All\_mem(cpu)\_consumed}{map(red)\_num} \tag{3}$$

$$spill\_size = javaOpt \times Per1 \times Per2 \tag{4}$$

$$merge\_file\_size = start\_merge\_file \times spill\_size \tag{5}$$

Table 6. BT template

Configuration items	Setting
Benchmark	Terasort & WordCount
Map task	$\lceil D/block\_size \rceil$
Reduce task	1

4.4 Environment 1: 35 Nodes

We randomly generate 25 TB and 50 TB data as the input data for two kinds of benchmarks by teragen and randomtextwriter. In order to guarantee the pressure of reduce&shuffle phase and "waves" could be made, the number of reduce task is set to 1000 for WordCount and 3000 for Terasort. The absolute normalization errors are show in Fig. 4(a) and (b).

RUPredHadoop model has a good predicting effect with the environment 1. The minimum error is 0.04% for network read/write rate of Terasort with 3000 reduces. The average error also come up to a receivable value, which is 2.4115%. Conversely, WordCount-disk read rate is relatively poor, whose ANE reaches 7.03% and 7.26%. The reason for that is merge runtime longer than theoretical $merge\_time$ extracted from BT by reviewing Hadoop logs.

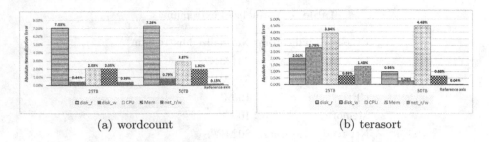

(a) wordcount   (b) terasort

Fig. 4. ANE of test cases in Environment 1

4.5   Environment 2: 80 Nodes

Different from the Environment 1, 80 nodes environment is a real industry data center. 100 TB data are generated for the first time to approximate the feasible production workloads, which is also a big trial ahead of the model. For together with such an order of magnitude, we decide to set 4500 and 6000 reduces for the two use cases respectively after many measurements and calculations. The prediction results are shown in Fig. 5(a) and (b).

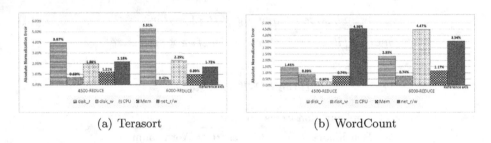

(a) Terasort   (b) WordCount

Fig. 5. ANE of test cases in Environment 2

For the more complex environment, the results still perform well in all test cases. The cpu error of WordCount-4500reduce reaches the minimum 0.30%, and the Terasort-4500reduce also shows 0.30% error for cpu utilization. On the contrary, there is something abnormal with cpu from 4500-reduce to 6000 in the Fig. 5(b). Depending on monitoring logs, it shows a drastic chang in cpu utilization, especially during reduce-sort and late-map phases. Combining benchmark configurations, one reason is that there is resource competition of cpu during reduce-sort stage, which leads to lower red_cpu.

4.6   Error Analysis with One Visualization Case

The prediction and the real resources utilization curves could be visualized by RUPredHadoop. In this section, we display a visual result (Terasort-50 TB-3000reduce- environment 1) and explain some information from Fig. 6.

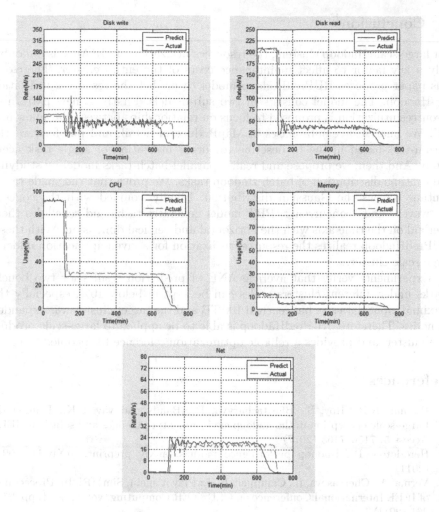

Fig. 6. Terasort-50 TB-3000reduce-Environment 1 (Color figure online)

From the time prediction point of view, the prediction curves can be very intuitive to see the prediction curves (blue) at the end of the horizontal axis is nearly coincident with the actual curve (red). From the RU point, Fig. 6 shows the prediction of disk read/write rate in the shuffle stage has larger fluctuations, but the actual monitoring of the curve is more gentle. This situation relates to complicated non-isomorphic production environment. Specifically, the reason is that there is a demand for data exchange at the same time between the other clusters, which causes the bandwidth available decrease of our target cluster. Hence, the data exchange at this time will become gentle, and disk write from memory during spill operation relatively longer delay. These situations lead to merge operation become relatively flat. For the other RU indicators, the forecast curves have a good fit with the actual curve.

5   Conclusion

For large-scale Hadoop clusters, the resources utilization of MRApps is undoubtedly a key factor to affect maintenance costs of the entire cluster. Therefore, this paper proposes RUPredHadoop model to predict the resources utilization for Hadoop. The model consists of two sub-models. Firstly, a new pattern of resources utilization is proposed to focus on relationships between data flow and RU. We analyze five sub-phases of MapReduce in detail, and put forward the datum-task-based model to describe resources utilization of single task accurately. And then, we propose and realize parallel-batch-tasks model by studying the runtime distribution of batch Hadoop tasks. We prove that the single task's runtime follows the Gaussian distribution. Finally, coupled with the concept of "waves" and "concurrent", PBT model generates tasks and schedules them overlaid on the time axis with the horizontal and vertical dimension. With these, RUPredHadoop realizes the resources utilization for a given application under a large-scale cluster.

Experiments verify that average ANE of prediction results for two benchmarks under different clusters scales can be kept in below 10%, especially the maximal input data size has reached 100 TB and the target cluster was expanded 80 nodes. Therefore, RUPredHadoop is able to be applied to large-scale production cluster, and provides a reliable optimization reference for people.

References

1. Parmar, R.R., Roy, S., Bhattacharyya, D., Bandyopadhyay, S.K., Kim, T.H.: Large-scale encryption in the hadoop environment: challenges and solutions. IEEE Access **5**, 7156–7163 (2017)
2. Herodotou, H.: Hadoop performance models. arXiv preprint. arXiv:1106.0940 (2011)
3. Verma, A., Cherkasova, L., Campbell, R.H.: Play it again, SimMR!. In: Proceedings of IEEE International Conference on CLUSTER Computing, vol. 8, no. 1, pp. 253–261 (2011)
4. Liu, N., Yang, X., Sun, X.H., Jenkins, J., Ross, R.: YARNsim: simulating hadoop YARN. In: Proceedings of the 15th IEEE/ACM International Symposium on Cluster, Cloud and Grid Computing (CCGrid), pp. 637–646 (2015)
5. Teng, F., Yu, L., Magoulès, F.: SimMapReduce: a simulator for modeling MapReduce framework. In: Proceedings of the 2011 Fifth FTRA International Conference on Multimedia and Ubiquitous Engineering (MUE 2011), pp. 277–282. IEEE Computer Society (2011)
6. Herodotou, H., et al.: Starfish: a self-tuning system for big data analytics. In: Proceedings of the 15th Biennial Conference on Innovative Data Systems Research, pp. 261–272 (2011)
7. Yigitbasi, N., Willke, T.L., Liao, G., Epema, D.: Towards machine learning-based auto-tuning of MapReduce. In: Proceedings of the 2013 IEEE 21st International Symposium on Modelling, Analysis & Simulation of Computer and Telecommunication Systems, pp. 11–20. IEEE Computer Society (2013)
8. Li, M., et al.: MRONLINE: MapReduce online performance tuning. In: Proceedings of the 23rd International Symposium on High-Performance Parallel and Distributed Computing, pp. 165–176 (2014)

9. Ganglia Monitoring System: Ganglia (2016). http://ganglia.sourceforge.net/. Accessed 10 Oct 2016

10. Nagios (2016). https://www.nagios.org/. Accessed 10 Oct 2016

11. Apache Ambari: Ambari (2016). https://ambari.apache.org. Accessed 07 Apr 2017

12. LinkedIn dr-elephant (2016). https://github.com/linkedin/dr-elephant. Accessed 07 Apr 2017

13. Wang, G., Butt, A.R., Pandey, P., Gupta, K.: A simulation approach to evaluating design decisions in MapReduce setups. In: IEEE International Symposium on Modeling, Analysis & Simulation of Computer and Telecommunication Systems, pp. 1–11 (2009)

14. Wang, G., Butt, A.R., Pandey, P., Gupta, K.: Using realistic simulation for performance analysis of MapReduce setups. In: Proceedings of the 1st ACM Workshop on Large-Scale System and Application Performance, pp. 19–26 (2009)

15. Apache: Mumak: Map-Reduce Simulator-ASF JIRA (2009). https://issues.apache.org/jira/browse/MAPREDUCE-728. Accessed 21 Apr 2017

16. Hammoud, S., Li, M., Liu, Y., Alham, N.K., Liu, Z.: MRSim: a discrete event based MapReduce simulator. In: Proceedings of the 2010 Seventh International Conference on Fuzzy Systems and Knowledge Discovery (FSKD), vol. 6, pp. 2993–2997 (2010)

17. Apache: Rumen: a tool to extract job characterization data from job tracker logs (2010). https://issues.apache.org/jira/browse/MAPREDUCE-751. Accessed 21 Apr 2017

18. Howell, F., McNab, R.: SimJava: a discrete event simulation library for Java. Simul. Ser. **30**, 51–56 (1998)

19. Buyya, R., Murshed, M.: GridSim: a toolkit for the modeling and simulation of distributed resource management and scheduling for grid computing. Concurr. Comput.: Pract. Exp. **14**(13–15), 1175–1220 (2002)

20. Herodotou, H., Dong, F., Babu, S.: MapReduce programming and cost-based optimization? Crossing this chasm with starfish. Proc. VLDB Endow. **4**(12), 1446–1449 (2011)

21. Herodotou, H., Babu, S.: Profiling, what-if analysis, and cost-based optimization of MapReduce programs. In: Encyclopedia of Database Systems, vol. 4, no. 11, pp. 1111–1122 (2011)

22. Apache: Apache hadoop (2017). http://hadoop.apache.org. Accessed 09 Oct 2016

23. Dean, J., Ghemawat, S.: MapReduce: simplified data processing on large clusters. In: Proceedings of the Sixth OSDI Symposium on Operating Systems Design and Implementation, pp. 137–150 (2004)

24. Shi, J., Zou, J., Lu, J., Cao, Z., Li, S., Wang, C.: MRTuner: a toolkit to enable holistic optimization for MapReduce jobs. Proc. VLDB Endow. **7**(13), 1319–1330 (2014)

25. Georges, A., Kotliar, G., Krauth, W., Rozenberg, M.J.: Dynamical mean-field theory of strongly correlated fermion systems and the limit of infinite dimensions. Rev. Mod. Phys. **68**(1), 13–125 (1996)

26. Pearson, K.: On the criterion that a given system of deviations from the probable in the case of a correlated system of variables is such that it can be reasonably supposed to have arisen from random sampling. In: Kotz, S., Johnson, N.L. (eds.) Breakthroughs in Statistics, pp. 11–28. Springer, New York (1992). https://doi.org/10.1007/978-1-4612-4380-9_2

27. Intel-Hadoop: HiBench-5.0 (2016). https://github.com/intel-hadoop/HiBench. Accessed 09 Oct 2016

Multi-keyword Parallel Search Algorithm for Streaming RDF Data

Jian Guan, Jingbin Wang$^{(\boxtimes)}$, and Long Yu

Fuzhou University, Fujian, China
{963016674,934198928,1095132907}@qq.com

Abstract. The existing keyword-based search algorithms based on streaming data are hard to meet the needs of users for real-time data processing. To solve this problem, multi-keyword parallel search algorithm for streaming RDF data (MPSASR) proposed in this paper combines the Spark and Redis frameworks to construct query subgraphs integrated with ontology based on the query keywords in real time. Associated with scoring function, regarding the high-priority query subgraph as a guide, parallel search is performed in the instance data, and finally the Top-k query results are returned. Of course, our algorithm uses a hash compression algorithm to compress RDF data, which reduces the space required. Moreover, our algorithm makes full use of historical data and effectively speeds up search efficiency. Our algorithm is experimentally verified to have great advantages in real-time search, response time, and search effects.

Keywords: RDF · Streaming data · Multi-keyword search · Real time

1 Introduction

With the rapid development of Internet technology and the rise of news, blogs, and social networks, the Internet has brought convenience to people's lives, but it has also brought about explosive growth of data. Streaming data real-time search has become a research hotspot [1–9]. The Internet produces a variety of streaming data. Because of the heterogeneity of data, RDF (Resource Description Framework) is widely used to provide a unified metadata representation in data streams [1]. RDF dynamic data streams are a source of great interest in the semantic network community, prompting growing demand for innovation in this area. In response to this demand, many scholars have studied and proposed their own RDF stream processing architectures. Among these, the dominant models are those of Barbieri et al. [2] who proposed C-SPARQL, and Le-Phuoc et al. [3] who proposed CQELS. Most of them feature SPARQL-like query language and operational semantics, through the definition of windowing technology in the data flow management system to achieve continuous queries. Borthakur [4] and others proposed a streaming data real-time search algorithm based on Hadoop and Hbase. Document [5] writes data in Nosql database in real time, and uses a three-level caching strategy to implement a streaming data search method. However, the above studies are directly conducted on large-scale data, and the real-time search response time is significantly affected by the data volume.

© Springer Nature Singapore Pte Ltd. 2018
Z. Xu et al. (Eds.): Big Data 2018, CCIS 945, pp. 494–511, 2018.
https://doi.org/10.1007/978-981-13-2922-7_33

As the RDF ontology information can reflect the classification and association of resources and attributes, under the premise of large data volume, using ontology as a guide for retrieval can greatly improve the efficiency of retrieval; the RDF ontology is usually determined by the quantity, with the scale of KB, building multiple-keywords ontology query subgraphs are very efficient and quick. Therefore, this paper proposes a multi-keyword parallel search algorithm (for streaming RDF Data) based on the Spark Streaming computing framework, which is abbreviated as the MPSASR algorithm. The algorithm comprises three parts: real-time data stream processing and distributed storage, query subgraph construction, and real-time data query. Our main contributions are summarized below.

1. Design of a data storage structure that combines Redis distributed storage clusters.
2. Mapping of the user-inquiry multi-keywords into categories, combining the ontology definitions to obtain the mapping class's association class diagrams, and then construction of ontology query subgraphs based on the operations of pruning, deduplication, and fusion.
3. Design of a scoring function that can reasonably and objectively score the ontology query subgraphs and sort them from high to low.
4. Design of a distributed search algorithm combining ontology query subgraphs and using the advantages of the iterative calculation of RDD (Elastic Distributed Data Set) combined with the stored historical retrieval result data to complete an efficient multiple keyword search.

The organization of this article is as follows. In Sect. 2, we summarize related work. In Sect. 3, we introduce the overall framework of the algorithm and related definitions. In Sect. 4, we introduce our algorithm in detail, and in Sect. 5, we evaluate it. Finally, in Sect. 6, we conclude and summarize our work.

2 Related Work

In this section, we will introduce some work related to streaming RDF data search. In general, streaming RDF data keyword search can be divided into two categories, one is a search based on formal query, and the other is a keyword-based search.

Searching based on formalized query statement is mainly through constructing formalized query statements to achieve keyword search. Among them, Barbieri et al. [2] proposed C-SPARQL and Le-Phuoc et al. [3] proposed CQELS, which were similar to SPARQL query language and operation semantics, and realized continuous inquiry through the definition of windowing technology in the data stream management system. Borthakur et al. [4] proposed a real-time search algorithm for streaming data based on Hadoop and Hbase. The literature [5] writes data into Nosql database in real time, and uses a three-level caching strategy to implement a search method for streaming data.

Searching based on keywords is mainly passing matching keywords directly on the data graph to obtain Top-k results. Huiying Li et al. proposed the RDF data keyword query method KREAG [10] based on entity triples association graph. Using the entity triples association graph as a model, it encapsulates text information into the vertices of the association graph and uses approximate algorithm of Steiner tree problem to solve the keyword search problem of RDF data. Virgilio et al. [11] proposed MapReduce-based RDF Distributed Keyword Search Scheme, which can use the MapReduce paradigm to search keywords by converting problems from graphic parallel to data parallel processing.

From the related work above, it can be found that most of the existing work focuses on searching directly on large-scale instance data, which will certainly hinder the real-time performance of search. Considering that ontology can represent the relationship between instances well, and the ontology file is relatively small, we propose a new keyword search scheme based on ontology construction query subgraphs.

3   The Overall Framework of the Algorithm and Related Definitions

3.1   The Overall Framework

The MPSASR algorithm includes three parts: data stream processing and distributed storage, construction of query subgraphs, and real-time data query. The algorithm combines the Spark Streaming framework of Spark platform with the distributed memory database Redis, and stores the newly added data in storage table designed by the algorithm to support the retrieval of multiple keywords by users. For the query keywords entered by the user, the algorithm first combines the associations of resources and attributes defined on the RDF ontology, maps the query keywords into multiple association classes, constructs multiple query subgraphs through pruning and fusion algorithms, and sorts by rating. The high-ranking query subgraph is used as a search basis to complete the real-time query calculation and return the query result. At the same time, the algorithm stores the query subgraph constructed for each query. For the keywords stored in the user's history, the construction query subgraph calculation will not be executed, and the existing query subgraph will be directly used to further improve the efficiency of the algorithm. The overall framework of the MPSASR algorithm is shown in Fig. 1.

3.2   Overall Process Description

Step 1. The RDF instance data is preprocessed and hash-coded by the Spark Streaming framework, and then the encoded and compressed data is stored in the Redis distributed database cluster in real time.

Step 2. Enter the keyword list Q and transfer it to the Spark framework for real-time search.

Step 3. In the real-time search and processing module, the keyword set Q is transmitted in the keyword mapping phase, and the keyword set Q is coded by hashing.

Step 4. Match the encoded keywords to obtain the corresponding ontologies for each keyword.

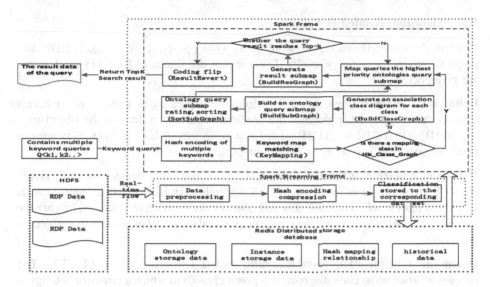

Fig. 1. The overall framework of the MPSASR algorithm

Step 5. For multiple ontology classes, query the historical ontology query subgraphs in the Redis database, judge whether these classes exist in the historical data; if so, take out corresponding sorted ontology query subgraphs and jump to Step 10; otherwise continue.

Step 6. Generate association class diagrams according to the ontology classes found in Redis, and continue.

Step 7. Prune each association class diagram, deleting the edges that do not affect the query results.

Step 8. Merge the pruned class diagrams to generate ontology query subgraphs.

Step 9. Use the correlation evaluation function in Definition 9 to score and sort the ontology query subgraphs; high scores give priority to such subgraghs distributed search.

Step 10. The unqueried ontology query subgraph with the highest priority in the set of ontology query subgraphs is passed to the RDD. The RDF instance data stored in the Redis database is iteratively matched, and this ontology query subgraph is removed from the sorted ontology query subgraphs set.

Step 11. The corresponding result subgraph is obtained in the distributed construction of the generated result subgraph stage. If the query results reach top-k, continue, otherwise back to Step 10.

Step 12. In the encoding inversion phase, the result subgraphs are coded in reverse. Spark's real-time search process ends.

Step 13. Write the inverted result subgraphs to the query result set, and return the top-k result subgraphs.

Step 14. End the algorithm.

3.3   Algorithm Related Definitions

Problem definition: Given the keyword query $Q = \{q_1, q_2, \ldots, q_i, \ldots, q_m\}$, RDF data graph G, return Top-k query results. The relevant definitions of RDF keyword search are given below.

Definition 1 (RDF triples). Let RDF triples be expressed as t <s, p, o>, where s denotes the subject of t, p denotes the predicate of t, and o denotes the object of t. $s \in (I \cup B), p \in (I \cup B), o \in (I \cup B \cup L)$, I is a collection of URI vertices, B is a set of blank vertices, and L is a vertex collection of text.

Definition 2 (RDF diagram). Let G = {t1, t2, ..., ti, ..., tn} denotes an RDF graph. An RDF graph can be defined by a set of RDF triples. An RDF graph is represented as a directed label graph. The subject si and object oi of each triple ti <si, pi, oi> are the vertices of the RDF graph, and the predicate pi is a directed label edge of the subject to the object.

Definition 3 (association class diagram, noted GSi). Let GSi = {T1, T2...Tn} denotes an association class diagram. In a given class Ci to which a keyword belongs, it is connected with its associated classes, that is, Ti <Si, Pi, Oi> is added to the GSi set, where Si = Ci or Oi = Ci.

Definition 4 (class diagram pruning). Multiple keywords are constructed into multiple GSs. In all GSs, the pattern triples <Si, Pi, Oi> that occur only once are loosely pending nodes, it will not affect our query result graphs if we delete them.

Definition 5 (class diagram fusion deduplication). Multiple keywords are constructed into multiple GSs. In all GSs, the pattern triple <Si, Pi, Oi> that occurs many times is regarded as closely related node, removing duplicated triples and maintaining one, forming a new graph association class diagram.

Definition 6 (ontology query subgraph). After multiple GSs are fused, the triple join operation is performed to form an ontology query subgraphs set Gsk.

Definition 7 (triples join). When constructing an ontology query subgraph or result subgraph, a triple join operation is performed for matched pattern triples or instance triples, where any two triples are connected by subjects, objects, or other triples. Formal representation of a triple join: For a pattern triple or instance triple set Set = {T1, T2, ..., Ti, ..., Tm}, given Ti(Si, Pi, Oi) and Tj(Sj, Pj, Oj), where $\exists i, j \in \{1, 2, \ldots, m\}$, if (Si = Sj&&Oi ≠ Oj) or (Si = Oj&&Oi ≠ Sj) or (Oi = Sj&&Si ≠ Oj) or (Oi = Oj&&Si ≠ Sj) says that Ti is adjacent to Tj, and can be connected by triple join.

Definition 8 (the query result, denoted as R). The known RDF data graph G and the keyword query Q. The query result is a set of connected subgraphs consisting of all the triplets of the query keyword, and any two triples are connected by subjects, objects, or other triples. Let R = {t1, t2, ..., tk, ..., tr}, where $\exists i, j \in \{1, 2, \ldots, r\}$, ti <si, pi, oi> and tj <sj, pj, oj>, then (si = sj&&oi ≠ oj) or (si = oj&&oi ≠ sj) or (oi = sj&&si ≠ oj) or (oi = oj&&si ≠ sj) or (ti − tk − tj). If the elements in the two triples set are not exactly the same, they are considered to be different query results.

Definition 9 (relevance score function, denoted as SE in score estimation). Enter the query $Q = \{q1, q2, \ldots, qi, \ldots, qm\}$, corresponding to the RDF ontology instance class $C = \{c1, c2, \ldots, ci, \ldots, cm\}$, assuming Q corresponds to an ontology query subgraph set $Gsk = \{g1, g2, \ldots gn\}$, where $gk \in C$.

$$SE(Gsk) = \alpha * \text{len}(Gsk) + (1 - \alpha) * \text{pageRanks}(Gsk) \qquad (1)$$

where $\text{len}(Gsk) = \frac{1}{Length(Gsk)}$, $Length(Gsk) = \sum_{i \in 1,2,\ldots,m} dis(ci, cj)$,

$$pageRanks(Gsk) = \sum_{i \in 1,2,\ldots,m} pageRank(ci)$$

The relevance evaluation function consists of two parts: the structural tightness score len(Gsk) and the content relevance score sim(C). Variable α is the adjustment parameter. In this paper, $\alpha = 0.5$ indicates that the two have the same degree of influence. The instance class nodes ci and cj on the ontology query subgraph are represented by dist(ci, cj); if the instance classes ci and cj are unreachable, the distance is

$$\text{dist}(ci, cj) = +\infty$$

$Length(Gsk)$ is equal to the sum of distances between vertices of two instances on the ontology query subgraph. The shorter the summation distance, the greater the value of 1/Length(Gsk), the closer the content contact is. The pageRank algorithm is a computing model proposed by Google to evaluate page rank (weight). In this paper, we refer to the pageRank algorithm to calculate the weight of each class node in the ontology query subgraph to evaluate the content relevance of the ontology query subgraph. The sum of the pageRank values of various classes on the ontology query subgraph is represented by pageRanks(Gsk). When the pageRanks(Gsk) value is larger, the content relevance of the Gsk is higher. An RDF instance vertex on the RDF data graph can be mapped to an instance class on the RDF ontology graph; the more closely related the instance vertices on the RDF data graph are, the closer the relationship between the ontology instance classes is, and the higher the score of result is.

4   MPSASR Algorithm

4.1   The Distributed Storage Scheme Design of MPSASR Algorithm

This section describes the distributed storage scheme of MPSASR algorithm. The algorithm uses a Redis in-memory database cluster as a medium for data storage. The number of Redis in-memory databases in the cluster can be dynamically increased or decreased as needed. In the Redis database, the average time complexity of adding, deleting, judging, and searching operations of Hash structure is O(1), and the Sorted Set is also an ordered set of convenient ordering. When the data is modified, the order of the Sorted Set will be automatically re-adjusted accordingly. To provide the capacity for MPSASR's data storage scheme, and meet the requirements for real-time streaming and real-time search of RDF data, this paper uses the three data structures in the Redis memory database to store RDF data in a distributed manner: Hash, Set, and Sorted Set. Considering the dynamic nature of streaming data, the ontology query subgraphs that have been searched can be recorded. Therefore, in addition to the increase in storage of ontology data and instance data, the storage of historical data is also increased. The storage schemes are listed in Tables 1, 2 and 3, and the storage scheme for hash mapping information is listed in Table 4.

4.2   Construction of Ontology Query Subgraphs

Because a keyword may map multiple instance classes and there are multiple different associations between classes, a group of query keywords will have multiple ontology query subgraphs, i.e., an ontology query subgraph set.

Construction of Association Class Diagrams. The association class diagram is the basis for constructing the ontology query subgraph. For each keyword, its corresponding classes can generate their own association class diagrams. The user enters multiple keywords, which may be classes, attributes, instances, or text. For each keyword, first determine whether the keyword is mapped to an property according to the C_C_Property table. If it is an property, it is stored for constructing the ontology query subgraph phase; if not, according to the Rdf_Ontology, C_C_Property, Instance_Class, and Literal_Triple, the type that the keyword may be mapped can be determined.

Due to the key-value storage form used by Redis, the time complexity in the query is O(1), so for each keyword, you can quickly locate its matched classes, while considering that a keyword may be mapped to multiple Classes, multiple keywords matching the classes will be repeated, then you need to remove the duplicate ontology classes. Given any class, the association class diagram GS corresponding to this class can be generated by Definition 3, and shown in Fig. 2. is the association class diagram of the FullProfessor class and the Course class in the LUBM ontology.

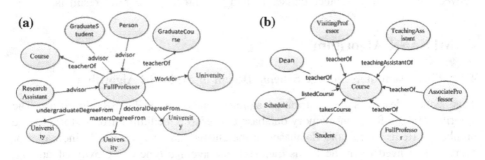

Fig. 2. (a) Association class diagram of the FullProfessor class (b) Association class diagram of the course class

Pruning of Associated Class Diagram. For any class, a generated association class diagram will contain a lot of pattern triples, but it will be doped with redundant pattern triples. In order to remove the edges that will not affect the query result, it is necessary to perform pruning. According to Definition 4, in the association class diagram, if only one of s, p, and o in a path <s, p, o> can be mapped to the class or attribute corresponding to a query keyword, the path can be considered not to affect the query result and can be pruned. Eventually, a pruned association class diagram GSSet is obtained, as shown in Algorithm 1.

Algorithm 1: CutGS pruning algorithm	Algorithm 2: Fusion algorithm
Function: Remove the edge that does not affect the query result	Function: multiple association class diagrams make connections to generate ontology query subgraph collection.
Input: Collection GSSet of association class diagrams, Pset of property set, Cset of class set	Input: a collection of association class diagrams, denoted as GSSet
Output: Pruned GS	Output: ontology query subgraph collection.
1. **For** $t_i < s_i, p_i, o_i > \in$ GS & $i = 1$, 2...,n	1. Set result = {};//Store the resulting ontology query subgraphs.
/*Traverse the association class diagram GS*/	2. Set allSet = {};//Store the pattern triples that have been checked.
//If the predicate of the pattern triple is in the attribute set Pset corresponding to the query keyword	3. Set temp = {};//Temporary variables of intersection.
2. **IF** Pset.contain(p_i)	4. For $Gs_i \in$ GSSet & $i = 1$, 2...,m
3. continue;	5. Gs_i.removeAll(result);
//If both the subject and object of the pattern triple are in the class set Cset corresponding to the query keyword	6. temp.clear();
4. ELSE IF Cset.contain(s_i) && Cset.contain(o_i)	7. temp.addAll(allSet);
5. continue;	8. temp.retainAll(Gs_i);
//If the above two conditions are not satisfied, the pattern triple is considered not to affect the query result.	9. temp.retainAll(Gs_i);
6. ELSE	10. result.addAll(temp);
7. GS.remove(t_i);	11. Gs_i.removeAll(temp);
8. END IF	12. allSet.addAll(Gs_i);
9. END FOR	13. End For
10. Return GS;	14. return result

The first line of codes starts traversing association class diagram set GS. The second line to the fifth line determine whether property pi of ti is present in query keywords, or subject and object of ti are both in the set Cset. If so, then ti can remain in the GS, and the algorithm continues; if not, in the scope of from the sixth line to the eighth line remove ti from GS.

Table 1. The storage scheme of ontology data

Name	Storage contents	Storage structure
Rdf_Ontology	Store RDF ontology information, which is stored as a key-value pair, where the key is a class name and the value is a class number	$Rdf\_Ontology = > \{Cindex1, Cindex2, \ldots Cindexi \ldots Cindexn\}$, $Cindexi$ is a class in the ontology, $Cindexi = \{Class\, i = > i\}$
SubPropertyOf	Store information about properties and their parent properties, which are stored in the form of key value pairs, where the key is the property and the value is the parent property of current property	$SubPropertyOf = > \{PS1, PS2, \ldots, PS\, i, \ldots, PS\, n\}$, $PSi = (Property i = > \{SubPropertyOf : []\})$
SubClassOf	Store information about classes and their parent classes, which are stored in the form of key value pairs, where the key is the class name and the value is the parent of current class	$Class\_Sup = > \{CS1, CS2, \ldots, CS\, i, \ldots, CS\, n\}$, $CSi = (Class i = > \{SubClass : []\})$
C_C_Property	Store the property and the classes to which the property is connected, where the key is the property and the value is the set of the class corresponding to current property	$C\_C\_Property = > \{P1, P2, \ldots, Pi, \ldots, Pn\}$, $Pi = (Property i = > \{[C1, C2], \ldots, [Cj, Ck]\})$, Cj is a Domain class and Ck is a Range class
Subject_Class	Store a pattern triple set whose subject is the same type, where the key is the name of the class that the subject belongs to, and the value is a combination of attributes and objects	$Subject\_Class = > \{C1, C2, \ldots Ci, \ldots Cn\}$, where $Ci = (Class i = > \{Property1 = > Class2, \ldots, Property j = > Classk\})$, $Property j$ is property, $Classk$ is the class to which the object belongs
Object_Class	Store a pattern triple set whose object is the same type, where the key is the name of the class to which the object belongs, and the value is the combination of the property and the class to which the subject belongs	$Object\_Class = > \{C1, C2, \ldots Ci, \ldots Cn\}$, where $Ci = (Class i = > \{[P1, C2], \ldots, [Pj, Ck]\})$, Pj is property, Ck is the class to which the subject belongs

Fusion of Association Class Diagrams. The purpose of the fusion is to connect the pruned association class diagrams. The association class diagrams belonging to the parent-child relationship should be separated and merged with other association class diagrams. In an association class diagram that needs to be fused, the algorithm iteratively determines whether there are common edges in two association class diagrams, and merges the two association class diagrams into one according to the common edge, to obtain one or more final association class diagrams. The input of the fusion process is a pruned GSSet, as shown in Algorithm 2.

Table 2. The storage scheme of instance data

Name	Storage contents	Storage structure
Literal_Triple	The data attribute and instance collection of the storage label are stored in the form of key value pairs, where the key is the label, and the value is the data attribute and instance	$Literal\_Triple = > \{L1, L2, \ldots, Lj, \ldots, Lm\}$, where $Lj = \{P1S1, P2S2, \ldots, PiSi, \ldots, PnSn\}$, Lj represents lable, $PiSi$ represents the pair of attribute instances corresponding to the Lj label, $PiSi = [Pi, Si]$
Instance_Class	The mapping relationship between the instance and the class of the instance is stored in the form of key-value pairs, where the key is the instance and the value is the class corresponding to the instance	$Instance\_Class = > \{IC1, IC2, \ldots, ICi, \ldots, ICn\}$, $ICi = (Instancei = > \{Class : []\})$
SC_OP_OC	Store the instance triples corresponding to the schema triples with object properties and store them as key-value pairs, where key is the subject of the instance triple and value is the object of the instance triple	$SC\_OP\_OC = > \{SO1, SO2, \ldots, SOi, \ldots, SOn\}$, $SOi = (Subjectk = > Objectk)$
OC_OP_SC	Store an inverted backup of SC_OP_OC, stored as a key-value pair, where the key is the object of the instance triple and the value is the subject of the instance triple	$OC\_OP\_SC = \{OS1, OS2, \ldots, OSi, \ldots, OSn\}$, $OSi = (Objectk = > Subjectk)$

Table 3. The storage scheme of historical data

Name	Storage contents	Storage structure
His_Classs_Graph	Store the query subgraphs that have been searched in the history, Where key is a subscript in the list of Rdf_Ontology, Subscripts are connected in descending order. The value is the query subgraph of the ontology	$His\_Classs\_Graph = \{K1, K2, \ldots, Ki, \ldots, Kn\}$, $Ki = (Index1\_Index2\_Index3\_\ldots\_Indexm = > [Gsk1, Gsk2, Gsk3, Gsk4\ldots])$, $Gski$ represents the ontology query subgraphs, sorted in descending order of scores

Lines 1–3 of the code define three storage variables. The result variable represents the resulting ontology query subgraphs, the allSet variable represents the collection of pattern triples that have been checked, and the temp variable represents a temporary variable for the intersection. Line 4 begins to traverse the set of association classes GSSet; Line 5 first deletes the pattern that has been queried by Gsi to avoid duplication; Lines 6 to 10 obtain the intersection of allSet and Gsi and add it to the result set; Lines 11 to 12 add non-intersections in Gsi to allSet.

Table 4. A storage scheme for hash mapping information

Name	Storage contents	Storage structure
Prefix_Hash	The form of a Hash stores the information of the prefix and its Hash value, which is stored in the form of key-value pairs, where the key is the prefix, and the value is the Hash value	Prefix_Hash = {PH1, PH2, ..., PHi, ..., PHn}, PHi = (Prefix = > Hash)
Hash_Prefix	The form of the Hash is stored as a reverse backup of the Prefix_Hash, stored in the form of key-value pairs, where the key is the Hash value, and the value is prefixed	Hash_Prefix = {HP1, HP2, ..., HPi, ..., HPn}, HPi = (Hash = > Prefix)
Conflict	The Set form stores the prefix for the conflict, where the value is the prefix string for the conflict	Conflict = {C1, C2, ..., Ci, ..., Cn}, Ci = (ConflictPrefix)

Construction of Ontology Query Subgraphs. Using the set of association class diagrams GSSet, the set of attributes Pset, and the set of classes Cset as input to construct the ontology query subgraph, an ontology query subgraph can be generated through two processes of pruning and fusion. The process of constructing the ontology query subgraphs is as shown in Algorithm 3.

4.3   Construction of Result Subgraphs

After completing the sorting of ontology query subgraphs (ScoreSubGraph), the ontology query subgraph with the highest priority is taken out to perform the Map stage to obtain the result subgraph. First, we combine the pattern triples in the ontology query subgraph to search the corresponding key-value pair records in the instance data; then we combine the predicates in the pattern triples and the subjects and objects in the key-value pairs to obtain matching instance triples; finally, according to the connection method of Definition 7, the matching instance triples are connected to obtain the corresponding result subgraph.

The keyword set may contain text-type keywords. You can search Keyword_Inf in $O(1)$ time to get the instance corresponding to the text type keyword. You can get matching instance triples based on pattern triples and text-type keywords (see Step 2 in Algorithm 4 for details). In addition, keyword sets may contain instance-type keywords. Searching for SC_OP_OC and OC_OP_SC within $O(1)$ time can obtain key-value pair records with this instance as the key. Matching instance triples can be

Algorithm 3: The ontology query construction algorithm subgraph

Input: the set of association class diagrams GSSet, the set of attributes Pset, and the set of classes Cset.

Output: Ontology query subgraph gsk collection.

For GS$_i$∈GSSet & i = 1, 2...,m /*Traverse GSSet*/

cutGS(GS$_i$,Pset,Cset) // Pruning operations on GSi.

End For

/*buildGSK merges pruned GSSets to generate an ontology query subgraph gsk*/

gsk = buildGS(GSSet)//Fusion operation for GSSet

Return gsk;

Algorithm 4: Generate result subgraph

Input: The highest priority ontology query subgraph, instance data

Output: Result subgraph set

Step 1. Get the instance corresponding to the text type keyword in Keywords_Inf and combine to get an instance triple matching pattern triple containing data attribute;

Step 2. Get the records with key of instance type in SC_OP_OC and OC_OP_SC, and get the triples of instances matching pattern triples containing object attributes;

Step 3. Get triples of instances matching other pattern triples in SC_OP_OC and OC_OP_SC;

Step 4. Connect the matching instance triples to get the result subgraph according to Definition 7;

Step 6. Add multiple result subgraphs to the result subgraph set;

Step 7. Output and pass the result subgraph set to the coding reversal phase (ResultReverse) to process, move to Step 8;

Step 8. End the algorithm.

obtained based on the pattern triples and key-value pair records (see Step 3 in Algorithm 4 for details). The specific process of the BuildResGraph stage is shown in Algorithm 4.

4.4 Time Complexity Analysis

The time complexity of the algorithm consists of two parts, namely the time complexity of constructing the ontology subgraphs and that of generating the result subgraphs. Among them, the construction of the ontology subgraphs stage includes constructing the association class diagram, the class diagram pruning and the class diagram fusion; and the result subgraphs generation phase includes the Map phase and the Reduce phase. Suppose the user inputs the number M of keywords, and each keyword generates m association class diagrams. Each association class diagram has n edges, and the number of generated ontology subgraphs is K, and the number of keyword-matched instance triples is p, then:

(1) The time complexity of constructing the ontology subgraphs stage: O(M) + O(m * n) + O(m);
(2) The time complexity of generating the result subgraphs stage: time complexity of the Map phase O(K), and time complexity of the Reduce phase O(p * K);

In summary, the time complexity of the MPSASR algorithm is (O(M) + O(m * n) + O(m)) + O(K) + O(p * K). Since the scale of the RDF ontology graph is relatively small, m and n and K are relatively small, so the overall time complexity of the algorithm is relatively small.

5 Experimental Analysis

5.1 Experimental Environment

The software environment used in the experiment was the Ubuntu operating system. Java was used as the programming language and the development environment was Eclipse. The number of Worker nodes in the Spark cluster determines the cluster computing performance. The size of the SPARK_WORKER_MEMORY configuration item in the Spark configuration file determines the available memory capacity of each Worker node in the cluster. One Master node and eight Worker nodes are deployed in the lab Spark cluster environment.

5.2 Experimental Data Sets and Settings

In order to verify the effectiveness of the MPSASR algorithm, this experiment uses the LUBM dataset as the simulation dataset and the Dbpedia dataset and the DBLP dataset as the real dataset. The basic parameters of the data set are described in Table 5.

The search examples used in the experiment are shown in Table 6. There are a total of 6 keyword sets Q1–Q6, and each keyword set contains 2–5 keywords. Among them, Q1–Q2 tests the LUBM dataset, Q3–Q4 tests the Dbpedia dataset, and Q5–Q6 tests the DBLP dataset. At the same time, Q1, Q3, and Q5 contain 4 types of keywords: instances, texts, classes, and properties. Q2, Q4, and Q6 contain only some types of keywords.

For different data sets, the MPSASR algorithm was compared with KREAG [10] and DKSRM [11] in the same experimental environment. In order to facilitate the comparison in experiment, the parameter α in the Eq. (1) of the correlation evaluation function of the Definition 9 was set to 0.5; the k value of the Top-k was set to 10.

Since the MPSASR algorithm implements real-time search of multiple keywords, the following method was used when testing the MPSASR algorithm: 8 MB data were flowed into the module of Spark in real time every 2 s, and performed real-time multi-keyword search continuously when the data sets flowed into it in real time. At the same time, when the inflow of data set was accomplished, multiple keyword search was performed on it. Finally, took the average of multiple keyword search results as the final result. The data for the other two algorithms in the comparison experiment was the average of 10 searches.

5.3 Experimental Results and Analysis

Analysis of Search Response Time Comparison. In order to test the search efficiency of the MPSASR algorithm, the experiment tested the search response time of the search examples belonging to different data sets, as listed in Table 7. The MPSASR algorithm utilizes historical data in the class generation association class diagram phase to skip unnecessary steps. In order to verify the role of historical data, the experiment tests the average search response time using the historical data method and the unused historical data method Q1–Q6, as shown in Fig. 3.

Table 5. Basic parameter description of the data set

Data set name	Number of attributes	Number of lasses	Number of pattern triples	Number of instance triples	File size
LUBM300	51	43	188	4135	3.19
LUBM500	51	43	188	6887	5.34
LUBM1000	51	43	188	13778	10.71
DBpedia3.8	1794	376	5795	1697	2.31
DBpedia3.9	2352	570	7766	2189	3.01
DBLP	43	18	487	1253	1.03

Table 6. Search example

Query	Keywords collection	Data set
Q1	Research12, teacherOf, Course, FullProfessor6, http://www. Department12.University0.edu/UndergraduateStudent95	LUBM
Q2	Lecturer4, UndergraduateStudent, takesCourse	LUBM
Q3	http://dbpedia.org/resource/Columbia_University, Columbia Lions, http://dbpedia.org/ontology/state, http://dbpedia.org/ontology/City	DBpedia
Q4	http://dbpedia.org/resource/Tom_Cruise, Rain Man	DBpedia
Q5	http://dblp.uni-trier.de/rec/bibtex/books/acm/kim95/DayalHW95, http://lsdis.cs.uga.edu/projects/semdis/opus#book_title, Modern Database Systems, http://lsdis.cs.uga.edu/projects/semdis/opus#University	DBLP
Q6	http://dblp.uni-trier.de/rec/bibtex/conf/ats/Zhang95, 1995, http://lsdis. cs.uga.edu/projects/semdis/opus#last_modified_date	DBLP

Table 7. MPSASR algorithm search response time under different data sets (unit:s)

Dataset	Q1	Q2	Q3	Q4	Q5	Q6
LUBM300	1.21	1.59	1.26	1.14	1.70	1.64
LUBM500	2.5	2.87	2.20	2.38	3.13	2.7
LUBM1000	3.79	3.28	3.54	3.36	4.0	3.81
DBpedia3.8	3.42	3.11	3.76	3.45	4.29	3.94
DBpedia3.9	3.12	3.59	3.68	3.62	4.47	4.17
DBLP	3.13	3.25	3.17	4.28	2.8	2.22

From Table 7, it can be seen that under different data sets, the MPSASR algorithm can search for keywords within 2 s, and the search response time is maintained at the same order of magnitude without obvious changes. High search efficiency satisfies the demand for real-time search of massive RDF data streams.

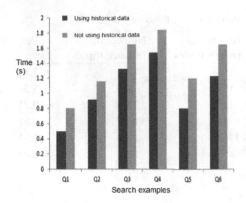

Fig. 3. Comparison of average search response times in used and unused historical data methods

From Fig. 3, we can see that in the method of using historical data, the average search efficiency of Q1–Q6 in each data set has been improved to varying degrees. Since the method that does not use historical data needs to re-map and match the keywords regardless of whether the current search keyword set was previously searched, it takes a certain amount of time; however, the method of using historical data is based on a variety of other methods. The situation judges the relationship between the current search keyword set and the previously searched keyword set, makes full use of historical data to reduce some unnecessary operations, and improves the efficiency of multiple keyword search to some extent.

Additionally, in order to reflect the advantages of the MPSASR algorithm in search efficiency, the average search response time of MPSASR, KREAG, and DKSRM algorithms for Q1–Q6 was respectively tested, as shown in Fig. 4. From Fig. 4, it can be seen that the average search efficiency of the MPSASR algorithm Q1–Q6 in the respective data sets has very obvious advantages over the other two algorithms. Because the MPSASR algorithm makes full use of Redis's optimization function for shaping numerical values, the RDF data is hash-coded and compressed, which reduces the storage space while speeding up the keyword search and matching, and stores the large-scale RDF data in Redis distributed. In the cluster, when reading and writing data is only O(1) time, while the Spark streaming processing framework performs real-time search on keywords, the generated intermediate data is stored in the memory, and the real-time flow is processed to the next stage. Both MKPSA and DKSRM algorithms use the MapReduce computing framework to perform a distributed multi-keyword search. When the MapReduce job is started, it takes a while to run; the intermediate data generated is stored on HDFS or on disk, and it takes time to read and write. The DKSRM algorithm stores a large number of path indexes distributed over different nodes of the cluster. When transmitting, it consumes a lot of network resources, increasing the network load, and leading to unsatisfactory keyword search efficiency.

Analysis of Search Effects. In order to verify the search effect of the MPSASR algorithm, evaluation is focused on precision and recall. Precision is used to measure the accuracy of top-k search results. It is the ratio of the number of correct instance triples in the search results to the total number of instance triples in the search results. Recall is the ratio of the number of correct instance triples in the search results to the total number of instance triples in the data set. For each data set, we create a set for each search example to store all relevant search results for that search example. The definitions of precision and recall are as follows.

$$precision = \frac{\text{Correct number of instance triples in search results}}{\text{Total number of instance triples in search results}}$$

$$recall = \frac{\text{Correct number of instances triples in search results}}{\text{Total number of instance triples in data set}}$$

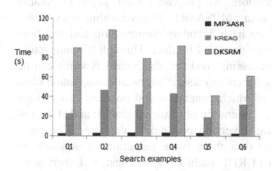

Fig. 4. The comparison in the average search response time of the three algorithms Q1–Q6

The precision and recall rates are used to evaluate performance of the MPSASR, KREAG, and DKSRM algorithms. Figure 5 shows average precision and recall of the three algorithms in Q1–Q6. From Fig. 5, it can be seen that precision and recall of the MPSASR algorithm in Q1–Q6 are high; higher on average than those of the DKSRM algorithm, and just slightly lower than those of the KREAG algorithm. The accuracy rate is between 84% and 93%, and the recall rate is between 81% and 94%. The MPSASR algorithm uses the Spark streaming framework to stream RDF data in real time, and so during a real-time multi-keyword search some of the RDF data directly related to the current keyword set may not completely flow into the Redis distributed cluster, causing the result submap to failed to build successfully, as it is missing a part of the correct result. The KREAG algorithm starts searching for keywords after all RDF data is written into the Redis distributed cluster, and therefor, almost all of the result submaps are built. The correct result is that the KREAG algorithm has a slightly higher precision and recall rate than the MPSASR algorithm.

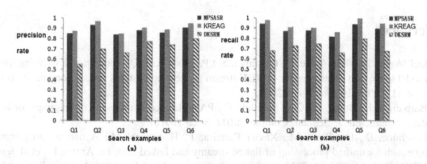

Fig. 5. The comparison in the precision and recall rates of the three algorithms Q1–Q6

6 Summary and Prospect

The keyword search of streaming RDF data has attracted more and more attention from academic and industrial circles. However, previous work is not ideal for real-time streaming data, and many queries based on formal languages are not directly based on the matching of RDF data graphs. Therefore, we propose a multi-keyword parallel search algorithm for streaming RDF data (MPSASR). The algorithm uses Spark Streaming framework to achieve real-time hash encoding compression and real-time inflow of RDF data, reducing the memory occupied by data. Through the interaction with the Spark Core real-time search processing modules, the input keywords can be streamed to complete mapping matching for entity classes; afterwards association class diagrams are constructed, on the basis of which generation and ordering of ontology query subgraphs are carried out as well as generation of result subgraphs; and at the last stage of the process, of course, encoding inversion operations; also should be mentioned is with exploitation of historical data there come reduction of unnecessary operations and efficiency improvement of RDF multi-keywords search. Experiments have proved that our algorithm has great advantages in real-time search, response time, and search effects.

Of course, our algorithm still has some deficiencies in some technical details. For example, when the number of search results does not reach Top-k, proper processing is not performed, and the expansion of the query range is not considered. In addition, the algorithm completes the processing of streaming data in memory, and the cached data needs to occupy memory. The algorithm performance and data size are limited by the cluster memory size. In the future research, the author will do further research from the following aspects: (1) In-depth study of the RDF semantic structure, expand the construction scope of the ontology query subgraphs to achieve a higher query accuracy; (2) Although the MPSASR algorithm has hashed and compressed the data, and the memory occupied by the data is relatively reduced, but the types of the stored data are diverse, so, a more reasonable and effective distributed storage solution and cache data clearing strategy should be studied.

References

1. Dell'Aglio, D., Della Valle, E., Calbimonte, J.P., et al.: RSP-QL semantics: a unifying query model to explain heterogeneity of RDF stream processing systems. Int. J. Semant. Web Inf. Syst. (IJSWIS) **10**(4), 17–44 (2014)
2. Barbieri, D.F., Braga, D., Ceri, S., et al.: C-SPARQL: a continuous query language for RDF data streams. Int. J. Semant. Comput. **4**(01), 3–25 (2010)
3. Le-Phuoc, D., Dao-Tran, M., Xavier Parreira, J., Hauswirth, M.: A native and adaptive approach for unified processing of linked streams and linked data. In: Aroyo, L., et al. (eds.) ISWC 2011. LNCS, vol. 7031, pp. 370–388. Springer, Heidelberg (2011). https://doi.org/10.1007/978-3-642-25073-6_24
4. Borthakur, D., Gray, J., Sarma, J.S., et al.: Apache Hadoop goes realtime at Facebook. In: ACM SIGMOD International Conference on Management of Data, SIGMOD 2011, Athens, Greece, pp. 1071–1080, June 2011

 5. Hou, R., Fang, J., Zhang, J.: Data query method for real-time streaming data protection. J. Comput. Appl. **31**(9), 2736–2740 (2014)
 6. Xu, W.: Research on streaming data real-time query method. Shandong University (2015)
 7. Jiang, C., Ji, Y., Sun, Y., et al.: Storm-oriented real-time streaming query system design for big data. J. Nanjing Univ. Posts Telecommun. **36**(3), 100–105 (2016)
 8. Zhu, M., Cheng, J., Bai, W.: An RDF data storage model based on HBase. J. Comput. Res. Dev. **50**(s1), 23–31 (2013)
 9. RDF concepts and abstract syntax. http://www.w3.org/TR/rdf-concepts/
10. Li, H., Ran, Y.: KREAG: RDF data keyword query method based on the relationship of entity triads. Chin. J. Comput. **34**(5), 825–835 (2011)
11. De Virgilio, R., Maccioni, A.: Distributed keyword search over RDF via MapReduce. In: Presutti, V., d'Amato, C., Gandon, F., d'Aquin, M., Staab, S., Tordai, A. (eds.) ESWC 2014. LNCS, vol. 8465, pp. 208–223. Springer, Cham (2014). https://doi.org/10.1007/978-3-319-07443-6_15

Efficient Memory Caching for Erasure Coding Based Key-Value Storage Systems

Jiajie Shen[1,2,3], Yi Li[1,2,3], Guowei Sheng[1,2,3], Yangfan Zhou[1,2,3(✉)], and Xin Wang[1,2,3]

[1] School of Computer Science, Fudan University, Shanghai, China
{14110240024,17210240148,17210240187,zyf,xinw}@fudan.edu.cn
[2] Shanghai Key Laboratory of Intelligent Information Processing, Shanghai, China
[3] Shanghai Institute of Intelligent Electronics Systems, Shanghai, China

Abstract. Erasure codes are widely advocated as a viable means to ensure the dependability of key-value storage systems for big data applications (*e.g.*, MapReduce). They separate user data to several data splits, encode data splits to generate parity splits, and store these splits in storage nodes. Reducing the disk Input and Output (I/O) latency is a well-known challenge to enhance the performance of erasure coding based storage systems. In this paper, we consider the problem of reducing the latency of read operations by caching splits in the memory of storage nodes. We find the key to solve this problem is that storage nodes need to cache enough splits in the memory, so that the application server can reconstruct the objects without reading data from disks. We design an efficient memory caching scheme, namely *ECCS*. The theoretical analysis verifies that ECCS can effectively reduce the latency of read operations. Accordingly, we implement a prototype storage systems to deploy our proposal. The extensive experiments are conducted on the prototype with the real-world storage cluster and traces. The experimental results show that our proposal can reduce the time of read operations by up to 32% and improve the throughput of read operations by up to 48% compared with current caching approaches.

Keywords: Key-value storage · Erasure codes · Caching approaches
Distributed storage · Read latency

1 Introduction

Object storage systems (*e.g.*, Ceph [1]) are widely used to store data objects of big data applications (*e.g.*, MapReduce [2]). Such storage systems are also known as the key-value storage systems, since they return the corresponding objects

This work was supported by the National Natural Science Foundation of China (Project Nos. 61571136 and 61672164), and a CERNET Innovation Project (No. NGII20160615).

Z. Xu et al. (Eds.): Big Data 2018, CCIS 945, pp. 512–539, 2018.
https://doi.org/10.1007/978-981-13-2922-7_34

according to their identifications (IDs). Recently, erasure codes are widely advocated as a viable means to ensure the data reliable, and enhance the efficiency of object storage systems [3]. They separate the original objects to data splits and encode data splits to generate parity splits, and store the splits in multiple storage nodes, so that a subset of these splits is sufficient to recover the original data. Instead of writing all data to a single storage node, erasure codes can enhance the performance of write operations by parallel writing data into multiple storage nodes [4].

However, such dependable systems suffer from high disk Input and Output (I/O) latency when reading objects, since storage nodes need to read the data from disks. To reduce such I/O latency, current storage systems typically cache all the splits for the objects with the high read frequency [5]. However, such caching schemes commonly waste the memory buffer to cache unnecessary splits, since erasure codes typically store redundant splits to ensure data reliable [3]. As a result, they cannot efficiently reduce the I/O overhead of read operations in erasure coding based object storage systems. How to reduce the disk I/O latency remains the key issue to enhance the efficiency of read operations in practical storage systems.

By analyzing the process of read operations, we find that the key to solve this problem is caching enough splits in the memory of storage nodes for the objects with high read frequency. In practice, storage nodes frequently fail. According to the measurement of Facebook cluster, there are up to 110 node failures which trigger repair jobs every day [6]. When the storage nodes failed, the cached data will also be unavailable. To ensure the application servers can read the objects from the memory, storage nodes need to cache the splits in the memory according to read frequency of objects and node failure rates. However, it is quite difficult to construct such caching scheme, since we face the following challenges.

- **Heterogeneous node failure rates.** Unlike traditional memory caching systems, storage nodes typically have diverse failure rates. For example, the storage nodes which are used for a long time typically have the higher failure rate than new ones. By analyzing the real-world node failure traces [7], we find that the largest failure rate of a given storage node is more than 10 times than the smallest value in the trace. Therefore, storage nodes need to carefully cache the splits according to such heterogeneous node failure rates, so that the caching approach can efficiently reduce the I/O latency of read operations.
- **Decentralized storage architecture.** Some object storage systems (*e.g.*, Ceph) are not centralized controlled by the master nodes. In other words, the storage nodes cannot obtain the global information of storage systems (*e.g.*, read frequency of objects) by accessing to the master nodes. However, such information is critical to determine the caching strategy of storage nodes. Therefore, the caching scheme needs to use an efficient approach to obtain the read frequency of objects according to the read frequency of splits and node failure rates.

To solve the above real-world challenges, we propose an efficient caching scheme, namely *Erasure Coding based Caching Scheme* (*ECCS*). ECCS can efficiently synchronize the information of storage nodes and calculate the performance gains of cached splits, so that storage nodes can cache the splits with the high performance gains. Accordingly, we implement a prototype storage system to deploy different caching approaches. We further conduct the extensive experiments with the real-world storage cluster and traces. The experimental results show that ECCS can effectively improve the performance of read operations under different application scenarios. Specifically, we make the following contributions.

- **Determining the caching strategy according to the failure rates of storage nodes.** According to cached splits and node failure rates, we formulate the I/O latency of read operation. We show that minimizing such I/O latency is an NP-hard problem. Based on this notion, ECCS defines the performance gains of splits, so that storage nodes caches the splits with high performance gains in the memory. Through theoretical analysis, we prove that ECCS can effectively improve the performance of read operations compared with current caching schemes.
- **Estimating the read frequency of objects without centralized master nodes.** Since the storage nodes cannot record the read frequency of splits when they fail, they cannot obtain the actual read frequency of objects. To solve this problem, we propose an approximate approach to estimate the read frequency of objects according to that of splits. Although storage nodes can receive the node failure rates from other storage nodes, ECCS can average the estimated results of storage nodes to ensure the correctness for the read frequency of objects.
- **Practical implementation.** We implement a real-world prototype storage system, namely *ECOS*. To verify the efficiency of our caching schemes, We further deploy different caching approaches in the prototype storage system. Accordingly, we conduct extensive experiments on the prototype storage system with the real-world storage cluster and traces. The experimental results show that our proposed scheme can reduce the time of read operations by up to 32% and improve the throughput by up to 48% compared with current caching approaches.

The remainder of the paper is organized as follows. In Sect. 2, we introduce the background and summarize the related work. In Sect. 3, we show the motivation of this work. In Sect. 4, we use an illustrative example to explain how our scheme works in practical storage systems. We formulate the problem in Sect. 5 and describe our proposed scheme in Sect. 6. We further implement the prototype storage system and deploy our scheme in Sect. 7. To verify the efficiency of our scheme, we evaluate the performance of different cache schemes with the prototype and real-world traces in Sect. 8. Finally, we conclude the paper in Sect. 9.

2   Background and Related Work

In this section, we introduce the preliminary of erasure codes. First, we introduce erasure coding based object storage systems. Then, we summarize the related work.

2.1   Current Architecture of Object Storage Systems

Considering the storage system shown in Fig. 1, four storage nodes and application servers are connected with the high speed data center networks in Fig. 1(a). To ensure the data reliability, the servers can use erasure codes to encode objects to several splits, and store these splits in storage nodes. In Fig. 1(b), the application server first separates object o_1 into data splits A and B, encodes data splits to generate parity splits $A + B$ and $A + 2B$, and stores these splits in storage nodes.

Then, the servers read splits from storage nodes and reconstruct objects even when some storage nodes failed. In Fig. 1(c), the server sends the read request for object o_1 to storage nodes. Suppose the storage nodes of split B is unavailable. Other available storage nodes return the split data. The server decode splits A and B by receiving any two splits. Then, the server aborts other transmission process and reconstruct object o_1. To clarify the contributions of our study, we summarize the related work.

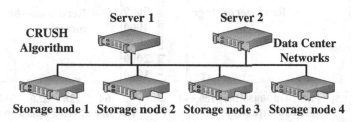

(a) The connection of storage nodes in storage systems.

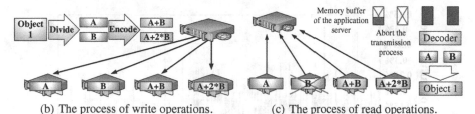

(b) The process of write operations.   (c) The process of read operations.

Fig. 1. The connection of storage systems, and the process of read and write operations. The parameters of the erasure code are $n = 4$, $k = 2$, and $m = 2$.

2.2   Related Work

Object storage systems are widely used in current applications. Weil *et al.* [1] propose an object system, namely Ceph. Yan *et al.* [8] propose a metadata

management scheme for object storage, namely Hashing Partition (HAP).
Dragojevic *et al.* [9] propose Fast Remote Memory (FaRM) to apply Remote
Direct Memory Access (RDMA) in object storage systems. Nishtala *et al.* [10]
propose Memcache in the Facebook cluster. Li *et al.* [5] propose an in-memory
storage system, namely Tachyon (*i.e.*, Alluxio). Similar storage systems include
KV-Direct [11], MICA [12], Scarlett [13], Amazon S3 [14], OpenStack Swift [15],
and Microsoft Azure Storage [16]. Since the storage nodes frequently fail, they
need to ensure the data reliability of storage systems.

Erasure codes [17] are widely used in storage systems to ensure the data reliability. Huang *et al.* [18] propose Local Reconstruction Codes (LRC) for Microsoft
Azure Storage. Sathiamoorthy *et al.* [6] use the regenerating code [19] to reduce
the network traffic of data recovery. Zhang *et al.* [20] first propose erasure coding
based object storage system, namely Cocytus. Rashmi *et al.* [3] propose erasure
code based caching approach, namely EC-Cache. Similar studies also can be
found in hitchhiker [21], STAIR [22], Carousel [4], MemEC [23], BCStore [24],
Agar [25], and Core [26]. Although these storage systems can ensure data reliability, they suffer from high read latency when reading data from the storage
nodes. To elaborate motivation of our work, we measure the performance of read
operations and analyze the real-world traces.

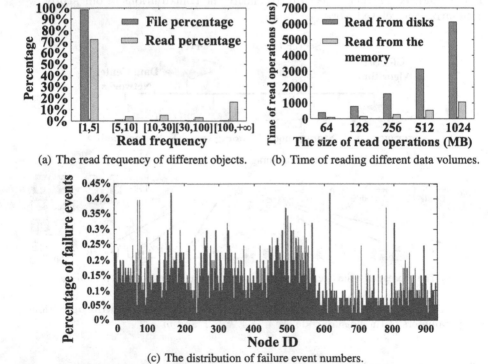

(a) The read frequency of different objects.   (b) Time of reading different data volumes.

(c) The distribution of failure event numbers.

Fig. 2. The frequency and latency of read operations and distribution of failure event
numbers under different traces.

3   Measurement and Motivation

In this section, we describe the preliminary and motivations of this paper. First, we introduce the setup of measurement environment and description of the traces. Then, we depict the key observations of measurement results.

3.1   Measurement Setup

To compare the performance of read data from disks and memory, we store the same data in the memory and disks of storage nodes. We use two Dell PowerEdge R730 equipped with Seagate Hard Disk Drive (HDD) disks as the application server and storage node. The server and storage node are connected with 10 Gigabit network. Such setting is widely used in practical storage systems [20]. To analyze the user behavior, we employ two real-world traces: Facebook SWIM trace and PNNL failure trace. The descriptions of these traces are listed as follows.

- **Facebook SWIM trace.** Facebook SWIM trace [27] is a production Hadoop MapReduce trace. The trace is collected from a 3000-machine cluster. The trace spans 1.5 months from October 2010 to November 2010, and contains roughly 1 million jobs which includes basis information and anonymized input path of objects.
- **PNNL failure trace.** PNNL failure trace [7] is a node failure trace. This trace includes the failure events of 980 nodes which is collected by Pacific Northwest National Laboratory (PNNL) from 2003 to 2007.

We use Facebook SWIM trace and PNNL failure trace to count the distribution of read frequency and node failure rates, respectively. We normalize the number of failure events for each storage node by dividing total number of failure events. We also measure the time of read operations, when the server reads different data volumes (64 MB, 128 MB, 256 MB, 512 MB, and 1024 MB) from the memory and disks of storage nodes, since current big data applications (*e.g.*, MapReduce) typically need to read large data volume [28]. Accordingly, we conduct the measurement analysis and obtain the key observations of these measurement results.

3.2   Key Observations

The read frequency of different objects, and node failure rates are shown in Fig. 2. By analyzing the measurement results, we obtain the following key observations.

The Read Frequencies of Different Objects Are Various. In Fig. 2(a), the number of read requests for 99% of objects is smaller than 5 times. However, 0.1% of objects read larger than 100 times incur more than 28% of read requests.

Disk I/O Operations Are Expensive. In Fig. 2(b), the server can reduce the time of read operations by up to 84% by caching data in the memory compared with reading data from disks, even the data are transmitted through the network.

Failure Rates of Storage Nodes Are Various. In Fig. 2(c), the largest number of failure events in a given node is more than 10 times than the smallest number of failure events. The failure rates of all storage nodes are different with each other.

To improve the performance of read operations, storage systems need to cache some splits with the high read frequency, so that these objects can be directly read from the memory. To depict how our scheme can improve the performance of read operations, we introduce an illustrative example for a given storage system.

4 Beyond Memory Caching Approaches: A Motivation Example

Consider the example in Fig. 3. There are four storage nodes n_1, n_2, n_3, and n_4, and four objects o_1, o_2, o_3, and o_4. For each timeslot, the number of read requests for object o_1, o_2, o_3, and o_4 are 400, 300, 200, and 100. Each object

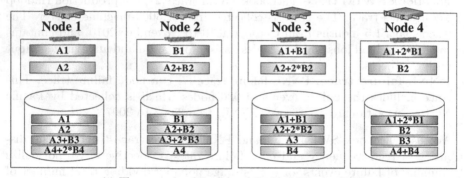

(a) The example of Traditional Caching Scheme (TCS).

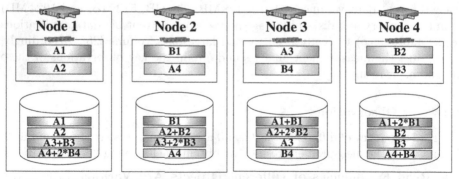

(b) The example of Erasure Coding based Caching Scheme (ECCS).

Fig. 3. Example of TCS and ECCS. The parameters of the erasure code are $n = 4$, $k = 2$, and $m = 2$. We use the splits with different colors to represent the splits with different read frequencies. Each node stores two data splits and two parity splits in disks and caches two splits. (Color figure online)

is separated to two data splits and encoded to two parity splits. For example, the application server separates object o_1 to data splits A_1 and B_1 and encodes them to generate parity splits $A_1 + B_1$ and $A_1 + 2*B_1$. Each storage node holds four splits and caches two splits in the memory. According to the measurement of Facebook cluster, there are up to 110 node failures in 3000 storage nodes every day [6]. Therefore, we set the node failure rate to $\alpha = 3\%$ (*i.e.*, storage nodes drop 3% read requests). If there are $a > k$ cached splits are available, the application server can read the object from the memory of storage nodes. When storage nodes cache $c = 2$, $c = 3$, and $c = 4$ splits, the failure rate of reading object o_i from the memory is $R_{o_i}^2 = 1 - (1 - \alpha)^2 = 1 - 0.97^2 = 5.91\%$, $R_{o_i}^3 = \alpha^3 + C_3^1 * (1 - \alpha) * \alpha^2 = 0.26\%$, and $R_{o_i}^4 \approx 0$.

Traditional Caching Scheme (TCS) typically caches the splits with the high read frequency in the memory. In Fig. 3(a), TCS caches splits A_1, B_1, $A_1 + B_1$, $A_1 + 2 * B_1$, A_2, B_2, $A_2 + B_2$, and $A_2 + 2 * B_2$ since these splits are read with high frequency. Suppose the application servers can parallel read splits from storage nodes. Since storage nodes cache $c = 4$ splits for objects o_1 and o_2, the application servers need to wait storage nodes to read $I_{o_1} = I_{o_2} = 0$ split to obtain objects o_1 and o_2, $I_{o_3} = 200$ splits for object o_3, and $I_{o_4} = 100$ splits for object o_4. Therefore, the total I/O latency of read operations is $I_{tot} = I_{o_1} + I_{o_2} + I_{o_3} + I_{o_4} = 300$ splits.

Since the difference of read frequencies for objects o_1, o_2, o_3, and o_4 are small, our scheme, namely ECCS, caches all data splits A_1, A_2, B_1, B_2, C_1, C_2, D_1, and D_2. The I/O latency of reading object o_1 is $I_{o_1} = R_{o_2}^2 * 400 = 24$ splits. Similarly, that of reading object o_2 is $I_{o_2} = 18$ splits, $I_{o_3} = 12$ splits for object o_3, and $I_{o_4} = 6$ splits for object o_4. The total I/O latency of read operations is $I_{tot} = I_{o_1} + I_{o_2} + I_{o_3} + I_{o_4} = 60$ splits. ECCS can effectively reduce the I/O latency of read operations to 20% compared with current caching scheme (*i.e.*, TCS).

Through this example, we notice that the I/O latency of read operations is affected by the cached splits. To reduce such I/O latency, storage nodes need to calculate the performance gains of splits according to the read frequency of objects and node failure rates. Moreover, some object storage systems are not centralized controlled. Instead of obtaining the read frequency of objects and node failure rates from the master nodes, storage nodes need to estimate the read frequency and receive node failure rates from other storage nodes. To elaborate how our scheme can reduce the I/O latency of read operations, we formulate the caching problem in storage systems.

5   Problem Formulation

In this section, we formulate the memory caching problem. First, we calculate the failure rates of read objects from the memory. Then, we obtain the I/O latency of read operations. To facilitate further discussions, we summarize the important notations for ease of reference in Table 1.

Table 1. Table of nomenclature

Notation	Definition
n_i	The ith storage node
o_i	The ith object in the storage system
s_i	The ith split for a given object
k	Number of data splits encoded from one object
m	Number of parity splits encoded from one object
n	Number of splits encoded from one object ($n = k + m$)
u	Number of unavailable splits in the memory for a given object
a	Number of available splits in the memory for object
c	Number of cached splits for a given object ($c = a + u$)
$S_{o_i}(t)$	The size of object o_i on timeslot t
$S_{s_i}(t)$	The size of split s_i on timeslot t
$S_{n_i}(t)$	The buffer size of storage node n_i on timeslot t
$N_{n_i}(t)$	Number of splits in storage node n_i on timeslot t
$N_o(t)$	Number of objects in the storage system on timeslot t
$I_{o_i}^{each}(t)$	The I/O latency of reading object o_i on timeslot t
$I_{o_i}^{read}(t)$	The total I/O latency of reading object o_i on timeslot t
$I_{tot}(t)$	The total I/O latency for reading objects on timeslot t
$\alpha_{n_i}(t)$	The failure rate of storage node n_i on timeslot t
$F_{o_i}(t)$	The read frequency of object o_i on timeslot t
$F_{s_i}(t)$	The read frequency of split s_i on timeslot t
$U_{o_i}(t)$	Unavailable nodes cached splits of object o_i on timeslot t
$A_{o_i}(t)$	Available nodes cached splits of object o_i on timeslot t
$C_{o_i}(t)$	Storage nodes cached splits of object o_i on timeslot t
$R_{o_i}^c(t)$	Failure rate of reading object o_i when storage nodes cache c splits of the object on timeslot t

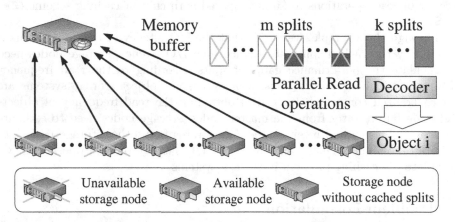

Fig. 4. Example of the caching process in a single node. We use different colors to represent the splits with different read frequencies. (Color figure online)

5.1 Failure Rate of Reading Objects from Memory

Let o_i represent the ith object in the storage system. Assume object o_i is separated to k data splits, and encoded to m parity splits. For object o_i, the storage system caches c splits in storage node set $C_{o_i}(t)$. When the server reads object o_i, there are a available nodes in set $A_{o_i}(t)$ and u unavailable nodes in $U_{o_i}(t)$. Specifically, $A_{o_i}(t)$, $U_{o_i}(t)$, and $C_{o_i}(t)$ satisfy

$$a = |A_{o_i}(t)|, u = |U_{o_i}(t)|, \text{ and } c = |C_{o_i}(t)| \tag{1}$$

where $|X|$ is the number of elements in the set X, $A_{o_i}(t) + U_{o_i}(t) = C_{o_i}(t)$, and $c = a + u$.

Let n_i, $n_j^a \in A_{o_i}(t)$, and $n_l^u \in U_{o_i}(t)$ represent the ith storage node, the jth available storage node, and the lth unavailable storage node. Suppose the failure rate of storage node n_i is $\alpha_{n_i}(t)$. When storage nodes cache c splits, we can calculate the possibility of a available splits $P_c^a(t)$ as

$$P_c^a(t) = \sum_{z=1}^{C_c^a} \prod_{n_j^u \in U_{o_i}(t)} \alpha_{n_j^u}(t) * \prod_{n_j^a \in A_{o_i}(t)} (1 - \alpha_{n_i^a}(t)) \tag{2}$$

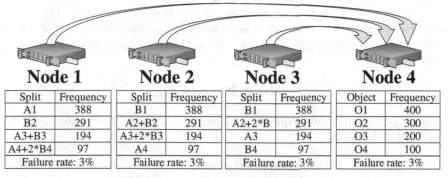

Node 1

Split	Frequency
A1	388
B2	291
A3+B3	194
A4+2*B4	97
Failure rate: 3%	

Node 2

Split	Frequency
B1	388
A2+B2	291
A3+2*B3	194
A4	97
Failure rate: 3%	

Node 3

Split	Frequency
B1	388
A2+2*B	291
A3	194
B4	97
Failure rate: 3%	

Node 4

Object	Frequency
O1	400
O2	300
O3	200
O4	100
Failure rate: 3%	

(a) Storage node n_4 generates the read frequencies.

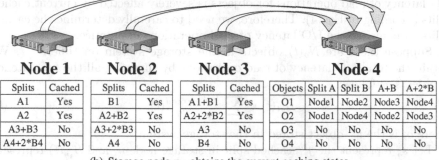

Node 1

Splits	Cached
A1	Yes
A2	Yes
A3+B3	No
A4+2*B4	No

Node 2

Splits	Cached
B1	Yes
A2+B2	Yes
A3+2*B3	No
A4	No

Node 3

Splits	Cached
A1+B1	Yes
A2+2*B2	Yes
A3	No
B4	No

Node 4

Objects	Split A	Split B	A+B	A+2*B
O1	Node1	Node2	Node3	Node4
O2	Node1	Node4	Node2	Node3
O3	No	No	No	No
O4	No	No	No	No

(b) Storage node n_4 obtains the current caching states.

Fig. 5. Example of the sync process for Fig. 3. The parameters of the erasure code are $n = 4$, $k = 2$, and $m = 2$.

where C_j^i is the permutation number that chooses i elements from j elements. For instance, if the number of cached splits and failure rates are $c = 5$ and $\alpha = 0.03$, the possibility that $a = 3$ available splits is $P_5^3 = C_5^3 * (1-\alpha)^2 * \alpha^3 = 8.21 * 10^{-3}$.

Since storage nodes can reconstruct object o_i with any k splits, the server can directly read object o_i from the memory if $a \geq k$ cached splits are available. Otherwise, storage nodes have to read splits from disks when the server reads object o_i. When storage nodes cache c splits, we obtain the failure rate of reading object o_i from the memory $R_{o_i}^c(t)$ as follow:

$$R_{o_i}^c(t) = \begin{cases} 1 & c < k \\ \sum_{a=0}^{k-1} P_c^a(t) & c \geq k \end{cases} \tag{3}$$

If the storage nodes read the splits from disks, they incur the high disk I/O latency of read operations in practice. According to the failure rates of reading objects from the memory, we formulate the I/O latency of read operations.

5.2   The I/O Latency of Read Operations

Since the servers can read these splits from these storage nodes in parallel, the I/O latency of read operations is determined by data volume which is read from disks of each storage node. The reading process is shown in Fig. 4. The servers reconstruct object o_i after receiving any k splits. Since the failure rates of reading object o_i from the memory is $R_{o_i}^c(t)$, we define the I/O latency of read operations $I_{o_i}^{each}(t)$ as

$$I_{o_i}^{each}(t) = R_{o_i}^c(t) * S_{s_j}(t) \tag{4}$$

For object o_i, the read frequency on timeslot t is $F_{o_i}(t)$. According to Eq. (4), the total I/O latency $I_{o_i}^{read}(t)$ is

$$I_{o_i}^{read}(t) = I_{o_i}^{each}(t) * F_{o_i}(t) \tag{5}$$

Given the failure rates of storage nodes and read frequency of objects, the total I/O latency of read operations for objects is severely affected by current cached splits according to Eq. (4). Therefore, we need to carefully determine the cached splits to minimize the I/O latency of read operations in practice.

Suppose there are $N_o(t)$ objects in the storage system on timeslot t. We obtain the total I/O latency of read operations by summing all the I/O latency of objects as follow:

$$I_{tot}(t) = \sum_{i=1}^{N_o(t)} I_{o_i}^{read}(t) \tag{6}$$

where o_i is the ith object stored in the storage system.

Let s_j^i represent the jth cached split in storage node n_i and $S_{s_j^i}^c(t)$ represent the size of split s_j^i. Suppose storage node n_i can buffer $S_{n_i}(t)$ bytes. For timeslot t, we minimize the I/O latency of read operations by the following expression:

$$\min I_{tot}(t) \tag{7}$$
$$s.t. \ \texttt{For all} \ s_j^i \in n_i,$$
$$\sum_{j=1}^{N_{n_i}(t)} S_{s_j^i}^c(t) \le S_{n_i}(t)$$

where $N_{n_i}(t)$ is the number of splits cached in storage node n_i. Since the buffer size of storage nodes is limited, the storage node needs an efficient caching scheme to cache splits in the memory. To reduce the I/O latency of read operations in practice, our scheme caches data according to read frequency and node failure rates.

6 Memory Caching for Erasure Codes

In this section, we propose our caching scheme to reduce the latency of read operations. First, we synchronize the information of storage nodes. Second, we propose the caching scheme, namely ECCS. Then, we depict the sync and caching process in practice. Finally, we analyze the performance of our proposal.

6.1 Storage Node Sync Scheme

We construct the scheme to synchronize the information of storage nodes, if storage systems are not centralized controlled by master nodes. Since storage nodes can record the read frequency for splits, the sync scheme estimates the read frequency of objects according to that of splits. Let s_j represent the jth split of object o_i in storage node n_l. For split s_j, storage node n_l can record read frequency $F_{s_j}(t)$ when storage node n_l is available and the failure rate of storage node n_l is $\alpha_{n_l}(t)$. Storage nodes estimate read frequency of object $F_{o_i}(t)$ as

$$F_{o_i}(t) = \max \left\{ \frac{F_{s_j}(t)}{1 - a_{n_l}(t)} \right\}, s_j \in o_i \ \text{and} \ s_j \in n_l \tag{8}$$

In Fig. 5, storage node n_4 obtains the information from storage nodes n_1, n_2, and n_3. Specifically, storage node n_4 estimates the read frequency of objects $F_{o_1}(t)$, $F_{o_2}(t)$, $F_{o_3}(t)$, and $F_{o_4}(t)$ are 400, 300, 200, and 100 according to Eq. (8) in Fig. 5(a), and receives the caching state of splits in Fig. 5(b).

Suppose storage nodes cache $c - 1$ splits. When storage nodes cache the cth split $s_j \in o_i$, we define expected decreased I/O latency $\Delta \bar{I}_{o_i}^j(t)$ by the following expression:

$$\Delta \bar{I}_{o_i}^j(t) = \begin{cases} \frac{(1 - R_{o_i}^k(t)) * F_{o_i}(t) * S_{s_j}(t)}{k}, c \ \text{or} \ j \le k \\ \Delta R_{o_i}^j(t) * F_{o_i}(t) * S_{s_j}(t), j, c > k \\ 0, \texttt{Otherwise} \end{cases} \tag{9}$$

where $\Delta R_{o_i}^j(t) = R_{o_i}^{j-1}(t) - R_{o_i}^j(t)$. We call $\Delta R_{o_i}^j(t)$ as decreased failure rate of reading object o_i from memory. In other words, the application servers have

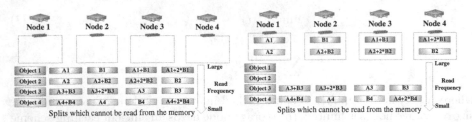

(a) Sort objects and splits with descend order. (b) Cache splits A_1, B_1, and C_1 in the memory.

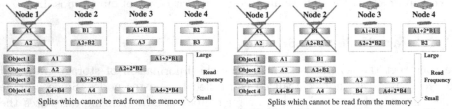

(c) Cache split $A_1 + B_1 + C_1$ in the memory. (d) Cache splits A_2, B_2, and C_2 in the memory.

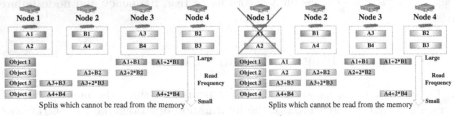

(e) Remove objects o_3 and o_4 from candidates. (f) Finish caching process when no candidate.

Fig. 6. Example of caching process for ECCS. The parameters of the erasure code are $n = 4$, $k = 2$, and $m = 2$. We use the splits with different colors to represent the splits with different performance gains. Each node stores two data splits and two parity splits. (Color figure online)

$\Delta R_{o_i}^j(t)$ more possibility to read object o_i from the memory of storage nodes, if storage node n_j caches split $s_j \in o_i$.

For split s_i, we define performance gain $G_{s_i}^j(t)$ as decreased I/O latency $\Delta \bar{I}_{o_i}^j(t)$ divided by split size $S_{s_i}(t)$ as follow:

$$G_{s_i}^j(t) = \frac{\Delta \bar{I}_{o_i}^j(t)}{S_{s_i}(t)} = \begin{cases} \frac{(1-R_{o_i}^k(t))*F_{o_i}(t)}{k}, c \text{ or } j \le k \\ \Delta R_{o_i}^j(t) * F_{o_i}(t), j, c > k \\ 0, \texttt{Otherwise} \end{cases} \tag{10}$$

Accordingly, we can reduce the I/O latency of read operations by caching the splits with high performance gains. To finish the caching process, our scheme need to cache data according to the performance gains of the splits.

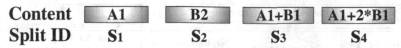

Content	A1	B2	A1+B1	A1+2*B1
Split ID	s_1	s_2	s_3	s_4

(a) Splits s_1, s_2, s_3, s_4 for object o_1 when all nodes are available.

Content	⊠A1	B2	A1+B1	A1+2*B1
Split ID		s_1	s_2	s_3

(b) Splits s_1, s_2, s_3 for object o_1 when storage node n_1 failed.

Content	⊠A1	⊠B2	A1+B1	A1+2*B1
Split ID			s_1	s_2

(c) Splits s_1 and s_2 for object o_1 when storage nodes n_1 and n_2 failed.

Fig. 7. Example of identifying the splits of object o_1 in Fig. 6. The parameters of the erasure code are $n = 4$, $k = 2$, and $m = 2$.

6.2   Heuristic Memory Caching Approach

We show the example of caching process in Fig. 6. In Fig. 6(a), there are four storage nodes n_1, n_2, n_3, and n_4, and four objects o_1, o_2, o_3, and o_4. The node failure rates are $\alpha = 3\%$ and read frequencies $F_{o_1}(t)$, $F_{o_2}(t)$, $F_{o_3}(t)$, and $F_{o_4}(t)$ are 400, 300, 200, and 100. Each object is divided to two data splits, and encoded to two parity splits. When caching two splits, failure rates $R_{o_1}^2(t)$, $R_{o_2}^2(t)$, $R_{o_3}^2(t)$, and $R_{o_4}^2(t)$ are

$$R_{o_1}^2(t) = R_{o_2}^2(t) = R_{o_3}^2(t) = R_{o_4}^2(t)$$
$$= \sum_{a=0}^{1} P_2^a = C_2^0 * \alpha^2 + C_2^1 * (1 - \alpha) * \alpha$$
$$= 0.03^2 + 2 * 0.97 * 0.03 = 5.91\% \tag{11}$$

Similarly, failure rates $R_{o_1}^3(t)$, $R_{o_2}^3(t)$, $R_{o_3}^3(t)$, and $R_{o_4}^3(t)$ are

$$R_{o_1}^3(t) = R_{o_2}^3(t) = R_{o_3}^3(t) = R_{o_4}^3(t)$$
$$= \sum_{a=0}^{1} P_3^a = C_3^0 * \alpha^3 + C_3^1 * (1 - \alpha) * \alpha^2$$
$$= 0.03^3 + 3 * 0.97 * 0.03^2 = 0.26\%. \tag{12}$$

According to Eqs. (11) and (12), we obtain decreased failure rates of objects o_1, o_2, o_3, and o_4 as

$$\Delta R_{o_i}^3(t) = R_{o_i}^2(t) - R_{o_i}^3(t) = 5.65\%, 1 \leq i \leq 4. \tag{13}$$

Before the caching process, ECCS first allocates the split IDs in Fig. 7. In Fig. 7(a), when cached splits are available, storage nodes mark splits A_1, B_2, $A_1 + B_1$, and $A_1 + 2 * B_2$ as s_1, s_2, s_3, and s_4. According to Eqs. (11), (12), and (13), ECCS caches these splits in the memory by the following steps.

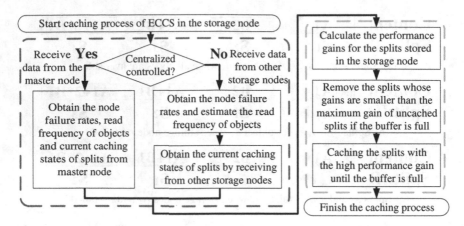

Fig. 8. The flow chart of ECCS. The storage node executes ECCS to cache the splits, when the number of read requests from application servers exceeds a threshold. After the caching process, the storage node records the number of read requests to calculate the read frequency for each split and object. We mark two main steps of ECCS (*i.e.*, the sync and caching process) by the rectangles with the red and green colors. (Color figure online)

First, storage nodes cache all splits of objects o_1 and o_2 in Fig. 6(b), since the memory buffer is empty. In Fig. 6(c), storage nodes replace the splits, according to the performance gains of splits. According to Eq. (10), storage node n_2 calculates the performance gain as $G_{A_4} = \frac{(1-R_{o_4}^2(t))*F_{o_4}(t)}{2} = 47$, and that of parity split $A_2 + B_2$ is $G_{A_2+B_2} = \Delta R_{o_2}^3(t) * F_{o_2}(t) = 16.95$. Accordingly, storage node n_2 replaces cached splits $A_2 + B_2$ by split A_4. Similarly, storage nodes n_3 and n_4 replace parity splits $A_1 + B_1$, $A_2 + 2*B_2$, and $A_1 + 2*B_1$ by data splits A_3, B_4, and B_3, respectively.

In Fig. 6(d), ECCS modifies the caching strategy when storage node n_1 failed. Since other nodes cannot receive the data from storage node n_1, they reallocate the split IDs in Fig. 7(b). Storage nodes mark splits B_1, A_1+B_1, and A_1+2*B_1 as s_1, s_2, and s_3, since split A_1 is unavailable. In Fig. 6(e), storage node n_2 replaces split A_4 by split $A_2 + B_2$ since the performance gains satisfy $G_{A_2+B_2} > G_{A_4}$. Similarly, storage node n_4 replaces split B_4 by split $A_1 + B_1$. In Fig. 6(f), we also show the storage node replace the cached splits when storage nodes n_1 and n_2 failed. For object o_1, they mark splits $A_1 + B_1$ and $A_1 + 2*B_1$ as s_1 and s_2 in Fig. 7(c), since splits A_1 and B_1 are unavailable. Storage nodes n_3 and n_4 replace cached splits A_3 and B_3 by splits $A_2 + 2*B_2$ and $A_1 + 2*B_1$. According to the sync and caching schemes, we construct erasure coding based caching scheme to cache the splits in practical storage systems.

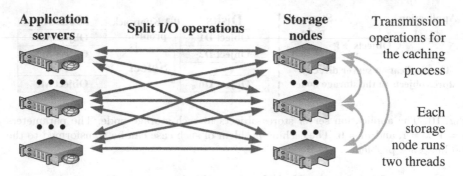

Fig. 9. The network traffic of ECOS. We mark the network traffic caused by the I/O operations of the application servers with the red color and that caused by the sync process with the green color. (Color figure online)

6.3 Erasure Coding Based Caching Scheme

We show the flow chart of ECCS in Fig. 8 and mark two main steps (*i.e.*, the sync and caching process) by the rectangles with the red and green colors.

First, the storage node obtains the read frequency of objects, caching states of splits, and the failure rates of storage nodes. If the storage system is centralized controlled, the storage node directly obtains the information of objects and storage nodes from the master node. Otherwise, the storage node estimates the read frequency of objects according to Eq. (8).

Then, the storage nodes sort the splits according to the performance gains and cache the splits with the high performance gains in memory. To execute the I/O operations and caching process simultaneously, the storage nodes run a daemon thread to cache the splits and transfer the data. We show the network traffic of split I/O operations and caching process in Fig. 9. To verify the efficiency of ECCS, we analyze the caching problem and efficiency of different caching approaches.

6.4 Performance Analysis

First, we show this caching problem is an NP-hard problem. Consider a simple case for $(k, m) = (1, 0)$ in Fig. 10. The application server stores $N_o(t)$ objects in an available storage node. The storage node can cache c objects. While keeping the memory without overflow. Such caching problem is well known NP-completeness problem (*i.e.*, the 0-1 knapsack problem), our caching problem with multiple storage nodes is an NP-hard problem. Since storage nodes cache the splits with the high performance gains, they can quickly finish caching process by using ECCS. Suppose the buffer size of storage nodes is far larger than the split size (*i.e.*, $S_{n_i}(t) \gg S_{s_j}(t)$). We can ensure the performance of read operations for ECCS as follows.

Theorem 1. *For given read frequency of objects, buffer size, and node failure rates, the total disk I/O latency of ECCS $I_{tot}^{ECCS}(t)$ is no more than that of TCS $I_{tot}^{TCS}(t)$ after several timeslots (i.e., $I_{tot}^{ECCS}(t) \leq I_{tot}^{TCS}(t)$).*

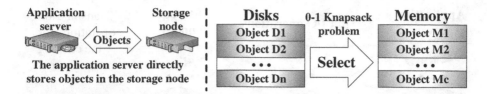

Fig. 10. The application server stores objects in each storage node. The parameters $n = 1$, $k = 1$, and $m = 0$. The caching problem of such case can be transformed to the 1-0 knapsack problem.

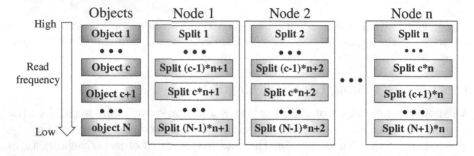

Fig. 11. The results of caching process for TCS. There are N objects and each object is encoded to n splits. We mark the objects, cached splits, and uncached splits with the red, purple, and yellow colors. (Color figure online)

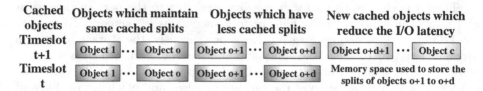

Fig. 12. The memory buffers of storage nodes for TCS. The storage node caches objects which have more than k cached splits. We mark the cached splits and objects with the red and purple colors. (Color figure online)

Proof. The servers send read requests to all storage nodes which store the splits. TCS caches all the splits of the objects with high read frequency. In Fig. 11, storage nodes store N objects and cache the splits of c objects. Since the memory buffer is empty on timeslot t_1, ECCS will also cache all splits of the objects and the I/O latencies of TCS and ECCS are the same (*i.e.*, $I_{tot}^{TCS}(t_1) = I_{tot}^{ECCS}(t_1)$). Then, storage nodes determine whether replacing the cached splits or not.

If the cached splits are not replaced, the I/O latency remains the original value (*i.e.*, $I_{tot}^{ECCS}(t) = I_{tot}^{ECCS}(t-1)$). Otherwise, storage nodes replace cached splits by high performance gain ones. According to Eq. (10), the performance gain decreases with the increase of split ID i if $i > k$. ECCS caches $c > k$ splits for object o_i, if read frequency $F_{o_i}(t)$ is large enough. When storage nodes

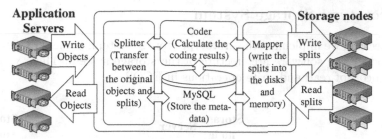

(a) The software architecture of the I/O interface in application servers.

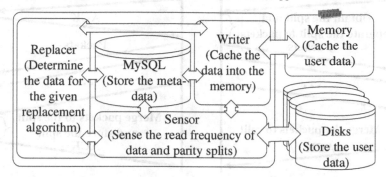

(b) The software architecture of caching component in storage nodes.

Fig. 13. The software architecture of ECOS, including the I/O interface in application servers and caching component in storage nodes.

cache $c > k$ splits, the servers can read object o_i from memory with success rate $V_{o_i}^c(t) = 1 - R_{o_i}^c(t)$. For given read frequency and node failure rates., storage nodes cache more objects after each replacement operation. In Fig. 12, storage nodes maintain the splits of o objects, remove the splits of d objects, and cache w new objects.

Since the buffer size of storage nodes is far larger than the split size (*i.e.*, $S_{n_i}(t) \gg S_{s_j}(t)$), the total size of removed splits $S_{tot}^{rem}(t)$ and that of added splits $S_{tot}^{add}(t)$ satisfy $S_{tot}^{rem}(t) \approx S_{tot}^{add}(t)$. For a given object o_i and read frequency $F_{o_i}(t)$, the application servers can read $I_{o_i}^{read} = (1 - R_{o_i}^c(t)) * F_{o_i}(t) = \sum_{j=1}^{c} G_{o_i}^j(t)$ splits from the memory. Since ECCS caches the splits with high performance gain, these new cached objects can reduce more I/O latency than the original ones by using the same memory space. The total I/O latency will decrease after each replacement process. Since TCS does not modify its cached splits, after the storage nodes execute the caching operations for l timeslots, the total latencies of TCS and ECCS satisfy $I_{tot}^{TCS}(t) = I_{tot}^{ECCS}(t_1) \geq I_{tot}^{ECCS}(t_2) \geq ... \geq I_{tot}^{ECCS}(t_l) = I_{tot}^{ECCS}(t)$.

According to Theorem 1, ECCS can reduce the I/O latency of read operations by replacing cached splits after several caching operations. To deploy different caching approaches in practice, we implement the prototype storage system and evaluate the performance of these approaches with the real-world storage cluster.

7   Prototype Storage System

In this section, we implement the prototype storage system, namely ECOS. First, we introduce the software architecture of the storage system. Then, we depict the protocols of read and write operations in ECOS.

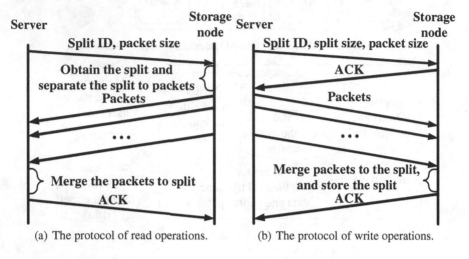

Fig. 14. The protocols of the read and write operations. The server use multi-thread to execute these read and write operations.

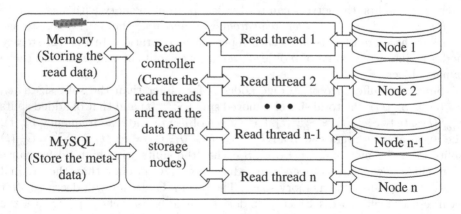

Fig. 15. The interface of read operations in application servers. Each thread reads the split from one storage node.

7.1   Software Architecture of ECOS

The software architecture of ECOS is shown in Fig. 13. For application servers in Fig. 13(a), we implement the splitter, coder, and mappers to separate objects to

Table 2. Current caching approaches and storage systems

Approach	Description	Storage systems
LRU	Replace the split which is least used after storage nodes execute each read operation	EC-Cache [3], Memcached [10], Tachyon [5]
LFU	Replace the split with the smallest read frequency after storage nodes execute each read operation	PACMan [30], Caffeine [31], TinyLFU [32]
TCS	Replace the splits whose read frequency is lower than a threshold after several read operations	HACFS [33], GDSF [34], Zerba [28]

splits, encode the splits and store encoded splits. We use Jerasure [29], an open source coding library, to finish encoding and decoding operations, and MySQL database to store the metadata. For storage nodes in Fig. 13(b), we code recorder, replacer, and writer to record the read frequency of splits, determine cached splits, and write these splits to the memory. To simply access memory space, we mount the memory as RAMDisk of the local storage [10]. In TCS, the storage node averages the number of read requests on timeslot t and read frequency on timeslot $t - 1$ to update the read frequency of split s_i by

$$F_{s_i}(t) = \alpha * D_{s_i}(t) + (1 - \alpha) * F_{s_i}(t - 1) \tag{14}$$

where $D_{s_i}(t)$ is the number of read requests for split s_i on timeslot t. We call parameter α as frequency update ratio. We apply Eq. (14) for ECCS and use the MySQL database to save metadata (e.g., read frequency and saving path of splits). Since TCP (Transmission Control Protocol) uses the slow-start scheme which severely affects the performance of high speed network [9], we do not directly apply TCP to transmit the splits, and implement an efficient transmission protocol. To improve the performance of transmission operations, we construct our protocol of read and write operations.

7.2 The Protocols of Read and Write Operations

Since data center networks are much more reliable than Internet, we can simplify the transmission protocol to improve the performance of transmission operations in Fig. 14.

For read operations in Fig. 14(a), the storage node separates the splits into packets and returns the packets to the server after receiving the read request, so that the server merges packets to the split and sends acknowledgement (ACK) to finish the read operation. For write operations in Fig. 14(b), the server sends write requests to the storage node and receives the ACK to confirm its availability. Then, the server divides the split to packets and sends the packets to the storage node.

To efficiently execute the I/O operations, we use the multi-thread model to parallelize the read operations. We show the interface of read operations in Fig. 15. Specifically, the read controller allocates the memory buffer in the server, and use different points to identify different splits when reading the splits.

To avoid such small traffic affects the performance of I/O operations, we use 1 Gbps and 10 Gbps networks to transmit the control information and split data, respectively. To compare the efficiency of caching approaches, we evaluate the performance of read operations under different caching schemes with our prototype storage systems and real-world traces.

8   Evaluation

In this section, we evaluate the performance of caching schemes. First, we describe our experimental environments. Then, we depict the advantages of our proposed scheme. Finally, we evaluate the overhead of caching approaches.

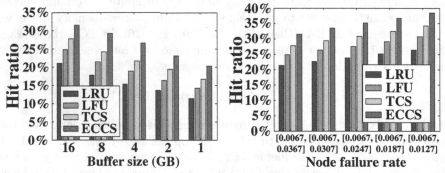

(a) Hit ratios under different memory buffer sizes.

(b) Hit ratios under different node failure rates.

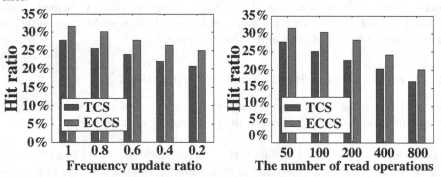

(c) Hit ratios under different frequency update ratios.

(d) Hit ratios under different replacement intervals.

Fig. 16. Hit ratios of caching approaches under different buffer sizes, node failure rates, frequency update ratios, and replacement intervals.

8.1   Experimental Setup

We use DELL PowerEdge R730 servers as the application server, and real-world storage cluster with 5 storage nodes to deploy our storage system. Such setting is widely used in current storage systems [20]. We compare our proposed scheme with several state-of-the-art caching approaches (*i.e.*, LRU [35], LFU [36], and TCS [33]). We list the descriptions of these approaches in Table 2. For TCS and ECCS, we replace some cached splits after the storage nodes execute given number of read operations. In this paper, we call such number of read operations as replacement intervals (*i.e.*, intervals of replacement operations). To verify the efficiency of our scheme, we compare the performance of read operations for different memory buffer sizes, node failure rates, frequency update ratios, and replacement intervals.

We evaluate three performance metrics: the hit ratio, time of read operations, and throughput of read operations. Moreover, we compare two overheads of storage systems: disk I/O overhead of read operations and the time of replacement operations. The hit ratio is obtained by the number of read operations directly

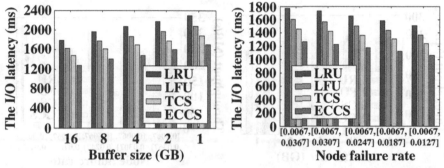

(a) The I/O latency of read operations under different memory buffer sizes.

(b) The I/O latency of read operations under different node failure rates.

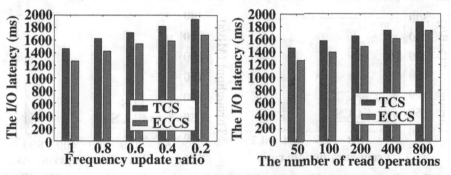

(c) The I/O latency of read operations under different frequency update ratios.

(d) The I/O latency of read operations under different replacement intervals.

Fig. 17. The I/O latency of read operations under different buffer sizes, node failure rates, frequency update ratios, and replacement intervals.

read objects from the memory divided by the total number of read requests. The time of read operations is measured by the duration of read operations. The throughput is the ratio obtained by dividing the size of read requests by the time of these read operations. The disk I/O overhead of read operations is obtained by the data volume that read from disks for each read operation. The time of replacement operations is measured by the average duration of replacement operations for each read operation.

We use RS code [29] to encode the data splits to generate parity splits. We set the default code parameters as $(k, m) = (3, 2)$, memory buffer size as 16 GB [20]. For TCS and ECCS, we set the default frequency update ratio to $\alpha = 1$, and replace cached splits after every 50 read operations [28]. According to the measurement of Facebook cluster, there are 30 to 110 failed nodes in 3000 storage nodes every day [6]. Accordingly, we set nodes failure rates as [0.0067, 0.0367], and use Facebook SWIM trace [27] to simulate the read requests. We run the trace for 100 times and average the results. These experimental results are shown in Figs. 16, 17, 18, 19, and 20. We elaborate our experimental studies for performance metrics and the overhead of caching approaches.

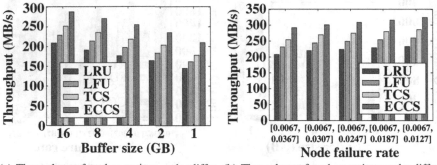

(a) Throughput of read operations under different memory buffer sizes.

(b) Throughput of read operations under different node failure rates.

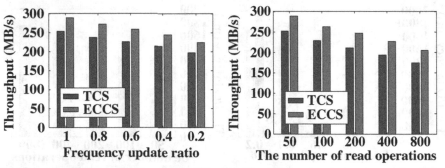

(c) Throughput of read operations under different frequency update ratios.

(d) Throughput of read operations under different replacement intervals.

Fig. 18. Throughput of read operations under different buffer sizes, node failure rates, frequency update ratios, and replacement intervals.

8.2   Performance Analysis of Caching Approaches

We compare the performance of read operations under different caching approaches, and analyze experimental results as follows.

First, we compare the performance of read operations under different memory buffer sizes 16 GB, 8 GB, 4 GB, 2 GB, 1 GB, and distributions of node failure rates [0.0067, 0.0367], [0.0067, 0.0307], [0.0067, 0.0247], [0.0067, 0.0187], and [0.0067, 0.0127]. These experimental results are shown in Figs. 16(a), (b), 17(a), (b) and 18(a), (b). In short, ECCS improve the performance of read operations in practice. Specifically, ECCS can improve the hit ratios by up to 65% compared with LRU, 53% compared with LFU, and 27% compared with TCS, reduce the time of read operations by up to 32% compared with LRU, 29% compared with LFU, and 21% compared with TCS, and improve the throughput of read operations by up to 48% compared with LRU, 32% compared with LFU, and 24% compared with TCS.

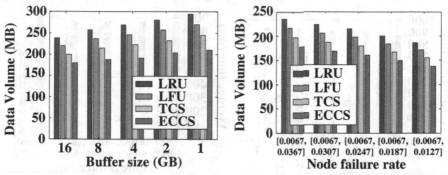

(a) The I/O overhead under different memory buffer sizes.

(b) The I/O overhead under different node failure rates.

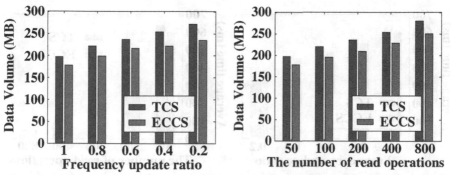

(c) The I/O overhead under different frequency update ratios.

(d) The I/O overhead under different replacement intervals.

Fig. 19. The I/O overhead of read operations under different buffer sizes, node failure rates, frequency update ratios, and replacement intervals.

Then, we test the performance of read operations under different frequency update ratios and intervals of replacement operations. These experimental results are shown in Figs. 16(c), (d), 17(c), (d) and 18(c), (d). Since both TCS and ECCS need current read frequency to cache the splits, they can cache correct splits to achieve the high performance of read operations when the frequency update ratio and replacement interval are small. In practice, the storage systems can modify these parameters to improve the performance of read operations. Because these parameters also affect the overhead of caching approaches, we compare the overhead of different caching approaches.

8.3 The Overhead Analysis of Caching Approaches

Since caching approaches incur the overhead of storage systems, we analyze how such overhead affects the efficiency of read operations. We summarize these overhead of different caching approaches as follows.

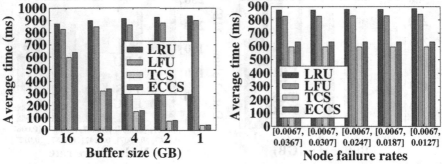

(a) Time of replacement operations under different memory buffer sizes.

(b) Time of replacement operations under different node failure rates.

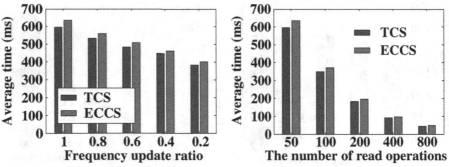

(c) Time of replacement operations under different frequency update ratios.

(d) Time of replacement operations under different replacement intervals.

Fig. 20. Time of replacement operations under different buffer sizes, node failure rates, frequency update ratios, and replacement intervals.

First, ECCS can efficiently reduce the I/O overhead of read operations compared with current caching approaches. These results are shown in Fig. 19. Specifically, ECCS can reduce the I/O overhead of read operations by up to 30% compared with LRU, 28% compared with LFU, and 25% compared with TCS. Since the disk I/O overhead severely affects the I/O latency of read operations, ECCS can improve the performance of read operations by reading the splits from the memory.

Moreover, we compare the time of replacement operations under different application scenarios. These experimental results are shown in Fig. 20. Since LRU and LFU execute the replacement operation after each read operation, such caching approaches suffer from high overhead when replacing the cached splits. Although both TCS and ECCS need to update the cached splits in the memory during the process of replacement operations, the average time of replacement operations for each read operation is not quite large, since these approaches execute replacement operations after several read operations. When buffer size is small, the storage nodes need to replace less splits in the memory and hit ratios of read operations are small. Accordingly, the storage nodes read more splits from disks when they are not cached in the memory.

9 Conclusion

Erasure codes are widely used as a viable means to ensure the data reliability in key-value storage systems. They encode objects to several splits and store these splits in multiple storage nodes, so that the application servers can use some splits to reconstruct original data. Since the storage nodes need to read splits from disks if they failed to cache these data in the memory, the application servers suffer from the high I/O latency when reading stored objects. In this paper, we consider how to improve the performance of read operations by caching the data in the memory. We propose an efficient caching scheme, namely ECCS. Through theoretical analysis, ECCS can effectively reduce the I/O latency of read operations compared with current caching schemes. We implement a prototype storage system and conduct extensive experiments with real-world storage cluster and traces. The experimental results show that ECCS can reduce the time of read operations by up to 32% and improve throughput by 48% compared with the state-of-the-art approaches, i.e., LRU, LFU, and TCS.

References

1. Weil, S.A., Brandt, S.A., Miller, E.L., et al.: Ceph: a scalable, high-performance distributed file system. In: Proceedings of Symposium on Operating Systems Design and Implementation, OSDI 2006, pp. 307–320 (2006)
2. Dean, J., Ghemawat, S.: MapReduce: simplified data processing on large clusters. Commun. ACM 51(1), 107–113 (2008)
3. Rashmi, K., Chowdhury, M., Kosaian, J., et al.: EC-Cache: load-balanced, low-latency cluster caching with online erasure coding. In: Proceedings of the 12th USENIX Symposium on Operating Systems Design and Implementation, OSDI 2016, pp. 401–417 (2016)

4. Li, J., Li, B.: On data parallelism of erasure coding in distributed storage systems. In: Proceedings of IEEE International Conference on Distributed Computing Systems, ICDCS 2017, pp. 1–12 (2017)
5. Li, H., Ghodsi, A., Zaharia, M., et al.: Tachyon: reliable, memory speed storage for cluster computing frameworks. In: Proceedings of the ACM Symposium on Cloud Computing, SoCC 2014, pp. 1–15 (2014)
6. Sathiamoorthy, M., et al.: XORing elephants: novel erasure codes for big data. In: Proceedings of IEEE Conference on International Conference on Very Large Data Bases, VLDB 2013, pp. 325–336 (2013)
7. Javadi, B., Kondo, D., Iosup, A., Epema, D.: The failure trace archive: enabling the comparison of failure measurements and models of distributed systems. J. Parallel Distrib. Comput. **73**(8), 1208–1223 (2013)
8. Yan, J., Zhu, Y.L., Xiong, H., et al.: A design of metadata server cluster in large distributed object-based storage. In: Proceedings of IEEE Conference on MASS Storage Systems and Technologies, MSST 2004, pp. 199–205 (2004)
9. Dragojević, A., Narayanan, D., Castro, M., Hodson, O.: FaRM: fast remote memory. In: Proceedings of the 11th USENIX Symposium on Networked Systems Design and Implementation, NSDI 2014, pp. 401–414 (2014)
10. Nishtala, R., Fugal, H., Grimm, S., et al.: Scaling memcache at Facebook. In: Proceedings of the 10th USENIX Symposium on Networked Systems Design and Implementation, NSDI 2013, pp. 385–398 (2013)
11. Li, B., Ruan, Z., Xiao, W., et al.: KV-Direct: high-performance in-memory key-value store with programmable NIC. In: Proceedings of ACM Symposium on Operating Systems Principles, SOSP 2017, pp. 137–152 (2017)
12. Kaminsky, M., Andersen, D., Lim, H.: MICA: a holistic approach to fast in-memory key-value storage, pp. 429–444 (2014)
13. Ananthanarayanan, G., et al.: Scarlett: coping with skewed content popularity in MapReduce clusters. In: Proceedings of the 6th European Conference on Computer Systems, EUROSYS 2011, pp. 287–300 (2011)
14. Amazon S3 storage, 14 September 2017. http://aws.amazon.com/s3
15. OpenStack Swift, 14 September 2017. http://swift.openstack.org
16. Calder, B., Wang, J., Ogus, A., et al.: Windows Azure Storage: a highly available cloud storage service with strong consistency. In: Proceedings of the 23rd ACM Symposium on Operating Systems Principles, pp. 143–157 (2011)
17. Reed, I.S., Solomon, G.: Polynomial codes over certain finite fields. J. Soc. Ind. Appl. Math. **8**(2), 300–304 (1960)
18. Huang, C., Simitci, H., Xu, Y., et al.: Erasure coding in Windows Azure Storage. In: Proceedings of the USENIX Annual Technical Conference, ATC 2012, pp. 15–26 (2012)
19. Dimakis, A.G., Godfrey, P.B., Wu, Y., et al.: Network coding for distributed storage systems. IEEE Trans. Inf. Theory **56**(9), 4539–4551 (2010)
20. Zhang, H., Dong, M., Chen, H.: Efficient and available in-memory KV-store with hybrid erasure coding and replication. In: Proceedings of the 14th USENIX Conference on File and Storage Technologies, FAST 2016, pp. 167–180 (2016)
21. Rashmi, K.V., Shah, N.B., Gu, D., et al.: A "Hitchhiker's" guide to fast and efficient data reconstruction in erasure-coded data centers. In: Proceedings of the Annual Conference on ACM Special Interest Group on Data Communication, SIGCOMM 2014, pp. 331–342 (2014)

22. Li, M., Lee, P.P.: STAIR codes: a general family of erasure codes for tolerating device and sector failures in practical storage systems. In: Proceedings of the 12th USENIX Conference on File and Storage Technologies, FAST 2014, pp. 147–162 (2014)
23. Yiu, M.M.T., Chan, H.H.W., Lee, P.P.C.: Erasure coding for small objects in in-memory KV storage. In: Proceedings of ACM International Systems and Storage Conference, SYSTOR 2017, pp. 1–12 (2017)
24. Li, S., Zhang, Q., Yang, Z., Dai, Y.: BCStore: bandwidth-efficient in-memory KV-store with batch coding. In: Proceedings of International Conference on Massive Storage Systems and Technology, MSST 2017, pp. 1–13 (2017)
25. Halalai, R., Felber, P., Kermarrec, A.M., Taiani, F.: Agar: a caching system for erasure-coded data. In: Proceedings of IEEE International Conference on Distributed Computing Systems, ICDCS 2017, pp. 23–33 (2017)
26. Li, R., Lin, J., Lee, P.P.: CORE: augmenting regenerating-coding-based recovery for single and concurrent failures in distributed storage systems. In: Proceedings of IEEE Conference on Mass Storage Systems and Technologies, MSST, pp. 1–6. IEEE (2013)
27. Facebook SWIM traces, 17 June 2016. https://github.com/SWIMProjectUCB/SWIM/wiki/Workloads-repository
28. Li, J., Li, B.: Zebra: demand-aware erasure coding for distributed storage systems. In: Proceedings of IEEE Symposium on Quality of Services, IWQoS 2016, pp. 1–10 (2016)
29. Plank, J.S., Luo, J., Schuman, C.D., Xu, L., Wilcox-O'Hearn, Z.: A performance evaluation and examination of open-source erasure coding libraries for storage. In: Proceedings of the 7th USENIX Conference on File and Storage Technologies, FAST 2009, pp. 253–265 (2009)
30. Ananthanarayanan, G., Ghodsi, A., Borthakur, D., et al.: PACMan: coordinated memory caching for parallel jobs. In: Proceedings of the 9th USENIX Symposium on Networked Systems Design and Implementation, NSDI 2012, pp. 1–20 (2012)
31. Manes, B.: Caffeine: a high performance caching library for JAVA 8 (2016). https://github.com/benmanes/caffeine
32. Einziger, G., Friedman, R.: TinyLFU: a highly efficient cache admission policy. In: Proceedings of Euromicro International Conference on Parallel, Distributed and Network-Based Processing, PDP 2014, pp. 146–153 (2014)
33. Xia, M., Saxena, M., Blaum, M., Pease, D.A.: A tale of two erasure codes in HDFS. In: Proceedings of the 13th USENIX Conference on File and Storage Technologies, FAST 2015, pp. 213–226 (2015)
34. Cherkasova, L.: Improving WWW proxies performance with greedy-dual-size-frequency caching policy. Hp Technical report (1998)
35. Dan, A., Towsley, D.: An approximate analysis of the LRU and FIFO buffer replacement schemes. ACM SIGMETRICS Perform. Eval. Rev. 18(1), 143–152 (1990)
36. Robinson, J.T., Devarakonda, M.V.: Data cache management using frequency-based replacement. In: Proceedings of ACM Conference on Measurement and Modeling of Computer Systems, SIGMETRICS 1990, pp. 134–142 (1990)

Data Quality Control and Data Governance

A Task-Driven Reconfigurable Heterogeneous Computing Platform for Big Data Computing

Pengfei Yang (ID), Quan Wang(✉) (ID), Peiheng Zhang, Zhike Wang, Lu Fan, and Caihong Huang

School of Computer Science and Technology, Xidian University, Xi'an 710126, China
qwang@xidian.edu.cn

Abstract. Big data computing and analysis can uncover hidden patterns, correlations and other insights by examining large amounts of data. Comparing with the traditional processor, the new types of processors, just like digital signal processor (DSP), Field Programmable Gate Array (FPGA), graphics processing unit (GPU), could improve the speed of data analysis significantly. Heterogeneous multicores systems have become the primary architecture as devices are tasked to do more complicated functions faster. While, in most cases, these heterogeneous resources cannot be utilized sufficiently because the system software is provided by vendors, loaded pre-sale and doesn't change. The cloud computing offers the capability of distributing infrastructures according to the requirements. We build a cloud-like heterogeneous computing platform which including PowerPC, DSP, GPU and FPGA. A task-driven dynamic loading scheme is proposed by making use of the virtualization and middleware technologies. The system can manage the entire lives of allocating, loading, using, and recovering. Taking this as a guide, a private cloud principle verification system including web application layer, main control layer, and computing service layer is designed and verified, which proves the feasibility of the computing platform. According to the test results of web system, the platform can well meet the design intention of acquiring the computing resources according to the task requirements.

Keywords: Task-driven · Heterogeneous system · Big data · IoT

1 Introduction

The traditional embedded systems were dominated by hardware constraints, they used to be relatively simple and less consider the issues of computing speed and efficiency. The problems such as Storage walls, Power walls and Algorithm walls restrict the improvement of the system performance. To solve this problem, on the one hand, many dedicated embedded devices are manufactured to increase the compute efficiency of missions in specific areas significantly. Such

© Springer Nature Singapore Pte Ltd. 2018
Z. Xu et al. (Eds.): Big Data 2018, CCIS 945, pp. 543–557, 2018.
https://doi.org/10.1007/978-981-13-2922-7_35

as DSP owns powerful floating-point computing capability for its internal hardware multipliers and particular instruction sets [1], FPGA has high algorithm parallelism because of its property of hardware-level parallel [2–5]. On the other hand, more simple and flexible interconnection methods, just like optical fibres, configurable buses and network on the chip are designed, make it possible to achieve free combination and efficient communication between multi/many different resources [6,7]. With the development of these related technologies, most embedded systems are taking on the characteristics of the general-purpose computing system. The performance of the embedded system is so high that in most cases, it only needs little part of resources could meet the requirement of task execution. Therefore, improving the resource utilization of embedded systems becomes a new research focus.

At the same time, cloud computing has emerged as a popular computing model. Cloud computing has sufficient resources all the time for its clients as resource pools, efficiently and dynamically allocates or deallocates these resources, is considered suitable for its clients. It must be efficient if supervise and schedule the abundant embedded resources using cloud computing model [8,9].

"Virtualization" is a most important and widely used technology in cloud computing. It takes a bounded set of system hardware and makes them act like multiple virtual environments. For heterogeneous resources such as DSPs and FPGA that can significantly improve system performance, we can use virtualization management technology and middleware technology to connect these independently embedded devices through a network or "cloud", make them a computing resource similar to a virtual machine, implemented according to user needs. Dynamic reconfiguration and flexible capacity expansion will further leverage the performance of these high-performance heterogeneous hardware and enhance the system scalability [10–12].

This paper proposes a task-driven, high-performance heterogeneous dynamic loading program, and builds their private cloud for prototype verification. The results show that the embedded computing platform has the capacity of reconstruction and expansion, the performance of DSP and FPGA is fully utilized and has significant flexibility and efficiency improvement over traditional methods.

The rest of this paper is organized as follows. The system architectures of hardware and software are introduced in Sect. 2. Section 3 describes detailly the implementation of dynamic loading for two kinds of heterogeneous resources, DSP and FPGA. A web system test scenario is presented in Sect. 4. Finally, a summary of related work is made in Sect. 5.

2   System Architecture

The traditional system architecture of embedded computing platform is shown in Fig. 1(a). Heterogeneous computing resources just like DSP and FPGA make up the underlying hardware layer. The customized program is developed according to user's specific requirements. After debugging and verification, the program is solidified into the chip and can't be changed. This architecture has many draw-backs: firstly, a special development always need to be made by programmers for different tasks and data input due to the tight integration of software and hard-ware, this increases the difficulty of system development. Secondly, the system has poor versatility and openness. It is difficult for third-party software to be ported to hardware systems. It is also tricky to adjust system functions dynam-ically and cannot meet the requirements for higher complexity control, resource management, and multitasking. Last but not least, the overall system load is low for it always runs specific tasks. In most scenarios, the high-performance com-puting components in the system can't adequately exert their own performance, which also results in a waste of computing power of the platform.

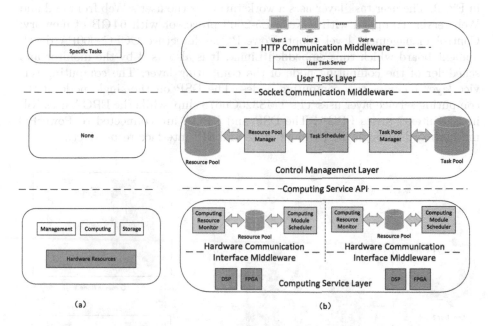

Fig. 1. System architecture comparison

The flexible expansion and dynamic allocation features of cloud computing can improve the efficiency of server usage and enhance the computing power of the system. Based on the concept of hierarchical organization and central-ized scheduling in the cloud platform, the overall architecture of the optimized platform is shown in Fig. 1(b). On the basis of the original system, the control management layer is added, and the system functions is further refined: The user task layer is responsible for collecting the task requests submitted by the terminal users. The control management layer is responsible for maintaining the virtual resource pool of the underlying computing resources and schedul-ing the user task requests. The computing service layer is responsible for the specific virtualization and dynamic loading of the underlying hardware. According to the scheduling result of the control management layer, it is responsible for delivering the specific task data and task code to the DSP or FPGA and realising the monitoring and management of the entire lifecycle of the distribution, loading, claiming, and registering/updating of the heterogeneous hardware resources. The middleware technology is used to make communication between different layers. The lower layer services only provide service interfaces to the upper layer applications, shielding specific implementation details of the lower layer services, and sending messages to the upper layer applications to notify service execution status.

Regarding the specific hardware specifications of the private cloud computing platform used in this paper, the specific hardware architecture is built as shown in Fig. 2. The user task layer uses a workstation as the user's Web front end and Web server, equipped with a Xeon E5-2603 processor with 64 GB of memory; Control management level adopts PowerPC architecture CCFC9000PA development board which runs embedded Linux. It is also used by the monitor and scheduler of the computing node of the computing layer. The computing service layer includes two computing nodes. The DSP on the single node of the computing service layer uses TI's TMS320C6678 chip, while the FPGA uses Xilinx's Kintex-7 series FPGA. The DSP and FPGA are connected to PowerPC motherboard through the PCIe interface and SPI interface respectively.

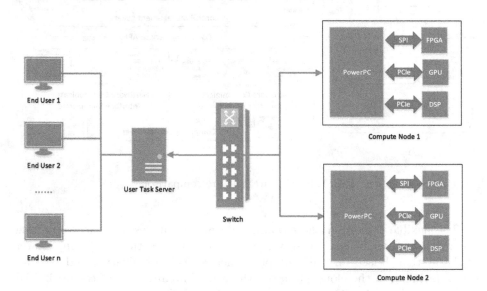

Fig. 2. Hardware constitution

For the whole system, socket communication middleware is responsible for the communication between the user task server and the control management layer. By creating two processes at the client and the server respectively, both parties

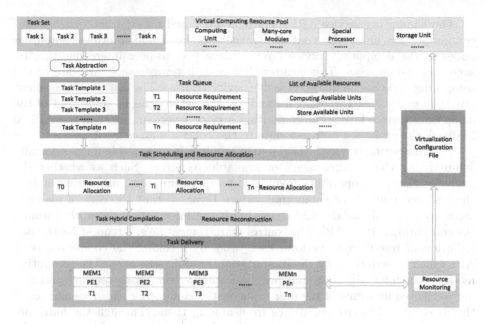

Fig. 3. Software solution

can ensure that anyone can initiate communication requests autonomously. At the same time, IO multiplexing technology ensures that the client process and server process at one end can responses timely after receiving data. The middleware automatically completes the encapsulation and decapsulation of user-defined protocols at both parties, and at the same time guarantees the stability of the communication function through means such as breakpoints retransmission.

The control management layer listens for the socket connection request of the user task layer, calls the socket communication middleware interface to decapsulate the user task information, and writes the task pool. At the same time, according to the current status of the hardware resource pool, the user task request is scheduled, and the scheduling result is fed back to the user task layer through the Socket communication middleware, so that, it can update the deployment status of the user task. After the user receives the task deployment status and confirms it is correct, the task is delivered and executed. The control management layer calls the computing API interface provided by the calculation service layer to perform the allocation of the heterogeneous hardware resources, the online loading of the user task code to the hardware, the execution of the user task, and the execution status report after the task is completed.

The resource pool stores the status information and access addresses of the computing resources. The information is maintained and updated by the resource pool manager. The task pool, which maintained by the task pool manager, stores user task requests, task images, and data. At the same time, the task pool manager is responsible for updating the status of user tasks according to the task execution status.

The computing service API is constructed based on ethernet switches and communication links. The control management layer and multiple computing nodes in the computing service layer achieve peer-to-peer interconnection by accessing the switches. This ensures that each heterogeneous hardware of the computing node has unique addressable device address. The control management layer can call the API to apply for and use the heterogeneous equipment of the computing service layer. The computing service layer can also call the API to feedback the processed user task data to the control management layer.

The computing service layer resource monitor is responsible for periodically summarizing the resource usage of available resources. Such as whether the device is occupied, operating frequency, temperature, etc., and writes it into the resource pool of the computing service layer, so that the control management layer can obtain the underlying hardware information promptly through the computing service API. The control management layer's request for the distribution of resources to perform user tasks by invoking the compute service API which is performed by the computation module scheduler. The computing module scheduler will filter and sort the available computing resources according to the requirements of the user tasks on the computing resources and select the most suitable hardware device from among them. Through the hardware communication interface middleware, the selected device is soft reset to wake the sleep device or reset the running device. Then, the user task code and data are loaded into the hardware device memory via the hardware communication interface middleware, and the soft reset state of the device is released after the loading is completed, and the hardware device is controlled to jump to the entry address of the user code to perform the user task. After the task is completed, the hardware device notifies the computing module scheduler with an external interrupt. The scheduler uses the hardware communication interface middleware to obtain the processed user data, release the device occupied by the task, and control the device to enter a low-power sleep state. The scheduler writes the processed user data to the resource pool and updates the state of the device to idle. At the same time, the scheduler uses the calculation service API to inform the control management layer's task pool manager to read the processed user data. The overall software architecture shown in Fig. 3 resolves the software application flow of the platform.

3   Dynamic Loading of Heterogeneous Resources

The core computing components of the computing platform are DSPs and FPGAs. In order to make both of them similar to the virtual machines in the cloud computing architecture that can be dynamical allocated and flexibly expanded according to user requirements, the key is to improve the development of these two devices from the traditionally loaded pre-sale to online task-driven load. In other words, to achieve the dynamic loading of heterogeneous resources. The following describes the DSP and FPGA dynamic loading implementation.

3.1 DSP Dynamic Loading

The traditional DSP development is written in the CCS integrated development environment provided by Texas Instruments. After debugging, it is downloaded to the DSP's internal FLASH through the simulator. After the DSP starts up, it takes the instruction from the FLASH by default and runs the fixed user task code. The online loading means that the DSP is used as a general-purpose computing device whose running code is specified by the user. With its PCIe bootstrap loader mode, the task code is loaded into the DSP's memory via the PCIe interface and the loaded user program is executed.

The online loading of TMS320C6678 DSP used in the computing platform of this paper includes DSP end boot loading and Linux driver modification. (unless otherwise specified, the DSP refers to the TMS320C6678 DSP). For boot loading, the 6678 supports multiple boot modes including I2C, SPI, EMIF, SRIO, EMAC, and PCIe. It is determined by the BOOTMODE[2:0] pins. After DSP power-on reset, it will sample the level of these three pins. The bootloader selects the corresponding boot mode based on the sampled value. When BOOTMODE[2:0]= 0b100, the DSP selects the PCIe boot mode which the host (Compute node's PowerPC) acts as the PCIe master, and the DSP serves as a slave. The host loads the user code into the DSP memory space through the PCIe bus to implement the boot loading process of the DSP.

Specifically, after the DSP is powered on or reset in the PCIe boot mode, the Bootloader will first initialize the DSP's associated registers and leave it in the IDEL state. The bootloader will control the boot flow based on the data in the DSP L2 SRAM (address space 0x0087 2DC0-0x0087 FFFF). The most critical register is the Boot magic address field located in the last 4 bytes of the L2 SRAM space, which is used to store the first address of applications. The bootloader determines the DSP state by checking whether this field is 0. If it is 0, the DSP is in the IDEL state; if it is not, it jumps to this address to execute the user code.

After the program is loaded, the Boot magic address is set to the first address of the loaded program in the DSP memory space. The host sends the MSI interrupt to the DSP. After the DSP receives the interrupt, it executes the interrupt program. The interrupt program is responsible for detecting whether the Boot magic address field is 0 and controlling the DSP to jump to the corresponding address when it is not. This completes the boot loading of PCIe.

The online loading of the DSP can be achieved through the PCIe bootstrap loading mode, but it also has specific requirements for the format of the user program. The .out file, which is generated by CCS compilation in the traditional mode, contains some positioning symbols and header information. The information is used to instruct the emulator to extract a useful program from the COFF file and load it into the L2 SRAM of the DSP. However, if the PCIe bootstrap loading method is adopted, the emulator instruction information contained in the COFF file is invalid data, and it must be filtered to generate a binary file that we call the firmware before it can be used as a PCIe loader.

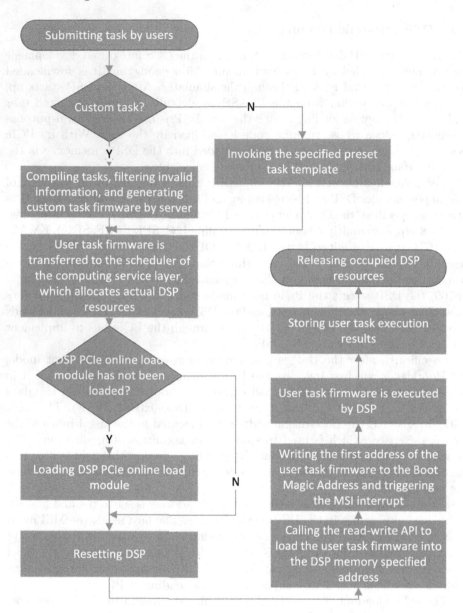

Fig. 4. Flowchart of DSP dynamic loading

For Linux, the PCIe device is a character device. We compile the code that loads the user program through the PCIe interface into a Linux driver module, which is one of the core components of the hardware communication interface middleware of the computing service layer. Based on the analysis above, the flow from the user submitting the task to the DSP is shown in Fig. 4.

3.2 FPGA Dynamic Loading

The dynamic loading principle of a computing node with an FPGA device is shown in Fig. 5. The FPGA computing module consists of PowerPC, CPLD and FPGA. PowerPC takes charge of hardware tasks transferring into FPGA source point. CPLD take charge of the communications between PowerPC and FPGA. FPGA take charge of performing computing tasks.

The PowerPC is connected to the switch of the underlying communication link of the computing service API through the Ethernet interface. It can communicate to CPLD with IIC. Several IO pins of the PowerPC are also directly connected to the FPGA as configuration pins for controlling the startup mode of the FPGA. CPLD connect with FPGA with IIC and SPI, and the connection consists of image line and task data line.

Among them, CPLD is used for changing the sending sequence. It converts the baud rate of MCU and baud rate of FGPA mutually, guarantee the communication between them.

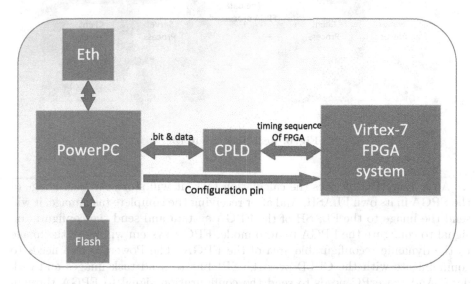

Fig. 5. Architecture of FPGA dynamic loading

When the PowerPC resets the FPGA through the control pin and reloads the user program into the FPGA, the PowerPC transfers the user program to the CPLD. The CPLD is used to generate the FPGA code writing sequence, and then erase the existing logic of the FPGA and load the user-specified task code.

The FPGA resources are divided into static parts and dynamically reconfigurable areas. Dynamically reconfigurable areas are provided for users. The tasks performed in the FPGA can be regarded as hardware processes, and the static area is responsible for I/O management and memory access management of the hardware processes. In cooperation with other resources of the user, the data needs to be sent to the PowerPC through the static area, further to be sent to the bottom device management module of the control management layer, and to be forwarded to the destination device.

As shown in Fig. 6, the network module in PowerPC contains socket-based server processes and client processes. The PowerPC needs to receive the task information sent from the control management node and send data to other nodes to work in harmony.

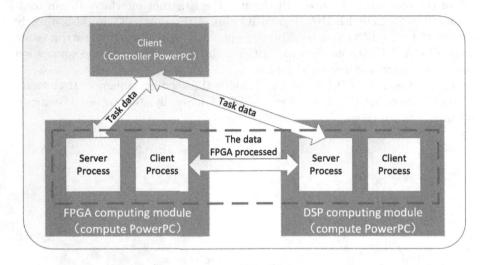

Fig. 6. Real-time monitoring

After PowerPC receives the calculation task, it will store the task image of the FPGA in its own FLASH. And after receiving the complete task image, it will send the image to the FLASH of the FPGA system and send the configuration signal to configure the FPGA to auto mode. FPGA system will write the image to the dynamic reconfigurable area of the FPGA. The PowerPC also needs to communicate with the CPLD over the SPI bus to send task images and task data. And PowerPC needs to send the configuration signal to FPGA through GPIO interface, to take charge of the work of FPGA.

4   Web System Testing Solution Based on User's View

According to the system software and hardware architecture and task-driven DSP and FPGA online loading scheme proposed above, the corresponding computing platform is built to verify the prototype. The architecture proposed in this paper is based on the use of the cloud platform. Therefore, we develop our own web application at the user task level and then call the functions of the control management layer and the computing service layer to achieve access to the platform's computing resources. The following gives the overall test plan and results of the system from the perspective of the user.

The use of computing platforms by users can be divided into five parts: real-time monitoring, task deployment, task delivery, results viewing, and system logging. We set out from these five parts to test the overall functionality of the system.

Real-time monitoring data comes from the virtual resource pool of the control management layer, which is updated by the control management layer periodically calling the computing service API. The computing platform consists of two DSP development boards and two FPGA development boards. The platform can display its working status in real time: whether it is idle, occupied or faulty, furthermore, the operating frequency of the control management's master PowerPC. The real-time monitoring interface is shown in Fig. 7.

During the task deployment phase, the user can select the specified task execution from the system-preset DSP, FPGA task templates, and specify the task attributes and task data. In this stage, the user task layer collects the user task request and sends it to the control management layer through the Socket communication middleware. The control management layer writes the user task request to the task pool, and schedules the user request according to the current virtual resource pool status, and returns the scheduling result to the user. Figure 8 shows the use of the task deployment phase.

In the delivery phase of the task, the user can optionally perform the submitted task, which is based on the control management's scheduling of the user task requests. As for task delivery, concurrency mechanism is supported, if the user currently has multiple scheduled tasks, the user can choose to perform a single specified task, or execute all tasks concurrently. After the user selects the execution task, the control management layer receives the user request, calls the computing service API to allocate the DSP and FPGA computing instances, and loads the user code on the DSP and the FPGA for execution. Occupancy of hardware resources during task execution is displayed in real-time on the right hardware topology. Hardware resources occupied by different tasks are identified in different colours. After the task is completed, the compute node will invoke the computing service API to notify the control management layer to read the processed user data and update the task status. These results will also be displayed on the interface. Figure 9 shows the task delivery interface.

In the result viewing stage, the task pool manager of the control management layer is responsible for retrieving all tasks belonging to the current user that have been executed in the task pool. The manager also needs to obtain the paths of the original data files and the processed data files of these tasks, compare and display the results of these tasks on the page. Figure 10 shows the result viewing page.

The system logging retrieves all the data related to the current user in the task pool, cascades and displays the user's operation behaviour on the platform. If it is an administrator authority user, it cascades and displays the operation behaviour of all the authenticated users on the platform so that the administrator performs user detection and problem tracing. Figure 11 shows the system logging.

Through user-view testing, the computing platform meets expectations and can support task-driven reconfigurable computing.

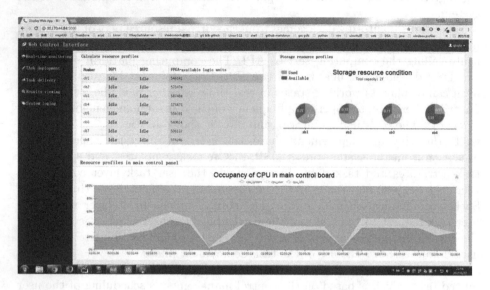

Fig. 7. Real-time monitoring

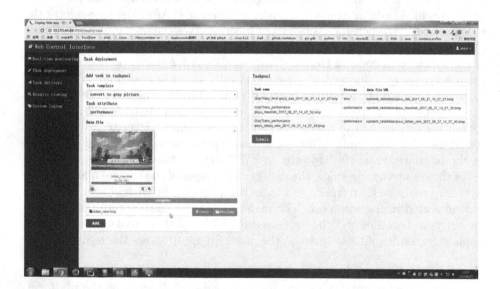

Fig. 8. Task deployment

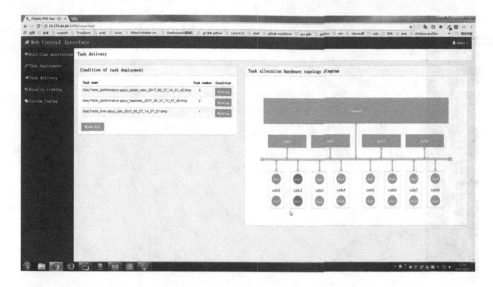

Fig. 9. Task delivery

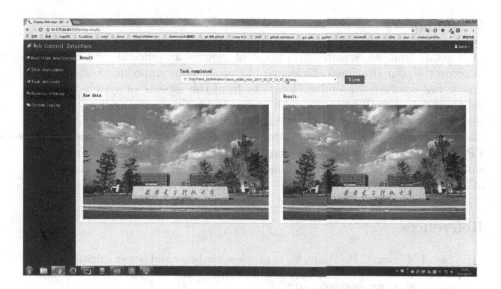

Fig. 10. Results viewing

Fig. 11. System logging

5 Summary

Optimizing from an architectural point of view is an efficient way to further improve the computational efficiency of heterogeneous embedded systems. This article has conducted useful explorations from the aspects of system hardware configuration, overall software architecture, and task processing flow, and completed the establishment of the verification system to verify the principle. In the next phase, GPUs will be introduced into the underlying resources, and the FPGA granularity will reach the intra-chip IP core level. At the same time, consider the introduction of intelligent methods for the management and scheduling of system task resources.References

References

1. Chen, J., Chang, C.H., Wang, Y., et al.: New hardware and power efficient sporadic logarithmic shifters for DSP applications. IEEE Trans. Comput.-Aided Des. Integr. Circuits Syst. **37**, 896–900 (2017)
2. Guo, K., Sui, L., Qiu, J., et al.: Angel-Eye: a complete design flow for mapping CNN onto embedded FPGA. IEEE Trans. Comput.-Aided Des. Integr. Circuits Syst. **37**(1), 35–47 (2018)
3. Angizi, S., He, Z., DeMara, R.F., et al.: Composite spintronic accuracy-configurable adder for low power digital signal processing. In: 2017 18th International Symposium on Quality Electronic Design (ISQED), vol. 2017, pp. 391–396. IEEE (2017)
4. Yang, P., Wang, Q., Zhang, J.: Parallel design and implementation of error diffusion algorithm and IP core for FPGA. Multimed. Tools Appl. **75**(8), 4723–4733 (2016)

5. Wang, C., Gong, L., Yu, Q., et al.: DLAU: a scalable deep learning accelerator unit on FPGA. IEEE Trans. Comput.-Aided Des. Integr. Circuits Syst. **36**, 513–517 (2017)

6. Adegbija, T., Rogacs, A., Patel, C., et al.: Microprocessor optimizations for the internet of things: a survey. IEEE Trans. Comput.-Aided Des. Integr. Circuits Syst. **37**(1), 7–20 (2018)

7. Yang, P., Wang, Q.: Heterogeneous honeycomb-like NoC topology and routing based on communication division. Int. J. Futur. Gener. Commun. Netw. **8**, 19–26 (2015)

8. Hashem, I.A.T., Yaqoob, I., Anuar, N.B.: The rise of "big data" on cloud computing: review and open research issues. Inf. Syst. **47**, 98–115 (2015)

9. Botta, A., Donato, W.D., Persico, V., et al.: Integration of cloud computing and internet of things: a survey. Futur. Gener. Comput. Syst. **56**, 684–700 (2016)

10. Tian, K., Dong, Y., Cowperthwaite, D.: A full GPU virtualization solution with mediated pass-through. In: USENIX Annual Technical Conference, vol. 2014, pp. 121–132 (2014)

11. Chen, F., Shan, Y., Zhang, Y., et al.: Enabling FPGAs in the cloud. In: Proceedings of the 11th ACM Conference on Computing Frontiers, vol. 2014, p. 3. ACM (2014)

12. Tarafdar, N., Lin, T., Fukuda, E., et al.: Enabling flexible network FPGA clusters in a heterogeneous cloud data center. In: Proceedings of the 2017 ACM/SIGDA International Symposium on Field-Programmable Gate Arrays, vol. 2017, pp. 237–246. ACM (2017)

Deduplication with Blockchain for Secure Cloud Storage

Jingyi Li[1], Jigang Wu[1,2(✉)], Long Chen[1], and Jiaxing Li[1]

[1] Guangdong University of Technology, Guangzhou, China
[2] Guangdong Key Laboratory of Big Data Analysis and Processing,
Guangzhou 510006, China
asjgwucn@outlook.com

Abstract. Secure deduplication techniques have been wildly used in cloud storage to save both disk space and network bandwidth. However, traditional schemes store only one copy in cloud which raises problems with data reliability. Meanwhile, the existence of any third party, which is adopted by many deduplication schemes, will bring the risk of being single-pointed. In order to solve the above problem, this paper proposes a deduplication scheme which distributes files to multi servers and records storage informations on blockchain. We present smart contract based protocols without the participants of the central authorities to provide secure deduplication. With safety analysis, we demonstrate that our deduplication scheme is secure and reliable under the definitions specified in the proposed security model with acceptable performance through simulation experiments.

Keywords: Deduplication · Blockchain · Cloud · Security

1 Introduction

Cloud storage has become popular in recent years give the credit to its cross platform, high scalability, high availability and almost unlimited capacity of data outsource service [5]. It is reported that the data in the network is growing fast and it is predicted to 10.4 zettabytes in 2019 while 83% of them will come from the cloud [1]. However, the explosive growth of data has caused problems such as insufficient bandwidth and tight storage space. In this case, data deduplication techniques are demanded to save disk space and bandwidth by storing a single copy of duplicate data in cloud.

But applying deduplication to cloud storage scenarios faces many challenges and difficulties. The most important point is that, due to the different needs of the application scenario, the commonly used method of deduplication usually cannot guarantee the confidentiality of data. Users are difficult to confirm whether the files they outsource to the cloud will be used or tampered by others arbitrary. Conventional encryption schemes is useless to solve the security problem because identical data copies of different users will lead to different

Z. Xu et al. (Eds.): Big Data 2018, CCIS 945, pp. 558–570, 2018.
https://doi.org/10.1007/978-981-13-2922-7_36

ciphertexts when they encrypt them by their private keys and making deduplication impossible.

The first attempt on this challenge is convergent encryption [17]. By generating a secret key based on the plaintext, such encryption scheme will return the
identifiable and coalescing supporting ciphertext. On the basis of this method,
the protection of metadata is concerned to improve the security in paper [17]
and the confidentiality of file auditing is also discussed in [12].

In spite of this, most of the secure deduplication schemes only consider the
cloud servers as a single one and ignore the trade off between data confidentiality
and reliability. With these schemes, it is expected to result in a negative effect on
a large number of users who store the same file when the only file stored in the
cloud is lost. Some survey considers providing higher reliability in deduplication
process is necessary. Here is an example that Li et al. [10] proposed a series of
schemes with the utility of secret sharing technique. But most of such deduplication schemes are relying on trusted third parties to provide key management or
auditing service and they are threatened by single point of failure in this setting.

Blockchain has received great attention in recent years due to its high credibility, security and efficiency. Consensus mechanism is well used to make decentralized system with superior fault tolerance. With such characteristic, blockchain
technique is suitable for constructing traceable distributed point-to-point transactions or collaboration systems. Decentralized storage systems, for example,
Filecoin [6] takes advantages of blockchain which can achieve security file storage
through an incentive mechanism and distributed construction. There are some
other research [8,9] pay attention on the efficiency of blockchain based storage
system and made some good results. However, existing similar systems use traditional distributed storage methods to achieve data reliability. By storing multiple
complete copies, they provide high fault tolerant but ignoring redundant waste of
system resources. In the case of distributed system mentioned above, the deduplication scheme will easily achieve the advatages of high reliability. But it must
tolerate node crashes, data loss, and even byzantine failures [2] at the same time.

To address these challenges, we present a cloud storage deduplication scheme
that can meet both security, reliability and integrity requirements. In the proposed construction, blockchain technique is utilized to protect system confidentiality and data integrity and which is also compatible with the distributed storage systems. In more details, file fingerprints will be included in the transaction
information and being showed in blockchain to provide trusted records. Every
data owner can make the duplicate check locally depend on these unchange-
able information. Our scheme not only reduces the communication frequency
between data owner and storage servers but also reduce the risk of communications being monitored and information leakage. Different from existing work, our
proposed scheme automatically perform deduplication tasks through smart contracts to ensure openness and credibility of the process towards all participants.
The information of files is displayed on blockchain and any eligible participant
can reconstruct the file with these information to gain remuneration. Our paper
has the following contributions.

- We have proposed a blockchain based distributed deduplication scheme to provide secure cloud storage with high reliability. File tags are published on blockchain and used for locally duplicate check or data integrity check. Combining the conventional encryption algorithm and an incentive mechanism, our construction can realize online missing file reconstruction through automatically executed smart contract.
- We demonstrate our scheme have high confidentiality and reliability defined in security analysis. Security models are presented to show the security goals. It is used to demonstrate that the outsource data can resist a certain degree of collusion attack or malicious tampering.
- Simulation results show that the proposed construction is efficient. Under the same degree of reliability, the redundancies of our scheme are optimized and comparable with the others.

2   Related Work

There are many previous researchs have discussed system models and schemes for reliable deduplication from many aspects. Meanwhile, distributed storage systems are gradually enabled to avoid single source download which will increase the risk of the attacking against server. It is worth noting that several decentralized storage schemes based on revenue incentives have proposed in these solutions to guarantee the higher efficiency and security of the distributed system.

2.1   Reliable Deduplication

Data deduplication techniques are wildly used in enterprise data storage service to save cost. As cloud storage services become popular, this technology has gained higher attention and broader application scenarios. At the same time, this technology faces more diverse challenges and difficulties like reliability under the premise of ensuring confidentiality. To address the reliability in deduplication, [11] showed how to achieve reliable key management in deduplication but it do not take into account the encrypted files. Paper [14] achieve reliability through the utility of ECC codes and distribute the encoded file among multi-servers. But this structure does not match the confidentiality requirement of cloud storage.

2.2   Decentralized Storage

Most cloud storage systems like Dropbox [4] and Google Drive [3] are the construction that map distributed heterogeneous storage devices to a single continuous storage space and managed by a centralized platform. In order to avoid being overly dependent on a single trusted entity, which is threatened by the single point of failure, blockchain based decentralized storage systems are concerned.

Blockchain is the underlying technique of Bitcoin, a distributed transaction ledger proposed in 2008 [15]. As the most important role, blockchain makes it possible to deal with the problem of double spending in a distributed network by generating a chained time-stamped data structure with transactions. Without the administration from the central manager, all the participants (miners) follow the consensus that only the longest chain is extended and the others are deemed invalid. The second generation of blockchain, such as Ethereum [18], uses the concept of smart contract to provide a general, programmable infrastructure. Smart contracts are deployed and running across the blockchain network which can express triggers, conditions, and even an entire business process automatically which brings more convenience to the application of blockchain technology. InterPlanetary File System(IPFS) [6] is a typical blockchain based decentralized storage platform which maintains the confidentiality of stored file by implementing an incentive mechanism to encourage servers provide better storage service. Platform developer claims that IPFS distributes all the files across the network and makes it accessible to all computers connected to the network at the same time. With bitcoin-blockchain-like block structure, the content is linked using hash key like referencer, that can be used to access them from client computers and all users will become miners with paid maintenance of this structure. But such systems keep their reliability with storing multiple copies of the same file across distributed servers. It brings about similar problems as traditional storage system that does not have deduplication scheme.

To achieve both higher security and reliability, we proposed a novel scheme combines the technology of secure deduplication and blockchain. In order to reduce space redundancy and communication consumption, deduplication technology is applied in blockchain-based storage systems On the other hand, to protect data from the risk of a single server, smart contract is fully utilized in the scheme.

3   Problem Formulation

3.1   System and Security Models

Our proposed scheme consists of two types of entities as shown in Fig. 1:

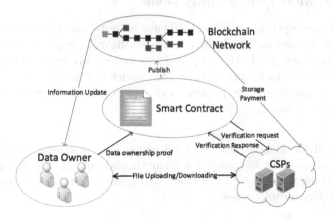

Fig. 1. System model

Cloud Service Provider (CSP) is an entity who provide data storage. The CSPs in network store only one copy of the same files or file blocks. We consider a series of CSPs to provide fault tolerance which means that the user data will be stored in multiple CSP servers in a distributed way.

Data owner upload their files to CSPs and accesses them in any time. Deduplication check is necessary to avoid duplicate data uploads. They also use a fault-tolerant method to encode the file before uploading for a higher level of reliability.

CSP and Data owner is ought to join the blockchain network as a node for related services. Each node can join or leave the blockchain network without any price and we assume that the blockchain is properly maintained. Unique private key is offered to every node which is used to generate transaction address to interact anonymously with each other. All the duplication and transaction information are recorded on blockchain to ensure that the data are authentic.

Notations. Table 1 Shows some notations used in our scheme. Among them, we will cover some methods about PoR [7] and deduplication scheme [10].

Table 1. System notations

Parameter	Description
f_h	A handle file to support the verification of PoR algorithm
c	The challenge parameter for PoR check
r	The response parameter for PoR check
(t, k, n)	The parameter of Ramp Secret Sharing Scheme
$Share()$	The algorithm to split files into shards
$Recover()$	The algorithm to restore the original file from file shards

3.2 Threat Model and Security Goals

As a generic assumption, we define CSPs are honest-but-curious. They prefer to follow designed protocols but desire to learn additional file information out of the scope of their privileges. In the proposed scheme, an adversary may plays a role of CSP. Collusion attacks among such adversaries are concerned in our security threat model. Specifically, we present brief security requirements as follows. For more detail, see paper [13].

Confidentiality. This requires that all users, i.e. those who do not hold the file, cannot obtain any meaningful information about the file. Specifically, no unauthorized users can get all the ciphertexts, and a certain number of servers can't restore the complete files.

Integrity. It is similar to previous work [16], but we require that data owner can verify the data integrity whether the it is stored in CSP or downloaded to local. Integrity requires that the data downloaded from the cloud storage can be verified by data owner that it has not been altered. To achieve this, the credible tag is required to support verification mechanism and it is not allowed to be changed by authorized adversaries which may lead to a wrong verification. In addition, data owner can detect whether data is maliciously-generated with the correct tag.

Reliability. The deduplication scheme should provide fault tolerance. The solution also needs to ensure the availability of data in the cloud, ensuring that files can be downloaded for a longer period of time. In our scheme, an incentive-based file recovery method was proposed to ensure file reliability.

4   Blockchain Based Deduplication Scheme

4.1   Blockchain

The essence of the blockchain is a time-stamped chained data structure. We use this structure and combine the consensus mechanism to achieve effective record of data tags, and automated information service processing through smart contracts. The main tool for implementing deduplication is ramp secret sharing scheme, and blockchain information will be used for localization duplicate check and integrity verification.

Blockchain Initialization. We set the blockchain network by Blockchain as a Service (BaaS) mode. That is, only the blockchain service is provided, the user do not need to concerned about the related content of the blockchain maintenance. The advantage of this mode is that there are always nodes of sufficient amount to maintain the security and reliability of the blockchain. It is not necessary to consider that the security changes when the number of users fluctuates. To initialize the blockchain service, we need to perform the following three steps.

A. Data owners and CSPs add in blockchain network as the participant.
B. Each participant gains a private key after step A. Data owner generates a public key$PubK_D$ from $PriK_D$ and the value $PubK_{CSPi}$ is also derived from the i-th CSP's private key $PriK_{CSPi}$. $Addr_D$ and $Addr_{CSPi}$ are generated by key pairs as the anonymous communication address.
C. Data owner needs to prepare some gas as a cost to storage transactions or set up smart contracts.

Algorithms with Smart Contract. Business Smart Contract (BSC) and Transaction Smart Contract (TSC) are proposed in our proposed scheme, both of them are implemented on blockchain network. At the begining of system setup, BSC is uploaded as a long-standing algorithm carrier. The algorithms set of BSC are responsible for verification and restore the missing file. We will introduce three of them in detail.

Verify: BSC sends challenge $c = POR_F(F)$ to a prover to ask for a response. Target participant return $r = POR_r(c)$ and BSC compute $POR_v(c, r)$ after received it. At the same time, all participants can use this algorithm to request the retrievability verification of a file.

Register: If the value $Tag(F, Addr_{CSPi})$,which is provided by data owner, is successfully passed the POR_F, BSC create an unsinged TSC script and send it to the data owner.

Reconstruct: With the Reconstruct function, it is allowed to restore the lost shard with metadata stored on the blockchain by using $Recover(b_i)$. $Tag(F, Addr_{CSPi})$ and a POR_F is necessary to start this function. When it is activated, Reconstruct function check the running TSC to find whether F is saved continuously. In particular, we define that if any one TSC is invalid, the corresponding data is not saved properly. Once the check is failed, BTC broadcast the requirements to related data owner for specific data and provide acceptable gas to whom uploading the correct file.

The TSC is generated by BSC and published after file uploading. In our scheme, the storage service demander lock some of their gas on blockchain as a deposit to set up the TSC which is agreed by both data owner and CSP. To be more specific, a part of the gas is seen as prepaid storage fee which is paid to the CSP who provide long term storage service via an installment contract. If the contract is expired or no valid verification information is received within a long period of time, it terminates itself and return the remaining gas.

To access to the file, data owner construct a TSC with related CSPs to make payment through piecewise payment channel. That is, there are many mini transactions for every unit of data, such as 32bit. These sub-transactions will continue until the transaction is completed or the contract is forcibly terminated.

4.2   Data Deduplication Management

File Upload. To upload a file F as Fig. 2. shows, the data owner first computes the file tag $Tag(F)$ and performs a local duplicate check with the downloaded blockchain tag information $Tag(F')$ which is updated in real time. If the file has duplicate in cloud, data owner computes and sends $Tag(F, Addr_{CSPi})$ to BSC via a secure channel to authorize participants with *Register*. Then a TSC is sent to data owner. Data owner signs the TSC and upload to CSP. CSP publish the TSC on blockchain and return a pointer for the share stored at server $Addr_{CSPi}$ to payer after the transaction created by TSC is conceded in blockchain network.

If no duplicate is found, data owner performs the following. Firstly, the PoR_F is computed to get handle $f_h = Gen_h(F)$, secret sharing algorithm over F is executed and output $b_i = Share(C)$, where b_i is the i-th shard of F. And then, the data owner runs the tag generation algorithm to get $Tag(b_i)$ and

$Tag(F, Addr_{CSPi})$ for each server with $Addr_{CSPi}$. Data owner uploads the set of values $\{Tag(F), Tag(F, Addr_{CSPi}), f_h\}$ to BTC to get unsigned TSC. Then the values $\{Tag(F), Tag(F, Tag(b_i), Addr_{CSPi})\}$ with the signed TSC script to the i-th CSP. Finally, each CSP stores these values and return a pointer to data owner after signs and broadcasts the TSC to blockchain network.

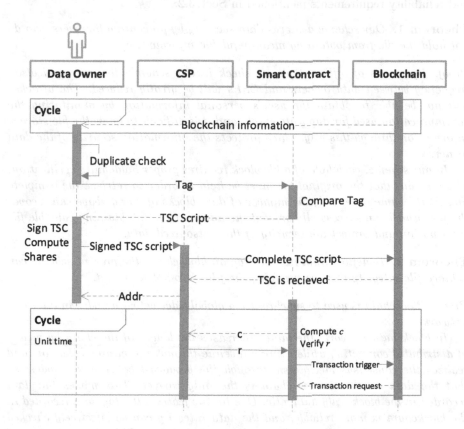

Fig. 2. File upload (no duplicate file)

File Download. Data owner finds available nodes by traversing blockchain data to download F. At least k shares of b_i are downloaded from CSP server then data owner restores the file by $F = Recover(b_i)$.

File Auditing. CSP can benefit from continuously storing data by continuing to pass BSC verification PoR_F. With transaction information which contains $Tag(F)$, $Tag(b_i)$, every participant can check the proof of ownership for specific data. The transaction is completed by off-chain micro-payment channel to avoid high cost of on-chain transaction. Through high-reliability chain information, we can compare the tag values $Tag(F)$ and $Tag(b_i)$ to determine whether the file has been maliciously tampered with.

5 Security Analysis

In this paper, we present a secure deduplication scheme with high reliability by the utility of blockchain techniques and secret sharing scheme. In this section, the proposed scheme would be testified that it could meet confidentiality, integrity and reliability requirements mentioned in Sect. 3.2.

Theorem 1. *Our scheme achieves data security by preventing the users who do not hold the file from obtaining meaningful file information.*

Proof. By means of offline duplicate check in our scheme, the communication frequency between data owners and CSPs will be greatly reduced. The attacker will not be able to obtain the user's personal information by monitoring the communication used for duplicate check, and it is difficult to know the file storage information through this way, which protects the information security of the data owner.

In our scheme, each hold of a file block requires proper holdability verification, which means that the original file must be held in order to retrieve the complete file. At the same time, only the number of data blocks of k and above can recover the file, which can successfully prevent the adversary from obtaining valuable file information and protect the security of the outsourced data.

Theorem 2. *Benefited from recorded tag on blockchain, the present scheme can achieve file integrity through verifying the file is complete or not.*

Proof. Blockchain is used to synchronize a global ledge between nodes in the same network.

In blockchain network, the miner increases the length of the chain by means of distributed computing while gaining revenue through the mining behavior, and realizes the consensus mechanism through the broadcast between the nodes, so that the data on the longest chain is the only correct. This makes the data recorded in the block truly authentic. Due to this feature, the tag data recorded in the blockchain is more reliable, and the data integrity can be effectively checked by comparing the tag values. Anyone who wants to successfully change the record needs to hold more than 51% full network computing power, which is considered impossible.

Theorem 3. *Through the utilization of secret sharing scheme, our scheme can easily achieve reliability requirement.*

Proof. The data can be restructed by Recover algorithm from any k pieces of file blocks. The reliability of the system rises as the $n - k$ rises. Besides, economic incentives are also applied to Reconstruct methods to increase participants' enthusiasm for maintaining system reliability.

6   Performance Analysis

6.1   Experiment Setup

Table 2 Shows the main part of the experiment environment.

Among them, we choose the hash function is SHA256, the default data block size is 4 KB.

Table 2. Experiment environment

Item	Parameter
CPU	AMD Ryzen 3 PRO 1200 Quad-Core Processor(3.10GHz)
Operating system	Ubuntu 16.04 LTS
Software toolkit	Jerasure Version 1.2

6.2   Experiment Analysis

The key technique of our scheme is RSSS. It's the main method to share the file among storage nodes and it has greater impact on the performance of our scheme. Thus, the computation cost of RSSS will be our primary consideration while the throughput is not the main influence factors and we have not done the simulation of it.

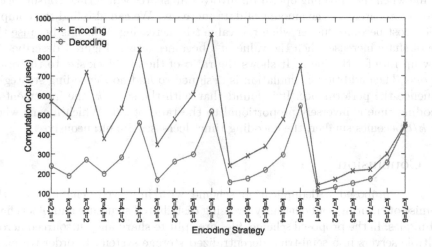

Fig. 3. $n = 8$, $2 <= k <= 7$

We make simulation experiments to explore how different combination of parameter t, k and n in RSSS affect the computation cost. As showed in Fig. 3, the encoding and decoding time of our proposed scheme are almost in the order

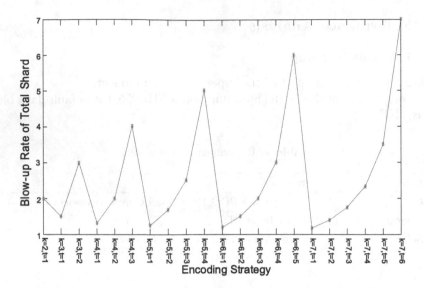

Fig. 4. Blow-up Rate

of microseconds under the defined parameters. We can see that the decoding time is less than the encoding time due to the use of less computing resources. In addition, the decoding time growth rate under all parameters is much smaller than the encoding time. It is because the decoding operation only use k shares of the file when the encoding operation involves all shares which time consumption becomes apparent as the amount of data increases. We can also find the computation cost becomes higher when the value t is increasing. Which is because the size of share increase when the value of t becomes larger. Figure 4 describe the blow-up rate for the file F. It shows the ratio of the file block size to the original one. Then additional simulation is designed to find how encoding strategies influence the performance. It is found that with the same k, the increment of encoding time is inversely proportional to the ratio of k/t. Which means when the k/t becomes smaller, the encoding time decreases Simultaneously.

7   Conclusion

We have proposed a deduplication scheme with improved reliability on the promise of ensuring confidentiality by using distributed structure and blockchain techniques. In the proposed scheme, files are split to shard and outsourced across multiple servers in a semi-trust decentralized storage system. In order to maintain system stability, we use an incentive mechanism, the storage payment model, to ensure that the vast majority of rational CSPs prefer to profit by following established protocals rather than malicious behavior. Payment will be automatically assigned through blockchain smart contracts which contain a time-stamp based authentication protocol. The presented scheme has also provided a reconstruct scheme that a missing file can be restored. Besides, auditing schemes have provided the security and integrity through smart contract without trusted third parties to avoid single point of failure. Severial kinds of attacks are proposed and discussed as the goal to build our security model and then we do the simulations with main tool. Although it has incurred a little higher computation cost than baseline, security analysis has demonstrated that our proposed scheme has achieved the presented goal and provided higher security and confidentiality than the previous work.

Acknowledgment. This work was supported by the National Natural Science Foundation of China under Grant Nos. 61702115 and 61672171, Natural Science Foundation of Guangdong, China under Grant No. 2018B030311007, and Major R&D Project of Educational Commission of Guangdong under Grant No. 2016KZDXM052, and China Postdoctoral Science Foundation Fund under Grant No. 2017M622632, and Opening Project of Guangdong Province Key Laboratory of Big Data Analysis and Processing under Grant No. 201805, Guangdong Pre-national project under Grant No. 2014GKXM054.

References

1. CISCO: Cisco global cloud index (2012–2017). http://www.cisco.com/c/en/us/solutions/service-provider/global-cloud-index-gci/index.html/
2. Douceur, J.R.: The sybil attack. In: Druschel, P., Kaashoek, F., Rowstron, A. (eds.) IPTPS 2002. LNCS, vol. 2429, pp. 251–260. Springer, Heidelberg (2002). https://doi.org/10.1007/3-540-45748-8_24
3. Google Drive: Google drive. http://drive.google.com/
4. Dropbox: Dropbox: A file-storage and sharing service. https://www.dropbox.com/
5. Elzeiny, A., Elfetouh, A.A., Riad, A.: Cloud storage: a survey. Int. J. Emerg. Trends Technol. Comput. Sci. (IJETTCS) **2**(4), 342–349 (2013)
6. Filecoin: A decentralized storage network. https://www.filecoin.io/
7. Juels, A., Kaliski, B.S.: PORs: proofs of retrievability for large files. In: ACM Conference on Computer and Communications Security, pp. 584–597 (2007)
8. Li, J., Liu, Z., Chen, L., Chen, P., Wu, J.: Blockchain-based security architecture for distributed cloud storage. In: 2017 IEEE International Symposium on Parallel and Distributed Processing with Applications and 2017 IEEE International Conference on Ubiquitous Computing and Communications (ISPA/IUCC), pp. 408–411. IEEE (2017)
9. Li, J., Wu, J., Chen, L.: Block-secure: blockchain based scheme for secure P2P cloud storage. Inf. Sci. **465**, 219–231 (2018). https://doi.org/10.1016/j.ins.2018.06.071. http://www.sciencedirect.com/science/article/pii/S0020025518305012
10. Li, J., et al.: Secure distributed deduplication systems with improved reliability. IEEE Trans. Comput. **64**(12), 3569–3579 (2015)
11. Li, J., Chen, X., Li, M., Li, J., Lee, P.P.C., Lou, W.: Secure deduplication with efficient and reliable convergent key management. IEEE Trans. Parallel Distrib. Syst. **25**(6), 1615–1625 (2014)
12. Li, J., Li, J., Xie, D., Cai, Z.: Secure auditing and deduplicating data in cloud. IEEE Trans. Comput. **65**(8), 2386–2396 (2016)
13. Li, J., Wu, J., Chen, L., Li, J.: Blockchain-based secure and reliable distributed deduplication scheme. In: 2018 International Conference on Algorithms and Architectures for Parallel Processing (ICA3PP) (2018, accepted)

14. Liu, C., Gu, Y., Sun, L., Yan, B., Wang, D.: R-ADMAD: high reliability provision for large-scale de-duplication archival storage systems. In: International Conference on Supercomputing. pp. 370–379 (2009)
15. Nakamoto, S.: Bitcoin: a peer-to-peer electronic cash system. (2008, consulted)
16. Shin, Y., Koo, D., Yun, J., Hur, J.: Decentralized server-aided encryption for secure deduplication in cloud storage. IEEE Trans. Serv. Comput. **PP**(99), 1 (1939)
17. Storer, M.W., Greenan, K., Long, D.D.E., Miller, E.L.: Secure data deduplication. In: ACM International Workshop on Storage Security and Survivability, pp. 1–10 (2008)
18. Wood, G.: Ethereum: a secure decentralised generalised transaction ledger (2014)

Big Data System and Management

Virtual Machine Live Migration Strategy in Big Data Information System

Juan Fang[✉], Lifu Zhou, and Mengxuan Wang

Faculty of Information Technology, Beijing University of Technology,
Beijing 100022, China
fangjuan@bjut.edu.cn

Abstract. In recent years, as an emerging technology, cloud computing has provided us with convenient services, and power consumption on issues have become increasingly prominent. Virtual machine live migration technology has become an important technology to reduce the power consumption of cloud computing centers. In the process of virtual machine migration, the performance of the virtual machine is inevitably degraded, which may violate service level agreement (SLA, Service Level Agreement). How to use virtual machine live migration technology to reduce power consumption as much as possible while ensuring a low SLA violation rate becomes a hot issue. This paper aims to optimize the light load detection and virtual machine redistribution in the virtual machine live migration model. Aiming at the problem that the existing virtual machine light load detection method is easy to cause "over-migration", this paper proposes a threshold-based minimum CPU utilization method for light load detection, which effectively avoids excessive virtual machine migration. Aiming at the problem that the current process of virtual machine re allocation algorithm is relatively simple, and there is a certain power loss space, we present power aware simulation annealing algorithm (PASA). The algorithm combines the simulated annealing algorithm based on the power aware best fit decreasing algorithm (PABFD), which largely avoids the disadvantage that the PABFD easily falls into the local optimal solution trap. The paper uses the CloudSim simulator as simulation platform. The results show that compared with the best algorithm combination proposed by the previous researchers, the power consumption of the new algorithm combination proposed in the paper is reduced by 16.79%, and the SLA violation rate is reduced by 85.37%. Combining the two algorithms together can lead to better energy efficiency, performance and quality of service than using the two algorithms.

Keywords: Cloud computing center · Virtual machine migration
Low load detection algorithm · Virtual machine reallocation algorithm

1 Introduction

The cloud computing concept emerged in 2006 and was developed on the basis of large-scale distributed computing technology. NIST defines cloud computing as a convenient model that can access the pool of computing resources on the network on demand and with less administrative effort [1]. The emergence of cloud computing

© Springer Nature Singapore Pte Ltd. 2018
Z. Xu et al. (Eds.): Big Data 2018, CCIS 945, pp. 573–588, 2018.
https://doi.org/10.1007/978-981-13-2922-7_37

provides a more feasible way for general enterprises to meet their own processing data requirements. Compared with building their own data centers, the price is low, and there are advantages such as rapid implementation, low maintenance cost, and low IT staff demand [2]. For the above reasons, cloud computing has become a hot spot in the industry. While providing us with convenient services,

tasks submitted by users are usually big data computing tasks, the power consumption of cloud computing centers has become increasingly prominent. Virtual machine live migration technology in cloud computing center has become a practical and effective solution. Virtualization technology is the foundation of big data. Virtualization technology enables multiple virtual machines (VMs, virtual machines) to share the resources of the same physical machine in parallel as real physical machines, while ensuring isolation between virtual machines [3]. The live migration of the virtual machine can be used to adjust the load in the large cloud computing center, dynamically adjust the load according to the current load level and migrate all the virtual machines in the light-loaded physical node to other physical nodes with normal load. Finally, the emptied physical node is adjusted to sleep mode, thereby achieving the purpose of saving energy and reducing carbon emissions. However, in the process of virtual machine live migration, the user can only be as transparent as possible. In the migration process, the performance of the virtual machine is reduced, which seriously affects the user experience and may even violate SLA. Based on the above reasons, how to make better use of the advantages brought by virtual machine live migration while avoiding some drawbacks has become one of the most popular and most meaningful research contents.

There were outstanding contributions previously in the four modules of the virtual machine live migration - Over Overloading Detection, Host Underloading Detection, Virtual Forwarder Selection, and VM redistribution.

The purpose of overload detection is to determine whether a physical node is in an overload state. However, if a general detection method is used to determine whether the node is overloaded, it is likely that the physical node has entered an overload state before taking the corresponding measures after the determination is completed. Therefore, in recent years, the academic community has mainly used various methods to predict the impending overload. Kim et al. [4] proposed a local weighted regression method to determine Host Overloading Detection. The method is to apply the mathematical local weighted regression method to the overload detection. The core idea is to periodically collect the physical node load status data and fit the CPU utilization data for a period of time into a curve and use this curve to predict the CPU utilization for the next time period. The mathematical model of the method is relatively simple, and the calculation process is relatively simple, so it has a low time complexity. At the same time, it is also ideal in predicting the effect. This model is directly used in our experiment. Arianyan et al. [5] proposed the minimum migration time strategy, the strategy takes full account of the impact of virtual machine live migration on performance and service quality and can effectively avoid the rise of SLA violation rate but does not consider the problem of effectively reducing virtual machine load. Buyya et al. [6] proposed the maximum relational coefficient method to determine the VM migration selection problem. This method is able to determine whether a physical node in light load condition, but due to the process of safety factor need extra operation time, so the method may lead to detection of extra time overhead. To solve the Host Under-loading Detection, Ferreto et al. [7] proposed the CPU arithmetic mean method. Virtual machine redistribution is to reallocate the migrated virtual machine to other physical nodes. After redistribution, on the one hand, it is required that the other physical nodes cannot be overloaded, and on the other hand, the energy consumption of all physical nodes is required to be as small as possible.

Beloglazov et al. [8, 9] proposed the optimal adaptive descending algorithm for energy perception, which is the application of the best adaptive descending algorithm in virtual machine redistribution. The general idea of the algorithm is to arrange all virtual machines in reverse order of CPU utilization, and then find out one physical node with sufficient resources in all physical nodes to carry the virtual machine and minimize the power consumption increment after migration. The experimental data of Beloglazov et al. show that the algorithm has lower energy consumption than the energy-aware first-time adaptive descending algorithm, but the algorithm structure and algorithm are relatively simple, and there is still room for further improvement in energy efficiency.

2 Host Underloading Detection Based on Threshold

2.1 Minimum CPU Utilization

The core idea of the Host Underloading Detection method is to periodically traverse all the hosts in the cloud data center, and to calculate the hardware usage of all the hosts. Then, the hosts are sorted according to the CPU utilization rate. Determine under-loading host according to the inequality (1).

$$h \in H | \forall a \in H, h_u \leq a_u \tag{1}$$

Where h, a is a single host, H is the physical host list of the entire cloud data center, h_u is the CPU utilization of the h node, and a_u is the CPU utilization of the node.

The MCU (MCU, Minimum CPU Utilization) is one of the lowest power and SLA violation algorithms based on the results of a large number of experiments in [4]. But there is no scholar to propose a virtual machine live migration overhead. This gives the MCU a big flaw.

The results of the study [10] turned out that virtual machine live migration process will produce a certain power. On this basis, Zhou et al. [11, 12] proposed over-migration. Over-migration causes the VM moved to sleep mode when the host moved out. And this may lead to the situation described in (2).

$$Power < Power - Power_{saved} + Power_{migration} \tag{2}$$

Where Power is the total power consumption before the migration occurs. $power_{saved}$ is the power saved by adjusting some physical nodes to sleep mode after migration, and it is also the part we want to maximize. $Power_{migration}$ is the migration power overhead caused by moving some physical nodes out of VM. In the event of this, the migration of VMs is not worth the loss, because power consumption is even higher than before migration. Due to virtual machine live migration will cause a certain degree of performance degradation, which has violated the risk of SLA, so these unnecessary transfers can also lead to an increase in SLA violation. This is over-migration.

2.2   Minimum CPU Utilization Based on Threshold

Overview of the Algorithm

To solve the problem above, the paper proposes minimum CPU utilization based on threshold (MUT, Minimum Utilization with Threshold). The core idea of this method is to use a large number of experiments to find the best value of the threshold to make further restrictions on the determination of the original MCU on the light load. Determine whether the threshold of light load host CPU utilization as a constraint or not. If the threshold is lower than the threshold, it is determined that the host is currently in the light load state. If it is higher than the threshold value, it is determined that the load is normal. It is not difficult to predict that the algorithm can effectively avoid the over-migration problem mentioned in the previous section if the appropriate threshold is taken.

In summary, from the theoretical level of the algorithm has the following advantages:

1. Reduce unnecessary light load decisions that can lead to over-migration problems, thereby reducing unnecessary migration costs and performance degradation.
2. Compared with the original algorithm is likely to produce lower power and lower SLA violation.

Algorithm Structure

The specific flow of the MCU based on the threshold is to periodically traverse all the hosts in the cloud data center, statistics the hardware usage of all the hosts, and then sort the hosts according to the utilization rate of the CPU, and then determine the underloading according to the inequality (3).

$$h \in H | \forall a \in H, h_u \leq a_u | h_u < threshold \tag{3}$$

Where h, a is a single host, H is the host list of the entire cloud data center, h_u is the CPU utilization of the h node, a_u is the CPU utilization of a node, and the threshold is the final threshold determined by the experiment.

If there is a host h, try to migrate all the VMs on the physical machine to other hosts without causing overload to other hosts. If it can be implemented, the VM on the host will be moved out of the scheme.

The pseudo code of MCUT is as follows (Table 1):

Table 1. pseudo code of MCUT

Algorithm1: Minimum CPU Utilization with Threshold
1 Input: host List, threshold **Output**: feasibility of allocation **2** min Utilization ← MAX **3** allocated Host ← NULL **4 foreach** host in host List **do** **5**   CPU Utilization ← host.getUtilization() **6**   **if** CPU Utilization < min Utilization **then** **7**   min Utilization← CPU Utilization **8**   allocated Host ← host **9 if** allocated Host ≠ NULL **then** **10**   vm List ← getVMsFromHost(allocated Host) **11**   host List ← deleteHostFromList(host List, allocated Host) **12**   **if** min Utilization < threshold **then** **13**   **if** host List *has enough resources for* vm List **then** **14**   **return** True **15**   **else** **16**   **return** False

Algorithm Summary

Through the minimum CPU utilization method based on the threshold and the description of the specific algorithm structure above, the following judgment can be made:

1. After obtaining a suitable threshold, the threshold-based minimum CPU utilization method can effectively avoid the over-migration problem, so the focus of the follow-up work is to find a suitable threshold by a large number of experiments.

2. It can be inferred that in the process of finding the appropriate threshold, if the threshold is reduced from 100% (that is, for the Host Underloading Detection process does not make the second constraint) to 0% (that is, if all the host if the CPU utilization rate of 0%, it is determined that the load is normal, under normal circumstances this means that all host load is normal) in the process. Over-migration problem will gradually reduce or even disappear with the constraints of the gradual tightening. In the process the unnecessary power consumption due to over-migration is gradually reduced and the total power consumption is reduced. Then because of the tightening of CPU utilization limits, the normal light-load decision will be affected, and more and more hosts that are really in a light-load state will be judged to be normally loaded. The total power consumption will rise and the SLA violation rate is declining throughout this process because the virtual

machine live migration is decreasing which caused a performance degradation. Therefore, the optimal threshold is then taken at the minimum power consumption, where the over-migration problem is minimized due to threshold constraints and it do not affect the normal light load determination process.

3. After using the appropriate threshold, MUT is likely to be superior to the original CPU utilization method in both total power consumption and SLA violation rate.
4. In summary, the next section of the paper is about finding the appropriate threshold through a lot of scientific experiments and comparing it with the minimum CPU utilization method in terms of power consumption and SLA violation rate.

2.3   Experiments and Results Analysis

Experimental Design

The overall experimental idea is that building a simulation of the cloud computing center by using a cloud computing center simulator. In order to control the experimental variables, variables in this simulation of the cloud computing center will not be changed. After the simulation of the MCU and the other three matching algorithms in the simulation of the cloud computing center run the situation, which repeat 10 experiments. And determine the power consumption in this case and SLA violation data. Next, the Host Underloading Detection algorithm is replaced by a threshold-based MCU. The 20 sets of thresholds are taken from 100% to 0% of the difference. Each set of thresholds is repeated ten times. The power consumption and SLA violation data are also determined. And compare the power consumption under the optimal threshold value and SLA violation data with MCU experimental data. Finally, it proves that the proposed minimum CPU utilization method can be made improvements on the basis of predecessors.

Experimental Environment

This paper uses CloudSim 3.0.2 as a cloud computing center simulation platform. The simulator is one of the most powerful and powerful cloud computing platform simulators currently favored by researchers. The simulator is currently one of the most popular and most powerful cloud computing platform simulators. The simulator comes with an energy consumption and SLA violation rate monitoring module that automatically generates an operational report with these two data after each simulation run. In the experiment simulation of a 800 host with a medium-sized cloud computing center. Of which 50% of the host is Huawei Fusion Server Rh2288H. Each server is equipped with two Intel Xeon E5_2609 processors. The server model memory size holds 65G. Hard disk size holds 1 TB. Another 50% of the host model is equipped with the processing Switch to Intel Xeon E5_2699. The available bandwidth per host is 1Gbit/s.

Workload

In order to make the results of the simulation in this paper more realistic and effective, it is necessary to use the workload data of the real system environment, so we use some of the real data provided by the CoMon project. The specific workload data is the ten-day operational data randomly selected by PlanetLab from March to April 2011 recorded by the CoMon project. The specific data characteristics of the workload are shown in Table 2.

Table 2. Workload data characteristics (CPU utilization)

Date	Number of VMs	Average (%)	Sample estimation deviation
03/03/2011	1052	12.31	17.09
06/03/2011	898	11.44	16.83
09/03/2011	1061	10.70	15.57
22/03/2011	1516	9.26	12.78
25/03/2011	1078	10.56	14.14
03/04/2011	1463	12.39	16.55
09/04/2011	1358	11.12	15.09
11/04/2011	1233	11.56	15.07
12/04/2011	1054	11.54	15.15
20/04/2011	1033	10.43	15.21

Since the load data is derived from the real environment, and each group contains the entire plant running data of the entire PlantLab throughout the day, the user requests and tasks of different characteristics are evenly distributed among the ten groups of workloads. The experimental data is close to the real environment. Theoretically, it can be inferred that the experimental data obtained by applying these workloads and the conclusions based on experimental data are highly scalable.

Experimental Data and Results Analysis
Figure 1 is a graph of the final data obtained from the experiment for determining the threshold. The circular coordinate point uses the left ordinate for the power data and the X coordinate point for the SLA violation rate data using the right ordinate.

Fig. 1. Energy consumption and SLA violation with different threshold.

It is not difficult to see that the power at different thresholds and the SLA violation data are consistent with the predictions made in the previous section. As the threshold decreases, the power consumption decreases firs. And the minimum value is minimized when the over-migration problem is minimized. Then the power consumption is limited due to the normal host underloading detection. SLA violation is declining due to the gradual reduction of virtual machine live migration. When the threshold is 0.45, the power consumption is the minimum, and the SLA violation is at a relatively low point. The optimal threshold is 0.45.

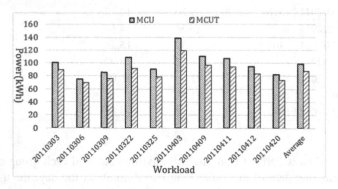

Fig. 2. Energy consumption and SLA violation with different threshold.

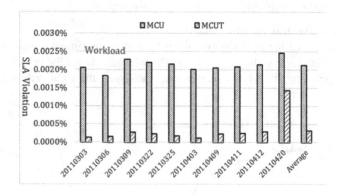

Fig. 3. SLA violation of two algorithms with different workloads

Figures 2 and 3 are compared with the algorithm of the minimum CPU utilization method when the threshold is 0.45. From the above two graphs, it can be seen that the minimum CPU utilization method based on threshold is significantly lower than the control group with the minimum CPU utilization method in terms of power consumption and SLA violation rate, especially in terms of SLA violation rate It is significantly reduced. From the average point of view, the algorithm proposed in this paper has a significant reduction in power consumption compared with the previous algorithm, the reduction rate is 11.88%, SLA violation is more obvious, the average decline of 84.47%. This shows that the threshold-based MCU proposed in this paper has improved the predecessor's algorithm and has made great progress.

3 Power Aware Simulation Annealing

3.1 Power Aware Best Fit Decreasing

VM reallocation is reassigned VM to other host up which host is identified as overload by Host Overloading Detection and egress selection module chooses to move out of the VM, and which host identified as light load through Host Underloading Detection. On the one hand, it is required that redistribution should not lead to the overload of the host. On the other hand, power of all the host moving in should increase as little as possible.

PABFD is a constructive heuristic algorithm to solve the packing problem. After a series of experiments, it is proved that the algorithm can give a smaller feasible solution in a tiny time complexity. But each time the allocation of VM only to find the a single VM optimal allocation of power under current circumstances, but not consider of the best overall direction.

For the above reasons, the paper aims to find an algorithm which can give a better solution than PABFD for the VM allocation algorithm. Because the VM reallocation process will increase the SLA violation at a certain extent if the waiting time is too long, which reduce the quality of service, so the new algorithm will control time complexity in a lower range.

3.2 Power Aware Simulation Annealing

I apply the simulation annealing algorithm to the VM reallocation module. Under the controlling of a cool down scheduler, the simulation annealing algorithm can be achieved in the acceptable time complexity. It can avoid falling into the local optimal "trap", so as to obtain a better perspective from the global point of view.

In this paper, PASA use solution given by PABFD as the initial solution. The state is recorded as the initial state i. According to the power model, we calculate the power of the entire cloud computing center, which recorded as E(i). Then we assign a VM randomly selected from the queue to be allocated to a randomly selected host with sufficient resources to carry the VM, which denote the state at that time as j. We calculate the whole power consumption of cloud computing center based on the power model, denoted by E (j). Calculate the power difference between the two states, denoted as ΔE. The specific formula as shown in (4).

$$\Delta E = E(j) - E(i) \tag{4}$$

If $\Delta E < 0$, it is shown that the current solution provides a VM reallocation scheme with which power equal to or because of the current solution. So we accept j as a new solution; if $\Delta E > 0$, it is shown that the solution is higher than the current solution

power. However, for the possibility of jumping out of the local optimal solution trap, we will be accept this difference solution in a certain probability, where the probability is recorded as ζ. The specific probability formula as shown in (5).

$$\zeta = \exp(\frac{-\Delta E}{T}) \tag{5}$$

If $\zeta >$ random $(0, 1)$, then accept the lower solution j as a new solution. If $\zeta \leq$ random $(0, 1)$, then we give up j. And then continue to cool down and cycle the implementation of the steps until meet the termination conditions of cooling coefficient table set. Then calculate the entire cloud computing center power consumption of the final solution. And compare it with the power consumption of initial solution given by the PABFD. If the power consumption is lower than the initial solution, it is shown that PASA successfully found a better solution. Then we accept the final solution for the PASA VM reallocation program. If the power consumption is higher than initial solution, it is indicated that there requires more algorithm execution time to accept the solution. The power consumption is higher than PABFD, so we accept its initial solution as the final VM reallocation scheme.

3.3 Experimental Results and Analysis

The General Idea of the Experiment
There are three goals in this experiment. First, it is necessary to determine the value of each parameter in the cool down scheduler to obtain the best combination of parameters to achieve the best PASA effect. Then, in the same experimental environment, using the optimal combination of parameters obtained in the previous step, the most efficient one is selected from the three deployment scenarios through a large number of experiments as the final deployment plan. Finally, on the basis of a large amount of experimental data, we compare the power to SLA breaches by PABFD, which is aim to demonstrate the performance of PASA advantage through the mature algorithm proposed by the previous.

Due to the random search characteristics of the simulation annealing algorithm, the results given by PASA is likely to fluctuate within a certain interval. Therefore, all the experiments in this section will use a large number of experiments with taking the average of each sets of data to ensure the reliability of results.

Parameter Determination
To begin with the experiment, the parameters of the PASA are determined by using the parameter range of which is larger difference. In this experiment, the range of the initial temperature T is selected by {300, 600, 900}. The value of the iterations per time L is {200, 400, 600}. And the cooling coefficient(β) is in the range of {0.65, 0.8, 0.95}. Then we will choice 27 parameters randomly to combine with in every range of value. And take the average of ten experimental power and SLA violation data. The specific data is shown in Fig. 4.

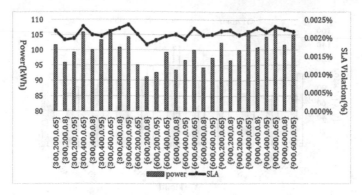

Fig. 4. Average power and SLA violation under different coarse-grained parameters

The combination of {600, 200, 0.8}, which is clearly visible, is the relative minimum of the power and SLA violation. The reference value for the next set of experiments is {600, 200, 0.8}. The specific value is based on the range of fine-grained parameters. The resulting average experimental data is shown in Fig. 5.

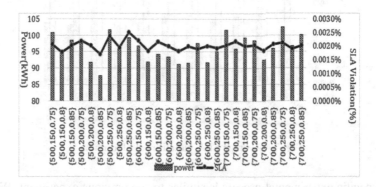

Fig. 5. Average power and SLA violation under different fine grain parameters

It can be seen from Fig. 5 that the combination of parameters {500, 200, 0.85} has the minimum average power and average SLA violation rate. There is a significant advantage over the other 26 sets of parameter combinations. The final combination is finalized {500, 200, 0.85}.

Experimental Results Analysis
It is significant to compare of the PASA and PABFD performance proposed in this paper. The reference index is still the power and SLA violation rate, which can measure the power efficiency of the two algorithms and the performance impact to the cloud computing center.

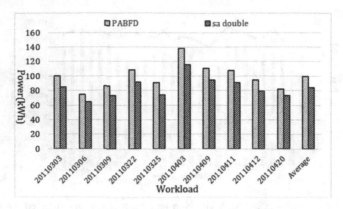

Fig. 6. Average power consumption of algorithms

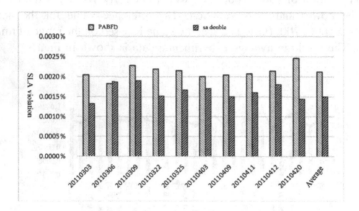

Fig. 7. Average SLA violation of algorithms

The specific experimental process is similar to the previous experiment. We test a number of times and take the average of the experimental results as the final reference data.

The specific power consumption data shown in Fig. 6. From the figure we can clearly see that the power under the method proposed PASA is lower than the control group by using PABFD. From the average of the 10 power groups under workloads, the algorithm proposed in this paper is 11.66% lower than that of the PABFD algorithm.

Figure 7 shows the SLA violation rate comparison of the two algorithms. It can be seen that under most workloads, the SLA violation of the energy-aware simulated annealing algorithm proposed in this paper is lower than PABFD. "20110306" is slightly higher under a set of workloads. From the average point of view, the average SLA violation rate of the energy-aware simulated annealing algorithm is 0.0015%, while the average SLA violation rate of PABFD is 0.0021%, which is 28.57% compared with the latter. The above data can be used to illustrate that the energy-aware simulated annealing algorithm is better than the previous ones in terms of performance and quality of service.

In summary, the PASA proposed in this paper is superior to the PABFD proposed by predecessors in the same experimental conditions, both in power and SLA violation, which is using the best combination of parameters obtained from the large number of experiments in the above two subsections. It can be explained that the algorithm proposed in this paper has some improvement on the basis of previous research in power efficiency and service quality.

4   Combination of Virtual Machine Live Migration Algorithm

4.1   Experimental Design

The overall experimental idea is to experiment with the same cloud computing center configuration and the same workload in the simulator. First, we get the power and SLA violation of the virtual machine live migration system using the threshold-based MCU and PASA matching algorithm. Then three groups of control data were obtained and compared with each other to demonstrate whether the two algorithms proposed in this paper can play the proper role in the same cloud computing center.

The first group of the three sets of algorithm combinations used as the control group is the best combination of a group of algorithms in the previous combination of algorithms. The second group of algorithms is a virtual machine based on the matching algorithm based on MCU. Live migration system, the third set of algorithm combinations is a separate use of PASA with matching algorithm composed of virtual machine live migration system. In order to guarantee the control variables in the experiment, the matching algorithm of the virtual machine live migration system proposed in this paper is consistent with the control group.

4.2   Experimental Configuration and Workload

Since the experimental process in the previous section is successful and there is no obvious problem and the resulting data is reliable, the experimental environment and the workload in this section are consistent with the experimental environment in chapter 3.4

4.3   Experimental Data and Analysis

We experiment multiple times with a set of target groups and three groups of control group algorithm combination. Taking the average of each group algorithm combination under each group of workload data.

Figure 8 is the combination of four algorithms of the power comparison. The figure of the four columnar data is legend from left to right, respectively, said the best proposed combination of the previous algorithm, a separate application based on the threshold of the MCU combination of algorithms, PABFD algorithm combination and combination of threshold-based MCU and PASA algorithm combination. From the final average data, we can see that the two algorithms combination proposed in this article decreases 16.70% compared to the previous combination of the best combination. It declines 7.36% compared to a separate application based on the threshold of the MCU power average. It decreases 5.70% compared to the average application of PABFD power alone.

Figure 9 for the combination of four algorithms SLA violation comparison. It can be seen that MCUT proposed to reduce over-migration and improve service quality is superior to PASA in performance and quality of service. In the combination of these two algorithms, the SLA violation rate is significantly lower than PASA alone and is basically the same as the control group applying MCUT.

From the specific data point of view, the proposed combination of the two algorithms in the average SLA violation compared to the previous combination of the best combination of the average algorithm SLA violation decreased by 84.43%. It compared to the average SLA violation of thresholds based on thresholds decreased by 3%. It

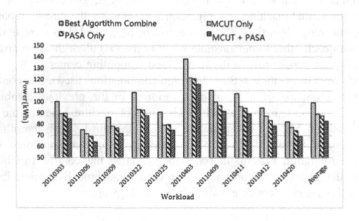

Fig. 8. Average power consumption of sets of algorithms

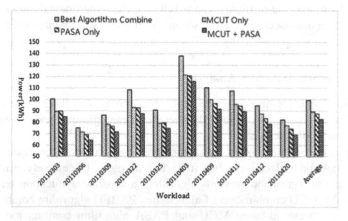

Fig. 9. Average SLA violation of sets of algorithms

compared with the average SLA violation of the control group with PASA alone reduced by 78.00%.

In summary, the two algorithms applied to a virtual machine live migration system giving full play to the threshold-based MCU which improves performance and quality of service features. The PASA reduce the power, and the combination of the two algorithms compared with the alone is a certain gain. There is a significant decline in the case of two algorithms at the same time power and SLA violation compared with the previous study of the best combination of the algorithm.

5 Conclusion

In this paper, we research on reducing power of virtual machine live migration system. Host Underloading Detection and VM reallocation module, which are related to the purpose, are the main module we aim to. We improved the new Host Underloading Detection method and VM Reallocation algorithm on the basis of previous research.

In the case of Host Underloading Detection, this paper proposes a minimum CPU utilization method disposed of judging light load carelessly when the minimum CPU utilization and leading to over-migration at a extent. This method limits the decision of the light load host by using an experimentally obtained optimal threshold as a quadratic constraint to ensure that only the host. That is really out of light load and does not raise the over-migration problem. The experimental data show that the proposed algorithm shows a significant reduction in power consumption compared with the previous algorithm. And the decrease rate is 11.88%. The SLA violation rate is more obvious, which decreases of 84.47% in average. It shows that the threshold-based minimum CPU utilization method proposed in this paper can effectively identify the light load state of the host. It can effectively avoid the occurrence of over-migration. On the basis of predecessors there is higher energy efficiency and better performance and quality of service.

In the case of VM reallocation, this paper presents PASA, which is a concrete application of simulation annealing algorithm VM reallocation. Using the VM reallocation scheme proposed by PABFD as the initial solution of the algorithm, the random change VM reallocation scheme adopts the new scheme. If the total power consumption of the cloud computing center is lower than the total power consumption of the initial solution, the total power dissipation is higher than the total power consumption of the initial solution. Then the new scheme is adopted with the probability of using the metropolis criterion. It can avoid the greedy algorithm in some ways which is easy to fall into the local optimal solution trap shortcomings. The experimental data show that the average power obtained by the algorithm proposed in this paper is 11.66% which is lower than that of PABFD and the average SLA violation rate is 28.57%. It shows that the PASA proposed in this paper can be better in ensuring performance and quality of service based on the lower power.

Finally, this paper demonstrates the effect of the two algorithms proposed in this paper in the same VM migration system. The experimental results show that the minimum CPU utilization method works well with PASA in the same cloud computing center. And both algorithms run at the same time compared to running one of the algorithms with better energy efficiency and better Performance and service quality.

Acknowledgement. This work is supported by the National Natural Science Foundation of China (Grant No. 61202076), along with other government sponsors. The authors would like to thank the reviewers for their efforts and for providing helpful suggestions that have led to several important improvements in our work. We would also like to thank all teachers and students in our laboratory for helpful discussions.

References

1. Mell, P.: The NIST definition of cloud computing. Commun. ACM **53**(6), 50 (2011)
2. Yang, H., Tate, M.: A descriptive literature review and classification of cloud computing research. Commun. Assoc. Inf. Syst. **31**(2), 35–60 (2012)
3. Barham, P., Dragovic, B., Fraser, K., et al.: Xen and the art of virtualization. In: ACM SIGOPS Operating Systems Review, vol. 37, no. 5, pp. 164–177. ACM (2003)
4. Kim, N., Cho, J., Seo, E.: Energy-credit scheduler: an energy-aware virtual machine scheduler for cloud systems. In: Future Generation Computer Systems, pp. 128–137 (2014)
5. Arianyan, E.: Multi objective consolidation of virtual machines for green computing in Cloud data centers. In: 2016 8th International Symposium on Telecommunications (IST), pp. 654–659. IEEE (2016)
6. Buyya, R., Yeo, C.S., Venugopal, S., et al.: Cloud computing and emerging IT platforms: vision, hype, and reality for delivering computing as the 5th utility. Futur. Gener. Comput. Syst. **25**(6), 599–616 (2009)
7. Ferreto, T.C., Netto, M.A.S., Calheiros, R.N., et al.: Server consolidation with migration control for virtualized data centers. Futur. Gener. Comput. Syst. **27**(8), 1027–1034 (2011)
8. Beloglazov, A., Buyya, R.: Optimal online deterministic algorithms and adaptive heuristics for energy and performance efficient dynamic consolidation of virtual machines in cloud data centers. Concurr. Comput. Pract. Exp. **24**(13), 1397–1420 (2012)
9. Beloglazov, A., Abawajy, J., Buyya, R.: Energy-aware resource allocation heuristics for efficient management of data centers for cloud computing. Futur. Gener. Comput. Syst. **28** (5), 755–768 (2012)
10. Strunk, A., Dargie, W.: Does live migration of virtual machines cost energy. In: Advanced Information Networking and Applications (AINA), pp. 514–521. IEEE (2013)
11. Fang, J., Zhou, L., Hao, X.: Energy and performance efficient underloading detection algorithm of virtual machines in cloud data centers. In: Cluster Computing (CLUSTER), pp. 134–135. IEEE (2016)
12. Sun, X., Ansari, N., Wang, R.: Optimizing resource utilization of a data center. IEEE Commun. Surv. Tutor. **18**(4), 2822–2846 (2016)

Author Index

Printed in the United State
by Bookmasters

Printed in the United States
By Bookmasters